T0202798

# Lecture Notes in Artificial Intelligence     13070

Subseries of Lecture Notes in Computer Science

Series Editors

Randy Goebel
*University of Alberta, Edmonton, Canada*

Wolfgang Wahlster
*DFKI, Berlin, Germany*

Zhi-Hua Zhou
*Nanjing University, Nanjing, China*

Founding Editor

Jörg Siekmann
*DFKI and Saarland University, Saarbrücken, Germany*

More information about this subseries at https://link.springer.com/bookseries/1244

Lu Fang · Yiran Chen · Guangtao Zhai ·
Jane Wang · Ruiping Wang ·
Weisheng Dong (Eds.)

# Artificial Intelligence

First CAAI International Conference, CICAI 2021
Hangzhou, China, June 5–6, 2021
Proceedings, Part II

 Springer

*Editors*
Lu Fang 🆔
Tsinghua University
Beijing, China

Yiran Chen 🆔
Duke University
Durham, NC, USA

Guangtao Zhai 🆔
Shanghai Jiao Tong University
Shanghai, China

Jane Wang 🆔
University of British Columbia
Vancouver, BC, Canada

Ruiping Wang 🆔
Institute of Computing Technology
Chinese Academy of Sciences
Beijing, China

Weisheng Dong 🆔
Xidian University
Xi'an, China

ISSN 0302-9743          ISSN 1611-3349 (electronic)
Lecture Notes in Artificial Intelligence
ISBN 978-3-030-93048-6          ISBN 978-3-030-93049-3 (eBook)
https://doi.org/10.1007/978-3-030-93049-3

LNCS Sublibrary: SL7 – Artificial Intelligence

This Springer imprint is published by the registered company Springer Nature Switzerland AG
The registered company address is: Gewerbestrasse 11, 6330 Cham, Switzerland

# Preface

The present book includes extended and revised versions of papers selected from the 1st CAAI International Conference on Artificial Intelligence (CICAI 2021), held in Hangzhou, China, on June 6, 2021.

CICAI is a summit forum in the field of artificial intelligence and the 2021 forum was hosted by Chinese Association for Artificial Intelligence (CAAI). CICAI aims to establish a global platform for international academic exchange, promote advanced research in AI and its affiliated disciplines, and promote scientific exchanges among researchers, practitioners, scientists, students, and engineers in AI and its affiliated disciplines in order to provide interdisciplinary and regional opportunities for researchers around the world, enhance the depth and breadth of academic and industrial exchanges, inspire new ideas, cultivate new forces, implement new ideas, integrate into the new landscape, and join the new era.

The conference program included invited talks delivered by two distinguished speakers, Harry Shum and Song-Chun Zhu, as well as a panel discussion, followed by an oral session of 15 papers and a poster session of 90 papers. Those papers were selected from 307 submissions using a double-blind review process, and on average each submission received 3.2 reviews. The topics covered by these selected high-quality papers span the fields of machine learning, computer vision, natural language processing, and data mining, amongst others.

This book contains 101 papers selected and revised from the proceedings of CICAI 2021. We would like to thank the authors for contributing their novel ideas and visions that are recorded in this book.

June 2021

Lu Fang
Yiran Chen
Guangtao Zhai
Jane Wang
Ruiping Wang
Weisheng Dong

# Organization

## General Chairs

| | |
|---|---|
| Lu Fang | Tsinghua University, China |
| Yiran Chen | Duke University, USA |
| Guangtao Zhai | Shanghai Jiao Tong University, China |

## Program Chairs

| | |
|---|---|
| Jane Wang | University of British Columbia, Canada |
| Ruiping Wang | Chinese Academy of Sciences, China |
| Weisheng Dong | Xidian University, China |

## Publication Chairs

| | |
|---|---|
| Yuchen Guo | Tsinghua University, China |
| Le Wu | Hefei University of Technology, China |

## Presentation Chairs

| | |
|---|---|
| Xia Wu | Beijing Normal University, China |
| Jian Zhao | AMS, China |

## International Liaison Chair

| | |
|---|---|
| Chunyan Miao | Nanyang Technological University, Singapore |

## Advisory Committee

| | |
|---|---|
| C. L. Philip Chen | University of Macau, China |
| Xilin Chen | Institute of Computing Technology, Chinese Academy of Sciences, China |
| Yike Guo | Imperial College London, UK |
| Ping Ji | City University of New York, USA |
| Licheng Jiao | Xidian University, China |
| Ming Li | University of Waterloo, Canada |
| Chenglin Liu | Institute of Automation, Chinese Academy of Sciences, China |
| Derong Liu | University of Illinois at Chicago, USA |

| Hong Liu | Peking University, China |
| Hengtao Shen | University of Electronic Science and Technology of China |
| Yuanchun Shi | Tsinghua University, China |
| Yongduan Song | Chongqing University, China |
| Fuchun Sun | Tsinghua University, China |
| Jianhua Tao | Institute of Automation, Chinese Academy of Sciences, China |
| Guoyin Wang | Chongqing University of Posts and Telecommunications, China |
| Weining Wang | Beijing University of Posts and Telecommunications, China |
| Xiaokang Yang | Shanghai Jiao Tong University, China |
| Changshui Zhang | Tsinghua University, China |
| Lihua Zhang | Fudan University, China |
| Song-Chun Zhu | Peking University, China |
| Wenwu Zhu | Tsinghua University, China |
| Yueting Zhuang | Zhejiang University, China |

## Area Chairs

| Badong Chen | Xi'an Jiaotong University, China |
| Peng Cui | Tsinghua University, China |
| Weihong Deng | Beijing University of Posts and Telecommunications, China |
| Yang Feng | Institute of Computing Technology, Chinese Academy of Sciences, China |
| Yulan Guo | National University of Defense Technology, China |
| Di Huang | Beihang University, China |
| Gao Huang | Tsinghua University, China |
| Qing Ling | Sun Yat-sen University, China |
| Qi Liu | University of Science and Technology of China, China |
| Risheng Liu | Dalian University of Technology, China |
| Deyu Meng | Xi'an Jiaotong University, China |
| Jinshan Pan | Nanjing University of Science and Technology, China |
| Xi Peng | Sichuan University, China |
| Chao Qian | Nanjing University, China |
| Boxin Shi | Peking University, China |
| Dong Wang | Dalian University of Technology, China |
| Jie Wang | University of Science and Technology of China, China |

# Contents – Part II

**Explainability, Understandability, and Verifiability of AI**

Reducing Adversarial Examples Through Boundary Methods . . . . . . . . . . . . . . . . 3
    *Xinyi Hu, Zhengming Zhang, Zhouyong Liu, Zhiwei Han, and Luxi Yang*

Explainable AI for Classification Using Probabilistic Logic Inference . . . . . . . . . 16
    *Xiuyi Fan and Siyuan Liu*

A Consistency Regularization for Certified Robust Neural Networks . . . . . . . . . . 27
    *Mengting Xu, Tao Zhang, Zhongnian Li, and Daoqiang Zhang*

Fooling Neural Network Interpretations: Adversarial Noise to Attack
Images . . . . . . . . . . . . . . . . . . . . . . . . . . . . . . . . . . . . . . . . . . . . . . . . . . . . . . . . . . . . . 39
    *Qianqian Song, Xiangwei Kong, and Ziming Wang*

**Machine Learning**

BPN: Bidirectional Path Network for Instance Segmentation . . . . . . . . . . . . . . . . . 55
    *Fan Xu, Lijuan Duan, and Yuanhua Qiao*

Claw U-Net: A UNet Variant Network with Deep Feature Concatenation
for Scleral Blood Vessel Segmentation . . . . . . . . . . . . . . . . . . . . . . . . . . . . . . . . . . . 67
    *Chang Yao, Jingyu Tang, Menghan Hu, Yue Wu, Wenyi Guo, Qingli Li,
    and Xiao-Ping Zhang*

Attribute and Identity Are Equally Important: Person Re-identification
with More Powerful Pedestrian Attributes . . . . . . . . . . . . . . . . . . . . . . . . . . . . . . . . 79
    *Shuangye Chen and Kai Xu*

Disentangled Variational Information Bottleneck for Multiview
Representation Learning . . . . . . . . . . . . . . . . . . . . . . . . . . . . . . . . . . . . . . . . . . . . . . . . 91
    *Feng Bao*

Real-Time Collision Warning and Status Classification Based Camera
and Millimeter Wave-Radar Fusion . . . . . . . . . . . . . . . . . . . . . . . . . . . . . . . . . . . . . . 103
    *Lei Fan, Qi Yang, Yang Zeng, Bin Deng, and Hongqiang Wang*

AD-DARTS: Adaptive Dropout for Differentiable Architecture Search . . . . . . . . 115
    *Ziwei Zheng, Le Yang, Liejun Wang, and Fan Li*

An Improved DDPG Algorithm with Barrier Function for Lane-Change
Decision-Making of Intelligent Vehicles .................................. 127
*Tianshuo Feng, Xin Xu, Xiaochuan Zhang, and Xinglong Zhang*

Self-organized Hawkes Processes ........................................ 140
*Shen Yuan and Hongteng Xu*

Causal Inference with Heterogeneous Confounding Data: A Penalty
Approach .............................................................. 152
*Zhaofeng Lu and Bo Fu*

Optimizing Federated Learning on Non-IID Data Using Local Shapley
Value ................................................................. 164
*Zuoqi Tang, Feifei Shao, Long Chen, Yunan Ye, Chao Wu, and Jun Xiao*

Boosting Few-Shot Learning with Task-Adaptive Multi-level Mixed
Supervision ........................................................... 176
*Duo Wang, Qianxia Ma, Ming Zhang, and Tao Zhang*

Learning Bilevel Sparse Regularized Neural Network ..................... 188
*Xin Xu, Liangliang Zhang, and Qi Kong*

**Natural Language Processing**

DGA-Net: Dynamic Gaussian Attention Network for Sentence Semantic
Matching .............................................................. 203
*Kun Zhang, Guangyi Lv, Meng Wang, and Enhong Chen*

Disentangled Contrastive Learning for Learning Robust Textual
Representations ....................................................... 215
*Xiang Chen, Xin Xie, Zhen Bi, Hongbin Ye, Shumin Deng,
Ningyu Zhang, and Huajun Chen*

History-Aware Expansion and Fuzzy for Query Reformulation .............. 227
*Wei Pang and Ruixue Duan*

Stance Detection with Knowledge Enhanced BERT ........................ 239
*Yuqing Sun and Yang Li*

Towards a Two-Stage Method for Answer Selection and Summarization
in Buddhism Community Question Answering ............................. 251
*Jiangnan Du, Jun Chen, Suhong Wang, Jianfeng Li, and Zhifeng Xiao*

Syllable Level Speech Emotion Recognition Based on Formant Attention ..... 261
*Abdul Rehman, Zhen-Tao Liu, and Jin-Meng Xu*

Judging Medical Q&A Alignments in Multiple Aspects ..................... 273
   *Pengda Si, Qiang Deng, Yiru Wang, Bin Zhong, Jin Xu, and Yujiu Yang*

DP-BERT: Dynamic Programming BERT for Text Summarization ........... 285
   *Shiyun Cao and Yujiu Yang*

**Robotics**

Research on Obstacle Avoidance Path Planning of Manipulator Based
on Improved RRT Algorithm .......................................... 299
   *Tianying Hu*

Visual Odometer Algorithm Based on Dynamic Region Culling .............. 311
   *Hongwei Mo and Xifeng Zhang*

Viewing Angle Generative Model for 7-DoF Robotic Grasping .............. 323
   *Xiang Gao, Wei Li, and Zhiqing Wen*

RGB-D Visual Odometry Based on Semantic Feature Points in Dynamic
Environments ....................................................... 334
   *Hao Wang, Yincan Wang, and Baofu Fang*

**Other AI Related Topics**

Robust Anomaly Detection from Partially Observed Anomalies
with Augmented Classes ............................................. 347
   *Rundong He, Zhongyi Han, Yu Zhang, Xueying He, Xiushan Nie,
   and Yilong Yin*

A Triple-Pooling Graph Neural Network for Multi-scale Topological
Learning of Brain Functional Connectivity: Application to ASD Diagnosis .... 359
   *Zhiyuan Zhu, Boyu Wang, and Shuo Li*

HierarIK: Hierarchical Inverse Kinematics Solver for Human Body
and Hand Pose Estimation ........................................... 371
   *Xinyu Yi, Yuxiao Zhou, and Feng Xu*

A Novel Conditional Knowledge Graph Representation and Construction ...... 383
   *Tingyue Zheng, Ziqiang Xu, Yufan Li, Yuan Zhao, Bin Wang,
   and Xiaochun Yang*

Unlocking the Potential of MAPPO with Asynchronous Optimization ......... 395
   *Wei Fu, Chao Yu, Yunfei Li, and Yi Wu*

A Random Opposition-Based Sparrow Search Algorithm for Path
Planning Problem ...................................................... 408
  *Guangjian Zhang and Enhao Zhang*

Communication-Efficient Federated Learning with Multi-layered
Compressed Model Update and Dynamic Weighting Aggregation ............ 419
  *Kaiyang Zhong and Guiquan Liu*

**Author Index** ........................................................ 431

# Contents – Part I

**Applications of AI**

Comparative Sharpness Evaluation for Mobile Phone Photos ................ 3
  *Qiang Lu, Guangtao Zhai, Yucheng Zhu, Xiongkuo Min, Tao Wang,*
  *and Xiao-Ping Zhang*

*DiffGNN*: Capturing Different Behaviors in Multiplex Heterogeneous
Networks for Recommendation ......................................... 15
  *Tiankai Gu, Chaokun Wang, and Cheng Wu*

Graph-Based Exercise- and Knowledge-Aware Learning Network
for Student Performance Prediction ................................... 27
  *Mengfan Liu, Pengyang Shao, and Kun Zhang*

Increasing Oversampling Diversity for Long-Tailed Visual Recognition ....... 39
  *Liuyu Xiang, Guiguang Ding, and Jungong Han*

Odds Estimating with Opponent Hand Belief for Texas Hold'em Poker
Agents ............................................................... 51
  *Zhenzhen Hu, Jing Chen, Wanpeng Zhang, Shaofei Chen, Weilin Yuan,*
  *Junren Luo, Jiahui Xu, and Xiang Ji*

Remote Sensing Image Recommendation Using Multi-attribute
Embedding and Fusion Collaborative Filtering Network .................... 65
  *Boce Chu, Jinyong Chen, Meirui Wang, Feng Gao, Qi Guo, and Feng Li*

Object Goal Visual Navigation Using Semantic Spatial Relationships ......... 77
  *Jingwen Guo, Zhisheng Lu, Ti Wang, Weibo Huang, and Hong Liu*

Classification of COVID-19 in CT Scans Using Image Smoothing
and Improved Deep Residual Network ................................... 89
  *Changzu Chen, Zhongyi Hu, Shan Jin, Lei Xiao, Mingzhe Hu, Qi Wu,*
  *Jingjing Shao, Zhenzhen Luo, and Mianlu Zou*

Selected Sample Retraining Semi-supervised Learning Method for Aerial
Scene Classification ................................................. 101
  *Ye Tian, Ju Li, Liguo Zhang, Jianguo Sun, and Guisheng Yin*

Knowledge Powered Cooperative Semantic Fusion for Patent Classification .... 111
  *Zhe Zhang, Tong Xu, Le Zhang, Yichao Du, Hui Xiong, and Enhong Chen*

Diagnosis of Childhood Autism Using Multi-modal Functional
Connectivity via Dynamic Hypergraph Learning ......................... 123
  Zizhao Zhang, Jian Liu, Baojuan Li, and Yue Gao

CARNet: Automatic Cerebral Aneurysm Classification in Time-of-Flight
MR Angiography by Leveraging Recurrent Neural Networks ............... 136
  Yan Hu, Yuan Xu, Xiaosong Huang, Deqiao Gan, Haiyan Huang,
  Liyuan Shao, Qimin Cheng, and Deng Xianbo

White-Box Attacks on the CNN-Based Myoelectric Control System .......... 149
  Bo Xue, Le Wu, Aiping Liu, Xu Zhang, and Xun Chen

MMG-HCI: A Non-contact Non-intrusive Real-Time Intelligent
Human-Computer Interaction System .................................. 158
  Peixian Gong, Chunyu Wang, and Lihua Zhang

DSGSR: Dynamic Semantic Generation and Similarity Reasoning
for Image-Text Matching ............................................. 168
  Xiaojing Li, Bin Wang, Xiaohong Zhang, and Xiaochun Yang

Phase Partition Based Virtual Metrology for Material Removal Rate
Prediction in Chemical Mechanical Planarization Process .................. 180
  Wenlan Jiang, Chunpu Lv, Tao Zhang, and Huangang Wang

SAR Target Recognition Based on Model Transfer and Hinge Loss
with Limited Data ................................................... 191
  Qishan He, Lingjun Zhao, Gangyao Kuang, and Li Liu

Neighborhood Search Acceleration Based on Deep Reinforcement
Learning for SSCFLP ................................................ 202
  Zonghui Zhang, Zhangjin Huang, and Lu Zou

GBCI: Adaptive Frequency Band Learning for Gender Recognition
in Brain-Computer Interfaces ......................................... 213
  Pengpai Wang, Yueying Zhou, Zhongnian Li, and Daoqiang Zhang

**Computer Vision**

Hybrid Domain Convolutional Neural Network for Memory Efficient
Training ............................................................ 227
  Bochen Guan, Yanli Liu, Jinnian Zhang, William A. Sethares, Fang Liu,
  Qinwen Xu, Weiyi Li, and Shuxue Quan

Brightening the Low-Light Images via a Dual Guided Network .............. 240
  Jianing Sun, Jiaao Zhang, Risheng Liu, and Fan Xin

Learning Multi-scale Underexposure Image Correction ..................... 252
    Wei Zhong, Xiaodong Zhang, Long Ma, Risheng Liu, Xin Fan,
    and Zhongxuan Luo

Optimizing Loss Function for Uni-modal and Multi-modal Medical
Registration ......................................................... 264
    Zi Li, Fan Xin, Risheng Liu, and Zhongxuan Luo

Registration of 3D Point Clouds Based on Voxelization Simplify
and Accelerated Iterative Closest Point Algorithm ......................... 276
    Jiayu Wang and Hongjun Li

Few-shot Weighted Style Matching for Glaucoma Detection ................ 289
    Jinhui Liu and Xin Yu

Lightweight Convolutional SNN for Address Event Representation Signal
Recognition .......................................................... 301
    Zhaoxin Liu, Bangbo Huang, Jinjian Wu, and Guangming Shi

In-the-Wild Facial Highlight Removal via Generative Adversarial Networks ... 311
    Zhibo Wang, Ming Lu, Feng Xu, and Xun Cao

A Cross-Layer Fusion Multi-target Detection and Recognition Method
Based on Improved FPN Model in Complex Traffic Environment ............. 323
    Cuijin Li, Dewei Chen, Junji Chen, and Hongying Dai

Various Plug-and-Play Algorithms with Diverse Total Variation Methods
for Video Snapshot Compressive Imaging ............................... 335
    Xin Yuan

EEG Signals Classification in Time-Frequency Images by Fusing
Rotation-Invariant Local Binary Pattern and Gray Level Co-occurrence
Matrix Features ...................................................... 347
    Zhongyi Hu, Zhenzhen Luo, Shan Jin, and Zuoyong Li

Reduced-reference Perceptual Discrepancy Learning for Image
Restoration Quality Assessment ....................................... 359
    Leida Li, Bo Hu, Yipo Huang, and Hancheng Zhu

EFENet: Reference-Based Video Super-Resolution with Enhanced Flow
Estimation .......................................................... 371
    Yaping Zhao, Mengqi Ji, Ruqi Huang, Bin Wang, and Shengjin Wang

Multi-label Aerial Image Classification via Adjacency-Based Label
and Feature Co-embedding .................................................... 384
  *Xiangrong Zhang, Shouping Shan, Jing Gu, Xu Tang, and Licheng Jiao*

Coarse-to-Fine Attribute Editing for Fashion Images ....................... 396
  *Qinghu Wang, Jianjun Qian, Xingxing Zou, Jian Yang,
  and Waikeung Wong*

PSS: Point Semantic Saliency for 3D Object Detection ..................... 408
  *Jiajing Cen, Pei An, Gaojie Chen, Junxiong Liang, and Jie Ma*

Image Segmentation Based on Non-convex Low Rank Multiple Kernel
Clustering ................................................................. 420
  *Xuqian Xue, Xiao Wang, Xiaoqian Zhang, Jing Wang, and Zhigui Liu*

Novel View Synthesis of Dynamic Human with Sparse Cameras ............. 432
  *Xun Lv, Yuan Wang, Feiyi Xu, Jianhui Nie, Feng Xu, and Hao Gao*

Attention Guided Retinex Architecture Search for Robust Low-light
Image Enhancement ........................................................ 444
  *Xiaoke Shang, Jingjie Shang, Long Ma, Shaomin Zhang, and Nai Ding*

Dual Attention Feature Fusion Network for Monocular Depth Estimation ...... 456
  *Yifang Xu, Ming Li, Chenglei Peng, Yang Li, and Sidan Du*

A Strong Baseline Based on Adaptive Mining Sample Loss for Person
Re-identification .......................................................... 469
  *Yongchang Gong, Liejun Wang, Shuli Cheng, and Yongming Li*

Unsupervised Domain Adaptation via Attention Augmented Mutual
Networks for Person Re-identification ..................................... 481
  *Hui Tian and Junlin Hu*

MPNet: Multi-scale Parallel Codec Net for Medical Image Segmentation ...... 492
  *Bin Huang, Jian Xue, Ke Lu, Yanhao Tan, and Yang Zhao*

Part-Aware Spatial-Temporal Graph Convolutional Network for Group
Activity Recognition ...................................................... 504
  *Qi Wang, Xianglong Lang, Ye Xiang, and Lifang Wu*

A Loop Closure Detection Algorithm Based on Geometric Constraint
in Dynamic Scenes ........................................................ 516
  *Cheng Hang, Bo Zhao, and Baoyun Wang*

Unsupervised Deep Plane-Aware Multi-homography Learning for Image
Alignment . . . . . . . . . . . . . . . . . . . . . . . . . . . . . . . . . . . . . . . . . . . . . . . . . . . . . . . . .  528
  *Tao Cai, Yunde Jia, Huijun Di, and Yuwei Wu*

3D Hand Pose Estimation via Regularized Graph Representation Learning . . . . .  540
  *Yiming He and Wei Hu*

Emotion Class-Wise Aware Loss for Image Emotion Classification . . . . . . . . . . .  553
  *Sinuo Deng, Lifang Wu, Ge Shi, Heng Zhang, Wenjin Hu, and Ruihai Dong*

Image Style Recognition Using Graph Network and Perception Layer . . . . . . . . .  565
  *Quan Wang and Guorui Feng*

Arable Land Change Detection Using Landsat Data and Deep Learning . . . . . . .  575
  *Mei Huang and Wenzhong Yang*

Attention Scale-Aware Deformable Network for Inshore Ship Detection
in Surveillance Videos . . . . . . . . . . . . . . . . . . . . . . . . . . . . . . . . . . . . . . . . . . . . . . .  589
  *Di Liu, Yan Zhang, Yan Zhao, and Yu Zhang*

Context-BMN for Temporal Action Proposal Generation . . . . . . . . . . . . . . . . . . . .  601
  *Baoqing Tang, Shengye Yan, Yihua Ni, Yongjia Yang, and Kang Pan*

Revisiting Knowledge Distillation for Image Captioning . . . . . . . . . . . . . . . . . . . .  613
  *Jingjing Dong, Zhenzhen Hu, and Yuanen Zhou*

Enhanced Attribute Alignment Based on Semantic Co-Attention
for Text-Based Person Search . . . . . . . . . . . . . . . . . . . . . . . . . . . . . . . . . . . . . . . . . .  626
  *Hao Wang and Zhenzhen Hu*

ARShape-Net: Single-View Image Oriented 3D Shape Reconstruction
with an Adversarial Refiner . . . . . . . . . . . . . . . . . . . . . . . . . . . . . . . . . . . . . . . . . . . .  638
  *Hao Xu and Jing Bai*

Training Few-Shot Classification via the Perspective of Minibatch
and Pretraining . . . . . . . . . . . . . . . . . . . . . . . . . . . . . . . . . . . . . . . . . . . . . . . . . . . . . . .  650
  *Meiyu Huang, Yao Xu, Wei Bao, and Xueshuang Xiang*

Classification Beats Regression: Counting of Cells from Greyscale
Microscopic Images Based on Annotation-Free Training Samples . . . . . . . . . . . .  662
  *Xin Ding, Qiong Zhang, and William J. Welch*

Adaptive Learning Rate and Spatial Regularization Background Perception
Filter for Visual Tracking . . . . . . . . . . . . . . . . . . . . . . . . . . . . . . . . . . . . . . . . . . . . . .  674
  *Kai Lv, Liang Yuan, L. He, Ran Huang, and Jie Mei*

**Data Mining**

Estimating Treatment Effect via Differentiated Confounder Matching ......... 689
  *Zhao Ziyu, Kun Kuang, and Fei Wu*

End-to-End Anomaly Score Estimation for Contaminated Data
via Adversarial Representation Learning ................................. 700
  *Daoming Li, Jiahao Liu, and Huangang Wang*

Legal Judgment Prediction with Multiple Perspectives on Civil Cases ......... 712
  *Lili Zhao, Linan Yue, Yanqing An, Ye Liu, Kai Zhang, Weidong He,
  Yanmin Chen, Senchao Yuan, and Qi Liu*

Multi-view Relevance Matching Model of Scientific Papers Based
on Graph Convolutional Network and Attention Mechanism ................ 724
  *Jie Song, Zhe Xue, Junping Du, Feifei Kou, Meiyu Liang, and Mingying Xu*

A Hierarchical Multi-label Classification Algorithm for Scientific Papers
Based on Graph Attention Networks .................................... 735
  *Changwei Zheng, Zhe Xue, Junping Du, Feifei Kou, Meiyu Liang,
  and Mingying Xu*

HNECV: Heterogeneous Network Embedding via Cloud Model
and Variational Inference .............................................. 747
  *Ming Yuan, Qun Liu, Guoyin Wang, and Yike Guo*

DRPEC: An Evolutionary Clustering Algorithm Based on Dynamic
Representative Points ................................................... 759
  *Peng Li, Haibin Xie, and Zhiyong Ding*

User Reviews Based Rating Prediction in Recommender System ............. 771
  *Wenchuan Shi, Liejun Wang, Shuli Cheng, and Yongming Li*

Exploiting Visual Context and Multi-grained Semantics for Social Text
Emotion Recognition ................................................... 783
  *Wei Cao, Kun Zhang, Hanqing Tao, Weidong He, Qi Liu, Enhong Chen,
  and Jianhui Ma*

**Author Index** ........................................................ 797

# Explainability, Understandability, and Verifiability of AI

# Reducing Adversarial Examples Through Boundary Methods

Xinyi Hu$^{(\boxtimes)}$, Zhengming Zhang, Zhouyong Liu, Zhiwei Han, and Luxi Yang

School of Information Science and Engineering, Southeast University,
Nanjing 210096, China
{icedomain,zmzhang,liuzhouyong,zhw_h,lxyang}@seu.edu.cn

**Abstract.** At present, deep neural networks are widely used on a variety of tasks in computer vision, machine translation, speech recognition, etc. Unfortunately, this inexplicable black-box structure lacks robustness. In previous work, adversarial examples are proposed to describe the phenomenon that neural networks are vulnerable to be attacked. Interestingly, in addition to the widely accepted "noise" or "bugs", recent research has shown that the adversarial examples are "non-robust features", because the classifier trained on adversarial examples retains the ability to generalize to the original test set. In this paper, we link the relationship between large margin methods and the capabilities to defend against adversarial attacks, and further link the relationship to non-robust features. We compare the defense capabilities of the models trained by large margin loss function and general cross-entropy loss function against Fast Gradient Sign Method (FGSM) attack and Project Gradient Descent (PGD) attack and evaluate non-robust features extracted by the trained models. It is proved that *the model trained with large margin loss function is more resistant to adversarial perturbation and it gets fewer non-robust features*. This further indicates a direction for training robust networks: *to balance model test accuracy and defense capabilities*. Based on the margin method, we combined thickness to strengthen the description of the decision boundary. Through the feature space visualization, the effect of the boundary methods on the robustness of the model is intuitively illustrated.

**Keywords:** Large margin loss · Adversarial examples · Non-robust features · Boundary thickness · Feature clustering

## 1 Introduction

Deep neural network (DNN) [3] and its related research, especially convolutional neural network (CNN) [16], have been closely followed by many scholars. However, humans still have difficulty in understanding neural networks and can not give convincing explanations. Unfortunately, this black-box structure which is

---

This work was supported by the National Natural Science Foundation of China under Grant 61971128.

L. Fang et al. (Eds.): CICAI 2021, LNAI 13070, pp. 3–15, 2021.
https://doi.org/10.1007/978-3-030-93049-3_1

difficult to explain is severely non-robust. Neural networks are extremely sensitive to input data, and small perturbation can completely interfere with the correctness of the model. Research [19] proposes *adversarial examples* for the first time, i.e. minimal perturbation can fool models. Even modifying very few pixels can mislead models in the classification task [18].

Although people have proposed many methods for model defense, trying to resist malicious attacks, enhance the generalization ability and robustness of the model, humans have very little understanding of adversarial examples. More people believe that adversarial examples are "bugs" just like random noise for boundary smoothness exhibits certain defensive performance. Recent research [6] has expressed a novel point of view: the adversarial examples are intrinsic features. Unlike ordinary features, they may be the features of target labels in the dataset (not the label of this sample). They are called *non-robust features*. If the generation of adversarial examples is not omnidirectional in the high-dimensional input space, it may not be reasonable to treat adversarial examples as "bugs" or "noise". This is a new perspective for model defense. Perhaps we have to evaluate non-robust features in advance.

The large margin methods are widely used in traditional machine learning e.g. support vector machine. In this kind of method, the classifier is trained by maximizing the margin between samples. This suggests that large margin methods will affect the model's defense capabilities. In this case, greater perturbation is required to invalidate the model. In this work, we will combine the method of maximizing the margin for model training and verify that this will defend against adversarial attacks. Intuitively, large margin methods will optimize the decision boundary and extract features that are more relevant to its label, which will significantly reduce non-robust features. Boundary thickness [21] supplements the boundary margin, and we can get thick boundary by data augmentation method—mixup (a sufficient condition proved in [21]). It is a way to train more robust classifiers.

In this paper, we will compute the test accuracy of the model trained by large margin loss against FGSM [4] and PGD [13] attack to represent the defense capabilities. Next, we measure the non-robust feature scores according to the protocol in [6]. At the same time, observe the effect of different boundary methods in the training process, i.e. the discriminative feature distribution in the feature space. The main contributions of this paper are as follows:

1. Use the large margin loss function to train classification models, compare non-robust feature scores extracted by the model trained by cross-entropy loss and models' defense ability against perturbation.
2. Verify the effect of large margin loss on datasets [7,8].
3. We introduce boundary thickness, compare the effects of the large margin method, thick boundary method, and joint method on feature distribution. Then we intuitively illustrate the effect of boundary methods on model robustness through feature space visualization.

## 2   Related Works

Research [4,19] shows that conventional deep learning methods lack robustness. Szegedy et al. [19] proposed adversarial examples for the first time. Immediately after, Goodfellow et al. [4] showed that the adversarial examples come from the linear factor of the neural network, i.e. ReLU. This activation function is locally linear. Then iterative methods were developed, such as PGD [13]. Even in extreme cases, only the change of one-pixel value can attack successfully [18]. But these attack methods do not indicate exactly what the adversarial examples are.

In NeurIPS competition on defense against adversarial attacks, Liao et al. [9] propose High-Level Representation Guided Denoiser to defend against perturbation. They agree that adversarial perturbation is random noise. Ilyas et al. [6] show a completely different point of view. They say that the adversarial examples are features. This indicates that the original sample in the dataset has enhanced correlation with target labels, and then adversarial examples are generated. Although this does not conform to our intuitive understanding, it is reasonable because feature squeezing [20] and model distillation [15] have been used for model defense. These methods adjust the features extracted by the deep models.

How to optimize the feature extraction? Geometrically, the robustness of the neural network is related to its decision boundary. Yousefzadeh et al. [22] refers to a regularization method called mixup [24], and observes the softmax output scores predicted by the points on the connecting segments of two samples with different labels (these points cross the decision boundary at least once), i.e. they investigate the paths between inputs and flip points. Besides, there is also a work that compares the SVM and classifiers trained by deep learning which indicates that cross-entropy loss leads to poor boundary and margin. Therefore, we decide to verify the relationship between large margin methods, the robustness of models, and non-robust feature scores in this paper.

There are many perfect methods for the processing of large margin methods. On the one hand, the geometric relationship can be directly considered for mathematical derivation [2], e.g. SVM. Or through indirect processing, i.e. reducing the loss values of samples with low confidence (softmax prediction scores), Focal Loss [10] is a successful work. In addition, the calculation of the loss function can also be converted from the distance space to the angular space [25]. But how to deal with large margin methods is not the focus of attention in our work, we only consider the geometric margin methods.

The methods of training and optimization (e.g. regularization, etc. ) will directly affect the distribution of decision boundary [23]. There are various descriptions of the boundary. Thickness [21] is an enhanced description of boundary margin. The main boundary methods considered in this paper are only boundary margin and boundary thickness.

## 3   Boundary Margin

The geometrical large margin methods can regularize the model. The larger the distance of each sample to the decision boundary, the larger the perturbations for successful attacks. Now we need to convert the large margin methods into an optimization problem, i.e. the loss function.

## 3.1  Large Margin Methods

Consider a $n$ classification problem. $(\boldsymbol{x}_i, y_i)_{i=1,\cdots,N}$ are samples in the dataset $\mathcal{D}$, where $\boldsymbol{x}_i \in \mathcal{X}, i = 1, \cdots, N$. We train a classifier $f(\cdot; \theta)$ which satisfy $f : \mathcal{X} \rightarrow \mathbb{R}$. Its input is the sample $\boldsymbol{x}_i$, and the output is the predicted class.

More specifically, $f = (f_1, f_2, \cdots, f_n)$, here $f_i : \mathcal{X} \rightarrow \mathbb{R}$, for $i = 1, \ldots, n$ which represents the softmax predicted score. Then, the output of $f(\cdot; \theta)$ is the index of the maximum prediction score. i.e. $f(\boldsymbol{x}; \theta) = \arg\max_i f_i(\boldsymbol{x}; \theta)$.

Then, we expect to describe the decision boundary. The decision boundary between the two classes $\{i, j\}$ is where the prediction score of $i$ equals the one of $j$, i.e.

$$\Omega_{i,j} = \{\boldsymbol{x} \mid f_i(\boldsymbol{x}; \theta) = f_j(\boldsymbol{x}; \theta)\} \tag{1}$$

Based on this, the distance from the sample $\boldsymbol{x}$ to the decision boundary $\Omega_{i,j}$ is the minimum perturbation that makes the two output scores of $\{i, j\}$ the same. i.e.

$$d_{\boldsymbol{x},\{i,j\}} = \min_{\boldsymbol{\delta}} \|\boldsymbol{\delta}\|_p$$
$$\text{s.t.}\quad f_i(\boldsymbol{x} + \boldsymbol{\delta}; \theta) = f_j(\boldsymbol{x} + \boldsymbol{\delta}; \theta) \tag{2}$$

For $\boldsymbol{\delta}$ is the perturbation added to the sample $\boldsymbol{x}$. $\|.\|_p$ is $l_p$ norm, including but not limited to $l_1$, $l_2$, $l_\infty$ norm. We will use the $l_2$ norm more in this paper. According to the mathematical derivation in [2], we can have the following solution, i.e. the large margin loss function.

$$\hat{\theta} = \arg\min_\theta \sum_{k=1}^{N} \mathscr{A}_{i \neq y_k} \max\{0, \gamma + \frac{f_i(\boldsymbol{x}_k; \theta) - f_{y_k}(\boldsymbol{x}_k; \theta)}{\|\nabla_{\boldsymbol{x}} f_i(\boldsymbol{x}_k; \theta) - \nabla_{\boldsymbol{x}} f_{y_k}(\boldsymbol{x}_k; \theta)\|_q}\} \tag{3}$$

Here, $\|.\|_q$ is the dual norm of $\|.\|_p$ ($q = \frac{p}{p-1}$). $\mathscr{A}$ is only an aggregation operator. $\gamma$ is a slack-variable (hyperparameter) to balance for the irregular distribution of data. If interested, see more details in [2].

We will use the Eq. 3 for model training. This equation has little to do with the network structure and input data format, and it can be used with other regularization methods [5,14]. For comparison, we will use the cross-entropy loss function for baseline model training.

## 3.2  Informal Definition of Non-robust Features

Ilyas et al. [6] found that the adversarial examples contain the features. The more acceptable explanation before it is "noise" or "bugs" because we believe that the direction of adversarial perturbation is omnidirectional, and its correlation with target labels are similar, without bias, just like random noise in signal processing.

The point of view in [6] is that adversarial perturbation is biased, and it has different connections with different labels, i.e. the features extracted by the deep network include not only the features of the correct label but also the features of other labels. The protocol is shown in Fig. 1, i.e. models can have a good generalization even training on the dataset generated by adversarial examples (attacked and relabeled), e.g. the test accuracy is more than 60% on CIFAR10 described in [6] (this value is not low, because CIFAR10 has 10 classes).

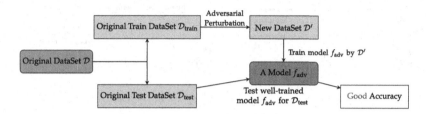

**Fig. 1. Illustration of non-robust features:** The model trained on the attacked and relabeled dataset generalize well on the original test set.

We further explain the process described in Fig. 1. Split the original dataset $\mathcal{D} = \{(\boldsymbol{x}_i, y_i)\}_{i=1}^{N}$ into the training set $\mathcal{D}_{\text{train}}$ and the testing set $\mathcal{D}_{\text{test}}$. In order to perform a white box attack on $\mathcal{D}_{\text{train}}$ and generate a new dataset $\mathcal{D}'$, we need to train a classifier $f(\cdot; \theta)$ in advance. Generally, Acc is not low, and the explanation in [6] is that adversarial examples contain "'non-robust features" of target labels, which is helpful for generalization. The accuracy (Acc) can be a good measure of non-robust features. Therefore, we can briefly give the definition of non-robust feature score:

**Definition 1 (Non-Robust Feature Score).** *The new network model $f_{adv}$ trained on the new dataset $\mathcal{D}'$ which is attacked and relabeled, test on the original clean test set $\mathcal{D}_{test}$ and get the test accuracy Acc. The generalization accuracy Acc is called* **Non-Robust Feature Score**[1]*.*

In the experimental part, we will compute the test accuracy in this way to measure non-robust features. Similar methods were used in recent works [21]. We intuitively get the less non-robust features extracted by a more ideal classifier, because these features reflect more information of target labels than the labels of the sample itself.

### 3.3   Experiments on MNIST and CIFAR10

As described in Sect. 3.1 and Sect. 3.2, the defense capability against adversarial attacks and non-robust feature scores are two important indicators for evaluating the decision boundary. In the experimental setup, we will use different adversarial

---

[1] The definition is not formal, it is limited to the research described in this paper.

8    X. Hu et al.

perturbations for model attacks. After that, the non-robust feature scores of the two methods are calculated for model comparison. Baseline models are trained by the cross-entropy loss function. We experiment on dataset [7,8], see Fig. 2.

We use Fast Gradient Sign Method (FGSM) and Project Gradient Descent(PGD) as attack methods, the number of iterations $t$ in the PGD attack is 10. If the maximum visual change accepted by humans is 32 pixels, then the maximum value in $x$-axis is roughly $32/255 \approx 0.125$.

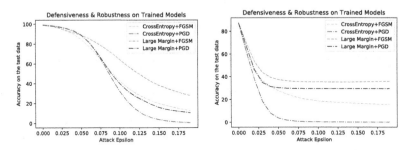

**Fig. 2. Comparison of the test accuracy of models against attack with different $\epsilon$:** Left is the curve for MNIST, right is the curve for CIFAR10. For the attacks of different amplitudes represented by the value in $x$-axis, the pixel value is normalized to $[0, 1]$, 0 means pixel value is 0, and 1 means pixel value is 255.

**Results:** the test accuracy will be reduced to a certain extent after being attacked. But FGSM attack and PGD attack show big differences. Compared with FGSM, PGD is stronger and can reduce the test accuracy to the greatest extent within a smaller $\epsilon$. The large margin loss can improve defensive performance. Under the basic FGSM attack, the large margin loss and cross-entropy loss show a significant difference with small $\epsilon$. And when $\epsilon = 0.1$, the difference of test accuracy exceeds 30%. Under the PGD attack, both of them did not perform well, and even the accuracy drop faster. With the increase of the attack amplitude, it can still show that the large margin loss has the greater ability.

*Remark 1.* In Fig. 2, the test accuracy tends to be stable during large attacks for the pixel value will be limited in the experiments. These are more like wrong samples or "noisy labels". Such values are meaningless.

Next, we evaluate the non-robust features as described in Sect. 3.2. Use the test accuracy to represent the non-robust feature score as shown in Fig. 3 during the training process of the new deep network $f_{adv}$ .

Although each time Acc will stabilize at different values, e.g. MNIST once stabilized at 40%, once stabilized at 45%, the trend of Acc and the difference between the two types of loss are obvious. In Fig. 3, the curves corresponding to the two models trained by different loss intuitively prove this.

**Results:** Models trained with large margin loss can have smaller non-robust feature scores, it is extremely impressive. This shows that a robust classifier tends

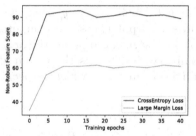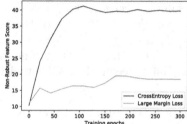

**Fig. 3. Non-robust feature scores:** Left are scores of MNIST which is stable at 90% and 60% for two loss functions, right are scores of CIFAR10 which is stable at 40% and 20%. In the early epochs, Acc increased significantly. Non-robust feature scores tend to stabilize at larger epochs. Models trained by large margin loss can have less non-robust features.

to be more geometrically ordered, and the large margin method has larger space to defense, which is consistent with the fact that we recognize during network training:*to balance model test accuracy and defense capabilities*. Research [23] shows a similar view. We are more inclined to reduce the proportion of non-robust features.

*Remark 2.* On the one hand, this part of the experiment needs to be observed repeatedly many times, because our scores are measured by the test accuracy(in the original test set) of the model $f_{adv}$, the skills of training (e.g. data augmentation, learning rate, weight decay, regularization, etc. ) will interfere with this evaluation result. On the other hand, we remove some points when drawing Fig. 3, this part of the scores more reflect the abnormality of the model during training in a single period.

Besides, MNIST and CIFAR10 show differences. More "easy" datasets have more non-robust features in the case of the same 10 classes, and its training is also easier. It shows that the difficulty of model training is related to the robustness of models.

## 4   Boundary Thickness

For the large margin methods, the decision boundary is just a set of thin spatial surfaces. e.g. the decision boundary is usually a subset that satisfies $f_0(x) = f_1(x) = 0.5$ in the binary classification problem. When crossing the boundary, the sample changes from $x_1$ to $x_2$, its confidence changes in various ways, and margin boundaries are not enough to describe. Here, we will discuss a new viewpoint of decision boundary: *thickness* [21]. It is equivalent to reducing a single subset of the boundary to a set of subsets that can better describe networks.

For a classifier $f(\cdot; \theta) = (f_1, f_2, \cdots, f_n)$, We pay attention to its maximum probability output $f(x; \theta) = \arg\max_i f_i(x; \theta)$, and use the difference of confidence to define boundary thickness. For the connection segment of two random samples $(x_i, y_i), (x_j, y_j)$ with different labels $(y_i \neq y_j)$, e.g. mixup. Then the confidence difference of the softmax output is easy to compute, i.e. $g_{ij}(x) = f_{y_i}(x) - f_{y_j}(x)$, where $x$ is on the segment of pairs $(x_i, x_j)$, i.e. $x(t) = tx_i + (1 - t)x_j, t \in [0, 1]$. The boundary thickness is the length of the confidence difference on the segment within a certain range, which is specifically defined as:

**Definition 2 (Boundary Thickness).** *For pairs* $(\alpha, \beta) \in (-1, 1) (\alpha < \beta)$, *a series of sample pairs* $(x_i, y_i), (x_j, y_j)$ *of different classes in the dataset* $\mathcal{D}$. *Boundary thickness of the classifier* $f(\cdot; \theta)$ *is:*

$$\Theta(f, \alpha, \beta) = \mathbb{E}_{(x_i, x_j) \sim \mathcal{D}} \left[ \int_{t \in [0,1]} \mathbf{I}\{\alpha < g_{ij}(x(t)) < \beta\} dx(t) \right] \qquad (4)$$

*where* $g_{ij}(x) = f_{y_i}(x) - f_{y_j}(x)$, $x(t) = tx_i + (1 - t)x_j, t \in [0, 1]$. $\mathbf{I}(\cdot)$ *output is 1 when the input is True.*

Intuitively, boundary thickness is related to robustness. The "thin" boundary fits the narrow space but is easy to attack between two different classes. On the contrary, a thicker boundary is difficult to fit the data, but it will be more robust, and it needs larger $\epsilon$ to successfully attacks. Therefore, boundary thickness leads to the following conjectures for general neural networks: *we need to maximize boundary margin and increase boundary thickness to achieve robustness.*

### 4.1   Comparison of Boundary Thickness and Margin

Here, we discuss the difference between boundary thickness and margin. The margin of $x$ is the minimum distance from $x$ to $f_i(x) = f_j(x)$, where $i, j$ represent the two classes in the multi-classification problem. The thickness is related to the confidence corresponding to the sample space under different $\alpha, \beta$. e.g. consider the one-dimensional two-classification problem on the $x$ axis, where the true label is $\text{sign}(x)$. Then the two prediction functions $f_1(x) = \arctan(x)$ and $f_2(x) = \arctan(100x)$ have the same margins for any $x$, but different boundary thicknesses. See Fig. 4 for the difference.

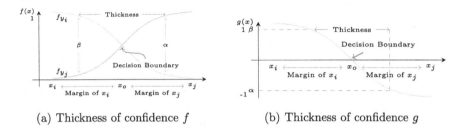

(a) Thickness of confidence $f$          (b) Thickness of confidence $g$

**Fig. 4.** Boundary margin vs. boundary thickness

The boundary thickness shown in Fig. 4 is consistent with our definition, but the margin is not completely correct. Most of the pairs $(x_i, x_j)$ are not orthogonal to the decision boundary. However, such an informal illustration can give an intuitive contrast. Generally, the boundary thickness is less than the margin if two margins (the distance of $x_i$ to the boundary add $x_j$ to the boundary) are considered together, i.e. the thickness is an enhanced version of the margin in Sect. 3.1. When $\alpha = f_{y_i}(x_j) - f_{y_j}(x_j)$ and $\beta = f_{y_i}(x_i) - f_{y_j}(x_i)$, boundary thickness degenerates to boundary margin.

## 4.2 Mixup and Boundary Thickness

Experiments prove that mixup will thicken the decision boundary, and it is a sufficient and unnecessary condition to maximize the boundary thickness [21]. The main idea of mixup is linear sampling, i.e.

$$f_{mixup}(\lambda x_i + (1 - \lambda)x_j) = \lambda y_i + (1 - \lambda)y_j \tag{5}$$

Train networks to maximize thickness:

$$f_{thick}(x) = \arg\max_{f(x)} \min_{(\alpha,\beta)} \Theta(f, \alpha, \beta) \tag{6}$$

**Theorem 1.** *In the binary classification problem, linear function $f_{lin}(tx_i + (1 - t)x_j) = [t, 1 - t]$ is a sufficient and unnecessary condition to maximize $\Theta(f)$, i.e.*

$$f_{lin}(x) = \arg\max_{f(x)} \min_{(\alpha,\beta)} \Theta(f, \alpha, \beta) \tag{7}$$

*where $(x_i, x_j)$ are sample pairs, and $(\alpha, \beta) \in (-1, 1)$ are parameter pairs.*

Research [21] gives proof of the Theorem 1 . If interested, see more details in their work. With the support of the Theorem 1, we can indirectly train a classifier with thicker boundaries by mixup.

## 5   Boundary Methods and Feature Space

We have previously described the relationship between boundary margin, boundary thickness, and the robustness of models. In the process of increasing the margin and the thickness, the proportion of adversarial examples caused by boundary will change, and non-robust features will reduce. In essence, the change of boundary margin and thickness is a tighter clustering in the high-dimensional sample/feature space, which is similar to the effect of the angle margin [25].

To observe the role of different boundary methods during training (i.e. the more discriminative feature distribution in the feature space), we use the large margin loss, data augmentation method mixup, and both of them to train different models.

It is difficult to visualize the distribution of high-dimensional features. It can be processed by some classic data dimensionality reduction methods, e.g. PCA [17], LDA [1], T-SNE [11,12], etc. Besides, we can directly set the output feature dimension of the last layer(Fully Connected Layer) to 3, corresponding to the $x, y, z$-axis. Due to the large feature span, to better observe the distribution, it is normalized by the $\ell_2$ norm before visualization(features are distributed on the unit ball), Fig. 5 shows the feature distribution of multiple methods (The results on MNIST and CIFAR10 are similar).

Figure 5(a) is the distribution trained by cross-entropy loss. Without additional methods, the feature distribution is scattered and contains a large number of crosses, which will lead to errors. In a sense, some of these samples are abnormal samples. While the classification task has a bottleneck, it is particularly easy to cause adversarial examples. Figure 5(b) is the distribution trained by large margin loss. It enlarges the margin between different classes and has a good clustering effect, therefore adversarial examples can be greatly reduced. Figure 5(c) uses the data augmentation method mixup to indirectly obtain a thick boundary classifier. Compared with Fig. 5(b), its feature distribution is slightly superior, which is mainly reflected in the reduction of intra-class variance. But mixup has the problem of instability during training, which directly leads to the disadvantage of insufficient inter-class spacing. It is obviously between Fig. 5(a) and Fig. 5(b), the contradiction between model robustness and accuracy is more prominent. Figure 5(d) combines the advantages of the large margin methods and the thick boundary methods, i.e. superimposes mixup and large margin loss function. Smaller intra-class variance and larger inter-class spacing, and then we can get less adversarial examples, further, it proves the fact that "multiple boundary methods help improve the model's defense capabilities (i.e. robustness)".

The comparison of Fig. 5 intuitively shows the direct way to reduce adversarial examples caused by decision boundary through the feature space distribution. Other data augmentation methods, network, feature, and other regularization methods are not inconsistent with this, and they can be used in combination to help model security. In addition, more in-depth descriptions of decision boundary are also guiding directions for improving network performance.

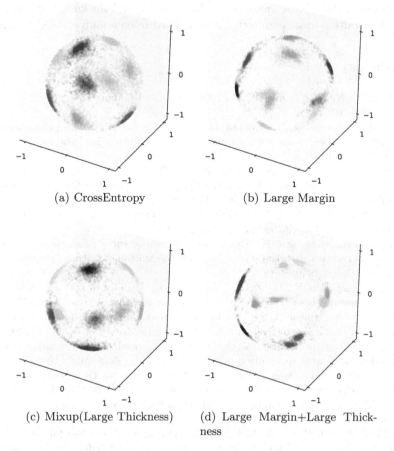

(a) CrossEntropy

(b) Large Margin

(c) Mixup(Large Thickness)

(d) Large Margin+Large Thickness

**Fig. 5. Boundary methods and feature space visualization.** Subfig a is the distribution trained by cross-entropy loss. Subfig b is the distribution trained by large margin loss. Subfig c is the distribution trained by mixup. Subfig d is the distribution trained by large margin loss and mixup.

## 6   Conclusions

Deep learning methods seriously lack robustness and are susceptible to interference from adversarial perturbation. We first train more robust classifiers with large margin loss. Compared with the standard cross-entropy loss, it can reduce non-robust features while enhancing the network defense capabilities. The experimental results on the dataset prove that *the model trained with large margin loss function is more resistant to adversarial perturbation and it gets fewer non-robust features.* This further indicates a direction for training the network:*to balance model test accuracy and defense capabilities.* Based on the margin methods, we combined the thickness to strengthen the description of the boundary. The effect of the boundary methods on the robustness is intuitively explained through the visualization of the feature space. Boundary margin and thickness can make the feature distribution more compact, and it is hard to cause adversarial examples.

# References

1. Blei, D.M., Ng, A.Y., Jordan, M.I.: Latent dirichlet allocation. J. Mach. Learn. Res. **3**, 993–1022 (2003)
2. Elsayed, G., Krishnan, D., Mobahi, H., Regan, K., Bengio, S.: Large margin deep networks for classification. In: Advances in Neural Information Processing Systems, pp. 842–852 (2018)
3. Glorot, X., Bengio, Y.: Understanding the difficulty of training deep feedforward neural networks. In: Proceedings of the 13th International Conference on Artificial Intelligence and Statistics, pp. 249–256 (2010)
4. Goodfellow, I.J., Shlens, J., Szegedy, C.: Explaining and harnessing adversarial examples. arXiv preprint arXiv:1412.6572 (2014)
5. Hinton, G.E., Srivastava, N., Krizhevsky, A., Sutskever, I., Salakhutdinov, R.R.: Improving neural networks by preventing co-adaptation of feature detectors. arXiv preprint arXiv:1207.0580 (2012)
6. Ilyas, A., Santurkar, S., Tsipras, D., Engstrom, L., Tran, B., Madry, A.: Adversarial examples are not bugs, they are features. arXiv preprint arXiv:1905.02175 (2019)
7. Krizhevsky, A., Hinton, G., et al.: Learning multiple layers of features from tiny images. In: Handbook of Systemic Autoimmune Diseases, vol. 1, no. 4 (2009)
8. LeCun, Y., Bottou, L., Bengio, Y., Haffner, P., et al.: Gradient-based learning applied to document recognition. Proc. IEEE **86**(11), 2278–2324 (1998)
9. Liao, F., Liang, M., Dong, Y., Pang, T., Hu, X., Zhu, J.: Defense against adversarial attacks using high-level representation guided denoiser. In: Proceedings of the IEEE Conference on Computer Vision and Pattern Recognition, pp. 1778–1787 (2018)
10. Lin, T.Y., Goyal, P., Girshick, R., He, K., Dollár, P.: Focal loss for dense object detection. In: Proceedings of the IEEE International Conference on Computer Vision, pp. 2980–2988 (2017)
11. Maaten, L., Hinton, G.: Visualizing data using t-sne. J. Mach. Learn. Res. **9**, 2579–2605 (2008)
12. Maaten, L.: Barnes-Hut t-SNE. Comput. Sci. **1301**, 3342 (2013)
13. Madry, A., Makelov, A., Schmidt, L., Tsipras, D., Vladu, A.: Towards deep learning models resistant to adversarial attacks. arXiv preprint arXiv:1706.06083 (2017)
14. Murugan, P., Durairaj, S.: Regularization and optimization strategies in deep convolutional neural network. CoRR abs/1712.04711 (2017). http://arxiv.org/abs/1712.04711
15. Papernot, N., McDaniel, P., Wu, X., Jha, S., Swami, A.: Distillation as a defense to adversarial perturbations against deep neural networks. In: 2016 IEEE Symposium on Security and Privacy (SP), pp. 582–597. IEEE (2016)
16. Sainath, T.N., Mohamed, A., Kingsbury, B., Ramabhadran, B.: Deep convolutional neural networks for LVCSR. In: 2013 IEEE International Conference on Acoustics, Speech and Signal Processing, pp. 8614–8618. IEEE (2013)
17. Shlens, J.: A tutorial on principal component analysis. Int. J. Remote Sens. **51**(2) (2014)
18. Su, J., Vargas, D.V., Sakurai, K.: One pixel attack for fooling deep neural networks. IEEE Trans. Evol. Comput. **23**, 828–841 (2019)

19. Szegedy, C., et al.: Intriguing properties of neural networks. Computer Science (2013)
20. Xu, W., Evans, D., Qi, Y.: Feature squeezing: detecting adversarial examples in deep neural networks. arXiv preprint arXiv:1704.01155 (2017)
21. Yang, Y., et al.: Boundary thickness and robustness in learning models. In: Advances in Neural Information Processing Systems, vol. 33, pp. 6223–6234. Curran Associates, Inc. (2020)
22. Yousefzadeh, R., O'Leary, D.P.: Investigating decision boundaries of trained neural networks. CoRR abs/1908.02802 (2019). http://arxiv.org/abs/1908.02802
23. Zhang, H., Yu, Y., Jiao, J., Xing, E., El Ghaoui, L., Jordan, M.: Theoretically principled trade-off between robustness and accuracy. In: International Conference on Machine Learning, pp. 7472–7482. PMLR (2019)
24. Zhang, H., Cisse, M., Dauphin, Y.N., Lopez-Paz, D.: mixup: beyond empirical risk minimization. arXiv preprint arXiv:1710.09412 (2017)
25. Zhang, X., Zhao, R., Qiao, Y., Wang, X., Li, H.: AdaCos: adaptively scaling cosine logits for effectively learning deep face representations. In: Proceedings of the IEEE Conference on Computer Vision and Pattern Recognition, pp. 10823–10832 (2019)

# Explainable AI for Classification Using Probabilistic Logic Inference

Xiuyi Fan$^{(\boxtimes)}$ and Siyuan Liu

Department of Computer Science, Swansea University, Swansea SA2 8PP, UK
{xiuyi.fan,siyuahn.liu}@swansea.ac.uk

**Abstract.** The overarching goal of Explainable AI is to develop systems that not only exhibit intelligent behaviours, but also are able to explain their rationale and reveal insights. In explainable machine learning, methods that produce a high level of prediction accuracy as well as transparent explanations are valuable. In this work, we present an explainable classification method, which works by first constructing a symbolic Knowledge Base from the training data, and then performing probabilistic inferences on such Knowledge Base with linear programming. Our approach achieves a level of learning performance comparable to that of traditional classifiers such as random forests, support vector machines and neural networks. It identifies decisive features that are responsible for a classification as explanations and produces results similar to the ones found by SHAP, a state-of-the-art Shapley Value based explainable AI method.

**Keywords:** Explainable AI · Classification · Probabilistic logic inference

## 1 Introduction

The need for building AI systems that are explainable has been raised [4]. The ability to make machine-led decision making transparent, explainable, and therefore accountable is critical in building trustworthy systems. Producing explanations is at the core of realising explainable AI. Two main approaches for explainable machine learning have been explored in the literature: (1) intrinsically interpretable methods [18], in which prediction and explanation are both produced by the same underlying mechanism, and (2) model-agnostic methods [12], in which explanations are treated as a post hoc exercise and are separated from the prediction model. In the case for methods (1), while many intrinsically interpretable models, such as short decision trees, linear regression, Naive Bayes, k-nearest neighbours and decision rules [20] are easy to understand, they can be weak for prediction and suffer from performance loss in complex tasks. As for methods (2), model agnostic approaches such as local surrogate [15], global surrogate [1], feature importance [7] and symbolic Bayesian network transformation [19] leave the prediction model intact and use interpretable but presumably weak models to "approximate" the more sophisticated prediction model. However, it has been argued that since model agnostic approaches separate explanation from prediction, explanation modules cannot be faithful representations of their prediction counterpart [18]. In this context, we present a classification approach that produces accurate predictions and explanations.

© Springer Nature Switzerland AG 2021
L. Fang et al. (Eds.): CICAI 2021, LNAI 13070, pp. 16–26, 2021.
https://doi.org/10.1007/978-3-030-93049-3_2

Classification is a set of machine learning problems described as follows. Given a set of data instances, whose class membership is known, classification is the problem of identifying to which of a set of classes a new instance belongs. Each instance is characterised by a set of features $\mathcal{F}$. For some data $\mathcal{D}$, there exists a labelling function $L : \mathcal{D} \mapsto \{POS, \neg POS\}$.[1] Let $D \subseteq \mathcal{D}$ be the training set s.t. for each $d \in D$, $L(d)$ is known. For $x \in \mathcal{D}$, we would like to know:

**Q1:** *whether* $L(x) = POS$;
**Q2:** *if so, which features* $f \subseteq \mathcal{F}$ *make* $L(x) = POS$.

Standard supervised learning techniques answer Q1 but not Q2, which asks for *decisive* features. Understanding *"what causes a query instance $x$ to be classified as in some class $C$?"* is as important as *"does $x$ belong to $C$?"*. For instance, for a diagnostic system taking patients' medical records as the input and producing disease classifications as the output, pinpointing symptoms that lead to the diagnosis is as important as the diagnosis itself. In this paper, we propose algorithms answering both questions. In a nutshell, we solve classification as inference on probabilistic Knowledge Bases (KBs) learned from data. Specifically, given training data $D$ with features $F$, we define a function $\mathcal{M}$ that maps $D$ to a probabilistic KB. Then, for a query $x$, we check whether $\mathcal{M}(D)$ and $x$ together entail POS.

We present two algorithms for probabilistic KB construction. The first one constructs KBs from decision trees and the second constructs KBs directly from data. Query classification is modelled with probabilistic logic inference carried out with linear programming. The main contributions are: (i) a method of performing classification with probabilistic logic inference; (ii) a polynomial time inference algorithm on KBs; and (iii) algorithms for identifying decisive features as explanations.

## 2    Training as Knowledge Base Construction

KB construction is at the core of our approach. Specifically, a KB contains a set of disjunction clauses and each clause has a probability, defined formally as follows.

**Definition 1.** *A Knowledge Base (KB)* $\{\langle p_1, c_1 \rangle, \ldots, \langle p_m, c_m \rangle\}$ *is a set of pairs of clauses* $c_i$ *and probability of clauses* $p_i = P(c_i)$, $1 \leq i \leq m$. *Each clause is a disjunction of literals and each literal is a propositional variable or its negation.*

*Example 1.* With two propositional variables $\alpha$ and $\beta$, $\{\langle 0.6, \neg\alpha \vee \beta \rangle, \langle 0.8, \alpha \rangle\}$ is a simple KB containing two clauses with probabilities 0.6 and 0.8, respectively.

Generating logic clauses from decision trees or random forests has been studied [3,11,14]. Unlike the existing approaches where, due to their use of strict inference methods, non-probabilistic rules are generated, our KBs consist of probabilistic rules. Specifically, from a decision tree constructed from the training data, we create a clause

---

[1] POS stands for *positive*. For presentation simplicity, we only consider binary classification problems in this paper. Our approach generalises to multi-category classification by replacing POS with class labels for each candidate class accordingly.

$c$ from each path from the root to the leaf of the tree. The probability of $c$ is the ratio between the positive samples and all samples at the leaf. Formally, we define the KB $\mathcal{K}_T$ drawn from a decision tree T as follows.

**Definition 2.** *Let* T *be a decision tree, each non-root node in* T *labelled by a feature-value pair $a\_v$, read as feature $a$ having value $v$. Let $\{\rho_1,\ldots,\rho_k\}$ be the set of root-to-leaf paths in* T, *where each $\rho_i$ is of the form $\langle root, a_1\_v_1,\ldots,a_n\_v_m\rangle$ and $a_n\_v_m$ labels a leaf node in* T. *Then, the KB drawn from* T *is $\mathcal{K}_T = \{\langle p_1,c_1\rangle,\ldots,\langle p_k,c_k\rangle\}$ s.t. for each $\rho_i$, $\langle p_i,c_i\rangle \in \mathcal{K}_T$, where $c_i = POS \vee \neg a_1\_v_1 \vee \ldots \vee \neg a_n\_v_m$, and $p_i$ is the ratio between positive and the total samples in the node labelled by $a_n\_v_m$.*

Algorithms 1 and 2 construct $\mathcal{K}_T$ from data $D$. Specifically, Algorithm 1 takes a root-to-leaf path from a decision tree to generate a clause. The path with features $a_1,\ldots,a_n$, s.t. each feature has a value in $\{v_1,\ldots,v_m\}$, is interpreted as $a_1\_v_1 \wedge \ldots \wedge a_n\_v_m \rightarrow POS$, and read as, *a sample is positive if its feature $a_1$ has value $v_1$, ..., feature $a_n$ has value $v_m$*. As a disjunction, the clause is then written as $POS \vee \neg a_1\_v_1 \vee \ldots \vee \neg a_n\_v_m$. Algorithm 2 builds a tree and then constructs clauses from paths in the tree. Example 2 illustrates how to build a KB from a decision tree.

---

**Algorithm 1.** Clause from Tree Path

1: **procedure** CLAUSEFROMPATH(*path*)
2:      *clause* ← POS
3:      **for** each edge $e$ in *path* **do**
4:          $a$ ← feature of $e$
5:          $v$ ← value of $e$
6:          *clause* ← *clause* $\vee \neg a\_v$
7:      **return** *clause*

---

**Algorithm 2.** Construct KB with Decision Tree

1: **procedure** DECSIONTREEKB(*D*)
2:      $\mathcal{K}_T$ ← {}; Use ID3 to compute a tree T from $D$
3:      *allPaths* ← all paths from the root to leaves in $T$
4:      **for** each *path* in *allPaths* **do**
5:          $n$ ← end node in *path*
6:          $r$ ← ratio between positive and total samples in $n$
7:          add $[r]$ CLAUSEFROMPATH(*path*) to $\mathcal{K}_T$
8:      **return** $\mathcal{K}_T$

---

*Example 2.* Given a data set with four strings, *0000, 1111, 1010, 1100*, labelled positive, and four strings, *0010, 0100, 1110, 1000*, labelled negative. There are four features, bits 1–4, each feature takes its value from $\{0,1\}$. The decision tree constructed is shown in Fig. 1. There are eight leaves, thus eight root-to-leaf paths and clauses. E.g., *root* $\rightarrow a_4\_0 \rightarrow a_1\_0 \rightarrow a_2\_0 \rightarrow a_3\_0$ gives the clause $POS \vee \neg a_4\_0 \vee \neg a_1\_0 \vee \neg a_2\_0 \vee \neg a_3\_0$. The probability of the clause being the number of positive samples over the total samples at the leaf. There is only one sample, 0000, at this leaf, since it is positive, the clause probability is 1. The KB $\mathcal{K}_T$ is shown in Table 1.[2]

---

[2] Henceforth, $[p]$ $z_1 \vee \ldots \vee z_l$ denotes an $l$-literal clause in a KB with probability $p$.

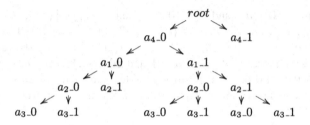

**Fig. 1.** Decision tree learned from data in Example 2. A node $a_X\_Y$ is read as "bit $X$ has value $Y$".

**Table 1.** $\mathcal{K}_\mathsf{T}$ from the tree in Fig. 1.

[0.0] POS $\vee \neg a_1\_0 \vee \neg a_2\_0 \vee \neg a_3\_1 \vee \neg a_4\_0$

[1.0] POS $\vee \neg a_1\_0 \vee \neg a_2\_0 \vee \neg a_3\_0 \vee \neg a_4\_0$

[0.0] POS $\vee \neg a_1\_0 \vee \neg a_2\_1 \vee \neg a_4\_0$

[1.0] POS $\vee \neg a_1\_1 \vee \neg a_2\_0 \vee \neg a_3\_1 \vee \neg a_4\_0$

[0.0] POS $\vee \neg a_1\_1 \vee \neg a_2\_0 \vee \neg a_3\_0 \vee \neg a_4\_0$

[0.0] POS $\vee \neg a_1\_1 \vee \neg a_2\_1 \vee \neg a_3\_1 \vee \neg a_4\_0$

[1.0] POS $\vee \neg a_1\_1 \vee \neg a_2\_1 \vee \neg a_3\_0 \vee \neg a_4\_0$

[1.0] POS $\vee \neg a_4\_1$

Algorithm 2 constructs clauses from root-to-leaf paths in a decision tree. We can also use paths from the root to all nodes, not just the leaves, to construct clauses, i.e., replacing line 3 in Algorithm 2 with

*allPaths* ← all paths from the root to **all nodes** in $T$.

As random forests have been introduced to improve the stability of decision trees, we can apply the same idea to obtain more clauses from a forest, i.e., repeatedly generating different decision trees, and for each tree, we construct clauses for each path originated at its root, in the spirit of [11]. If we further take the above idea of "generating as many clauses as possible" to its limit, we realise that constructing KBs from trees is a special case of selecting clauses constructed from all $k$-combinations of feature-value pairs, for $k = 1 \ldots n$, where $n$ is the total number of features in the data. Formally, we define the KB $\mathcal{K}_\mathsf{D}$ drawn directly from data D as follows.

**Definition 3.** *Given data* D *with features* $F = \{a_1, \ldots, a_n\}$ *taking values from* $V = \{v_1, \ldots, v_m\}$, *for each* $F_k = \{a'_1, \ldots, a'_k\} \in 2^F \setminus \{\}$, *let* $C_k^1 = \{a'_1\_v | v \in V\}, \ldots, C_k^k = \{a'_k\_v | v \in V\}$. $C_k = C_k^1 \times \ldots \times C_k^k$. *For each* $c = \{a''_1\_v'_1, \ldots, a''_k\_v'_k\} \in C_k$, $S_i \subseteq$ D *is the set of samples s.t. feature* $a''_i$ *having value* $v'_{i'}$ *for all* $i \in \{1, \ldots, k\}$. *If* $|S_i| \neq 0$, *then let* $p_i$ *be the ratio between positive samples in* $S_i$ *and* $|S_i|$, $\langle p_i, \mathsf{POS} \vee \neg a''_1\_v'_1 \vee \ldots \vee \neg a''_k\_v'_k \rangle$ *is in the KB* $\mathcal{K}_\mathsf{D}$ *drawn directly from data. There is no other clause in* $\mathcal{K}_\mathsf{D}$ *except those constructed as above.*

Definition 3 can be illustrated with the following example.

*Example 3.* Let $F = \{a_1, a_2\}$ and $V = \{0, 1\}$. Then $2^F \backslash \{\} = \{\{a_1\}, \{a_2\}, \{a_1, a_2\}\}$. For illustration, let us choose $F_k = \{a_1, a_2\}$. Then $C_k^1 = \{a_1\_0, a_1\_1\}$, $C_k^2 = \{a_2\_0, a_2\_1\}$, and $C_k = \{\{a_1\_0, a_2\_0\}, \{a_1\_0, a_2\_1\}, \{a_1\_1, a_2\_0\}, \{a_1\_1, a_2\_1\}\}$. Then, suppose we choose $c = \{a_1\_0, a_2\_0\}$ and add $\langle p_i, \text{POS} \vee \neg a_1\_0 \vee \neg a_2\_0 \rangle$ to $\mathcal{K}_\text{D}$, where $p_i$ is the ratio between positive samples with both features $a_1, a_2$ having value 0 and total samples with these feature-values. $\mathcal{K}_\text{D}$ can be constructed by choosing different $F_k$ and $c$ iteratively.

Finally, Algorithm 3 gives a procedural construction for $\mathcal{K}_\text{D}$.

---

**Algorithm 3.** Construct KB Directly

---

1: **procedure** DIRECTKB(*data*)
2:     *counts* $\leftarrow \{\}, \mathcal{K}_\text{D} \leftarrow \{\}$
3:     **for** each *entry* in *data* **do**
4:         *feaVals* $\leftarrow \{a\_v|$ feature $a$ has value $v$ in *entry*$\}$
5:         *label* $\leftarrow$ binary label of *entry* as integer
6:         $S \leftarrow$ POWERSET(*feaVals*) $\backslash \{\}$
7:         **for** each *key* as an element of $S$ **do**
8:             **if** *key* is in *counts* **then**
9:                 *counts*[*key*] $\leftarrow$ *counts*[*key*] $+ [1, label]$
10:            **else**
11:                *counts*[*key*] $\leftarrow [1, label]$
12:     **for** each *key* in *counts* **do**
13:         $r \leftarrow$ *counts*[*key*][1]/*counts*[*key*][0]
14:         Insert "[$r$] POS $\vee \neg key$" to $\mathcal{K}_\text{D}$
15:     **return** $\mathcal{K}_\text{D}$

---

In Line 14, $\neg\{s_1, \ldots, s_n\}$ is $\neg s_1 \vee \ldots \vee \neg s_n$, e.g. for *key* $= \{a_1\_v_1, a_2\_v_2\}$, insert "[$p$] POS $\vee \neg a_1\_v_1 \vee \neg a_2\_v_2$" to $\mathcal{K}_\text{D}$. *counts* is a dictionary with keys being sets of feature-value pairs and values being two-element arrays. *label* is either 0 or 1. Line 9 is an element-wise addition, e.g., $[1, 0] + [1, 1] = [2, 1]$. At the end of the first loop, *counts*[*key*][0] is the number of samples containing *key* and *counts*[*key*][1] is the number of positive ones.

## 3   Querying as Probabilistic Inference

Our KB construction methods produce clauses with probabilities. Intuitively, for a query that asserting some feature-value pairs, we want to compute the probability of POS under these feature-value pairs and predicting the query being positive when the probability is greater than 0.5. To introduce our inference method for computing such probabilities, we first review a few concepts in probabilistic logic [13], which pave the way for discussion.

Given a KB $\mathcal{K}^3$ with clauses $c_1, \ldots, c_m$ composed from $n$ propositional variables, the *complete conjunction set*, denoted as $\mathcal{W}$, over $\mathcal{K}$ is the set of $2^n$ conjunctions s.t.

---

[3] From this point on, we use $\mathcal{K}$ to denote a KB constructed using either of the two approaches ($\mathcal{K}_\text{T}$ or $\mathcal{K}_\text{D}$).

each conjunction contains $n$ distinct propositional variables. A *probability distribution* $\pi$ (wrt. $\mathcal{K}$) is the set of $2^n$ probabilities $\pi(w) \geq 0 (w \in \mathcal{W})$, s.t. $\sum_{w \in \mathcal{W}} \pi(w) = 1$. $\pi$ *satisfies* $\mathcal{K}$ iff for each $i = 1, \ldots, m$, the sum of $\pi(w)$ equals $P(c_i)$ for all $w$ s.t. the truth assignment satisfying $w$ satisfies $c_i$. A KB $\mathcal{K}$ is *consistent* iff there exists a $\pi$ satisfying $\mathcal{K}$. With a consistent KB, Nilsson suggested that one can derive *literal probabilities* from $\pi$, i.e., for all literals $z$ in the KB, $P(z)$ is the sum of $\pi(w)$ for all $w \in \mathcal{W}$ containing $z$, e.g., for a consistent KB with two literals $\alpha$ and $\beta$, $P(\alpha) = P(\alpha \wedge \beta) + P(\alpha \wedge \neg \beta)$ [13]. In short, to compute literal probabilities, one first computes probability assignments over the complete conjunction set, and then adds up all relevant probabilities for the literal.

At first glance, since POS is an literal in our knowledge base, it might be possible to perform our inference with the above approach for computing $P(\text{POS})$: all clauses in a KB are of the form $\text{POS} \vee \neg a_1\_v_1 \vee \ldots \vee \neg a_n\_v_m$, each with an associated probability; a query is a set of feature-value pairs, e.g., $a'_1\_v'_1, \ldots, a'_n\_v'_m$, each with an assigned probability 1; $P(\text{POS})$ computed as the sum of $P(\text{POS} \wedge a_1\_v_1 \wedge \ldots \wedge a_n\_v_m)$, $P(\text{POS} \wedge a_1\_v_1 \wedge \ldots \wedge \neg a_n\_v_m), \ldots, P(\text{POS} \wedge \neg a_1\_v_1 \wedge \ldots \wedge \neg a_n\_v_m)$ estimates the likelihood of POS. However, this idea fails for the following two reasons. Firstly, this approach requires solving the probability distribution $\pi$, which has been shown to be NP-hard wrt. the number of literals in the KB [8], thus the state-of-the-art approaches only work for KB with a few hundred of variables [6].

Secondly, putting a KB and a query together introduces inconsistency, so there is no solution for $\pi$. For instance, for the KB in Example 2, let the query be 0000, which translates to four clauses, $a_1\_0, a_2\_0, a_3\_0$ and $a_4\_0$, each with $P(a_i\_0) = 1$. Consequently, $P(\neg a_i\_0) = 0$. Together with $P(\text{POS} \vee \neg a_1\_0 \vee \neg a_2\_0 \vee \neg a_3\_0 \vee \neg a_4\_0) = 1$, we infer $P(\text{POS}) = 1$. However, $P(\text{POS}) = 1$ is inconsistent with $P(\text{POS} \vee \neg a_1\_0 \vee \neg a_2\_0 \vee \neg a_3\_1 \vee \neg a_4\_0) = 0$, as for any $\alpha, \beta$, we must have $P(\alpha) \leq P(\alpha \vee \beta)$. In this case, $\mathcal{K}$ is inconsistent with the query thus there is no solution for $\pi$.

Therefore, we formulate the computation as an optimization problem so that inconsistency is tolerated. This is the core of our inference method. More specifically, we propose to use linear programming to estimate the literal probabilities.

**Definition 4.** *Given a KB* $\mathcal{K} = \{\langle p_1, c_1 \rangle, \ldots, \langle p_m, c_m \rangle\}$ *with clauses* $\mathcal{C} = \{c_1, \ldots, c_m\}$ *over literals* $\mathcal{Z}$, *a linear program* $L_{\mathcal{K}}$ *of* $\mathcal{K}$ *with unknowns* $\omega(\sigma), \sigma \in \mathcal{C} \cup \mathcal{Z}$, *is the following.*
minimise:

$$\sum_{i=1}^{m} |\omega(c_i) - p_i| \tag{1}$$

subject to: for each clause $c_i = z_1 \vee \ldots \vee z_l$,

$$\omega(c_i) \leq \omega(z_1) + \ldots + \omega(z_l); \tag{2}$$

for $z_j = z_1 \ldots z_l$ in clause $c_i$:

$$\omega(c_i) \geq \omega(z_j); 1 = \omega(z_j) + \omega(\neg z_j); 0 \leq \omega(z_j) \leq 1. \tag{3}$$

Definition 4 estimates literal probabilities from clause probabilities without computing the distribution over the complete conjunction set, i.e., for any literal $z$ in the

KB, $\omega(z)$ approximates $P(z)$. We illustrate probability computation with the following example.

*Example 4.* (Example 1 cont.) Given these two clauses, $c_1 = \neg\alpha \vee \beta; c_2 = \alpha$, and their probabilities, $P(c_1) = 0.6, P(c_2) = 0.8$, the complete conjunction set $\mathcal{W} = \{\neg\alpha \wedge \neg\beta, \neg\alpha \wedge \beta, \alpha \wedge \neg\beta, \alpha \wedge \beta\}$. Truth assignments satisfying $\alpha \wedge \beta, \neg\alpha \wedge \beta$, and $\neg\alpha \wedge \neg\beta$ satisfy $c_1$ and truth assignments satisfying $\alpha \wedge \beta$ and $\alpha \wedge \neg\beta$ satisfy $c_2$. $\mathcal{K}$ is consistent iff $\pi_1 = \pi(\alpha \wedge \beta)$, $\pi_2 = \pi(\alpha \wedge \neg\beta)$, $\pi_3 = \pi(\neg\alpha \wedge \beta)$, and $\pi_4 = \pi(\neg\alpha \wedge \neg\beta)$ s.t. $\sum_{j=1}^{4} \pi_j = 1$, $\pi_1 + \pi_3 + \pi_4 = 0.6$ and $\pi_1 + \pi_2 = 0.8$. $L_{\mathcal{K}}$ is:
minimise:

$$|\omega(c_1) - 0.6| + |\omega(c_2) - 0.8|$$

subject to:

$$\omega(c_1) \leq \omega(\neg\alpha) + \omega(\beta); \quad \omega(c_2) \leq \omega(\alpha); \quad \omega(c_1) \geq \omega(\neg\alpha); \quad \omega(c_1) \geq \omega(\beta);$$
$$\omega(c_2) \leq \omega(\alpha); \quad \omega(c_2) \geq \omega(\alpha); \quad 1 = \omega(\alpha) + \omega(\neg\alpha); \quad 1 = \omega(\beta) + \omega(\neg\beta);$$
$$0 \leq \omega(\alpha) \leq 1; \quad 0 \leq \omega(\beta) \leq 1.$$

A solution to $L_{\mathcal{K}}$ is: $\omega(\neg\alpha \vee \beta) = 0.6$; $\omega(\alpha) = 0.8$; $\omega(\neg\alpha) = 0.2$; $\omega(\beta) = 0.6$; $\omega(\neg\beta) = 0.4$.

With a means to reason with KBs, we are ready to answer queries. Algorithm 4 defines the query process. Let $L_{\mathcal{K}}$ be the linear system constructed from $\mathcal{K}$. Given a query $\mathcal{Q}$ with feature-value pairs $a_1\_v_1, \ldots, a_n\_v_m$, we amend $L_{\mathcal{K}}$ by inserting $\omega(a_i\_v_j) = 1$ and $\omega(a_i\_v'_j) = 0$, where $v_j$ is a possible value of $a_i$, $v'_j \neq v_j$, for all $a_i, v_j$ in $\mathcal{Q}$. $\omega(\texttt{POS})$ computed in $L_{\mathcal{K}}$ answers whether $\mathcal{Q}$ is positive. Since the solution of $\omega(\texttt{POS})$ can be a range, we compute the upper and lower bounds of $\omega(\texttt{POS})$ by maximising and minimising $\omega(\texttt{POS})$ subject to minimising Eq. (1), respectively, and use the average of the two. It returns *positive* when the average is greater than 0.5. The intuition of our approach is that, for a query $x$, to evaluate whether $\mathcal{K}$ and $x$ entail POS, we compute $\omega(\texttt{POS})$ in $L_{\mathcal{K}}$. Example 5 illustrates the query process.

---

**Algorithm 4.** Query Knowledge Base

---

1: **procedure** QUERYKB(*query*, $L_{\mathcal{K}}$)
2:      **for** each feature $a$ in *query* **do**
3:          **for** each possible value $v$ of $a$ **do**
4:              **if** $a$ has value $v$ in *query* **then**
5:                  Add $\omega(a\_v) = 1$ to $L_{\mathcal{K}}$
6:              **else**
7:                  Add $\omega(a\_v) = 0$ to $L_{\mathcal{K}}$
8:      **return** $\omega(\texttt{POS})$ computed in $L_{\mathcal{K}}$

---

*Example 5.* (Example 2 cont.) For query 0101, we add the following equations as constraints to $L_{\mathcal{K}}$:

$$\omega(a_1\_0) = 1, \quad \omega(a_1\_1) = 0, \quad \omega(a_2\_0) = 0, \quad \omega(a_2\_1) = 1,$$
$$\omega(a_3\_0) = 1, \quad \omega(a_3\_1) = 0, \quad \omega(a_4\_0) = 0, \quad \omega(a_4\_1) = 1.$$

The computed $\omega(\text{POS})$ is no greater than 0.5, representing a negative classification.

The proposed querying mechanism differs fundamentally from that of decision trees. A decision tree query can be viewed as finding the longest clause in the KB that matches with the query in and checking whether its probability is greater than 0.5. For instance, for query *0101*, a decision tree query returns positive as the longest matching clause in "POS $\vee \neg a_4\_1$" has probability 1. However, our approach considers probabilities from other clauses in the $\mathcal{K}$ and produces a different answer.

## 4  Explanations

One advantage of the presented classification method is that it supports *partial queries*, which are queries with missing values, as the probability of POS can be computed without values assigned to all features. Explanation computation can be supported with partial queries in our approach.

Algorithm 5 outlines one approach. Given a query $Q$ with $n$ features, to find the $k$ most decisive features, we construct *sub-queries* s.t. each sub-query contains exactly $k$ feature-value pairs in $Q$. If $Q$ yields a positive classification, then the sub-query that maximises $\omega(\text{POS})$ is an explanation; otherwise, the sub-query that minimises $\omega(\text{POS})$ is. Since we know that there are $\binom{n}{k}$ different sub-queries in total, the order of sub-query evaluation can be strategised with methods such as hill climbing for more efficient calculation. Although in principle, Algorithm 5 could work with any classification technique supporting partial queries, our proposed method does not require reconstructing the trained model for testing each of the sub-queries, making the explanation generation convenient. The explanation approach is illustrated in Example 6.

---

**Algorithm 5.** Explanation Computation

1: **procedure** COMPUTEEXPLANATION$(Q, L_{\mathcal{K}}, k)$
2:     $S \leftarrow \{sQ | sQ \in 2^Q, \text{SIZEOF}(sQ) = k\}$
3:     **if** QUERYKB$(Q, L_{\mathcal{K}}) > 0.5$ **then**
4:         **return** $\arg\max_{sQ \in S}$ QUERYKB$(sQ, L_{\mathcal{K}})$
5:     **else**
6:         **return** $\arg\min_{sQ \in S}$ QUERYKB$(sQ, L_{\mathcal{K}})$

---

*Example 6.* (Example 5 cont.) To compute the single most decisive feature, we let $k = 1$. $S$ contains four feature-value pairs: $q_1 = \{a_1\_0\}$, $q_2 = \{a_2\_1\}$, $q_3 = \{a_3\_0\}$, $q_4 = \{a_4\_1\}$. Let $\omega_i, i = 1 \ldots 4$ be $\omega(\text{POS})$ computed with $q_1 \ldots q_4$, respectively. We have $\omega_1 = 0.33, \omega_2 = 0.5, \omega_3 = 0.5$, and $\omega_4 = 1$. Thus, the computed explanation for the classification is $a_1\_0$. We read this as:

> *0 - - - is responsible for 0101 being negative.*

This matches with our intuition well as for each of the other choices, there are at least as many positive samples as negative ones.

## 5 Performance Analysis

We study the accuracy of the proposed approach in classification and finding explanations over synthetic data sets. The major reason to use synthetic data sets is that there is no ground truth for explanations in real data sets. For example, even if symptoms are found as explanations for diagnosis results using explainable AI techniques, there is no ground truth to verify whether these symptoms explain the results. Therefore, we performed experiments with synthetic data sets with known explanation ground truth. Specifically, we created four synthetic data sets of integer strings, Syn 10/4, Syn 10/8, Syn 12/4, and Syn 12/8, with the following rules. For each data set, we set a (random) *seed* string of the same length as strings in the data set from the same alphabet. For instance, for the "Syn 10/4" data set with 10 bits strings where each bit can take 4 possible values, 3232411132 is the seed. Here, the size of the alphabet is 4. Each 10-bit string denotes a data instance with 10 features s.t. each feature takes its value from $\{1, 2, 3, 4\}$. A string $s$ in the data set is labelled positive iff $s$ match bits in the seed for exactly five places. E.g., $3\underline{133}4\underline{21}24\underline{2}^4$ is positive and $3\underline{1334}\underline{21}2\underline{32}$ is negative (it shares 6 bits as the seed rather than 5). For each string classified as positive, we compute a $k$-bit explanation. An explanation is *correct* iff the seed string has the same values for the bits identified as the explanation. The accuracy of an explanation is defined as the number of correct bits over the length of explanation. For instance, for $k = 5$, we have

| Query | Explanation | Seed | Accuracy |
|-------|-------------|------|----------|
| 3233112143 | 323–1-1– | 3232411132 | 1.0 |
| 3244341112 | -2—411-2 | 3232411132 | 0.8 |

The 2nd query contains an incorrect explanation 4. On our synthetic data sets with a 70% to 30% split on training and testing, the classification result is shown in Table 2. Our approaches are *Tree* (Algorithm 2) and *Direct* (Algorithm 3). We use CART (a decision tree algorithm), multi-layer perceptron (MLP) neural networks (with two hidden layers with 12 and 10 nodes, respectively), random forest (with 100 trees) and support vector machine as our comparison baselines. To evaluate our explanation approach, we compare *Direct* with the state of the art Shapley Value based approach SHAP [10], one of the latest approaches for explaining classification results from trees. The explanation accuracy is shown in Table 2.

**Table 2.** Experiment results (F$_1$ scores)

|  | Syn 10/4 | Syn 10/8 | Syn 12/4 | Syn 12/8 |
|--------|----------|----------|----------|----------|
| Tree | 0.71 | 0.78 | 0.62 | 0.70 |
| Direct | **0.92** | 0.95 | **0.89** | **0.94** |
| CART | 0.79 | 0.87 | 0.70 | 0.84 |
| MLP | 0.77 | 0.83 | 0.73 | 0.80 |
| Forest | 0.90 | **0.96** | 0.85 | 0.93 |
| SVM | 0.85 | 0.86 | 0.81 | 0.81 |

**Table 3.** Explanation accuracy on four syntactic data sets and various explanation lengths $k$.

|  |  | $k = 1$ | $k = 2$ | $k = 3$ | $k = 4$ | $k = 5$ |
|------|-------|---------|---------|---------|---------|---------|
| 10/4 | Direct | 1 | 1 | 1 | 0.995 | 0.972 |
|  | SHAP | 1 | 1 | 0.996 | 0.993 | 0/962 |
| 10/8 | Direct | 1 | 1 | 0.997 | 0.980 | 0.976 |
|  | SHAP | 0.996 | 0.995 | 0.972 | 0.967 | 0.951 |
| 12/4 | Direct | 1 | 0.982 | 1 | 0.997 | 0.901 |
|  | SHAP | 0.993 | 0.980 | 0.973 | 0.942 | 0.856 |
| 12/4 | Direct | 1 | 1 | 0.998 | 0.975 | 0.964 |
|  | SHAP | 1 | 0.990 | 0.977 | 0.929 | 0.918 |

---

[4] The underlined bits are identical to the seed.

Table 2 shows that the classification accuracy of our approach is competitive comparing to the baseline approaches. This further validates our approach for classification. Table 3 shows that although our approach (Direct) and SHAP both can identify part of the seed string from each query instances, hence computing correct explanations, ours gives higher accuracy across the board.

## 6   Related Work

Performing probabilistic logic inference with mathematical programming has been studied recently in [9] with its NonlInear Probabilistic Logic Solver (NILS) approach. NILS either assumes independence amongst its variables or expand probability of conjunctions as the product of the probability of a literal and some conditionals. Thus NILS produces non-linear systems and rely on gradient descent methods for finding solutions. Consequently, NILS is unsuitable for classification as the independence assumption does not hold between the class labels and feature values or, in general, values across different features. When independence cannot be assumed, systems constructed with NILS are difficult to solve numerically.

In explainable machine learning, there has been significant interest in providing explanations for classifiers; see e.g., [2] for an overview. Works have been proposed to use simpler thus weaker classifiers to explain results from stronger ones, e.g., [5]. Recent works on *model-agnostic* explainers [15,16] focus on adding explanations to existing (black-box) classifiers. [1] use KB based classifiers to explain results obtained from MLP and random forests. LIME [15] augment the data with randomly generated samples close to the instance to be explained and then construct a simple thus explainable classifier to generate explanations. [17] works by decomposing a model's predictions based on individual contributions of each feature. [19] explains Bayesian network classifiers by compiling naive Bayes and latent-tree classifiers into Ordered Decision Diagrams. [10] provides explanations for decision trees based on the game-theoretic Shapley values.

## 7   Conclusion

We present a non-parametric classification technique that gives explanations to its predictions. Our approach is based on approximating literal probabilities in probabilistic logic by solving linear systems corresponding to KBs, which are either directly learned from data or augmented with additional knowledge. Our linear program construction is efficient and our approaches tolerate inconsistency in a KB. As a stand-alone classifier, our approach matches or exceeds the performance of existing algorithms.

There are three research directions that we plan to explore. Firstly, this work focuses on developing the underlying explainable classification techniques. We will apply techniques developed practical applications and perform user studies in the future. Secondly, we will study semantics for inconsistent KBs. Thirdly, we will study richer explanation generation with (probabilistic) logic inference.

**Acknowledgements.** The work is supported with a funding contribution from the Welsh Government Office for Science, Sêr Cymru III programme – Tackling Covid-19.

# References

1. Alonso, J., Ramos-Soto, A., Castiello, C., Mencar, C.: Hybrid data-expert explainable beer style classifier. In: Proceedings of the IJCAI-17 Workshop on Explainable AI (2018)
2. Biran, O., Cotton, C.V.: Explanation and justification in machine learning: a survey. In: Proceedings of the IJCAI-17 Workshop on Explainable AI (2017)
3. Chiang, D., Chen, W., Wang, Y., Hwang, L.: Rules generation from the decision tree. J. Inf. Sci. Eng. **17**(2), 325–339 (2001)
4. Doran, D., Schulz, S., Besold, T.R.: What does explainable AI really mean? A new conceptualization of perspectives. CoRR abs/1710.00794 (2017)
5. Féraud, R., Clérot, F.: A methodology to explain neural network classification. Neural Netwo. **15**(2), 237–246 (2002). https://doi.org/10.1016/S0893-6080(01)00127-7. http://www.sciencedirect.com/science/article/pii/S0893608001001277
6. Finger, M., Bona, G.: Probabilistic satisfiability: logic-based algorithms and phase transition. In: Proceedings of the 22nd International Joint Conference on Artificial Intelligence, IJCAI 2011, Barcelona, Catalonia, Spain, 16–22 July 2011, vol. 17, pp. 528–533 (2011)
7. Fisher, A., Rudin, C., Dominici, F.: All models are wrong but many are useful: variable importance for black-box, proprietary, or misspecified prediction models, using model class reliance, pp. 237–246. arXiv preprint arXiv:1801.01489 (2018)
8. Georgakopoulos, G., Kavvadias, D., Papadimitriou, C.H.: Probabilistic satisfiability. J. Complex. **4**(1), 1–11 (1988)
9. Henderson, T.C., et al.: Probabilistic sentence satisfiability: an approach to PSAT. Artif. Intel. **278**, 103199 (2020)
10. Lundberg, S.M., et al.: From local explanations to global understanding with explainable AI for trees. Nat. Mach. Intel. **2**(1), 56–67 (2020)
11. Mashayekhi, M., Gras, R.: Rule extraction from decision trees ensembles: new algorithms based on heuristic search and sparse group lasso methods. Int. J. Inf. Technol. Decis. Mak. **16**(06), 1707–1727 (2017)
12. Molnar, C.: Interpretable Machine learning. Lulu.com (2020)
13. Nilsson, N.J.: Probabilistic logic. Artif. Intel. **28**(1), 71–87 (1986)
14. Quinlan, J.R.: Generating production rules from decision trees. In: IJCAI, vol. 87, pp. 304–307. Citeseer (1987)
15. Ribeiro, M.T., Singh, S., Guestrin, C.: "Why should i trust you?" explaining the predictions of any classifier. In: Proceedings of the 22nd ACM SIGKDD International Conference on Knowledge Discovery and Data Mining, pp. 1135–1144 (2016)
16. Ribeiro, M.T., Singh, S., Guestrin, C.: Anchors: high-precision model-agnostic explanations. In: Proceedings of the AAAI Conference on Artificial Intelligence, vol. 32 (2018)
17. Robnik-Šikonja, M., Kononenko, I.: Explaining classifications for individual instances. IEEE Trans. Knowl. Data Eng. **20**(5), 589–600 (2008)
18. Rudin, C.: Stop explaining black box machine learning models for high stakes decisions and use interpretable models instead. Nat. Mach. Intel. **1**(5), 206–215 (2019)
19. Shih, A., Choi, A., Darwiche, A.: A symbolic approach to explaining Bayesian network classifiers. arXiv preprint arXiv:1805.03364 (2018)
20. Yang, H., Rudin, C., Seltzer, M.: Scalable Bayesian rule lists. In: International Conference on Machine Learning, pp. 3921–3930. PMLR (2017)

# A Consistency Regularization for Certified Robust Neural Networks

Mengting Xu, Tao Zhang, Zhongnian Li, and Daoqiang Zhang[✉]

College of Computer Science and Technology, Nanjing University of Aeronautics
and Astronautics, Nanjing 211106, China
{xumengting,dqzhang}@nuaa.edu.cn

**Abstract.** A range of provable defense methods have been proposed to
train neural networks that are certifiably robust to the adversarial examples. Among which, COLT [1] combined adversarial training and provable defense method that achieves state-of-the-art accuracy and certified
robustness. However, COLT treats all examples equally during training,
which ignores the inconsistent constraint of certified robustness between
correctly classified (natural) and misclassified examples. In this paper,
we explore this inconsistency and add a regularization to exploit misclassified examples efficiently. Specifically, we identified that the certified
robustness of networks can be significantly improved by refining inconsistent constraint on misclassified examples. Besides, we design a new
defense regularization called *Misclassification Aware Adversarial Regularization* (MAAR), which constrains the output probability distributions of all examples in the certified region of the misclassified example. Experimental results show that MAAR achieves the best certified
robustness and comparable accuracy on CIFAR-10 and MNIST datasets
in comparison with several state-of-the-art methods.

**Keywords:** Adversarial defense · Certified robustness · Misclassified
examples · Consistency regularization

## 1 Introduction

Despite the widespread success of neural network on diverse tasks such as image
classification [8], face and speech recognition [17]. Recent studies have highlighted
the lack of robustness in state-of-the-art neural network models, e.g., a visually

Supported by the National Key Research and Development Program of China (No.
2018YFC2001600, 2018YFC2001602, 2018ZX10201002), the National Natural Science
Foundation of China (Nos. 61876082, 61861130366, 61732006 and 61902183), and
the Royal Society-Academy of Medical Sciences Newton Advanced Fellowship (No.
NAF_R1_180371).

**Supplementary Information** The online version contains supplementary material
available at https://doi.org/10.1007/978-3-030-93049-3_3.

© Springer Nature Switzerland AG 2021
L. Fang et al. (Eds.): CICAI 2021, LNAI 13070, pp. 27–38, 2021.
https://doi.org/10.1007/978-3-030-93049-3_3

imperceptible adversarial image can be easily crafted to mislead a well-trained network [16]. Even worse, researchers have identified that these adversarial examples are not only valid in the image classification task [4] but also plausible in the object detection [23] and speaker recognition [11]. Considering the significance of adversarial robustness in neural network, a range of defense methods have been proposed. Adversarial training [4,22] which can be regarded as a data augmentation technique that trains neural networks on adversarial examples are highly robust against the strongest known adversarial attacks such as C&W attack [2], but it provides no guarantee—it is unable to produce a certificate that there are no possible adversarial attack which could potentially break the model. Recent line of work on provable defense [15,20] has been proposed to train neural networks that no attacks within a certain region will alter the networks prediction. Moreover, COLT [1] combines adversarial training and provable defense methods to train neural network with both high certified robustness and accuracy. However, recall that the formal definition of certified robustness is conditioned on natural examples that are correctly classified [20]. COLT treats both correctly classified and misclassified examples equally during training process while evaluating certified robustness just on correctly classified examples. From this perspective, the effect of misclassified example on certified robustness is unknown.

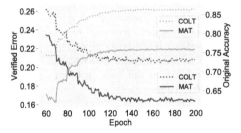

**Fig. 1.** Verified error (red lines) and original accuracy (blue lines) for COLT [1] and Misclassification Aware Training (MAT). The dataset is CIFAR-10 with $L_\infty$ maximum perturbation $\epsilon = 2/255$. (Color figure online)

Therefore, it is not clear for the following questions: *(1) Do misclassified examples have effectiveness for improving certified robustness? (2) If yes, how can we make better use of misclassified examples to improve the certified robustness?*

To address these issues, we explore the inconsistent constraint of certified robustness between the correctly classified and misclassified examples on COLT. For correctly classified examples, the objectives of robust constraint and original accuracy constraint are the same: *Constraining the correctly classified examples and all examples within their perturbations sets to be close enough to the correct labels, which will improve the original accuracy as well as certified robustness of the network.* However, for misclassified examples, the constraints on robustness and accuracy are different: *Making the misclassified examples and the examples in the perturbation sets close enough to the original labels can improve the original accuracy of the network, but it is undeniable that this will destroy the stability of the misclassified examples (the original label is the "wrong label" for misclassified example), thereby reducing the certified robustness of the network.* To deal with this problem, we firstly propose to use the output label (the output label is the "true label" for misclassified example) of the misclassified example during training process to keep the stability of misclassified examples (which we call *Misclassification Aware Training* (MAT) in Sect. 3.2). Interestingly, as shown in Fig. 1, we find that misclassified examples have a significant effect on the final certified robustness of network. Compared with COLT [1] (dashed red line), the verified error (the details of these metrics are in Appendix 1.3) of MAT (red line) drops drastically. However, the original accuracy of MAT (blue line) is extremely lower than standard COLT (dashed blue line).

In this paper, in order to make better use of misclassified examples, we propose a consistency regularization to constrain the output probability distributions of all examples in the certified region of the misclassified example. The regularization term called Misclassification Aware Adversarial Regularization (MAAR) aims to encourage the output of network to be stable against misclassified adversarial examples. In other words, MAAR focuses on solving the inconsistency of certified robustness on both correctly classified and misclassified examples, which improves the final certified robustness of network. Meanwhile, MAAR does not change the training label during the training process, which alleviates the decrease of model accuracy.

Our main contributions are:

- We investigate the inconsistency on constraint of certified robustness caused by misclassified examples by a proof-of-concept experiment (i.e., *Misclassification Aware Training* (MAT)).
- We propose a consistency regularization term called *Misclassification Aware Adversarial Regularization* (MAAR) which improves certified robustness by maintaining the stability of misclassified examples as well as relieving the degree of accuracy decline.
- We show the effectiveness of MAAR by different networks and perturbations on two datasets. Specifically, MAAR achieves the state-of-the-art certified robustness of 62.8% on CIFAR-10 with 2/255 $L_\infty$ perturbations on 4-layer convolutional network as well as 97.3% on MNIST dataset with $L_\infty$ perturbation 0.1 on 3-layer convolutional network.

## 2   Related Works

### 2.1   Empirical Adversarial Defense

The most successful empirical defense to date is adversarial training. It was first proposed in [16] and [4], where they showed that adding adversarial examples to the training set can improve the robustness against attacks. More recently, Madry et al. [12] formulated adversarial training as a min-max optimization problem and demonstrated that adversarial training with PGD attack leads to empirical robust models. However, it is equivalent to the minimization of the lower bound on the inner maximum worst-case loss that will lure and mislead an optimizer. Indeed, while adversarial training often provides robustness against a specific attack, it often fails to provide guarantee to generalize to new attacks.

## 2.2   Certified Adversarial Defense

Different from adversarial training, another approach is to compute the worst-case perturbation exactly, and most of the existing methods are based on Satisfiability Modulo Theories [6] or Mixed Integer Linear Programming [3]. Currently, these approaches can take up to several hours to compute the loss for a single example even for small networks. Computing an upper bound on the worst-case loss can avoid the intractability of exact computation. These approaches are typically based on linear [20], hybrid zonotope [13] or interval bound propagation [5]. While these approaches obtain robustness guarantees, accuracy of these networks is relatively low. COLT [1] combines adversarial training and convex relaxation method [20] layerwisely to train networks with high certified robustness and accuracy. Another line of work proposes to replace neural networks with a randomized classifier [10] which comes with probabilistic instead of exact guarantees on its robustness.

However, all the methods above treat natural examples equally during training process regardless whether the examples can be correctly classified or not, but only evaluate certified robustness on correctly classified examples. The effect of misclassified example on final certified robustness is "undefined". In this paper, we explore the impact of these misclassified examples, and provide a *Misclassification Aware Adversarial Regularization* (MAAR) as explained in next section.

## 3   Methods

### 3.1   Preliminaries

**Base Classifier.** For a $K$-class ($K \geq 2$) classification problem, denote a dataset $\{(\boldsymbol{x}_i, y_i)\}_{i=1,\cdots,n}$ with distribution $\boldsymbol{x}_i \in \mathbb{R}^d$ as natural input and $y_i \in \{1, \cdots, K\}$ represents its corresponding true label, a classifier $h_\theta$ with parameter $\boldsymbol{\theta}$ predicts the class of an input example $\boldsymbol{x}_i$:

$$h_\theta(\boldsymbol{x}_i) = \arg\max_{k=1,\cdots,K} \boldsymbol{p}_k(\boldsymbol{x}_i, \boldsymbol{\theta}) \tag{1}$$

$$\boldsymbol{p}_k(\boldsymbol{x}_i, \boldsymbol{\theta}) = \exp(\boldsymbol{z}_k(\boldsymbol{x}_i, \boldsymbol{\theta})) / \sum_{k'=1}^{K} \exp(\boldsymbol{z}_{k'}(\boldsymbol{x}_i, \boldsymbol{\theta})) \tag{2}$$

where $\boldsymbol{z}_k(\boldsymbol{x}_i, \boldsymbol{\theta})$ is the logits output of the network with respect to class $k$, and $\boldsymbol{p}_k(\boldsymbol{x}_i, \boldsymbol{\theta})$ is the probability (softmax on logits) of $\boldsymbol{x}_i$ belonging to class $k$.

**Adversarial Risk.** The adversarial risk [12] on dataset $\{(\boldsymbol{x}_i, y_i)\}_{i=1,\cdots,n}$ and classifier $h_\theta$ output probability $\boldsymbol{p}(\boldsymbol{x})$ can be defined as follows:

$$\mathcal{ADV}(\boldsymbol{p}(\boldsymbol{x}_i'), y) = \frac{1}{n} \sum_{i=1}^{n} \max_{\boldsymbol{x}_i' \in \mathcal{B}_\epsilon(\boldsymbol{x}_i)} \mathcal{L}(\boldsymbol{p}(\boldsymbol{x}_i'), y_i) \tag{3}$$

where $\mathcal{L}$ is the loss function such as commonly used cross entropy loss, and $\mathcal{B}_\epsilon(\boldsymbol{x}_i) = \{\boldsymbol{x} : ||\boldsymbol{x} - \boldsymbol{x}_i||_p \leq \epsilon\}$ denotes the $L_p$-norm ball centered at $\boldsymbol{x}_i$ with radius $\epsilon$. We will focus on the $L_\infty$-ball in this paper.

**Original Training Risk in COLT.** The original training risk in COLT [1] is defined as follows:

$$\mathcal{R}_{COLT}(h_\theta, \boldsymbol{x}_i) := \mathcal{L}_{ori}(\boldsymbol{p}(\boldsymbol{x}_i), y_i) + \mathcal{ADV}(\boldsymbol{p}(\boldsymbol{x}'_i), y_i) \tag{4}$$

where $\mathcal{L}_{ori}(\cdot)$ is the original training loss function such as cross entropy loss and $\boldsymbol{x}'_i \in \mathcal{B}_\epsilon(\boldsymbol{x}_i)$.

## 3.2 Misclassification Aware Training

Note that the training risk in Eq. (4) is defined on all natural examples, regardless of whether they are correctly classified ($h_\theta(\boldsymbol{x}_i) = y_i$) or misclassified ($h_\theta(\boldsymbol{x}_i) \neq y_i$) by the current model $h_\theta$. To differentiate and explore the effect of misclassified examples, we reformulate the training risk based on the prediction of the current network $h_\theta$. Specifically, we split the natural training examples into two subset according to $h_\theta$, with one subset of correctly classified examples ($\mathcal{C}_{h_\theta}^+$) and one subset of misclassified examples ($\mathcal{C}_{h_\theta}^-$):

$$\mathcal{C}_{h_\theta}^+ = \{i : i \in [n], h_\theta(\boldsymbol{x}_i) = y_i\} \tag{5}$$

$$\mathcal{C}_{h_\theta}^- = \{i : i \in [n], h_\theta(\boldsymbol{x}_i) \neq y_i\} \tag{6}$$

Considering the inconsistency on constraints of certified robustness on correctly classified and misclassified examples in COLT, we use output label for misclassified examples rather than original label during training, which we call *Misclassification Aware Training* (MAT). The training risk on misclassified examples is formulated as follows:

$$\mathcal{R}_{MAT}^-(h_\theta, \boldsymbol{x}_i) := \mathcal{L}_{ori}(\boldsymbol{p}(\boldsymbol{x}_i), y_i) + \mathcal{ADV}(\boldsymbol{p}(\boldsymbol{x}'_i), h_\theta(\boldsymbol{x}_i)) \tag{7}$$

As we observed in Fig. 1, although directly changing the training label of the misclassified example can achieve higher certified robustness, it leads to lower classification accuracy.

## 3.3 Misclassification Aware Adversarial Regularization

With the purpose of avoiding excessive reduction of the original accuracy as well as keeping the consistency of certified robustness on two subsets, we regularize misclassified examples by an additional term (a KL-divergence term that was used previously in [19,24]) rather than changing the training labels. The proposed consistency regularization aims to encourage the output probability distributions of neural network to be stable against examples in the perturbation region of misclassified adversarial examples, thus improving the certified robustness of network. The improved training risk of misclassified examples is formulated as follows:

$$\begin{aligned} \mathcal{R}^-(h_\theta, \boldsymbol{x}_i) := \mathcal{L}_{ori}(\boldsymbol{p}(\boldsymbol{x}_i), y_i) + \mathcal{ADV}(\boldsymbol{p}(\boldsymbol{x}'_i), y_i) \\ + \mathcal{KL}(\boldsymbol{p}(\boldsymbol{x}_i) \| \boldsymbol{p}(\boldsymbol{x}'_i)) \end{aligned} \tag{8}$$

where

$$KL(\boldsymbol{p}(\boldsymbol{x}_i) || \boldsymbol{p}(\boldsymbol{x}_i')) = \sum_{k=1}^{K} \boldsymbol{p}_k(\boldsymbol{x}_i, \theta) \log \frac{\boldsymbol{p}_k(\boldsymbol{x}_i, \theta)}{\boldsymbol{p}_k(\boldsymbol{x}_i', \theta)} \tag{9}$$

measures the difference of two distributions.

For correctly classified examples, we simply use original training risk, i.e.,

$$\mathcal{R}^+(h_\theta, \boldsymbol{x}_i) := \mathcal{L}_{ori}(\boldsymbol{p}(\boldsymbol{x}_i), y_i) + \mathcal{ADV}(\boldsymbol{p}(\boldsymbol{x}_i'), y_i) \tag{10}$$

Finally, by combining the two training risk terms (i.e., Eq. (10) and Eq. (10)), we train a network that minimizes the following risk:

$$\begin{aligned} \mathcal{R}(h_\theta, \boldsymbol{x}) &:= \frac{1}{n} \Big( \sum_{\boldsymbol{x}_i \in \mathcal{C}_{h_\theta}^+} \mathcal{R}^+(h_\theta, \boldsymbol{x}_i) + \sum_{\boldsymbol{x}_i \in \mathcal{C}_{h_\theta}^-} \mathcal{R}^-(h_\theta, \boldsymbol{x}_i) \Big) \\ &= \frac{1}{n} \sum_{i=1}^{n} \{ \mathcal{L}_{ori}(\boldsymbol{p}(\boldsymbol{x}_i), y_i) + \mathcal{ADV}(\boldsymbol{p}(\boldsymbol{x}_i'), y_i) \\ &\quad + KL(\boldsymbol{p}(\boldsymbol{x}_i) || \boldsymbol{p}(\boldsymbol{x}_i')) \cdot \mathbb{I}(h_\theta(\boldsymbol{x}_i) \neq y_i) \} \end{aligned} \tag{11}$$

where $\mathbb{I}(h_\theta(\boldsymbol{x}_i) \neq y_i)$ is the indicator function. $\mathbb{I}(h_\theta(\boldsymbol{x}_i) \neq y_i) = 1$ if $h_\theta(\boldsymbol{x}_i) \neq y_i$, and $\mathbb{I}(h_\theta(\boldsymbol{x}_i) \neq y_i) = 0$ otherwise.

**Optimization for Regularization Term.** As presented in Eq. (11), the new training risk is a regularized adversarial risk with regularization term $\frac{1}{n} \sum_{i=1}^{n} \{ KL(\boldsymbol{p}(\boldsymbol{x}_i) || \boldsymbol{p}(\boldsymbol{x}_i')) \cdot \mathbb{I}(h_\theta(\boldsymbol{x}_i) \neq y_i) \}$. However, the indicator function cannot be directly optimized if we conduct a hard decision during the training process. In this study, we propose to use a soft decision scheme by replacing $\mathbb{I}(h_\theta(\boldsymbol{x}_i) \neq y_i)$ with the output probability $1 - \boldsymbol{p}_{y_i}(\boldsymbol{x}_i, \boldsymbol{\theta})$. The output probability will be large for misclassified examples and small for correctly classified examples, by which we could provide a approximate solution for 0–1 optimization problem .

**The Overall Objective.** Based on the regularization optimization, the objective function of our proposed *Misclassification Aware Adversarial Regularization* (MAAR) is formulated as:

$$\mathcal{R}^{MAAR}(\boldsymbol{\theta}) = \frac{1}{n} \sum_{i=1}^{n} \mathcal{L}^{MAAR}(\boldsymbol{x}_i, y_i, \boldsymbol{\theta}) \tag{12}$$

where $\mathcal{L}^{MAAR}(\boldsymbol{x}_i, y_i, \boldsymbol{\theta})$ is defined as:

$$\begin{aligned} \mathcal{L}^{MAAR}(\boldsymbol{x}_i, y_i, \boldsymbol{\theta}) &= \mathcal{L}_{ori}(\boldsymbol{p}(\boldsymbol{x}_i), y_i) + \mathcal{ADV}(\boldsymbol{p}(\boldsymbol{x}_i'), y_i) \\ &\quad + \lambda \cdot KL(\boldsymbol{p}(\boldsymbol{x}_i) || \boldsymbol{p}(\boldsymbol{x}_i')) \cdot (1 - \boldsymbol{p}_{y_i}(\boldsymbol{x}_i, \boldsymbol{\theta})) \end{aligned} \tag{13}$$

Here, $\lambda$ is the tunable scaling paremeters and fixed for all training examples. Our training process is shown in Algorithm 1.

---

**Algorithm 1.** Misclassification Aware Adversarial Regularization (MAAR)

---

**Input**  $d$-layer network $h_\theta$, training set $(\mathcal{X}, \mathcal{Y})$, learning rate $\eta$, step size $\alpha$, inner steps $n$, tunable scaling parameters $\lambda$, perturbation $\epsilon$.

**for** $l \leq d$ **do**

  **for** $j \leq n_{epochs}$ **do**

    Sample mini-batch: $(\boldsymbol{x}_1, y_1), (\boldsymbol{x}_2, y_2) \cdots, (\boldsymbol{x}_b, y_b)\} \sim (\mathcal{X}, \mathcal{Y})$;

    Compute convex relaxations: $\mathbb{C}_l(\boldsymbol{x}_1), \cdots, \mathbb{C}_l(\boldsymbol{x}_b)$;

    Initialize: $\boldsymbol{x}_1' \sim \mathbb{C}_l(\boldsymbol{x}_1), \cdots, \boldsymbol{x}_b' \sim \mathbb{C}_l(\boldsymbol{x}_b)$;

    **for** $i \leq b$ **do**

      Update in parallel $n$ times: $\boldsymbol{x}_i' \leftarrow \Pi_{\mathbb{C}_l(\boldsymbol{x}_i)}(\boldsymbol{x}_i' + \alpha \nabla_{\boldsymbol{x}_i'} \mathcal{ADV}(h_\theta^{l+1:d}(\boldsymbol{x}_i'), y_i))$;

    **end for**

    $\mathcal{L}(h_\theta^{l+1:d}(\boldsymbol{x}_i'), y_i) \leftarrow \mathcal{L}_{ori}(\boldsymbol{p}(\boldsymbol{x}_i), y_i) + \mathcal{ADV}(\boldsymbol{p}(\boldsymbol{x}_i'), y_i) + \lambda \cdot \mathcal{KL}(\boldsymbol{p}(\boldsymbol{x}_i) || \boldsymbol{p}(\boldsymbol{x}_i')) \cdot (1 - \boldsymbol{p}_{y_i}(\boldsymbol{x}_i, \boldsymbol{\theta}))$

    Update parameters: $\boldsymbol{\theta} \leftarrow \boldsymbol{\theta} - \eta \cdot \frac{1}{b} \sum_{i=1}^{b} \nabla_\theta \mathcal{L}(h_\theta^{l+1:d}(\boldsymbol{x}_i'), y_i)$;

  **end for**

  Freeze parameters $\boldsymbol{\theta}_{l+1}$ of layer function $h_\theta^{l+1}$.

**end for**

**Output** Certified robust neural network $h_\theta$

---

# 4 Experiments

In this section, we first introduce the experimental settings used in our experiments. Then we investigate the sensitivity of regularization parameter $\lambda$, and choose the best parameter $\lambda$ to evaluate the effectiveness of our proposed MAAR compared with COLT and MAT. Finally, we show the experimental results of our MAAR under different network architectures and different perturbations on two datasets (i.e., CIFAR-10 [7] and MNIST [9]) in comparison with several other defense methods.

## 4.1 Experimental Settings

Experiments are conducted on CIFAR-10 and MNIST datasets. We use a four-layer convolutional network with $L_\infty$ perturbations $\epsilon = 2/255, 8/255$ on CIFAR-10 dataset, a three-layer convolutional network with $L_\infty$ perturbations $\epsilon = 0.1$ on MNIST dataset. The layerwise training fashion[1] has been adopted in our MAAR's training. We use four metrics to evaluate our training models, i.e., original accuracy (ACC), certified robustness (CR), verified error (VE), and latent robustness (LR). The detailed information of these settings can be found in Appendix 1 and 2.

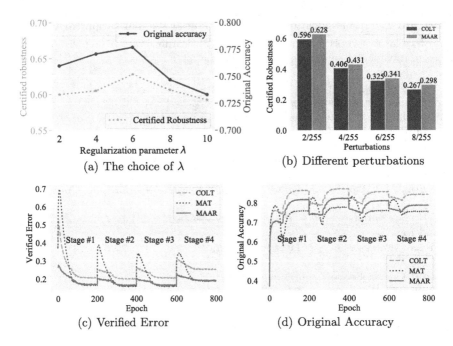

**Fig. 2.** The effectiveness of our proposed MAAR. (a) Sensitivity of regularization parameter $\lambda$. Layerwise verified error (b) and layerwise original accuracy (c) of COLT, MAT and proposed MAAR on different stages, respectively. (d) Certified robustness of COLT and MAAR with different perturbations. (Color figure online)

## 4.2  Sensitivity of Regularization Parameter $\lambda$

We investigate the parameter $\lambda$ with MAAR defined in Eq. (13) which controls the contribution of the regularization term. We present the results in Fig. 2(a) for different $\lambda \in \{2, 4, 6, 8, 10\}$. By explicitly setting different impact parameter of misclassified examples, the network achieves good stability and robustness across different choices of $\lambda$. According to the experimental results, we choose $\lambda = 6$ for our following experiments.

## 4.3  the Effectiveness of MAAR

**Comparison with COLT and MAT.** In order to verify the effectiveness of our proposed MAAR, we firstly compare MAAR with COLT and MAT. Note that all experiment settings of these three methods are the same except for the constraint on misclassified examples. The verified error and original accuracy evaluated at every epoch during training process has been shown in Fig. 2(c), Fig. 2(d). As shown in Fig. 2(c), the verified error of MAAR (green line) decreases more rapidly in comparison with COLT (red line) during each stage, which indicates that MAAR can reduce the proportion of potential adversarial examples in each layer. On the other hand, MAAR maintains the stability of misclassified examples by

**Table 1.** The final certified robustness (CR) and latent robustness (LR) of network trained on MAAR and COLT with the parameters of different stages. $LR^{3^{rd}}$ represents the latent adversarial attack is performed on the 3-rd ReLU layer.

| | Method | CR(%) | $LR^{3^{rd}}$(%) |
|---|---|---|---|
| Stage #1 | MAAR (Our work) | **54.1** | **58.3** |
| | COLT | 40.0 | 47.5 |
| Stage #2 | MAAR (Our work) | **57.5** | **60.1** |
| | COLT | 48.2 | 54.5 |
| Stage #3 | MAAR (Our work) | **60.7** | **62.0** |
| | COLT | 57.7 | 60.8 |
| Stage #4 | MAAR (Our work) | **62.8** | **64.7** |
| | COLT | 59.6 | 62.1 |

an additional regularization constraint rather than replacing the training label as MAT, which mitigates the decrease of original accuracy. As shown in Fig. 2(d), the accuracy of MAAR (green line) is obviously improved when compared with MAT (blue line).

**Table 2.** Comparison with the pior work. Accuracy and certified robustness evaluated with $L_\infty$ perturbation 2/255 and 8/255 on CIFAR-10 dataset, $L_\infty$ perturbation 0.1 on MNIST dataset. ACC: Accuracy, CR: Certified robustness.

| Method | CIFAR-10 | | | | MNIST | |
|---|---|---|---|---|---|---|
| | $\epsilon = 2/255$ | | $\epsilon = 8/255$ | | $\epsilon = 0.1$ | |
| | ACC (%) | CR (%) | ACC (%) | CR (%) | ACC (%) | CR (%) |
| Xiao et al. [21] | 61.1 | 45.9 | 40.5 | 20.3 | 99.0 | 95.6 |
| Mirman et al. [14] | 62.3 | 45.5 | 46.2 | 27.2 | 98.7 | 96.8 |
| IBP [5] | 58.0 | 47.8 | 47.8 | 24.9 | 98.8 | 95.8 |
| CROWN-IBP [25] | 61.6 | 48.6 | 48.5 | 26.3 | 98.7 | 96.6 |
| COLT [1] | 80.0 | 58.6 | 51.3 | 26.7 | 99.2 | 97.1 |
| Our work(MAAR) | 77.7 | **62.8** | 47.6 | **29.8** | 99.1 | **97.3** |

**The Consistent Promotion of MAAR in Layerwise Training Mechanism.** In addition, we evaluate the final certified robustness of the network on the checkpoint saved after each training stage. As shown in Table 1, we can observe that the final certified robustness of our network has been significantly improved in layerwise training fashion (from 54.1% on Stage #1 to 62.8% on Stage #4). Furthermore, the final certified robustness of our proposed MAAR is obviously higher than COLT when evaluated on all stages. Meanwhile, we investigate the latent robustness (LR) of the model. Generally, we run latent adversarial attack (i.e., PGD attack with 150 steps and step size of 0.01) on 3-rd ReLU layer with parameters of each stage. Table 1 indicates the LR of our proposed MAAR improves from 58.3% on Stage #1 to 64.7% on Stage #4, which is also obviously outperforming COLT on all stages. These results demonstrate that MAAR can bring the consistent promotion in layerwise training.

## 4.4   Certification Under Different Perturbations

We then evaluate the effectiveness of our proposed MAAR on certified robustness with different perturbations. As shown in Fig. 2(b), when perturbations $\epsilon \in \{2/255, 4/255, 6/255, 8/255\}$, the certified robustness of MAAR (orange bar) is obviously higher than COLT (blue bar). Specifically, our method MAAR achieves the state-of-the-art certified robustness (i.e., 62.8%) compared with COLT (59.6%) when $\epsilon = 2/255$.

## 4.5   Comparison with Prior Work on Different Datasets

We compare our MAAR with COLT [1], CROWN-IBP [25], and IBP [5] in the same network architecture and parameter settings. Furthermore, we list the results reported in literature of Xiao et al. [21] and Mirman et al. [14]. Table 2 shows the results.

**CIFAR-10.** For the $L_\infty$ perturbation 2/255. Experiment results show that MAAR substantially outperforms its competitors by certified robustness (i.e., 62.8%). Besides, the accuracy of our method also outperforms other works except COLT. This is because one side-effect of our regularization is that it will maintain the distribution around misclassified examples, which will decrease the accuracy in comparison with COLT. Actually, the accuracy–robustness trade-off has been proved to exist in predictive models when training robust models [18,24]. We also run the same experiment for $L_\infty$ perturbation 8/255, where MAAR also achieves the best certified robustness (i.e., 29.8%).

**MNIST.** To futher evaluate the effectiveness of our method, we also conduct experiments on MNIST dataset with $L_\infty$ perturbation 0.1. We report the full results in Table 2, MAAR also achieve the state-of-the-art certified robustness (i.e., 97.3%) comparable with best results from prior work (i.e., 97.1%).

## 5   Conclusion

In this paper, we investigated the inconsistent constraint of certified robustness between correctly classified and misclassified examples and find that misclassified examples have a recognizable impact on the final certified robustness of network. Based on this observation, we designed a consistency regularization which constrains the output probability distributions of examples in the certified region of the misclassified example. Our method, named Misclassification Aware Adversarial Regularization (MAAR), achieves the state-of-the-art certified robustness of 62.8% on CIFAR-10 with $2/255$ $L_\infty$ perturbation as well as 97.3% on MNIST dataset with $L_\infty$ perturbation 0.1. The method is general and can be instantiated with most of training risk.

In the future, we plan to investigate the association between accuracy and certified robustness among different neural networks, and apply our method to more provable defense frameworks.

# References

1. Balunovic, M., Vechev, M.: Adversarial training and provable defenses: bridging the gap. In: ICLR (2020)
2. Carlini, N., Wagner, D.: Towards evaluating the robustness of neural networks. In: SP, pp. 39–57. IEEE (2017)
3. Cheng, C.-H., Nührenberg, G., Ruess, H.: Maximum resilience of artificial neural networks. In: D'Souza, D., Narayan Kumar, K. (eds.) ATVA 2017. LNCS, vol. 10482, pp. 251–268. Springer, Cham (2017). https://doi.org/10.1007/978-3-319-68167-2_18
4. Goodfellow, I.J., Shlens, J., Szegedy, C.: Explaining and harnessing adversarial examples. In: ICLR (2015)
5. Gowal, S., Dvijotham, K., Stanforth, R., et al.: On the effectiveness of interval bound propagation for training verifiably robust models. arXiv preprint arXiv:1810.12715 (2018)
6. Katz, G., Barrett, C., Dill, D.L., Julian, K., Kochenderfer, M.J.: Reluplex: an efficient SMT solver for verifying deep neural networks. In: Majumdar, R., Kunčak, V. (eds.) CAV 2017. LNCS, vol. 10426, pp. 97–117. Springer, Cham (2017). https://doi.org/10.1007/978-3-319-63387-9_5
7. Krizhevsky, A., Hinton, G., et al.: Learning multiple layers of features from tiny images (2009)
8. Krizhevsky, A., Sutskever, I., Hinton, G.E.: Imagenet classification with deep convolutional neural networks. In: NeurIPS, pp. 1097–1105 (2012)
9. LeCun, Y., Bottou, L., Bengio, Y., et al.: Gradient-based learning applied to document recognition. Proc. IEEE **86**(11), 2278–2324 (1998)
10. Lecuyer, M., Atlidakis, V., Geambasu, R., et al.: Certified robustness to adversarial examples with differential privacy. In: SP, pp. 656–672. IEEE (2019)
11. Li, J., Zhang, X., Jia, C., et al.: Universal adversarial perturbations generative network for speaker recognition. In: ICME, pp. 1–6. IEEE (2020)
12. Madry, A., Makelov, A., Schmidt, L., et al.: Towards deep learning models resistant to adversarial attacks. arXiv preprint arXiv:1706.06083 (2017)
13. Mirman, M., Gehr, T., Vechev, M.: Differentiable abstract interpretation for provably robust neural networks. In: ICML, pp. 3578–3586 (2018)
14. Mirman, M., Singh, G., Vechev, M.: A provable defense for deep residual networks. arXiv preprint arXiv:1903.12519 (2019)
15. Raghunathan, A., Steinhardt, J., Liang, P.: Certified defenses against adversarial examples. In: ICLR (2018)
16. Szegedy, C., Zaremba, W., Sutskever, I., et al.: Intriguing properties of neural networks. In: ICLR (2014)

17. Taigman, Y., Yang, M., Ranzato, M., et al.: Deepface: closing the gap to human-level performance in face verification. In: CVPR, pp. 1701–1708 (2014)
18. Tsipras, D., Santurkar, S., Engstrom, L., et al.: Robustness may be at odds with accuracy. arXiv preprint arXiv:1805.12152 (2018)
19. Wang, Y., Zou, D., Yi, J., et al.: Improving adversarial robustness requires revisiting misclassified examples. In: ICLR (2019)
20. Wong, E., Kolter, Z.: Provable defenses against adversarial examples via the convex outer adversarial polytope. In: ICML, pp. 5286–5295 (2018)
21. Xiao, K.Y., Tjeng, V., Shafiullah, N.M.M., et al.: Training for faster adversarial robustness verification via inducing relu stability. In: ICLR (2018)
22. Zhang, H., Jia, F., Zhang, Q., et al.: Two-way feature-aligned and attention-rectified adversarial training. In: ICME, pp. 1–6. IEEE (2020)
23. Zhang, H., Zhou, W., Li, H.: Contextual adversarial attacks for object detection. In: ICME, pp. 1–6. IEEE (2020)
24. Zhang, H., Yu, Y., Jiao, J., et al.: Theoretically principled trade-off between robustness and accuracy. arXiv preprint arXiv:1901.08573 (2019)
25. Zhang, H., Chen, H., Xiao, C., et al.: Towards stable and efficient training of verifiably robust neural networks. arXiv preprint arXiv:1906.06316 (2019)

# Fooling Neural Network Interpretations: Adversarial Noise to Attack Images

Qianqian Song[1], Xiangwei Kong[2(✉)], and Ziming Wang[2]

[1] School of Information and Communication Engineering,
Dalian University of Technology, Dalian 116024, China
[2] School of Management, Zhejiang University, Hangzhou 310058, China
kongxiangwei@zju.edu.cn

**Abstract.** The accurate interpretation of neural network about how the network works is important. However, a manipulated explanation may have the potential to mislead human users not to trust a reliable network. Therefore, it is necessary to verify interpretation algorithms by designing effective attacks to simulate various possible threats in the real world. In this work, we mainly explore how to mislead interpretation. More specifically, we optimize the noise added to the input, which aims to highlight a certain area that we specify without changing the output category of network. With our proposed algorithm, we demonstrate that the state-of-the-art saliency maps based interpreters, e.g., Grad-CAM, Guided-Feature-Inversion, Grad-CAM++, Score-CAM and Full-Grad can be easily fooled. We propose two situations of fooling, Single-target attack and Multi-target attack, and show that the fooling can be transfered to different interpretation methods as well as generalized to the unseen samples with the universal noise. We also take image patches to fool Grad-CAM. Our results are proved in both qualitative and quantitative ways and we further propose a quantitative metric to measure the effectiveness of algorithm. We believe that our method can serve as an additional evaluation of robustness for future interpretation algorithms.

**Keywords:** Neural network · Interpretation · Attack

## 1 Introduction

As deep neural networks (DNNs) are increasingly being deployed in domains such as healthcare and biology [20], there is growing emphasis on building tools and techniques that can explain them to ensure that the decision-making mechanism is transparent and easily interpretable.

As a consequence, there has been a recent surge in post hoc techniques for explaining DNNs in an interpretable manner. Saliency map is a type of visualisation method that provides an intuitive explanation of the output of model by

---

Supported by National Natural Science Foundation of China (61772111).

L. Fang et al. (Eds.): CICAI 2021, LNAI 13070, pp. 39–51, 2021.
https://doi.org/10.1007/978-3-030-93049-3_4

highlighting the input regions which contributed the most to the final output, including Grad-CAM [1], Grad-CAM++ [3], Score-CAM [4], Full-Grad [5] and Guided-Feature-Inversion [2]. However, there has been very few analysis that a manipulated explanation of a DNN may not accurately reflect the accuracy of the network, especially in a controlled and adversarial setting, making human user trust in a reliable model influenced.

In this work, we demonstrate significant vulnerabilities in local explanation techniques that can be disturbed by the visually hardly noticeable noise to generate disturbed images whose explanations can be arbitrarily controlled. The perturbation does not change the label of the classification model. Another contribution of our work is to evaluate "goodness" of the fooling. We evaluate whether the visual area is aligned with a certain area we have predetermined. To sum up, our key contributions are summarized as follows:

- We introduce a novel algorithm to optimize noise added to images which fools the interpretation of the input without changing the label predicted by the classification model. We demonstrate its effectiveness for five explanation methods and we show that our method (a) attacks both single-target and multi-target interpretations (b) generalizes to unseen images and (c) transfers to different interpretation methods.
- We are also the first to fool interpretations with image patches [23]. Recently, [23] showed that image patches can be created for person detection algorithms. This is a complementary attacking form of noise, considering the possibility that the attacks in real world is not limited to noise.
- Our results are proved in both qualitative and quantitative ways and we further propose one quantitative metric that measures the effectiveness of the adversarial noise generated by our algorithm.

## 2   Related Work

### 2.1   Local-level Interpretation Methods

In order to apply DNNs into real world application, their results should be interpretable. The saliency map based interpreters, visualizing DNN internal representations, is a more straightforward and important way to understand the way networks interpret images [21,22]. Various algorithms have been proposed in this direction. Although these methods have shown great performance in explaining the decision of network, it is challenging to evaluate whether the explanation is reliable. We believe our method can serve as an additional evaluation for future interpretation algorithms.

### 2.2   Fooling Network Interpretation

[10] used the adversarial image to alter the attribution map while maintaining the predicted class unchanged which shows that some gradient-based explanation methods can be highly sensitive to small perturbations in the input. However, as

is mentioned in the discussion section in [10], the adversarial image after perturbation can lead to an unstructured change in the explanation map in this setting, i.e., completely replace the explanations. To mitigate this problem, we change only a small region of the image with our controlled setting. Here, we clearly know that the manipulated interpretation should be within this region. [8] also altered explanations of original images. Compared to [10], which was untargeted manipulations, [8] focused on targeted manipulations, i.e., to reproduce a given target map. [12] showed how saliency methods are unreliable by adding a constant shift to the input image and checking against different explanation maps. [11] showed that explanation maps are changed by randomization of (some of) the network weights. This is different from our method as it modifies the weights of the network. [9] also propose a novel theoretical framework for understanding and generating misleading explanations based on MUSE framework [17], but it can not extend well to other interpretation algorithms. [7] introduce a threat model by modifying the parameters of the model to fool the interpretations without hurting the accuracy of the original models. However, in a practical setting, the adversary requires computationally expensive training and might not always be able to modify the parameters. Our work is interested in modifying only the pixels in a small image area without altering the model. The paper most relevant to our work is [6] whose attacker based on adversarial patch [19] modifies the pixels in a small image area and leaves the rest unchanged to fool explanation. However, they need to change the label to create the adversarial patches.

## 3   Methods

We propose algorithms to optimize the noise so that when added to the image, the interpretation of the perturbed image would tend to a specific area without changing the label predicted by the classification model. In addition, we propose to use the approach similar to [16] to constraint the noise so that it is not to be perceived by the human eye. Figure 1 illustrates the overview of our method. We will mainly experiment on the following explanation methods:

- Grad-CAM (GCAM): This method calculates global average gradient of the target classification as a weight vector of each pixel belonging to a convolutional layer, then gets the interpretation as the weighted sum of activations of the convolutional layer discarding the negative values. We refer to the original publication for more details [1].
- Grad-CAM++ (GCAM++): This method has a more complicated weight vector calculation than Grad-CAM. This method thinks each pixel on the gradient contributes differently, so an additional weight is added to weight the pixel on the gradient [3].
- Guided-Feature-Inversion (GFI): This framework inverts the representations at higher layers of CNN to a synthesized image, while simultaneously encodes the location information of the target object in a mask [2].

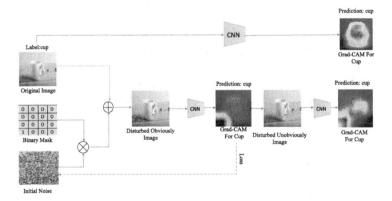

**Fig. 1. The overview of our attack on interpretation.** Top row shows that Grad-CAM [1] highlights the image location. Our attack algorithm goes beyond fooling the Grad-CAM visualization while keeping the final prediction unchanged.

- Score-CAM (SCAM): This method firstly extractes feature maps. Each feature map then works as a mask on original image, and obtain its forward-passing score on the target class. Finally, the explanation can be generated by linear combination of score-based weights and feature maps [4].
- Full-Grad (FGrad): This method provides attribution to both inputs and neurons. They alternate representation of the neural network output in terms of input-gradients and bias-gradients. See the original reference [5] for more details.

### 3.1  Fooling Interpretation with the Noise

Our method misleads the explanation methods to give out wrong interpretation by optimizing the noise with a new location importance loss, $l_{loim}(h, m)$. Here $h$ is the interpretation representing the important features of an input image of spatial dimension $D * D$ for the final prediction score for the class $c$, and $m \in \{0, 1\}^{D*D}$ is a predefined constant binary mask which take value of 0 corresponding to the location of the local area and 1 otherwise. We want to find an optimal noise $z \in R^{D*D*3}$ that changes the explanation of the perturbed image $\widetilde{x} \in R^{D*D*3}$ when added to the area of the original input $x$, so, assuming the perturbed image is generated by:

$$\widetilde{x} = x + z \odot (1 - m) \tag{1}$$

where $\odot$ is the element-wise product. The total loss function on each training image is a weighted sum of the cross-entropy loss $l_{ce}(\widetilde{x}; c)$ for the class $c$, and the location importance loss:

$$L(\widetilde{x}, m, h, c) = l_{ce}(\widetilde{x}; c) + \lambda l_{loim}(h, m) \tag{2}$$

$$l_{loim}(h, m) = \|h(\widetilde{x}) - (1 - m)\|_2 \tag{3}$$

in which $\lambda$ is the hyper-parameter to trade-off the effect of the above two loss terms, and $l_{loim}(h, m)$ is the value of $l_2 - norm$.

However, it makes no sense to change the interpretation of only one object if there are multiple objects in the image. In order to solve this problem, we predefine multiple binary masks to obtain respective desired regions simultaneously. Given an image with $n$ objects, the final perturbed image can be expressed as:

$$\tilde{x} = x + \sum_{j=1}^{n} (z \odot (1 - m_j)) \tag{4}$$

And we modify (2) as follows:

$$L(\tilde{x}, m_j, h_j, c_j) = \sum_{j=1}^{n} (l_{ce}(\tilde{x}; c_j) + \lambda_j l_{loim}(h_j, m_j)) \tag{5}$$

We take just two objects for example, namely $n = 2$ in our experiment. In addition to the single-target attack of interpretation, this work focuses on a more challenging task, multi-target attack of their interpretations, which can assign a corresponding different area that we want to highlight to each object, while keep the class of each object unchanged by a single noise without repeated training.

### 3.2 Fooling Interpretation with the Universal Noise

Universal attack of interpretation is a stronger and more practical form of attack wherein we optimize a noise just once that generalizes across images of original category. Such an attack is possibly strong enough to fool the explanation of an unknown test image using the noise learned by the training data. To do this, we adopt the batch processing technology to optimize the summation of losses for batch images for each category:

$$L(\tilde{x}_i, c, h_i, m) = \frac{1}{N} \sum_{i=1}^{N} (l_{ce}(\tilde{x}_i; c) + \lambda \| h_i(\tilde{x}_i) - (1 - m) \|_2) \tag{6}$$

Here, $N$ represents the batch size (32 in our experiment), and each disturbed image is synthesized from the corresponding input image, binary mask and the general noise. Each input image shares the same disturbed noise.

### 3.3 Fooling Interpretation with the Image Patches

In our paper, we also take image patches [23] to fool interpretations. These patches can be printed and 'pasted' on top of an image to attack person detection. We improve this by ensuring that the patches fool network interpretation. We first resize an image into the predefined area, and then fill the remained area with zeros. The image patch $I_{patch}$ will be processed, such as adding noise, modifying brightness and contrast. The perturbed image is generated by:

$$\tilde{x} = x \odot m + I_{patch} \odot (1 - m) \tag{7}$$

We optimize the $I_{patch}$ using the total loss function same as formula (2).

---

**Algorithm 1:** Fooling interpretation through disturbed and invisible noise.

---

**Input:** $X_0$, target label $c$, binary mask $m$, interpretation method $h$.
**Output:** $Z$, $X$.

1 Initialize the parameter $z$, iteration numbers $max\_iter$, $I_r$, $\lambda$, $\partial$, $t = 0$;
2 **while** $t \leq max\_iter$ **do**

3   $X_t = X_0 + Z_t \odot (1 - m)$;

4   $l_{loim}(h, m) = \|h(X_t) - (1 - m)\|_2$;

5   $L(Z_t, m, c) = l_{ce}(Z_t; c) + \lambda l_{loim}(Z_t, m)$;

6   $\widetilde{Z}_{t+1} = Adam(L(Z_t, m, c), I_r)$;

7   $Z_{t+1} \leftarrow clip(\widetilde{Z}_{t+1}, -\partial, \partial)$;

8   $t = t + 1$;

9 **end**
10 $X = X_0 + Z \odot (1 - m)$;
11 **return** $Z$, $X$.

---

### 3.4 Constraining the Noise

The generated image is unnatural if we just use the formula (2). Our goal is to inject a small amount of perturbation in the local area of the given input image so that the perturbation is not visually perceivable but result in significant misinterpretations. To find an $Z$ to minimize (2) with bounded $l_\infty - norm$ constraint ($\|Z\|_\infty \leq \partial$), we develop an algorithm based on the idea of I-FGSM [16,18].

Let $X_0$ denote the original input image and $X$ denote the perturbed version of $X_0$, so we adopt the I-FGSM [16,18] update rule to iteratively update $Z$ by (see Algorithm 1):

$$\widetilde{Z}_{n+1} = Z_n + \epsilon sgn(\nabla L(X_n, h, c, m)) \tag{8}$$

$$Z_{n+1} = clip_{-\partial, \partial}\left(\widetilde{Z}_{n+1}\right) \tag{9}$$

$$X_{n+1} = X_0 + Z_{n+1} \odot (1 - m) \tag{10}$$

Where $sgn\nabla$ is the sign of the gradient of the loss function, $\epsilon$ controls the amount of contribution that the calculated gradient provides at each iteration, and

$$clip_{a,b}(Z) = \min(\max(Z, a), b) \tag{11}$$

The term $\partial$ limits the maximum amount of perturbation to prevent noticeable changes of the perturbed image. The final perturbed image is obtained by $X = X_T$, where $T$ is the number of iterations.

## 4 Experiments

We use VGG19, ResNet18 and ResNet50 to experiment with various attack in all of our experiments. More special, we fool GCAM [1] and GCAM++ [3] on

ResNet50, SCAM [4] and FGrad [5] on ResNet18, and GFI on VGG19 (as used in [2]). We consider the ImageNet [14] validation set in our experiments. This set consists of a total of 50k images. We compare the changes before and after the interpretation algorithms are attacked.

## 4.1 Evaluation

In this section, we discuss the detailed experimental evaluation of our algorithm. More specially, we test how well the heatmap is focused on the area. We use the Localization metric from the object localization challenge of ImageNet competition. We draw a bounding box around the tresholded heatmap (0.15 in all of our experiments), and IOU (intersection over union) will be calculated as the overlap of the local area annotations and the bounding boxes. The number of IOUs less than 0.5 are used in our experiment. In this study, we propose and formulate the metric MPI, which we will refer to as the Mean Position Importance. We first sum the interpretation heatmap and normalize it to the range $[0, 1]$ for every image: $\widehat{h} = \frac{h}{\|h\|_1}$. Based on this, let the relationships between $\widehat{h}$, the mask $m$, the $i_{th}$ image, and the number of the images $N$ be expressed by $MPI = \frac{1}{N} \sum_{i=1}^{N} \left\| \widehat{h} \odot (1 - m) \right\|_1$. It will be 1 if the heatmap is completely focused on the area and 0 if the area is not highlighted at all. We also analyze the effectiveness of the optimized noise generated by our algorithm. We mainly compare the similarities between saliency maps of the original and the disturbed image with the Histogram comparison (HC) to calculate histogram of the heatmap respectively then normalize each, and finally compare the similarities between them. The value of this metric, a value of 1 indicates that the heatmap has not been changed, is the smaller, the better. For another similarity metric, we calculate the Spear-man's correlation coefficient (SC) [13] between saliency maps for quantifying the fooling effectiveness.

## 4.2 Single-target Attack of Interpretation

In this part, we show the attack performance on the single-target attack of interpretation task. We fool five interpretation methods, and we choose noise location along top right area of the image because they have the least probability to cover the salient object. The adversarial noise mask with size $60 \times 60$ ( 7% of the image) is applied. We use 50,000 images of the validation set and Adam [15] optimization algorithm for per image. The hyperparameter choices used in our single-target attack experiments are summarized in Table 1. The GFI [2] is trained on VGG19 along with 60 iteration steps, $10^{-2}$ learning rate. Figure 3 corresponding to four metrics presents summary of all the results. Figure 2 shows the qualitative attack results.

**Table 1.** Hyperparameters used in our analysis.

| Methods | Iterations | $I_r$ | $\lambda$ | $\partial$ |
|---------|-----------|-------|-----------|-----------|
| GCAM [1] | 50 | $10^{-1}$ | $10^{-1}$ | 70/255 |
| GFI [2] | 50 | $10^{-5}$ | $8 * 10^{-2}$ | 80/255 |
| GCAM++ [3] | 100 | $10^{-1}$ | $10^{-1}$ | 50/255 |
| SCAM [4] | 60 | $10^{-1}$ | $5 * 10^{-1}$ | 120/255 |
| FGrad [5] | 150 | $10^{-1}$ | $5 * 10^{-4}$ | 100/255 |

**Table 2.** Results for the image patches within the 7% of the input.

|  | MPI(%) | IOU(%) | HC(%) | SC(%) |
|--|--------|--------|-------|-------|
| No attack | 2.88 | 99.96 | 100 | 100 |
| Attack | **33.87** | **36.64** | **77.70** | **58.04** |

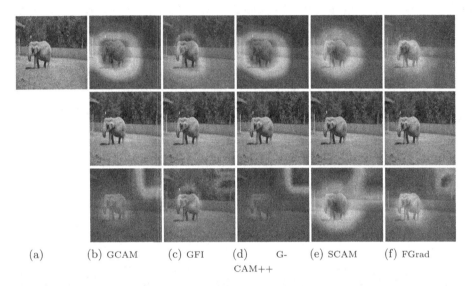

(a)    (b) GCAM    (c) GFI    (d)    G-    (e) SCAM    (f) FGrad
CAM++

**Fig. 2. Interpretations of the baseline and the fooled on an image from ImageNet validation set, of which the class label is 'African elephant' (shown in (a)).** The topmost row shows the baseline interpretations given the true class. The second row shows the disturbed images. The bottom row shows the fooled interpretations. See the baseline explanation results are changed dramatically when fooled

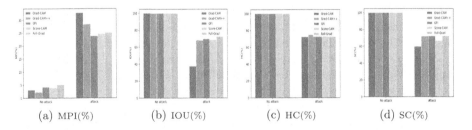

(a) MPI(%)    (b) IOU(%)    (c) HC(%)    (d) SC(%)

**Fig. 3. Results before and after five interpretation algorithms are attacked.** Note that for IOU, lower is better while for MPI, higher is better.

### 4.3  Multi-target Attack of Interpretation

With our algorithm, we can randomly decide the areas that we want the interpretation to output high importance of two objects as is shown in Fig. 4 for GCAM [1] and GCAM++ [3]. In our experiment, we set the area to the top left and right corner, respectively. This is a more evident form of fooling the interpretation, since that appearance that the explanation is inconsistent with the predicted category in an image containing more than one object indicates that the interpretation method is invalid.

**Table 3.** Comparison of MPI(%) for the single noise and our universal noise.

| Methods | Slug | Pier | Warplane | Stinkhorn | Green snake | Traffic light | Spiny lobster | Toilet tissue | European gallinule | Red-breasted merganser |
|---|---|---|---|---|---|---|---|---|---|---|
| Original | 2.22 | 2.76 | 2.00 | 2.62 | 2.98 | 2.94 | 1.98 | 2.50 | 3.57 | 2.95 |
| Universal | **38.04** | **22.65** | **25.25** | **31.52** | **36.90** | **37.79** | **32.38** | **35.24** | **32.56** | **31.47** |
| Single | 15.25 | 13.37 | 12.93 | 9.27 | 17.83 | 9.71 | 10.96 | 19.53 | 10.44 | 15.36 |

(a) GCAM                                (b) GCAM++

**Fig. 4. Interpretation of the 'tiger' and 'polecat' before and after their interpretations for ResNet50 are attacked.** The odd columns show original interpretations. The even columns show fooled interpretations. We set the area to the top left for 'polecat' and top right for 'tiger'.

### 4.4  Transfer to Different Interpretation Methods

The perturbed noise also possesses the transferability amoung different interpretation methods, which means the perturbed image generated for attacking one interpretation method can also mislead another method. For example, we observe that our noise optimized for FGrad [5] has varying effectiveness against GCAM [1], while noise that is optimized for GCAM is ineffective against FGrad, finding that the GCAM is more easy to fool. The similar phenomenon has been found in other explanations as is shown in Fig. 5.

### 4.5  Fooling Interpretation with the Universal Noise

Without loss of generality, we validate our method on a subset of 10 ImageNet categories, of which each category contains 260 test images and 1040 train images. For every fixed category, to learn an universal noise using train data, we fool GCAM [1] on ResNet50 along with 250 iterations, $I_r = 0.1$, $\lambda = 0.1$, $\partial = 70/255$ and then we evaluate it on test data. To prove the generalization of our noise, we calculate the effect of a single training image on each test image, and then evaluate 1040*260 attacks. The results are shown in Fig. 6 and Table 3 reflecting the universal noise has better generalization.

### 4.6   Fooling Interpretation with the Image Patches

In this paper we also consider image patches [23], which not just add noise to the particular area. Figure 7 shows examples of our patches to fool GCAM [1]. A lot of factors influence the appearance of the patch, so we do some transformations on the patch before applying it to the input: noise is put on the patch, the brightness and contrast of the patch is changed. In our experiment, We choose an image as our patch also placed on the top right corner of the input, and we fool GCAM on ResNet50 along with 150 iterations, $I_r = 0.01$, $\lambda = 0.1$ and then we evaluate it on 5,000 random images from the ImageNet validation set in Table 2.

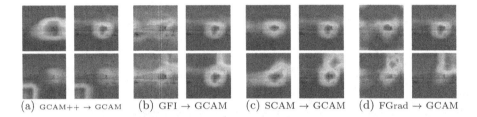

(a) GCAM++ → GCAM    (b) GFI → GCAM    (c) SCAM → GCAM    (d) FGrad → GCAM

**Fig. 5. Results showing transfers of our adversarial noise trained for GCAM++, GFI, SCAM, FGrad and evaluated on GCAM.** The first row shows original interpretations. The second row shows the effect of the transfers.

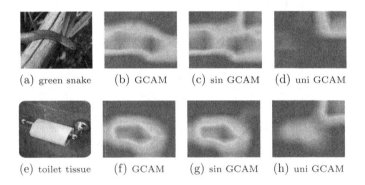

(a) green snake    (b) GCAM    (c) sin GCAM    (d) uni GCAM

(e) toilet tissue    (f) GCAM    (g) sin GCAM    (h) uni GCAM

**Fig. 6. GCAM visualization results comparing our universal noise vs the noise from a single training image.** (b), (f) are the original interpretations. As you can see from columns three and four, our universal noise reflects better attack effect.

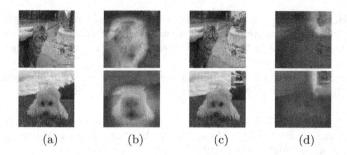

(a)              (b)              (c)              (d)

**Fig. 7. Example of image patches to fool GCAM.** (a) and (b) show the original images and interpretations, (c) and (d) show the disturbed images and fooled interpretations.

## 5   Conclusion

In this work, we introduce adversarial noise (small area, 7%, with restricted perturbations) which fool the interpretation of the unchanged category. We successfully design the adversarial noise that can highlight the small area that we specify. Compared with the existing attack algorithms, our proposed method is simpler and has better generalization performance. Moreover, we show that our attack works in various settings: (1) attacks multi-target interpretations, (2) generalizes across images of same category, (3) transfers from Grad-CAM++, GFI, Score-CAM and Full-Grad to Grad-CAM. We conclude that Grad-CAM is more vulnerable. We also fool interpretations with image patches, and some processes are taken to make the patches more robust. Our current noise do not transfer well to completely different interpretations, optimizing for different interpretations at the same time might improve upon this. In general, our work suggests that the community needs to develop more robust interpretation algorithms.

## References

1. Selvaraju, Ramprasaath, R., et al.: Grad-cam: visual explanations from deep networks via gradient-based localization. In: Proceedings of the IEEE International Conference on Computer Vision, pp. 618–626 (2017)
2. DU, M., et al.: Towards explanation of DNN-based prediction with guided feature inversion. In: Proceedings of the 24th ACM SIGKDD International Conference on Knowledge Discovery and Data Mining, pp. 1358–1367 (2018)
3. Chattopadhay, A., Sarkar, A., Howlader, P., Balasubramanian, V. N.: Grad-cam++: generalized gradient-based visual explanations for deep convolutional networks. In: IEEE Winter Conference on Applications of Computer Vision (WACV), pp. 839–847. IEEE (2018)

4. Wang, H., Wang, Z., Du, M., Yang, F., et al.: Score-CAM: score-weighted visual explanations for convolutional neural networks. In: Proceedings of the IEEE/CVF Conference on Computer Vision and Pattern Recognition Workshops, pp. 111–119 (2020)
5. Srinivas, S., Fleuret, F.: Full-gradient representation for neural network visualization. In: Advances in Neural Information Processing Systems, pp. 4124–4133 (2019)
6. Subramanya, A., Pillai, V., Pirsiavash, H.: Fooling network interpretation in image classification. In: Proceedings of the IEEE/CVF International Conference on Computer Vision, pp. 2020–2029 (2019)
7. Heo, J., Joo, S., Moon, T.: Fooling neural network interpretations via adversarial model manipulation. In: Advances in Neural Information Processing Systems, pp. 2921–2932 (2019)
8. Dombrowski, A.K., Alber, M., Anders, C., et al.: Explanations can be manipulated and geometry is to blame. In: Advances in Neural Information Processing Systems, pp. 13567–13578 (2019)
9. Lakkaraju, H., Bastani, O.: How do I fool you? manipulating user trust via misleading black box explanations. In: Proceedings of the AAAI/ACM Conference on AI, Ethics, and Society, pp. 79–85 (2020)
10. Ghorbani, A., Abid, A., Zou, J.: Interpretation of neural networks is fragile. In: Proceedings of the AAAI Conference on Artificial Intelligence, vol. 33, no. 01, pp. 3681–3688 (2019)
11. Adebayo, J., Gilmer, J., Muelly, M., et al.: Sanity checks for saliency maps. In: CoRR, abs/1810.03292 (2018)
12. Kindermans, P.-J., et al.: The (Un) reliability of saliency methods. In: Samek, W., Montavon, G., Vedaldi, A., Hansen, L.K., Müller, K.-R. (eds.) Explainable AI: Interpreting, Explaining and Visualizing Deep Learning. LNCS (LNAI), vol. 11700, pp. 267–280. Springer, Cham (2019). https://doi.org/10.1007/978-3-030-28954-6_14
13. Myers, L., Sirois, M. J.: S pearman correlation coefficients, differences between. In: Encyclopedia of statistical sciences (2004)
14. Russakovsky, O., et al.: Imagenet large scale visual recognition challenge. Int. J. Comput. Vis. **115**(3), 211–252 (2015)
15. Kingma, D.P., Ba, J.: Adam: a method for stochastic optimization. arXiv preprint arXiv:1412.6980 (2014)
16. Choi, J.H., Zhang, H., Kim, J.H., Hsieh, C.J., Lee, J.S.: Evaluating robustness of deep image super-resolution against adversarial attacks. In: Proceedings of the IEEE/CVF International Conference on Computer Vision, pp. 303–311 (2019)
17. Lakkaraju, H., Kamar, E., Caruana, R., Leskovec, J.: Faithful and customizable explanations of black box models. In: Proceedings of the 2019 AAAI/ACM Conference on AI, Ethics, and Society, pp. 131–138 (2019)
18. Kurakin, A., Goodfellow, I., Bengio, S.: Adversarial machine learning at scale. In: Proceedings of the International Conference on Learning Representations (2016)
19. Brown, T.B., Mané, D., Roy, A., Abadi, M., Gilmer, J.: Adversarial patch. In: Machine learning and Computer Security Workshop - NeurIPS (2017)
20. Ching, T., et al.: Opportunities and obstacles for deep learning in biology and medicine. J. R. Soc. Interface **15**(141), 20170387 (2017)
21. Yosinski, J., Clune, J., Nguyen, A., Fuchs, T., Lipson, H.: Understanding neural networks through deep visualization. In: CoRR, abs/1506.06579 (2015)

22. Zeiler, M.D., Fergus, R.: Visualizing and understanding convolutional networks. In: Fleet, D., Pajdla, T., Schiele, B., Tuytelaars, T. (eds.) ECCV 2014. LNCS, vol. 8689, pp. 818–833. Springer, Cham (2014). https://doi.org/10.1007/978-3-319-10590-1_53
23. Thys, S., Van Ranst, W., Goedemé, T.: Fooling automated surveillance cameras: adversarial patches to attack person detection. In: Proceedings of the IEEE/CVF Conference on Computer Vision and Pattern Recognition Workshops, pp. 49–55 (2019)

# Machine Learning

# BPN: Bidirectional Path Network for Instance Segmentation

Fan Xu[1,2], Lijuan Duan[1,3,4(✉)], and Yuanhua Qiao[5]

[1] Faculty of Information Technology, Beijing University of Technology, Beijing, China
ljduan@bjut.edu.cn
[2] Peng Cheng Laboratory, Beijing University of Technology, Beijing, China
[3] Beijing Key Laboratory of Trusted Computing, Beijing University of Technology, Beijing, China
[4] National Engineering Laboratory for Key Technologies of Information Security Level Protection, Beijing University of Technology, Beijing, China
[5] College of Mathematics and Physics, Beijing University of Technology, Beijing, China

**Abstract.** The feature pyramid network (FPN) has achieved impressive results in the field of object detection and instance segmentation by aggregating features of different scales, especially the detection of small objects. However, for some special large objects (such as tables, chairs, etc.), It is difficult to achieve good results for FPN. In this paper, we propose a new simple but effective network, the Bidirectional Path Network (BPN), for the problems that FPN cannot solve. In simple terms, it consists of a top-to-down FPN and bottom-to-up FPN. This bidirectional network structure can greatly enrich high-level semantic information and improve the detection effect of these large objects. And we also introduce dense connections to enrich the output features further. We tested our method on the COCO dataset. Firstly, on the object detection task, our method obtains comparable results with the state-of-the-art benchmark. Then, on the instance segmentation task, our method also achieved good results.

**Keywords:** Object detection · Instance segmentation · Feature pyramid network

## 1 Introduction

Instance segmentation that aims to identify and localize every object instance within an image while accurately segmenting each instance is no easy task. It is considered a combination of object detection and semantic segmentation. The interest in this field is driven by a broad set of applications, such as self-driving vehicles, medical imaging, and video surveillance. Recently, many powerful networks like Fast R-CNN [5], Faster R-CNN [25], Mask R-CNN [8], PANet [17] have been proposed with the development of convolutional neural networks. Many of them have achieved great rank or performance on the instance segmentation task.

The latest models that perform best on the coco dataset are basically based on FPN [13], like PANet [17], CenterNet [3]. Generally speaking, FPN uses high-level semantic

© Springer Nature Switzerland AG 2021
L. Fang et al. (Eds.): CICAI 2021, LNAI 13070, pp. 55–66, 2021.
https://doi.org/10.1007/978-3-030-93049-3_5

features to detect large objects, and low-level semantic features to detect small objects. Its top-down feature aggregation structure makes the output low-level features contain rich information. Therefore, FPN [13] has a very good effect on small object detection. However, the high-level semantic features output of FPN only includes information of one scale, which makes FPN not perform well on some special large objects (although large objects are usually thought to be well detected). Figure 1 shows the results of the winner of PASCAL VOC Challenge 2012, which backbone is FPN. We can see that its performance on class Table and class Plant is even worse than the bottle. It should be noted that the former belongs to large objects, and the latter belongs to small objects in the usual sense. We always pay too much attention to the detection of small objects and ignore the detection of these special large objects.

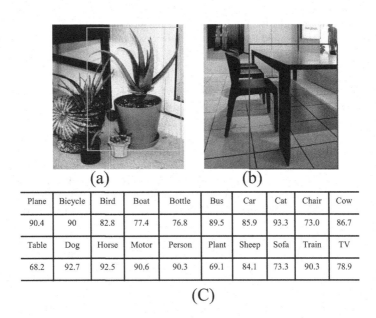

(a)                    (b)

| Plane | Bicycle | Bird | Boat | Bottle | Bus | Car | Cat | Chair | Cow |
|-------|---------|------|------|--------|-----|-----|-----|-------|-----|
| 90.4 | 90 | 82.8 | 77.4 | 76.8 | 89.5 | 85.9 | 93.3 | 73.0 | 86.7 |
| Table | Dog | Horse | Motor | Person | Plant | Sheep | Sofa | Train | TV |
| 68.2 | 92.7 | 92.5 | 90.6 | 90.3 | 69.1 | 84.1 | 73.3 | 90.3 | 78.9 |

(C)

**Fig. 1.** Table (c) lists the result of the winner of the PASCAL VOC Challenge 2012, whose backbone is FPN. The worst results come from categories Table and Plant. Figure (a) and figure (b) show the example images of those two classes, respectively. We can see that there are lots of irrelevant background information inside these bounding boxes.

In this paper, we intuitively consider this phenomenon caused by the extreme background-foreground information imbalance within the bounding box. For those instances bounding boxes with large background areas, it is essential to utilize the feature at the lower level, not just high-level features. Integrating information from several spatial scales is not uncommon. FPN [13] assign small proposal to low feature levels and vice versa. Simple but crude, this method ignores the case mentioned above. PANet [17] resolve the problem by augmenting a bottom-up path and cutting across the layers. Liao [1] proposed an enhanced-FPN architecture to lateral connecting feature maps. The above methods tried to maintain as much information undergoing the long path as

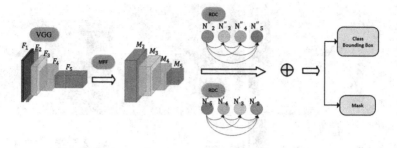

**Fig. 2.** Illustration of our framework. Note that we omit the channel dimension of feature maps for brevity.

possible. Inspired by bidirectional RNN, we present a novel Bidirectional Path Network (BPN) by exploiting FPN [13] in parallel to strengthen feature propagation and better fuse localized and semantic information to address this problem. The dense connection idea is implemented in two ways. When the primary features are extracting from a VGG backbone, they undergo a fully dense network directly connecting with each other and form intermediate ones. Then these intermediate features are sent to two parallel paths, a top-down path transferring semantic information gradually and a bottom-up path transferring detail and spatial information. At last, a set of high-quality feature maps are created for generating proposals. In addition, we also introduce a multi-level feature fusion (MFF) block to make each scale feature fuse the features of all scales, which thereby enhances the information amount of features. Similar to Mask R-CNN, the following of BPN is a class label prediction branch and a mask branch for mask prediction. They both take different scales of proposals as input. We evaluate the method on COCO [15] dataset, and the experimental results show the good performance of our method (Fig. 3). The contributions of this paper can be summarized as:

1. A multi-level feature fusion block is proposed to fuse features at multiple resolutions.
2. A bidirectional FPN network is proposed to boost the performance of FPN.
3. We validate the effectiveness of our method on the coco dataset.

## 2   Related Work

### 2.1   Proposal-Based and Proposal-Free Approaches

SDS [4] is built upon R-CNN [6]. It employs a method that combines multiple-scale regions into object candidates instead of SS. Hypercolumns [7] improved SDS via hypercolumns as pixel descriptors. In comparison, Fast R-CNN [5] followed SS [29] and proposed RoI Pooling, where each candidate is divided into $M \times N$ blocks, and then each block is max pooled so that the candidate regions are converted into uniform size feature vectors. Soon after, Faster R-CNN [25] and Mask R-CNN [8] is presented. The former replaces SS with RPN, while the later generating feature maps of different scales. Both of the above methods have improved performance a lot.

**Fig. 3.** Images in each row are visual results of our model (Top) and Mask R-CNN (Down) on COCO test-dev.

Although the proposal-based approaches are popular, many proposal-free methods come out one after the other. PFN [12] use off-the-shelf clustering method and for post-processing. [36] encodes instance on a pixel-level while [36] samples patches from the image and label each to have a patch-level prediction. [32] detect the instances from transformed maps. [26] detects and segments an object at a time by using recurrent neural network. [1] introduced inter-pixel Relation Network training with pseudo labels to estimate rough boundaries. [33] adopted keypoint-based representation.

### 2.2 Integrating Multi-level Knowledge

Local and global information are both required to achieve a good semantic segmentation result. Many approaches aim to make CNNs aware of multi-level information. [34] uses dilated convolutions instead of pooling layers to integrates multi-scale information. Feature fusion is another common way of aggregating context knowledge. PSPNet [37] proposed a pyramid pooling module to improve the capability of embedding global context information. ReNet [31], ReSeg [30] replace the omnipresent convolution and pooling layer with ResNet [9] layers, which composed of four recurrent neural networks that sweep the image. [23] trains different window sizes to predict labels. [22] makes global convolution practical through a large kernel. [4] described a method based on a Laplacian pyramid to combine both lower and higher resolution feature maps. Based on FCN [19], U-Net [27] combined high resolution feature with upsampled output, while [16,21] aggregates the results from all proposals. FPN [13] proposed feature pyramid networks and showed significant improvement.

## 3  BPN

The structure of BPN is shown in Fig. 2. It mainly consists of two parts, multi-level feature fusion (MFF) and bidirectional residual dense connection (BRDC). MFF enables

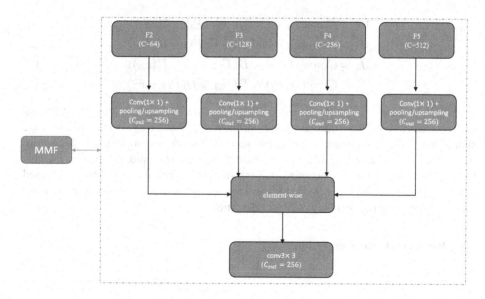

**Fig. 4.** Structure of the MFF. C denotes the channels of the feature.

the output of the network to aggregate features of different resolution sizes. BRDC combines the output from a top-to-down residual dense connection (RDC) block and a bottom-to-up RDC block. The structure of the top-to-down RDC is similar to the bottom-to-up RDC, and only the feature fusion direction of the former is opposite to the latter. At last, a set of high-quality features are created with those two blocks for generating proposals. Similar to Mask R-CNN [8], the following are a class label prediction branch and a mask branch for mask prediction.

## 3.1 Multi-level Feature Fusion

As mention in the Introduction, it is essential to integrate information from multiple spatial scales, especially for objects with a larger background context, which implies that we need to balance local and global information. So, we design a multi-level feature fusion block to combine both local and global contexts, which help to achieve good pixel-level accuracy and avoid local ambiguities. The structure of MFF is shown in Fig. 4.

The original features are extracted from a backbone, like VGG, with a scaling step of 2. The output of the backbone is denoted as $F_1, F_2, F_3, F_4, F_5$. Our multi-level feature fusion makes dense connection [10] on the original features. Every output of MFF can receive information from any other level through dense connection, not just preceding layers, as illustrated in Fig. 4. The output of MFF is calculated as:

$$M_2 = Conv(F_2 + U(F_3) + U(F_4) + U(F_5)) \tag{1}$$
$$M_3 = Conv(D(F_2) + F_3 + U(F_4) + U(F_5)) \tag{2}$$
$$M_4 = Conv(D(F_2) + D(F_3) + F_4 + U(F_5)) \tag{3}$$
$$M_5 = Conv(D(F_2) + D(F_3) + D(F_4) + F_5) \tag{4}$$

where $D(.)$ means a downsampling operation, $U(.)$ means a upsampling operation and $Conv$ means a $3 \times 3$ convolution operation. Each new feature $M_i$ is the element-wise sum of the all original features experienced several times downsampling with $3 \times 3conv$ or several times upsampling. In that way, the new feature $M_i$ combines features of multiple resolutions, so that $M_i$ contains rich information. When the object to be detected has a large background (Fig. 1), MFF can increase information of high-level feature, thereby improving the accuracy of object detection.

### 3.2 Residual Dense Connection

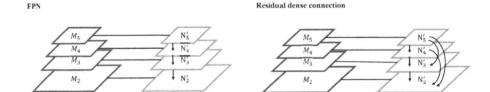

**Fig. 5.** Structure of the FPN(Left) and our residual dense connection(Right).

The powerful performance of FPN has been proven [8, 13]. In Fig. 5, the output features of each resolution only fuse the features of two resolutions except the highest resolution output (it is just one). We think it is not enough. In order to detect those objects whose bounding box is with a large background information, we should make the output features contain enough information. To this end, we designed a residual dense connection(RDC), it can be calculated as:

$$N_2' = M_2 \tag{5}$$
$$N_3' = Conv(M_2 + U(N_3')) \tag{6}$$
$$N_4' = Conv(M_2 + U(N_3') + U(N_4')) \tag{7}$$
$$N_5' = Conv(M_2 + U(N_3') + U(N_4') + U(N_5')) \tag{8}$$

where $Conv$ means a $3 \times 3$ convolutional operation and $U(.)$ means a upsampling operation. There are $\frac{L(L+1)}{2}$ connections in a network with L layers. In our case, it is six. Each feature in the bottom-up pathway is combined with all of the features from any other lower level. RDC makes the $N_2'$ used to detect large objects incorporate the other four resolution information, while for FPN, it is just two. For $N_3'$, in the same, RDC also contains more information than $N_3'$ in FPN. This allows RDC to help improve the detection of large objects with a lot of background information.

**Table 1.** Performance comparison (%) with the state-of-the-art methods on the MS-COCO test-dev dataset. MS means multi-scale training

| Method | $AP$ | $AP_{50}$ | $AP_{75}$ | $AP_S$ | $AP_M$ | $AP_L$ | Backbone |
|---|---|---|---|---|---|---|---|
| Faster R-CNN +++ [9] | 34.9 | 55.7 | 37.4 | 15.6 | 38.7 | 50.9 | ResNet-101 |
| Fitness R-CNN [28] | 41.8 | 60.9 | 44.9 | 21.5 | 45.0 | 57.5 | ResNet-101 |
| Cascade R-CNN [2] | 42.8 | 62.1 | 46.3 | 23.7 | 45.5 | 55.2 | ResNet-101 |
| Grid R-CNN w/ FPN [20] | 43.2 | 63.0 | 46.6 | 25.1 | 46.5 | 55.2 | ResNeXt-101 |
| Champion 2016[10] | 41.6 | 62.3 | 45.6 | 24.0 | 43.9 | 43.9 | ResNet-101 |
| Mask R-CNN [8] | 39.8 | 62.3 | 43.4 | 22.1 | 43.2 | 51.2 | ResNeXt-101 |
| RetinaNet800 [14] | 39.1 | 59.1 | 42.3 | 21.8 | 42.7 | 50.2 | ResNet-101 |
| YOLOv2 [24] | 21.6 | 44.0 | 19.2 | 5.0 | 22.4 | 35.5 | DarkNet-19 |
| SSD513 [18] | 31.2 | 50.4 | 33.3 | 10.2 | 34.5 | 49.8 | ResNet-101 |
| RefineDet512 [35] | 36.4 | 57.5 | 39.5 | 16.6 | 39.9 | 51.4 | ResNet-101 |
| RefineDet512 (MS) [35] | 41.8 | 62.9 | 45.7 | 25.6 | 45.1 | 54.1 | ResNet-101 |
| CornerNet511 (MS) [11] | 42.1 | 57.8 | 45.3 | 20.8 | 44.8 | 56.7 | Hourglass-104 |
| CenterNet511 (MS) [3] | 47.0 | 64.5 | 50.7 | 28.9 | 49.9 | 58.9 | Hourglass-104 |
| Ours[MS] | 42.8 | 63.6 | 47.0 | 25.9 | 45.1 | 52.1 | ResNet-50 |
| Ours[MS] | 46.9 | **68.2** | 51.5 | **30.5** | 49.6 | 56.6 | ResNeXt101 |

## 3.3 Bidirectional Paths Network

In the previous part, we introduce the RDC module, which uses dense connections to enrich the information of low-level semantic features like $N_2'$, $N_3'$. We think that high-level semantic features $N_4'$, $N_5'$ can be enriched in a similar way. Inspired by bidirectional RNN, we designe a bottom-up dense connection block, whose structure is shown in Fig. 6 and it can be calculated as:

$$N_5'' = M_5 \tag{9}$$

$$N_4'' = Conv(M_4 + D(N_5'')) \tag{10}$$

$$N_3'' = Conv(M_3 + D(N_4'') + D(N_5'')) \tag{11}$$

$$N_2'' = Conv(M_2 + D(N_3'') + D(N_4'') + D(N_5'')) \tag{12}$$

When the bottom-up augmentation output $N_2'$, $N_3'$, $N_4'$, $N_5'$ and the top-down augmentation output $N_2''$, $N_3''$, $N_4''$, $N_5''$ are ready, a combination process is conducted to fuse features in corresponding scale. $N_i'$ and $N_i''$ are fused through element-wise sum to $P_i$, which are the final feature maps to be fed into prediction networks. We name it as bidirectional residual dense block (BRDC).

$$P_i = N_i' + N_i'' \tag{13}$$

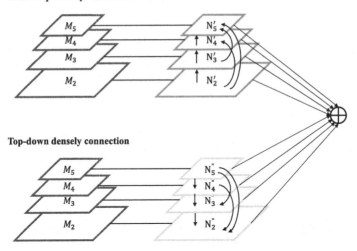

**Fig. 6.** Structure of our bidirectional paths network. It has a bottom-up dense connection block (Top) and a top-down dense connection block.

## 4    Experiments

We compare our method with state-of-the-art on challenging COCO [15]. Comprehensive ablation study is conducted on the COCO dataset.

### 4.1    Implementation Details

We follow the PANet, Mask R-CNN and FPN based to build BPN. We adopt image centric training [5]. For each image, we sample 512 region-of-interests (ROIs) with positive-to-negative ratio 1 : 3. Weight decay is 0.0001 and momentum is set to 0.9. Other hyper-parameters slightly vary according to datasets and we detail them in respective experiments. Following Mask R-CNN, proposals are from an independently trained RPN [13] for convenient ablation and fair comparison, i.e., the backbone is not shared with object detection/instance segmentation. For multi-scale training, we set longer edge to 1, 400 and the other to range from 400 to 1, 400. We calculate mean and variance based on all samples in one batch across all GPUs, do not fix any parameters during training, and make all new layers followed by a batch normalization layer, when using multi-GPU synchronized batch normalization.

### 4.2    Experiments on COCO

**Dataset and Metrics.** COCO [15] dataset is among the most challenging ones, for instance, segmentation and object detection due to the data complexity. It consists of 115k images for training and 5k images for validation (new split of 2017). 20k images are used in test-dev, and 20k images are used as test-challenge. Ground-truth

**Table 2.** Performance comparison (%) with the state-of-the-art methods on the MS-COCO val dataset. MS means multi-scale training

| Method | $AP^{bb}$ | $AP^{bb}_{50}$ | $AP^{bb}_{75}$ | $AP^{bb}_{S}$ | $AP^{bb}_{M}$ | $AP^{b}b_{L}$ | Backbone |
|---|---|---|---|---|---|---|---|
| Mask R-CNN [8] | 33.4 | 54.3 | 35.5 | 14.1 | 35.7 | 50.8 | ResNet-50 |
| PANet [17][MS] | 37.8 | 59.4 | 41 | 19.2 | 41.5 | 54.3 | ResNet-50 |
| Ours[MS] | 37.6 | 60.1 | 40.3 | 19.8 | 40.5 | 54.3 | ResNet-50 |
| Ours[MS] | 40.7 | 64.6 | 43.7 | 21.7 | 43.9 | 58.3 | ResNeXt-101 |

**Table 3.** Ablation study on COCO val dataset. The backbone of baseline method is FPN. "+" means we replace FPN with other block.

| Model | $AP^{bb}$ | $AP^{bb}_{50}$ | $AP^{bb}_{75}$ | $AP^{bb}_{S}$ | $AP^{bb}_{M}$ | $AP^{b}b_{L}$ |
|---|---|---|---|---|---|---|
| Baseline | 39.2 | 61.5 | 42.5 | 24.2 | 42.0 | 47.1 |
| +MFF | 39.8 | 59.1 | 42.3 | 21.8 | 42.7 | 50.2 |
| +BRDC | 39.9 | 61.6 | 43.4 | 24.5 | 42.8 | 47.7 |
| +MFF + BRDC | 40.2 | 61.9 | 43.4 | 24.5 | 42.7 | 48.2 |

labels of both test challenge and test-dev are not publicly available. There are 80 classes with pixel-wise instance mask annotation. We train our models on the train-2017 subset and report results on the val-2017 subset for ablation study. We also report results on test-dev for comparison. We follow the standard evaluation metrics, i.e., $AP, AP_{50}, AP_{75}, APS, APM$ and $APL$. The last three measure performance with respect to objects with different scales. Since our framework is general to both instance segmentation and object detection, we also train independent object detectors. We report mask AP, box ap APbb of an independently trained object detector, and box ap APbbM of the object detection branch trained in the multi-task fashion.

**Hyper-parameters.** We take 16 images in one image batch for training. The shorter and longer edges of the images are 800 and 1000, if not especially noted. For instance, segmentation, we train our model with learning rate 0.02 for 120k iterations and 0.002 for another 40k iterations. For object detection, we train one object detector without the mask prediction branch. Object detector is trained for 60k iterations with learning rate 0.02 and another 20k iteration with learning rate 0.002. These parameters are adopted from Mask R-CNN and FPN without any fine-tuning.

**Object Detection Results.** BPN is more like a backbone network. It can be used to replace FPN as a new benchmark backbone. To this end, we first test the performance of BPN on object detection tasks. The experimental results are shown in Table 1. We compared the effects of BPN and other detection networks on the COCO test dataset, including some one-stage detectors and two-stage detectors. Since BPN is based on mask-rcnn, it also belongs to two-stage detectors. We can see that BPN exceeds all one-step detectors and achieves comparable results with the best two-stage method. These things prove the effectiveness of BPN.

**Instance Segmentation Results.** FPN can be used as a backbone network for a variety of tasks, such as object detection and instance segmentation. In the previous section, we discuss the performance of BPN in object detection. Therefore, in this section, we mainly discuss the performance of BPN on instance segmentation task. Since COCO [15] does not currently provide test services for instance segmentation, we can only perform experiments on the val dataset. The results are shown in Table 2. We can see that the performance of BPN is much better than mask-rcnn. The former even exceeds the latter by 3%. In addition, BPN achieved comparable results with PANet. Note that PANet is the state-of-the-art method.

### 4.3  Ablation Study

In this section, we mainly analyze the impact of our proposed two modules, MFF and BRDC, on the performance of BPN. The backbone of baseline is FPN, we use MFF module or BRDC module to replace FPN to build a new network. The experimental results in the COCO val dataset for object detection are shown in Table 3. All methods are trained with multi-scale training. We can see that the performance of MFF block or BRDC block is better than FPN, and the combination of MFF and BRDC can achieve better experimental results. It means that both MFF and BRDC can be used as a backbone to improve the performance of a detector, and the combined use of MFF and BRDC can bring better results

## 5  Conclusion

In this article, we point out that the FPN network performs poorly on the detection of some special large objects and analyze the reasons behind this phenomenon. Based on this, we propose a new feature aggregation network, Bidirectional Path Network. It has two parts, a Multi-level feature fusion module (MFF) and a bidirectional dense connection module (BRDC). MFF can make each output feature contain all input information, thereby enhancing the information of the output features. The BRDC module is composed of a top-to-down residual dense connection (RDC) block and a bottom-to-up RDC block. The RDC block is similar to the FPN. It can be implemented by adding some extra connections to a FPN. The structures of MFF and BRDC are simple, and it is easy to implement those two blocks. We test the performance of BPN on the coco dataset, and the results show that BPN performs well on both the detection task and the instance segmentation task, which implies that BPN can be used as a new feature fusion backbone to replace FPN.

**Acknowledgements.** This work was supported in part by the Project of Beijing Municipal Education Commission Project (No.KZ201910005008), the National Natural Science Foundation of China (No. 62176009).

## References

1. Ahn, J., Cho, S., Kwak, S.: Weakly supervised learning of instance segmentation with inter-pixel relations. In: Proceedings of the IEEE Conference on Computer Vision and Pattern Recognition, pp. 2209–2218 (2019)

2. Cai, Z., Vasconcelos, N.: Cascade R-CNN: delving into high quality object detection. In: Proceedings of the IEEE Conference on Computer Vision and Pattern Recognition, pp. 6154–6162 (2018)

3. Duan, K., Bai, S., Xie, L., Qi, H., Huang, Q., Tian, Q.: Centernet: keypoint triplets for object detection. In: Proceedings of the IEEE International Conference on Computer Vision, pp. 6569–6578 (2019)

4. Ghiasi, G., Fowlkes, C.C.: Laplacian pyramid reconstruction and refinement for semantic segmentation. In: Leibe, B., Matas, J., Sebe, N., Welling, M. (eds.) ECCV 2016. LNCS, vol. 9907, pp. 519–534. Springer, Cham (2016). https://doi.org/10.1007/978-3-319-46487-9_32

5. Girshick, R.: Fast R-CNN. In: Proceedings of the IEEE International Conference on Computer Vision, pp. 1440–1448 (2015)

6. Girshick, R., Donahue, J., Darrell, T., Malik, J.: Rich feature hierarchies for accurate object detection and semantic segmentation. In: Proceedings of the IEEE Conference on Computer Vision and Pattern Recognition, pp. 580–587 (2014)

7. Hariharan, B., Arbeláez, P., Girshick, R., Malik, J.: Hypercolumns for object segmentation and fine-grained localization. In: Proceedings of the IEEE Conference on Computer Vision and Pattern Recognition, pp. 447–456 (2015)

8. He, K., Gkioxari, G., Dollár, P., Girshick, R.: Mask R-CNN. In: Proceedings of the IEEE International Conference on Computer Vision, pp. 2961–2969 (2017)

9. He, K., Zhang, X., Ren, S., Sun, J.: Deep residual learning for image recognition. In: Proceedings of the IEEE Conference on Computer Vision and Pattern Recognition, pp. 770–778 (2016)

10. Huang, G., Liu, Z., Van Der Maaten, L., Weinberger, K.Q.: Densely connected convolutional networks. In: Proceedings of the IEEE Conference on Computer Vision and Pattern Recognition, pp. 4700–4708 (2017)

11. Law, H., Deng, J.: Cornernet: detecting objects as paired keypoints. In: Proceedings of the European Conference on Computer Vision (ECCV), pp. 734–750 (2018)

12. Liang, X., Lin, L., Wei, Y., Shen, X., Yang, J., Yan, S.: Proposal-free network for instance-level object segmentation. IEEE Trans. Pattern Anal. Mach. Intell. **40**(12), 2978–2991 (2017)

13. Lin, T.Y., Dollár, P., Girshick, R., He, K., Hariharan, B., Belongie, S.: Feature pyramid networks for object detection. In: Proceedings of the IEEE Conference on Computer Vision and Pattern Recognition, pp. 2117–2125 (2017)

14. Lin, T.Y., Goyal, P., Girshick, R., He, K., Dollár, P.: Focal loss for dense object detection. In: Proceedings of the IEEE International Conference on Computer Vision, pp. 2980–2988 (2017)

15. Lin, T.-Y., et al.: Microsoft COCO: common objects in context. In: Fleet, D., Pajdla, T., Schiele, B., Tuytelaars, T. (eds.) ECCV 2014. LNCS, vol. 8693, pp. 740–755. Springer, Cham (2014). https://doi.org/10.1007/978-3-319-10602-1_48

16. Liu, C., et al.: Auto-deeplab: hierarchical neural architecture search for semantic image segmentation. In: The IEEE Conference on Computer Vision and Pattern Recognition (CVPR) (June 2019)

17. Liu, S., Qi, L., Qin, H., Shi, J., Jia, J.: Path aggregation network for instance segmentation. In: Proceedings of the IEEE Conference on Computer Vision and Pattern Recognition, pp. 8759–8768 (2018)

18. Liu, W., et al.: SSD: single shot multibox detector. In: Leibe, B., Matas, J., Sebe, N., Welling, M. (eds.) ECCV 2016. LNCS, vol. 9905, pp. 21–37. Springer, Cham (2016). https://doi.org/10.1007/978-3-319-46448-0_2

19. Long, J., Shelhamer, E., Darrell, T.: Fully convolutional networks for semantic segmentation. In: Proceedings of the IEEE Conference on Computer Vision and Pattern Recognition, pp. 3431–3440 (2015)

20. Lu, X., Li, B., Yue, Y., Li, Q., Yan, J.: Grid R-CNN. In: Proceedings of the IEEE Conference on Computer Vision and Pattern Recognition, pp. 7363–7372 (2019)
21. Noh, H., Hong, S., Han, B.: Learning deconvolution network for semantic segmentation. In: Proceedings of the IEEE International Conference on Computer Vision, pp. 1520–1528 (2015)
22. Peng, C., Zhang, X., Yu, G., Luo, G., Sun, J.: Large kernel matters-improve semantic segmentation by global convolutional network. In: Proceedings of the IEEE Conference on Computer Vision and Pattern Recognition, pp. 4353–4361 (2017)
23. Pinheiro, P.H., Collobert, R.: Recurrent convolutional neural networks for scene labeling. In: 31st International Conference on Machine Learning (ICML), No. CONF (2014)
24. Redmon, J., Farhadi, A.: Yolo9000: better, faster, stronger. In: Proceedings of the IEEE Conference on Computer Vision and Pattern Recognition, pp. 7263–7271 (2017)
25. Ren, S., He, K., Girshick, R., Sun, J.: Faster R-CNN: towards real-time object detection with region proposal networks. In: Advances in Neural Information Processing Systems, pp. 91–99 (2015)
26. Romera-Paredes, B., Torr, P.H.S.: Recurrent instance segmentation. In: Leibe, B., Matas, J., Sebe, N., Welling, M. (eds.) ECCV 2016. LNCS, vol. 9910, pp. 312–329. Springer, Cham (2016). https://doi.org/10.1007/978-3-319-46466-4_19
27. Ronneberger, O., Fischer, P., Brox, T.: U-Net: convolutional networks for biomedical image segmentation. In: Navab, N., Hornegger, J., Wells, W.M., Frangi, A.F. (eds.) MICCAI 2015. LNCS, vol. 9351, pp. 234–241. Springer, Cham (2015). https://doi.org/10.1007/978-3-319-24574-4_28
28. Tychsen-Smith, L., Petersson, L.: Improving object localization with fitness NMS and bounded IoU loss. In: Proceedings of the IEEE Conference on Computer Vision and Pattern Recognition, pp. 6877–6885 (2018)
29. Uijlings, J.R., Van De Sande, K.E., Gevers, T., Smeulders, A.W.: Selective search for object recognition. Int. J. Comput. Vis. **104**(2), 154–171 (2013)
30. Visin, F., et al.: Reseg: a recurrent neural network-based model for semantic segmentation. In: Proceedings of the IEEE Conference on Computer Vision and Pattern Recognition Workshops, pp. 41–48 (2016)
31. Visin, F., Kastner, K., Cho, K., Matteucci, M., Courville, A., Bengio, Y.: Renet: A recurrent neural network based alternative to convolutional networks. arXiv preprint arXiv:1505.00393 (2015)
32. Wu, Z., Shen, C., Hengel, A.V.D.: Bridging category-level and instance-level semantic image segmentation. arXiv preprint arXiv:1605.06885 (2016)
33. Yang, T.J., et al.: Deeperlab: Single-shot image parser. arXiv preprint arXiv:1902.05093 (2019)
34. Yu, F., Koltun, V.: Multi-scale context aggregation by dilated convolutions. arXiv preprint arXiv:1511.07122 (2015)
35. Zhang, S., Wen, L., Bian, X., Lei, Z., Li, S.Z.: Single-shot refinement neural network for object detection. In: Proceedings of the IEEE Conference on Computer Vision and Pattern Recognition, pp. 4203–4212 (2018)
36. Zhang, Z., Fidler, S., Urtasun, R.: Instance-level segmentation for autonomous driving with deep densely connected MRFs. In: Proceedings of the IEEE Conference on Computer Vision and Pattern Recognition, pp. 669–677 (2016)
37. Zhao, H., Shi, J., Qi, X., Wang, X., Jia, J.: Pyramid scene parsing network. In: Proceedings of the IEEE Conference on Computer Vision and Pattern Recognition, pp. 2881–2890 (2017)

# Claw U-Net: A UNet Variant Network with Deep Feature Concatenation for Scleral Blood Vessel Segmentation

Chang Yao[1], Jingyu Tang[1], Menghan Hu[1(✉)], Yue Wu[2], Wenyi Guo[2], Qingli Li[1], and Xiao-Ping Zhang[3]

[1] Shanghai Key Laboratory of Multidimensional Information Processing,
East China Normal University, Shanghai, China
`mhhu@ce.ecnu.edu.cn`
[2] Department of Ophthalmology, Ninth People's Hospital Affiliated to Shanghai Jiao Tong University, Shanghai, China
[3] Department of Electrical, Computer and Biomedical Engineering,
Ryerson University, Toronto, Canada

**Abstract.** Sturge-Weber syndrome (SWS) is a vascular malformation disease, and it may cause blindness if the patient's condition is severe. Clinical results show that SWS can be divided into two types based on the characteristics of scleral blood vessels. Therefore, how to accurately segment scleral blood vessels has become a significant problem in computer-aided diagnosis. In this research, we propose to continuously upsample the bottom layer's feature maps to preserve image details, and design a novel Claw UNet based on UNet for scleral blood vessel segmentation. Specifically, the residual structure is used to increase the number of network layers in the feature extraction stage to learn deeper features. In the decoding stage, by fusing the features of the encoding, upsampling, and decoding parts, Claw UNet can achieve effective segmentation in the fine-grained regions of scleral blood vessels. To effectively extract small blood vessels, we use the attention mechanism to calculate the attention coefficient of each position in images. Claw UNet outperforms other UNet-based networks on scleral blood vessel dataset. The robustness test also shows that the network structure has a better effect in resisting external Gaussian blur. The scleral blood vessel dataset can be downloaded by https://figshare.com/s/87d375bb37fd72912bee.

**Keywords:** Sturge-Weber syndrome · Scleral blood vessel · Medical image segmentation · Glaucoma · UNet

## 1  Introduction

Early diagnosis is significant for diseases such as glaucoma, hypertension, and diabetic retinopathy which lead to human vision deterioration [1]. Ophthalmol-

This work is sponsored by the Shanghai Education Development Foundation and Shanghai Municipal Education Commission (No. 19CG27).

L. Fang et al. (Eds.): CICAI 2021, LNAI 13070, pp. 67–78, 2021.
https://doi.org/10.1007/978-3-030-93049-3_6

ogists typically access the clinical condition of retinal blood vessels based on the retinal fundus images, and this is an effective indicator for the diagnosis of various eye diseases [2]. Sturge-Weber syndrome (SWS) is a vascular malformation disease, and it will cause glaucoma [3]. When the symptoms are severe, SWS can cause damage to the skin, brain, and eyes. Glaucoma caused by SWS has two onset peaks viz. onset at birth and onset in adolescence. Due to the seriousness of SWS, it has received great attention from ophthalmologists. Studies have found that the distribution of scleral blood vessels is abnormal for SWS patients. This abnormality may increase the outflow resistance, which in turn leads to glaucoma [4,5]. During trabeculotomy, ophthalmologists will divide the patients into two groups based on the degree of blood vessel expansion, whether there is a thick grid-like structure, and whether the blood vessel density of the surgical site is increased. Patients with diffuse vasodilatation and a thick grid-like vascular network have a lower surgical success rate, only 36% in 2 years, while the other group is as high as nearly 90% [6]. The artificial grouping is difficult to promote and requires a high clinical experience. Therefore, there is an urgent need to perform real-time automatic segmentation of scleral blood vessels. It is meaningful for ophthalmologists to take different surgical methods to improve the patient's prognosis and protect the optic nerve to the greatest extent.

The existing blood vessel segmentation approaches are mainly designed for fundus images. Unlike fundus images, in scleral vascular images, the vessels to be segmented are denser and have different scales. In addition, it may be not possible to obtain scleral vascular images with high quality during trabeculotomy. With the rapid development of convolutional neural networks (CNNs) [7], a variety of end-to-end segmentation models have been developed, such as fully convolutional neural networks (FCNs) [7], UNet [8], PSPNet [9] and DeepLab [10]. Among them, UNet shows a good segmentation effect on medical images. The decoder of UNet provides a high-level semantic feature map, and the encoder provides a low-level detailed feature map. These two phases are combined through skip connections. UNet++ [11] improves the strength of these connections by introducing nested and dense skip connections, reducing the semantic difference between encoder and decoder.

To obtain a good segmentation effect on scleral blood vessel images, we propose a novel UNet-based architecture called Claw UNet by adding skip connections between the deepest feature maps and the decoders. Each decoder part is connected with the upsampling of the bottom layer. By repeatedly using high-level semantic feature maps, the location information in the images can be captured from a complete scale, which helps to accurately segment, especially for the detailed areas of scleral blood vessels.

The objectives of this study are as follows:

1. A novel UNet-based network Claw UNet is proposed, which makes full use of the high-level semantics and can achieve effective segmentation in the fine-grained regions of scleral blood vessels.

2. The attention module and residual structure are added to make our network concentrate on the boundary segmentation of small blood vessels in the images.
3. We first establish scleral blood vessel image dataset, and the performance of Claw UNet is validated on this dataset. Gaussian blur and Gaussian noise are also added to the images to test the robustness of the network.

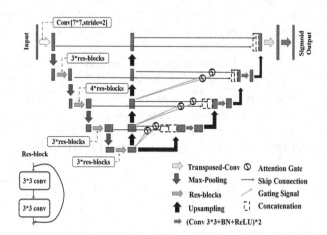

**Fig. 1.** The architecture of Claw UNet.

## 2  Claw UNet

Claw UNet combines the encoding and decoding structure of UNet, the nested and dense skip connections of UNet++ to achieve the ideal segmentation effect. Noteworthily, we respectively introduce a residual structure and an attention mechanism in the encoding part and the decoding part to further improve the network performance.

### 2.1  Claw UNet Architecture

The overall structure of Claw UNet is shown in Fig. 1. It contains three stages:

1. Encoding part: extract image features, learn image texture information, and use residual structure to realize shallow learning of deep information.
2. Bottom upsampling part: retain the deepest details in the scleral blood vessel images, and improve the detailed performance of the segmentation at the branch of the blood vessels.
3. Decoding part: upsample to restore the image size, and use the attention mechanism to combine the features of other parts for fusion to achieve an interpretable segmentation effect.

Claw UNet takes a scleral blood vessel image with the size of $512 \times 512$ as input. The first convolution operation uses kernels with the size of $7 \times 7$, and the stride is set to 2 to adjust the image size to $256 \times 256$. Max-pooling operation is used for downsampling to reduce the size of feature maps. For the following encoding parts, residual blocks are applied for all convolutional operations. The kernels in residual blocks are all set to $3 \times 3$ and the numbers of kernels are set to $32 \times 2^{i-1}$, where $i$ represents the layer number of the encoder. After this operation, we obtain feature maps $x_{En}^{i,0}$ with different depths corresponding to each layer. The size of feature maps of the $i - th$ layer is $512/2^i$. The downsampling part contains a total of four times, so the deepest feature maps are $16 \times 16$ with the depth of 512.

In the decoding part, the feature maps after convolution operation are upsampled. To fuse the extracted features, the decoder $x_{De}^{i,2}$ is combined with the corresponding encoder feature map $x_{En}^{i,0}$. The core of our Claw UNet is to send the $x_{Up}^{i,1}$ that is sampled $4 - i$ times from the bottom feature maps to the decoder. This operation allows us to more fully exploit deeper features to maintain more image details. The skip connection can be expressed by the following formula. when $i = 1, \cdots, N - 1$,

$$x_{De}^{i,2} = \left[ C \left( D \left( X_{En}^{k,0} \right) \right)_{k=1}^{i-1}, C \left( X_{Up}^{i,1} \right), C \left( U \left( X_{De}^{k,2} \right) \right)_{k=i+1}^{N} \right] \tag{1}$$

when $i = N$,

$$x_{De}^{i,2} = x_{En}^{i,0} \tag{2}$$

By concatenating the feature maps of encoding part, decoding part, and upsampling part of the bottom layer, the features of images can be fused together through a convolution operation. After each convolution operation, batch normalization (BN) [12], and ReLUs [13] are added. The BN layer can keep the training set and the test set independent and identically distributed, preventing the input distribution from gradually moving closer to both ends of the nonlinear function, resulting in a more obvious gradient in backpropagation and easier convergence. The linear activation function ReLU adds a threshold to the input to simplify the back propagation and improve the optimization effect.

It is worth mentioning that the attention mechanism is introduced between short connections. $x_{En}^{i,0}$ and $x_{Up}^{i,1}$ are combined with $x_{De}^{i,2}$ respectively to calculate the attention coefficient. This operation makes small areas easier to get more attention during segmentation. The first layer's concatenation does not use the attention mechanism, because the upsampling part has been completed. After that, the deconvolution operation restores the image size from $256 \times 256$ to the initial size of $512 \times 512$. Finally, the Sigmoid function after convolution with size of $3 \times 3$, BN and ReLUs for twice makes the output a binary image.

## 2.2   Feature Learning and Fusion

In this section, we will introduce how our network structure learns features and fuses different feature maps.

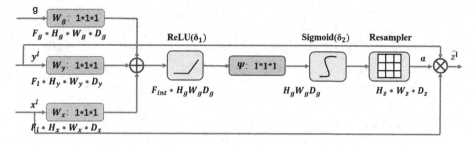

**Fig. 2.** Specific structure inside the attention block. A deep upsampling part is added as the input of the third attention.

To effectively learn the deep features, we modify the fundamental encoding block to the structure of the residual module and take Resnet-34 as the backbone of the downsampling part. The res-blocks can be divided into four parts corresponding to four layers in Claw UNet. In the same part, the same number of filters are used to maintain the same number of feature maps. As the number of encoding layers continues to increase, the number of filters also increases in multiples, so that the time complexity of each layer is similar. The residual part is connected by inserting shortcuts. It is worth noting that when inserting connections in the same dimension, other operations are not required. When inserting connections between different dimensions, additional zero entries can be added to achieve dimension matching. Four parts contain 3, 4, 6, 3 shortcuts respectively and each short connect is established between two layers of $3 \times 3$ convolution.

By integrating the decoding, upsampling and encoding parts, the segmentation of small areas may get better performance. The attention mechanism proposed in [14] uses the features of the next level to supervise the features of the upper level, and optimizes the segmentation by reducing the activation value of the background to achieve end-to-end output.

As shown in Fig. 2, the feature maps in the decoder and the feature maps in the corresponding encoder are sent to the module together. The attention coefficient is more targeted to the local small area, which helps improve performance [15]. To maintain more details in images, we also send the feature maps from the bottom upsampling part into the attention block together, and therefore the generated coefficients pay more attention to the deep features.

In the attention block shown in Fig. 2, $g$, $x^l$, and $y^l$ represent the feature matrix of the decoding part, the encoding part, and the bottom upsampling part respectively. Following, $x^l$ and $y^l$ are multiplied by a certain coefficient to achieve attention and then concatenate with $g$. The obtained feature maps are entered into the next layer of the decoding part. The resampler here resamples the feature map to the original size of $x^l$ and $y^l$. The calculation of the attention coefficient can be expressed by the following formula:

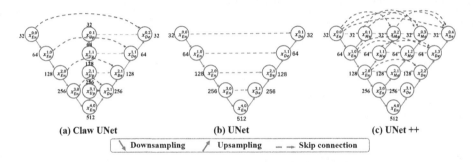

(a) Claw UNet          (b) UNet          (c) UNet ++

> ↘ **Downsampling**    ↗ **Upsampling**    — → **Skip connection**

**Fig. 3.** Comparison of (a) Claw UNet, (b) UNet, and (c) UNet++.

$$q_{att}^l = \psi^T \left( \sigma_1 \left( W_x^T x_i^l + W_y^T y_i^l + W_g^T g_i + b_g \right) \right) + b_\psi$$
$$\alpha_i^l = \sigma_2 \left( q_{att}^l \left( x_i^l, y_i^l, g_i; \Theta_{att} \right) \right) \tag{3}$$

where $g$, $y^l$, and $x^l$ are respectively multiplied by the weight matrix. The weight matrix can be learned through backpropagation to obtain the importance of each element of $g$, $y^l$, and $x^l$. That is to say, the purpose of introducing attention is to learn the importance of each element and target [14].

## 2.3   Comparison with UNet and UNet++

According to Fig. 3, we simply compare the similarities and differences between the proposed Claw UNet, UNet, and UNet++. Figure 3(b) shows that UNet uses an ordinary skip connection. The deep and shallow image features are captured by the encoding layers, and the precise localization on the symmetrical paths are realized by the decoding layers. The network realizes the end-to-end training, and can get better results on a small dataset. However, in this decoding process only uses the information in the corresponding encoder, and thus many details are often overlooked. Figure 3(c) shows that UNet++ uses nested and dense skip connections, and the redesigned skip connections aim to reduce the semantic gap between the feature maps of the encoder and decoder. Both of the above structures are short of exploring image information on a complete scale. Figure 3(a) shows that Claw UNet upsamples the feature maps of the bottom layer multiple times and has skip connections among the encoder, the decoder, and the feature maps upsampled from the bottom layer. We believe that when the feature maps from the decoder, the encoder, and the deepest layer are semantically similar, the optimizer will process the details in the image more effectively.

## 3   Experiments and Results

### 3.1   Experimental Protocol

**Dataset.** We establish the scleral blood vessel image dataset, and the images are taken from the actual surgery. Because it is more difficult to take images during

**Table 1.** Performance comparison of Claw UNet and other UNet-based networks on scleral blood vessel dataset. The best performer is highlighted in bold.

| Model | UNet | UNet++ | ResUNet | Channel-UNet | Attention-UNet | R2UNet | Claw UNet |
|---|---|---|---|---|---|---|---|
| MIoU (%) | 80.04 | 80.26 | 80.22 | 80.16 | 80.40 | 78.39 | **80.78** |
| Aver_hd | 12.63 | 12.49 | 12.15 | 12.29 | 12.54 | 12.88 | **12.02** |
| Dice (%) | 88.49 | 88.65 | 88.59 | 88.55 | 87.81 | 87.47 | **88.90** |

**Table 2.** Performance comparison of UNet++, ResUNet adding attention mechanism and Claw UNet without residual blocks and attention mechanism.

| Model | ResUNet + attention | ResUNet | UNet++ + attention | UNet++ | Claw UNet - residual - attention | Claw UNet |
|---|---|---|---|---|---|---|
| MIoU (%) | 80.20 | 80.22 | 79.03 | 80.26 | 79.64 | **80.78** |
| Aver_hd | 12.18 | 12.15 | 13.02 | 12.49 | 12.69 | **12.02** |
| Dice (%) | 88.57 | 88.59 | 87.85 | 88.65 | 88.28 | **88.90** |

operation, there are certain differences in the size, resolution, and perception field of the captured images. After discussing with experienced ophthalmologists, we intercepted specific blood vessel parts from the images. Ophthalmologists often judge this type of glaucoma caused by SWS based on the blood vessels in these areas. The dataset contains 51 images of scleral blood vessels taken from 51 different patients and covers two types of such diseases. To facilitate the subsequent network training, we set the size of all images to 512 × 512. The masks of the dataset are manually labeled by ophthalmologists from Department of Ophthalmology, Ninth People's Hospital Affiliated to Shanghai Jiao Tong University School of Medicine.

**Evaluation.** We divide the dataset for training and testing at a ratio of 4:1 and use binary cross-entropy as the loss function for optimization. Considering that the dataset is small, to avoid over-fitting, we adopt the method of 5-folded cross-validation [16], divide the total dataset into five parts, and use them as the validation set for training, separately. The final results are averaged. For image segmentation problems, Dice is more sensitive to the internal filling of the mask [17], and Hausdorff distance (Hd) is more sensitive to the segmented boundaries [18]. We finally adopt intersection over union (IoU) [19], Hd, and Dice as indicators to evaluate network performance. For a fair comparison, the parameters of all experiments are set to the same situation. We implement our model on NVIDIA GeForce RTX 2080 Ti using the PyTorch framework.

### 3.2 Comparison with Other UNet-Based Models

We choose a variety of derivative network structures based on UNet to compare with Claw UNet, including UNet [8], UNet++ [11], Resnet34-UNet [20], Channel-UNet [21], Attention-UNet [14], R2UNet [22]. Table 1 shows the cross-validation results of various networks. It can be clearly seen that Claw UNet has

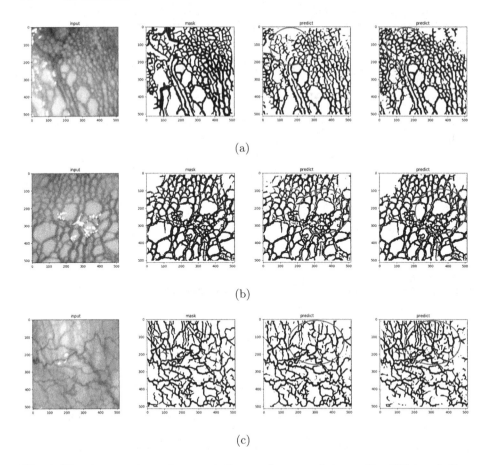

**Fig. 4.** Visual comparison of segmentation performance between Res-UNet and Claw UNet of three data from the scleral blood vessel dataset. The first column is the original images, the second column is the labeled images, the third column is the results of Res-UNet segmentation, and the fourth column is the results of Claw UNet segmentation.

achieved good results with MIoU, Dice, and Aver_hd of 80.78%, 88.90%, and 12.02, respectively. MIoUs of other network structures are below 80.40%. It is worth mentioning that Claw UNet has achieved the best performance in the first, third and fifth test sets in five-folded cross-validation. The MIoUs for all three are exceed 81%. UNet-based networks perform better than UNet. For the details of scleral blood vessels segmentation, ophthalmologists often judge the symptom type by blood vessels. Therefore, the segmentation of details is more important. In addition to focusing on the overall segmentation effect, it also requires higher performance for segmentation in details.

Figure 4 shows a detailed comparison of the segmentation results of our network and ResUNet. Claw UNet preserves more small vascular areas, indicating that Claw UNet is excellent in keeping image details. Due to the limited area of scleral blood vessels area in the image, the overall IoU difference between them is not big, but the performance of small region segmentation has been improved significantly. Small blood vessels are important for ophthalmologists to diagnose disease, and therefore, the sensitivity of Claw UNet to small blood vessels is necessary for computer-aided diagnostic systems.

### 3.3    Ablation Experiments

To illustrate the role of residual block and attention block, we conduct ablation experiments by adding such modules to the comparative network structures. Table 2 shows the comparison results of ResUNet and UNet++ combined with the attention mechanism. The MIoU of ResUNet + attention mechanism is basically the same as before. What's worse is that the MIoU of UNet++ drops more than 1% after adding the attention modules. In Claw UNet, we test the same network architecture without such mechanisms, and MIoU drops about 1%. The results show that the performance improvement not only depends on adding modules but also is greatly related to the network structure.

### 3.4    Robustness Experiments

The images taken in actual surgery often appear a lot of distortion such as noise, blur, and other unfavorable factors that affect the image quality. Therefore, improving the robustness of the network is of great help to practical applications. Although our data sets are all actual images during operation, we still artificially add a certain amount of Gaussian blur and Gaussian noise to compare the robustness of different networks to such effects.

We add Gaussian noise with a mean value of 0 and a variance of 0.0005, 0.001, 0.005 respectively to the images. Gaussian blur with a standard value of 3 and a template size of $3 \times 3$, $5 \times 5$, and $7 \times 7$ are also added respectively to the images. To compare the network robustness, we do not retrain the network, but directly use the images added to the trick to test.

Figure 5(a) illustrates the results of different models tested on the images adding Gaussian blur. It can be clearly seen that as the degree of blur increases, the segmentation results of all networks show a downward trend. Compared to other UNet-based networks, Claw Unet's results are relatively better, with the slowest rate of decline. When the template window size is $7 \times 7$, our network performance is more than 1.3% higher than other networks. Figure 5(b) illustrates the results of different models tested on the images adding Gaussian noise. It can be seen that when the noise level is low, Claw UNet still has a good segmentation performance. As the degree of noise continues to increase, the performance of the network begins to decline significantly. It is worth noticing that the performance of Attention-UNet is obviously due to all other networks, which also shows that the attention mechanism has better robustness in images with noise.

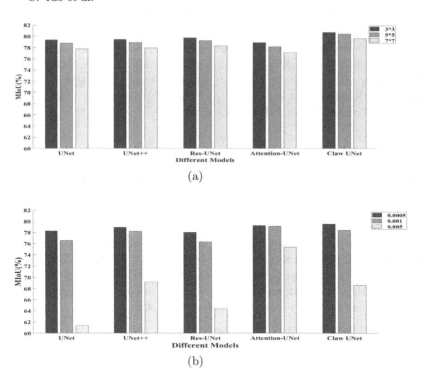

**Fig. 5.** Robustness test of different networks: a) Gaussian blur test, b) Gaussian noise test. Among them, Gaussian blur uses $3 \times 3$, $5 \times 5$, $7 \times 7$ windows, and Gaussian noise uses standard deviations of 0.0005, 0.001, and 0.005.

## 4    Conclusion

In this paper, we propose a novel UNet-based network called Claw UNet for the segmentation of scleral blood vessel images. There are no datasets available for segmenting scleral blood vessels, and hence, we first establish scleral blood vessel image dataset. To achieve more precise segmentation, we propose to continuously upsample from the bottom layer which provides rich detailed information. Residual structure and attention mechanism are introduced into Claw UNet to make our network more effective and extract features of an appropriate scale. The residual structure can effectively extract deep information, the deepest level features can retain more details of the original image, and the attention mechanism can produce accurate boundary perception. The scleral blood vessel image dataset is used to validate the performance of Claw UNet. The experimental results show that Claw UNet is more effective than and superior to previous work in detailed segmentation of scleral blood vessels. The robustness test also reflects the good anti-Gaussian blur capability of the network structure.

# References

1. Furtado, P., Travassos, C., Monteiro, R., Oliveira, S., Baptista, C., Carrilho, F.: Segmentation of eye fundus images by density clustering in diabetic retinopathy. In: 2017 IEEE EMBS International Conference on Biomedical & Health Informatics (BHI), pp. 25–28. IEEE (2017)
2. Jingfei, H., et al.: S-UNet: a bridge-style U-Net framework with a saliency mechanism for retinal vessel segmentation. IEEE Access **7**, 174167–174177 (2019)
3. Yue, W., et al.: Episcleral hemangioma distribution patterns could be an indicator of trabeculotomy prognosis in young SWS patients. Acta Ophthalmologica **98**, e685–e690 (2020)
4. Mantelli, F., Bruscolini, A., La Cava, M., Abdolrahimzadeh, S., Lambiase, A.: Ocular manifestations of Sturge-Weber syndrome: pathogenesis, diagnosis, and management. Clin. Ophthalmol. (Auckland NZ) **10**, 871 (2016)
5. Plateroti, A.M., Plateroti, R., Mollo, R., Librando, A., Contestabile, M.T., Fenicia, V.: Sturge-Weber syndrome associated with monolateral ocular melanocytosis, iris mammillations, and diffuse choroidal haemangioma'. Case Rep. Ophthalmol. **8**(2), 375–384 (2017)
6. Yue, W., et al.: Early trabeculotomy ab externo in treatment of Sturge-Weber syndrome. Am. J. Ophthalmol. **182**, 141–146 (2017)
7. Long, J., Shelhamer, E., Darrell, T.: Fully convolutional networks for semantic segmentation. In: Proceedings of the IEEE Conference on Computer Vision and Pattern Recognition, pp. 3431–3440 (2015)
8. Ronneberger, O., Fischer, P., Brox, T.: U-Net: convolutional networks for biomedical image segmentation. In: Navab, N., Hornegger, J., Wells, W.M., Frangi, A.F. (eds.) MICCAI 2015. LNCS, vol. 9351, pp. 234–241. Springer, Cham (2015). https://doi.org/10.1007/978-3-319-24574-4_28
9. Zhao, H., Shi, J., Qi, X., Wang, X., Jia, J.: Pyramid scene parsing network. In: Proceedings of the IEEE Conference on Computer Vision and Pattern Recognition, pp. 2881–2890 (2017)
10. Chen, L.-C., Papandreou, G., Kokkinos, I., Murphy, K., Yuille, A.L.: DeepLab: semantic image segmentation with deep convolutional nets, atrous convolution, and fully connected CRFs. IEEE Trans. Pattern Anal. Mach. Intell. **40**(4), 834–848 (2017)
11. Zongwei Zhou, Md., Siddiquee, M.R., Tajbakhsh, N., Liang, J.: UNet++: redesigning skip connections to exploit multiscale features in image segmentation. IEEE Trans. Med. Imaging **39**, 1856–1867 (2020)
12. Ioffe, S., Szegedy, C.: Batch normalization: accelerating deep network training by reducing internal covariate shift. arXiv preprint arXiv:1502.03167 (2015)
13. Nair, V., Hinton, G.E.: Rectified linear units improve restricted Boltzmann machines Vinod Nair. In: Proceedings of the 27th International Conference on Machine Learning (ICML 2010), Haifa, Israel, 21–24 June 2010 (2010)
14. Oktay, O., et al.: Attention U-Net: learning where to look for the pancreas. arXiv preprint arXiv:1804.03999 (2018)
15. Li, R., Liu, H., Zhu, Y., Bai, Z.: Arnet: attention-based refinement network for few-shot semantic segmentation. In: ICASSP 2020–2020 IEEE International Conference on Acoustics, Speech and Signal Processing (ICASSP), pp. 2238–2242. IEEE (2020)
16. Albadawy, E., Saha, A., Mazurowski, M.: Deep learning for segmentation of brain tumors: impact of cross-institutional training and testing. Med. Phys. **45**, 1150–1158 (2018)

17. Milletari, F., Navab, N., Ahmadi, S.-A.: V-Net: fully convolutional neural networks for volumetric medical image segmentation. In: 2016 Fourth International Conference on 3D Vision (3DV), pp. 565–571 (2016)
18. Karimi, D., Salcudean, S.: Reducing the Hausdorff distance in medical image segmentation with convolutional neural networks. IEEE Trans. Med. Imaging **39**, 499–513 (2020)
19. Chen, L.-C., Papandreou, G., Kokkinos, I., Murphy, K., Yuille, A.: Semantic image segmentation with deep convolutional nets and fully connected CRFs. CoRR, vol. abs/1412.7062 (2015)
20. Xiao, X., Lian, S., Luo, Z., Li, S.: Weighted Res-UNet for high-quality retina vessel segmentation. In: 2018 9th International Conference on Information Technology in Medicine and Education (ITME), pp. 327–331. IEEE (2018)
21. Chen, Y., et al.: Channel-Unet: a spatial channel-wise convolutional neural network for liver and tumors segmentation. Front. Genet. **10**, 1110 (2019)
22. Alom, Md.Z., Hasan, M., Yakopcic, C., Taha, T.M., Asari, V.K.: Recurrent residual convolutional neural network based on U-Net (R2U-Net) for medical image segmentation. arXiv preprint arXiv:1802.06955 (2018)

# Attribute and Identity Are Equally Important: Person Re-identification with More Powerful Pedestrian Attributes

Shuangye Chen and Kai Xu[✉]

Faculty of Information Technology, Beijing University of Technology,
Beijing 100124, China
xukai@emails.bjut.edu.cn

**Abstract.** Person re-identification (ReID) technology aims to identify characteristic people from different perspectives taken by different surveillance cameras. Due to the large changes in external light and viewing angle, the traditional ReID may completely fail. In response to this, we propose a joint recognition model with pedestrian attributes. Specifically, first, from the perspective of machine attention, we comprehensively consider the mutual exclusion and dependence between attributes, and propose a multi-task attribute recognition method that limits the attentional area of the neural network. Then we combine the pedestrian attribute and the person identity characteristics as the complete pedestrian feature information, to measure the similarity. In particular, we treat attribute information the same as identity information, instead of treating it as an attachment to identity information, in order to more effectively correct the errors of identity recognition. The experiments on the DukeMTMC-reID and Market1501 datasets prove the superiority of our method.

**Keywords:** Person re-identification · Pedestrian attribute recognition · Muti-task learning

## 1 Introduction

Person re-identification, is an artificial intelligence technology that uses computer vision technology to detect the target person in an image or video sequence. It is widely regarded as a sub-problem of image retrieval. In recent years, with the rise of deep learning, deep convolutional neural network has been widely used in ReID and have achieved encouraging performance [1]. At the same time, ReID is also used in many other fields, such as autonomous driving, video surveillance and activity analysis [2–4]. However, ReID still have certain problems, in the actual cross-camera scene, the appearance of the same person will change greatly in the surveillance video at different times. The changes of light and shadow or angle may make a person have two completely different appearance, and two different pedestrians may have similar visual appearance.

© Springer Nature Switzerland AG 2021
L. Fang et al. (Eds.): CICAI 2021, LNAI 13070, pp. 79–90, 2021.
https://doi.org/10.1007/978-3-030-93049-3_7

Zheng et al. [5] visualized the ReID clustering results and the activation map of pedestrian images, they found that the color of the clothes may be the most important clue to the ReID. The color of the clothes, as the most obvious feature, basically captures the most of the attention of the ReID model. However, the color of person's clothes can be changed easily by changes in illumination intensity. So, the robustness of the typical ReID model is insufficient. We need to use more robust feature information to improve this situation.

Pedestrian attribute (such as age, gender, hairstyle) is an abstract feature information that can be used to describe pedestrians. Different from low-level features such as color, stripes, or edges, pedestrian attributes can be regarded as high-level semantic features. They are a kind of robust semantic feature information, the most of them are not affected by color, so pedestrian attributes are not sensitive to changes in illumination. For example, in Fig. 1, if ReID model considers the attribute information of pedestrians, these similar pedestrians can be easily distinguished.

pants          age          gender          backpack

**Fig. 1.** Pairs of pedestrians with similar appearance and different attributes. Pairs of pedestrians in each column have similar visual appearance, but they can be easily distinguished by adding some specific attribute information to the judgment.

Considering the above issues comprehensively, we add pedestrian attributes to the ReID model. It uses identity information as global feature to identify pedestrians, and use attribute information as detailed features to correct pedestrian identification errors. The two complement each other to make the ReID model represents pedestrian's feature more comprehensively and robustly.

In conclusion, the main contribution of the proposed method is as following:

(1) We proposed a pedestrian attribute method with higher performance. Specifically, it includes multi-task pedestrian attribute grouping method, end-to-end model and auxiliary classification loss function. (2) Attribute information

is combined with identity information to improve the ReID model's ability to identify similar pedestrians. In particular, we put pedestrian attribute characteristics at the same priority as identity information, so that the role of pedestrian attributes can be exerted more effectively, and correct the wrong result of pedestrian identification. (3) The experimental results on the DukeMTMC-reID and Market1501 datasets prove the effectiveness of the framework. Whether it is the accuracy of pedestrian attribute recognition or the performance of reid, our method has achieved a good improvement.

## 2 Related Work

*Pedestrian Attribute Recognition.* DeepMAR [6] use the relationship between attributes, consider the global image only, and directly identify all attributes from a single image. The benefits are simple, intuitive and efficient. However, due to the lack of the considerations of fine-grained recognition, their performance is still limited. LGNet [7], PGDM [8] are based on local ideas, using local information to supplement the global feature. It can improve the overall recognition performance, but also brings the following shortcomings: Firstly, incorrect parts detection results will bring wrong features to the final classification; secondly, because human body parts are introduced, it is required more time to train or inference. Visual attention mechanism has been introduced in pedestrian attribute recognition [9,10], but the existing works are still limited. In this field, it is still necessary to explore how to design new attention models or learn from other fields [11]. Wu et al. [12] proposed a method based on sequence contextual relation learning (SCRL). The attribute relation sequence is regarded as a parallel branch, then the image sequence branch processed by CNN is merged to learn the context relation from the sequence to improve pedestrian attribute recognition.

*Person Re-identification.* Wu et al. [13] divided the image into five fixed-length regions, extracted the histogram of each region, and mixed it with the global depth feature. Although this cascade method extracts effective local features and obtains better performance, it ignores the problem of image misalignment, so its generalization ability is weak. In order to align pedestrians, Wei et al. [14] used Deeper Cut [15] to detect three non-overlapping body parts, and then learned feature vectors separately. Zheng et al. [16] used 14 key points located by the pose estimation model to assist pedestrian alignment. Wang et al. [17] used the attention mechanism to solve the problem of pedestrian dislocation.

ReID methods for adding pedestrian attributes. Lin et al. [18] re-weighting the attribute information, and then fused it with the global image feature for re-identification. Zhang et al. [19] calculated the appearance distance and attribute distance of two separate models, and merged these two distances together to obtain the final pedestrian ranking. Tay et al. [20] et al. designed an attribute attention network (AANet), which integrates the functions of part segmentation, attribute recognition and re-recognition, and performs these tasks in turn.

# 3   The Proposed Method

In the following part, we first describe the detail of pedestrian attribute recognition in Sect. 3.1 and then the Joint recognition model in Sect. 3.2.

## 3.1   Pedestrian Attribute Recognition

Combining the advantages of machine attention [21] and multi-task learning [22], we propose a multi-task attribute recognition method that limits the focus of neural networks. The method consists of two parts. The first part is to group the attributes according to the different attention area of different attributes by the neural network; the second part is the network model and loss function design based on the attention area limitation.

**Attribute Grouping.** Based on the research of Grad-cam [23], we visualize the area of interest of pedestrian attributes on the image. As shown in Fig. 2, different attributes have different areas of interest in the neural network. The regions of interest with different attributes may not affect each other, or may be large overlaps. For two different types of attributes that overlapped, their feature extraction will inevitably compete when performing in the overlapping image regions. In addition, attributes such as hats, backpacks, and boots have specific and local areas of interest, while abstract attributes such as age and gender usually do not involve fixed objects of interest and locations.

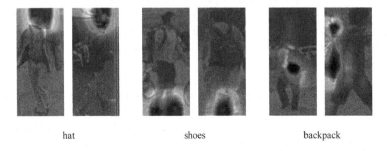

hat                              shoes                            backpack

**Fig. 2.** Class Activation Mapping (CAM) for some attributes.

Therefore, we divide the attributes into different groups according to the different area of interest of each attribute on the input image and the characteristics of specific attributes and abstract attributes, to ensure that the neural network pays attention to each attribute. At the same time, the attributes within the group will not cause interference, and the attributes of different groups promote each other through information sharing. The specific attribute groupings are shown in Table 1.

Pedestrian attribute recognition network. The network framework is shown in Fig. 3. The network mainly includes the following three parts:

*Input Network.* Because the Bottom layers of the deep neural network is used to extract low-level image features (such as texture, lines, etc.), we did not directly use the original image as the subsequent input, but chose the data processed by the base convolution block as the input of each sub-network. In this way, it can not only reduce the amount of network calculations and reduce the burden of hardware, but also enhance the connection between the sub-tasks of the network, and strengthen the information exchange between attributes.

**Table 1.** Market-1501-attribute grouping. *"up", "down", "clothes", "upcolor" and "downcolor" denote length of sleeve, length of lower-body clothing, style of clothing, color of upper-body clothing and color of lower-body clothing.*

| Group_1 | | Group_2 | | Group_3 | Group_4 | | | Group_5 |
|---------|------|-----------|---------|---------|-------|---------|----------|---------|
| Hair | Bag | Backpack | Handbag | Hat | Up | Young | Teenager | Gender |
| Upcolor | Down | Downcolor | | Clothes | Adult | Old | | |

*Feature Extraction Network.* The feature extraction part is divided into five feature extraction subtask networks, which are used to extract features of different attribute groups. The sub-network model uses VGG16 [24] as the backbone network, selects its three intermediate convolution blocks (ConvBlock_1, ConvBlock_2, ConvBlock_3 in Fig. 3), and then connects a dimension-raised convolution block (ConvBlock_4) to change the number of convolution kernels. Finally processed by Global Average Pooling(GAP) and put it to the subsequent classification layer. Using GAP instead of Fully Connected operation can greatly reduce the number of parameters of the model, and there is no limit to the size of the input image, which is more flexible and easily to transplant to Person re-identification.

*Attribute Prediction Network.* In this part, we input the feature information extracted by each feature extraction subtask network into two classification networks: an auxiliary classification network, which only used in the training phase and will be deleted during inference. Input the feature information of each sub-network into the respective sub-classification network for recognition; the other is the full classification network, which integrates the output information of each sub-feature extraction network, and then input into the full recognition classification network. Through the design of two classification networks, we can achieve three goals: (1) The machine attention areas between attributes in different subtasks do not overlap. (2) Fuse information of each subtask to realize joint identification of attributes. (3) Do not increase the amount of calculation in the inference stage.

**Pedestrian Attribute Recognition Loss Function.** Pedestrian attribute recognition is a multi-label classification problem. For these attributes, we regard it as multiple binary classification, using sigmoid and cross-entropy loss function.

84      S. Chen and K. Xu

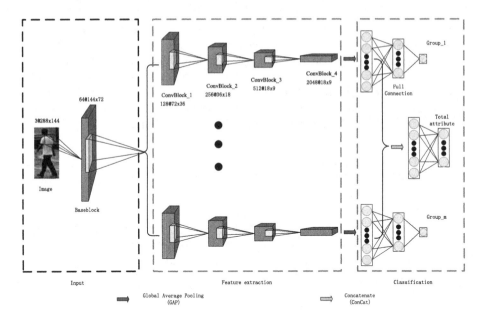

**Fig. 3.** Pedestrian attribute recognition framework.

For the full classification network loss function, *loss_full*, the cross-entropy loss function is used to measure the error between the ground truth and the predicted score:

$$loss\_full = -\frac{1}{N}\sum_{n=1}^{N}\left[Q_n \log \hat{P}_n + (1-Q_n)\log\left(1-\hat{P}_n\right)\right] \tag{1}$$

Where $N$ is the number of attributes, and $\hat{p}_n$ is the attribute prediction output by sigmoid. $Q_n$ is ground truth. The loss function of the auxiliary classification network *loss_aux*. Through the back propagation of this function, sub-networks are restricted to focus on the attributes in their own group:

$$loss\_aux = -\frac{1}{M}\sum_{m=1}^{M}\left[Q_m \log P_m + (1-Q_m)\log\left(1-P_m\right)\right] \tag{2}$$

Where $M$ is the number of auxiliary classification attributes. $P_m$ is the attribute prediction output by auxiliary classification network. $Q_m$ is ground truth. The attribute recognition loss function loss_total is the sum of the loss functions of the two classification networks:

$$loss\_total = loss\_full + loss\_aux \tag{3}$$

### 3.2   Joint Recognition Model

The structure of the joint recognition model is shown in Fig. 4. For the extraction of pedestrian identity features, we use ResNet-50 as the backbone. For the last sampling layer, we set its sampling rate to 1, so as to fully retain the pedestrian identity features. In order to ensure that pedestrian attribute information has sufficient influence on person re-identification, we set the same number and weight of attribute features and identity features.

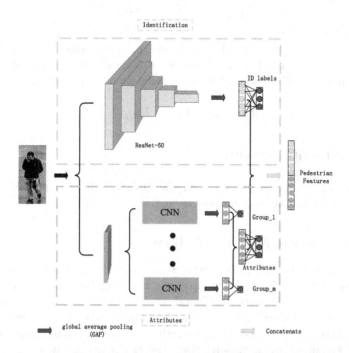

**Fig. 4.** The overview of the joint recognition model. The blue box is the pedestrian identity network, the green box is the pedestrian attribute network, and the gray is auxiliary layer for training, will not calculate in the inference stage. (Color figure online)

After a complete pedestrian image is processed by the joint recognition model, we can obtain a 1024-d pedestrian feature vector containing identity information and attribute information, and then the judgment of whether it is the same person is calculated by the similarity between different feature vectors, the closer the distance between the two feature vectors, the more similar the pedestrian images. In this paper, we select Euclidean Distance for the similarity measure of pedestrians.

### 3.3 Loss Function

Pedestrian identification part. We regard the task of pedestrian identification as a multi-classification problem, and use the cross-entropy loss to learn the identity classification. For a given training image X, joint model output the logits z of each identity label, $z = \{z_1, z_2, \ldots, z_k\}$, k is the number of identity labels. Then transform it into predicted probability of corresponding identity label by softmax: $p(k \mid x) = \frac{\exp(z_k)}{\sum_{i=1}^{K} \exp(z_i)}$. So, the identity loss function can be expressed as:

$$L_{ID} = -\sum_{i=1}^{K} p_i \log\left(\widehat{p_l}\right) \tag{4}$$

Where K is the number of identity labels, $p_i$ is true label of the i-th pedestrian's identity, $\hat{p}_i$ is the predicted output of the i-th pedestrian identity. The loss function of the pedestrian attribute classification Lattr refers to formula (1), (2).:

$$L_{total} = L_{ID} + L_{attr} \tag{5}$$

## 4  Experimental Result

In the following experiments, we used the DukeMTMCreID and Market1501 datasets to conduct our training and testing.

### 4.1  Pedestrian Attribute Recognition

For the experimental configuration of reference [5], we use the weights pre-trained by ImageNet for VGG16 backbone and use Kaiming Initialization to the new layer. The optimizer selects the SGD and sets weight decay $= 5e-4$ and Momentum $= 0.9$. And the learning rate of the pre-trained layer and the newly added layer is back-propagated according to 1:10. The initial learning rate is set to 0.01 and 0.001, after 20K iterations, the respective learning rates are attenuated to 1/10 of the original. The experimental platform uses Cloud Server (CPU: Intel(R) Xeon(R) CPU E5-2650 v3 @ 2.30 GHz, GPU: NVIDIA Tesla K80). The comparison results with other methods are shown in Table 2 and Table 3.

**Table 2.** Experimental results on the Market-1501 dataset. *"age\*"*, *"bp."*, *"up"*, *"down"*, *"clo."* denote Average accuracy of the four age attributes, whether backpack, length of sleeve, length of lower-body clothing, style of clothing. APR* denote using different backbone network, APR*1: Resnet18, APR*2: Resnet34, APR*3: Resnet50. The best performance is shown in black.

| Market-1501 | age* | bp. | bag | clo. | down | gender | hair | hb. | hat | up | avg. |
|---|---|---|---|---|---|---|---|---|---|---|---|
| ARN [18] | 85.80 | 86.60 | 78.60 | 93.60 | 93.60 | 87.50 | 84.20 | 88.10 | 97.00 | 93.50 | 88.85 |
| JVIA [25] | **91.60** | 86.70 | 80.20 | 80.90 | 84.70 | 88.90 | 84.90 | **92.30** | 97.60 | 78.30 | 86.61 |
| DeepMAR [6] | 82.60 | 88.10 | 79.30 | 94.60 | 93.50 | 91.70 | 87.10 | 88.70 | 97.30 | 93.20 | 89.61 |
| SCRL [12] | 86.80 | 86.20 | **89.90** | 90.70 | 91.70 | 87.10 | 86.00 | 86.30 | 96.40 | 93.20 | 89.43 |
| APR*1 [18] | 84.52 | 87.35 | 79.63 | 94.11 | 93.48 | 91.39 | 87.55 | 88.84 | 97.61 | 93.7 | 89.82 |
| APR*2 | 87.24 | 87.59 | 77.86 | **94.83** | 94.54 | 92.09 | 78.24 | 89.20 | **98.24** | 93.80 | 89.36 |
| APR*3 | 85.95 | 87.72 | 79.41 | 94.83 | 94.46 | 91.42 | 87.62 | 89.44 | 98.20 | 93.70 | 90.28 |
| Ours | 86.97 | **88.81** | 79.69 | 94.33 | 94.40 | **93.32** | **90.29** | 89.52 | 97.14 | 93.63 | **90.81** |

Through the analysis of Table 2 and Table 3, it can be found that although our method does not lead in every single attribute, it reaches the highest average accuracy rate. This shows that we succeeded in making the neural network pay enough attention to each attribute by limiting the focus area of the god machine, and thus the overall recognition performance was greatly improved.

## 4.2  Person Re-identification

In this section, we test the performance of the joint recognition model on the task of pedestrian re-recognition.

**Comparative Experiment.** For person re-identification, we use the joint recognition model to conduct experiments. The experimental settings are similar to the pedestrian attribute recognition experiments, using ImageNet pre-training weights and Kaiming initialization methods. The optimizer selects SGD and sets weight decay $= 5e-4$ and Momentum $= 0.9$. The initial learning rate for model training is set to 0.1. In addition, in order to speed up the model convergence, this paper uses CosineAnnealingWarmRestarts as the scheduler. The experimental platform is still Cloud Server. The experimental results are shown in Table 4. From the table, we can see that our joint recognition model can extract a more accurate representation of pedestrian features, making the distance measurement between pedestrian images more accurate, surpassing most of the algorithm models.

**Table 3.** Comparison of experimental results on the DukeMTMC dataset. *"shoes"*, *"top" denote the color of the shoes, length of upper-body clothing.*

| DukeMTMC | backpack | bag | boots | gender | handbag | hat | shoes | top | avg. |
|---|---|---|---|---|---|---|---|---|---|
| ARN [18] | 77.5 | 82.2 | 88.3 | 82 | 92.3 | 85.5 | 87.6 | 86.2 | 85.2 |
| JVIA [25] | 76.7 | 82 | 88.6 | 85.4 | 93.6 | **89.3** | 91.6 | 86.6 | 86.73 |
| DeepMAR [6] | 83.1 | 83.1 | 90.1 | 84.8 | 93.5 | 88.9 | 91.1 | **90.4** | 88.1 |
| Yin's [26] | **85.4** | 83.4 | 89.9 | 85.9 | 93.4 | 88.7 | 91.6 | 89.5 | 88.48 |
| APR*1 [18] | 81.47 | 82.69 | 89.96 | 85.24 | 93.74 | 88.9 | 91.73 | 89.67 | 87.92 |
| APR*2 | 83.01 | 82.76 | 90.26 | 85.66 | 93.45 | 89.47 | 91.37 | 89.39 | 88.17 |
| APR*3 | 82.97 | 82.99 | **90.47** | 86.24 | 93.27 | 88.95 | 91.74 | 89.99 | 88.33 |
| Ours | 83.35 | 84 | 90.34 | **86.94** | **93.49** | 89.04 | **91.79** | 89.47 | **88.55** |

**Self-contrast Experiment.** In this section, we set up a comparison with the benchmark experiment to verify the performance improvement of pedestrian re-identification after adding the pedestrian attribute verification module. The benchmark experiment is to use the standard ResNet-50 network to do pedestrian re-identification test. The experimental results are shown in Table 5. It can be seen from the results in the table that the addition of pedestrian attribute recognition can effectively enhance the model's accurate representation of pedestrian images, thereby improving the performance of pedestrian re-recognition.

**Table 4.** Comparison with other methods in Market-1501 and DukeMTMC-reID datasets. Rank-1 accuracy (%) and mAP (%) are shown. The best performance is shown in black.

|  | Market-1501 | | | | DukeMTMC-reID | | | |
|---|---|---|---|---|---|---|---|---|
|  | rank 1 | rank 5 | rank 10 | mAP | rank 1 | rank 5 | rank 10 | mAP |
| SAN [27] | 85.90 | 94.90 | 97.00 | 70.10 | 77.90 | – | – | 58.80 |
| APR [18] | 87.04 | 95.10 | 96.42 | 66.89 | 73.92 | – | – | 55.56 |
| PESR [28] | 85.60 | 94.80 | 97.50 | – | 79.40 | 91.30 | 92.10 | – |
| PSE [33] | – | – | – | – | 79.80 | 89.70 | 92.20 | 62.00 |
| DRAL [29] | 84.20 | 94.27 | 96.59 | 66.26 | 74.28 | 84.83 | 88.42 | 56.00 |
| DistributionNet [31] | 87.26 | 94.74 | 96.73 | 70.82 | 74.73 | 85.05 | 88.82 | 55.98 |
| Ours | **91.86** | **97.33** | **98.22** | **79.94** | **81.37** | **91.29** | **93.85** | **64.83** |

**Table 5.** Comparison results with baseline. Rank-1 accuracy (%) and mAP (%) are shown

|  | Market-1501 | | | | DukeMTMC-reID | | | |
|---|---|---|---|---|---|---|---|---|
|  | rank 1 | rank 5 | rank 10 | mAP | rank 1 | rank 5 | rank 10 | mAP |
| Baseline | 86.49 | 94.30 | 96.20 | 68.79 | 78.05 | 88.29 | 91.65 | 59.52 |
| Ours | 91.86 | 97.33 | 98.22 | 79.94 | 81.37 | 91.29 | 93.85 | 64.83 |

## 5    Conclusions and Future Work

In this paper, we propose a multi-task pedestrian attribute recognition method based on machine attention, and then add it to the person re-identification model to improve the recognition performance of similar pedestrians. For pedestrian attribute recognition, we grouped it first, and then an end-to-end network model is designed to achieve higher recognition accuracy by sharing the low-level features of the image and limiting the attention of the sub-network. For the joint recognition model, we give the attribute information greater influence weight, so that it can more effectively correct the error of the identity information. Finally, Experiments on the two public datasets have shown that our method achieves encouraging performance compared to the state-of-the-art approaches. For future work, we will study a more suitable pedestrian similarity measurement method.

## References

1. Siarohin, A., Lathuilière, S., Sangineto, E., et al.: Appearance and pose-conditioned human image generation using deformable GANs. IEEE Trans. Pattern Anal. Mach. Intell. **43**(4), 1156–1171 (2019)
2. Wojke, N., Bewley, A., Paulus, D.: Simple online and realtime tracking with a deep association metric. In: 2017 IEEE International Conference on Image Processing (ICIP), pp. 3645–3649. IEEE (2017)

3. Ristani, E., Tomasi, C.: Features for multi-target multi-camera tracking and re-identification. In: Proceedings of the IEEE Conference on Computer Vision and Pattern Recognition, pp. 6036–6046 (2018)

4. Li, W.H., Hong, F.T., Zheng, W.S.: Learning to learn relation for important people detection in still images. In: Proceedings of the IEEE/CVF Conference on Computer Vision and Pattern Recognition, pp. 5003–5011 (2019)

5. Zheng, Z., Zheng, L., Yang, Y.: A discriminatively learned CNN embedding for person reidentification. ACM Trans. Multimedia Comput. Commun. Appl. (TOMM) **14**(1), 1–20 (2017)

6. Li, D., Chen, X., Huang, K.: Multi-attribute learning for pedestrian attribute recognition in surveillance scenarios. In: 2015 3rd IAPR Asian Conference on Pattern Recognition (ACPR), pp. 111–115. IEEE (2015)

7. Liu, P., Liu, X., Yan, J., et al.: Localization guided learning for pedestrian attribute recognition. arXiv preprint arXiv:1808.09102 (2018)

8. Li, D., Chen, X., Zhang, Z., et al.: Pose guided deep model for pedestrian attribute recognition in surveillance scenarios. In: 2018 IEEE International Conference on Multimedia and Expo (ICME), pp. 1–6. IEEE (2018)

9. Liu, X., Zhao, H., Tian, M., et al.: HydraPlus-Net: attentive deep features for pedestrian analysis. In: Proceedings of the IEEE International Conference on Computer Vision, pp. 350–359 (2017)

10. Sarfraz, M.S., Schumann, A., Wang, Y., et al.: Deep view-sensitive pedestrian attribute inference in an end-to-end model. arXiv preprint arXiv:1707.06089 (2017)

11. Wang, X., Zheng, S., Yang, R., et al.: Pedestrian attribute recognition: a survey. arXiv preprint arXiv:1901.07474 (2019)

12. Wu, J., Liu, H., Jiang, J., et al.: Person attribute recognition by sequence contextual relation learning. IEEE Trans. Circuits Syst. Video Technol. **30**(10), 3398–3412 (2020)

13. Wu, S., Chen, Y.C., Li, X., et al.: An enhanced deep feature representation for person re-identification. In: 2016 IEEE Winter Conference on Applications of Computer Vision (WACV), pp. 1–8. IEEE (2016)

14. Wei, L., Zhang, S., Yao, H., et al.: GLAD: global-local-alignment descriptor for pedestrian retrieval. In: Proceedings of the 25th ACM International Conference on Multimedia, pp. 420–428 (2017)

15. Insafutdinov, E., Pishchulin, L., Andres, B., Andriluka, M., Schiele, B.: DeeperCut: a deeper, stronger, and faster multi-person pose estimation model. In: Leibe, B., Matas, J., Sebe, N., Welling, M. (eds.) ECCV 2016. LNCS, vol. 9910, pp. 34–50. Springer, Cham (2016). https://doi.org/10.1007/978-3-319-46466-4_3

16. Zheng, L., Huang, Y., Lu, H., et al.: Pose-invariant embedding for deep person re-identification. IEEE Trans. Image Process. **28**(9), 4500–4509 (2019)

17. Wang, C., Zhang, Q., Huang, C., Liu, W., Wang, X.: Mancs: a multi-task attentional network with curriculum sampling for person re-identification. In: Ferrari, V., Hebert, M., Sminchisescu, C., Weiss, Y. (eds.) ECCV 2018. LNCS, vol. 11208, pp. 384–400. Springer, Cham (2018). https://doi.org/10.1007/978-3-030-01225-0_23

18. Lin, Y., Zheng, L., Zheng, Z., et al.: Improving person re-identification by attribute and identity learning. Pattern Recogn. **95**, 151–161 (2019)

19. Zhang, X., Pala, F., Bhanu, B.: Attributes co-occurrence pattern mining for video-based person re-identification. In: 2017 14th IEEE International Conference on Advanced Video and Signal Based Surveillance (AVSS), pp. 1–6. IEEE (2017)

20. Tay, C.P., Roy, S., Yap, K.H.: AANet: attribute attention network for person re-identifications. In: Proceedings of the IEEE/CVF Conference on Computer Vision and Pattern Recognition, pp. 7134–7143 (2019)

21. Itti, L., Koch, C., Niebur, E.: A model of saliency-based visual attention for rapid scene analysis. IEEE Trans. Pattern Anal. Mach. Intell. **20**(11), 1254–1259 (1998)

22. Chennupati, S., Sistu, G., Yogamani, S., et al.: MultiNet++: multi-stream feature aggregation and geometric loss strategy for multi-task learning. arXiv preprint arXiv:1904.08492 (2019)

23. Chattopadhay, A., Sarkar, A., Howlader, P., et al.: Grad-CAM++: generalized gradient-based visual explanations for deep convolutional networks. In: 2018 IEEE Winter Conference on Applications of Computer Vision (WACV), pp. 839–847. IEEE (2018)

24. Simonyan, K., Zisserman, A.: Very deep convolutional networks for large-scale image recognition. arXiv preprint arXiv:1409.1556 (2014)

25. Zhang, S., He, Y., Wei, J., et al.: Person re-identification with joint verification and identification of identity-attribute labels. IEEE Access **7**, 126116–126126 (2019)

26. Yin, J., Fan, Z., Chen, S., Wang, Y.: In-depth exploration of attribute information for person re-identification. Appl. Intell. **50**(11), 3607–3622 (2020). https://doi.org/10.1007/s10489-020-01752-x

27. Shen, C., Qi, G.J., Jiang, R., et al.: Sharp attention network via adaptive sampling for person re-identification. IEEE Trans. Circuits Syst. Video Technol. **29**(10), 3016–3027 (2018)

28. Ha, Y., Tian, J., Miao, Q., et al.: Part-based enhanced super resolution network for low-resolution person re-identification. IEEE Access **8**, 57594–57605 (2020)

29. Liu, Z., Wang, J., Gong, S., et al.: Deep reinforcement active learning for human-in-the-loop person re-identification. In: Proceedings of the IEEE/CVF International Conference on Computer Vision, pp. 6122–6131 (2019)

30. Suh, Y., Wang, J., Tang, S., Mei, T., Lee, K.M.: Part-aligned bilinear representations for person re-identification. In: Ferrari, V., Hebert, M., Sminchisescu, C., Weiss, Y. (eds.) Computer Vision – ECCV 2018. LNCS, vol. 11218, pp. 418–437. Springer, Cham (2018). https://doi.org/10.1007/978-3-030-01264-9_25

31. Yu, T., Li, D., Yang, Y., et al.: Robust person re-identification by modelling feature uncertainty. In: Proceedings of the IEEE/CVF International Conference on Computer Vision, pp. 552–561 (2019)

32. Li, Y.J., Chen, Y.C., Lin, Y.Y., et al.: Recover and identify: a generative dual model for cross-resolution person re-identification. In: Proceedings of the IEEE/CVF International Conference on Computer Vision, pp. 8090–8099 (2019)

33. Sarfraz, M.S., Schumann, A., Eberle, A., et al.: A pose-sensitive embedding for person re-identification with expanded cross neighborhood re-ranking. In: Proceedings of the IEEE Conference on Computer Vision and Pattern Recognition, pp. 420–429 (2018)

# Disentangled Variational Information Bottleneck for Multiview Representation Learning

Feng Bao[✉]

University of California, San Francisco, USA
Feng.Bao@ucsf.edu

**Abstract.** Multiview data contain information from multiple modalities and have potentials to provide more comprehensive features for diverse machine learning tasks. A fundamental question in multiview analysis is what additional information can be brought by additional views and can we quantitatively identify this additional information. In this work, we try to tackle this challenge by decomposing the entangled multiview features into shared latent representations that are common across all views and private representations that are specific to each single view. We formulate this feature disentanglement in the framework of information bottleneck and propose disentangled variational information bottleneck (DVIB). DVIB explicitly defines the properties of shared and private representations using constrains from mutual information. By deriving variational upper and lower bounds of mutual information terms, representations are efficiently optimized. We demonstrate the shared and private representations learned by DVIB well preserve the common labels shared between two views and unique labels corresponding to each single view, respectively. DVIB also shows comparable performance in classification task on images with corruptions. DVIB implementation is available at https://github.com/feng-bao-ucsf/DVIB.

**Keywords:** Information bottleneck · Variational inference · Multiview representation learning · Information disentanglement

## 1 Introduction

With advances in the past decade, performances of major machine learning frameworks have reached their accuracy plateau in many tasks [1–3]. To further overcome the performance bottleneck, multiview learning methods are viewed as promising solutions [4,5]. By collecting additional views from samples, we expect to obtain more useful and task-relevant features, therefore enhancing the performance of methods through increasing the information abundance within the data [6–8].

In multiview data, each modality is collected using different technologies and approaches and contains different levels of corruptions, noises and/or missings.

© Springer Nature Switzerland AG 2021
L. Fang et al. (Eds.): CICAI 2021, LNAI 13070, pp. 91–102, 2021.
https://doi.org/10.1007/978-3-030-93049-3_8

92     F. Bao

One fundamental and critical question in multiview analysis is: compared with single view data, can additional views provide additional effective information to facilitate the learning tasks? If yes, can we explicitly identify the additional information to explain the view property and enhance the data interpretability?

To answer these questions, it requires us to decompose the entangled information embedded in multi-view data into view-shared and view-private (a.k.a view-specific) representations [9,10] (Fig. 1a). Based on the view decomposition, contributions from each single view can be explicitly quantified and analyzed. Besides that, the view-shared information exhibits the general and common features of the sample and can be used to reduce the effects of data corruption and noise [9,11]. Meanwhile view-private information represents the unique properties from single modality therefore can be used to evaluate its importance to specific tasks and reflect the strength and weakness of technologies that generate the view.

Learning disentangled representation from multiview data is challenging in terms of the modeling of the entanglement [7,9,10]. In this work, we formulate the disentangled representation learning in the framework of information bottleneck and propose disentangled variational information bottleneck (DVIB). In the optimization target (see Sect. 3), we aim to maximize the mutual information between shared latent representations generated from different views while minimizing the mutual information between private representations at the same time (Fig. 1b). With such constrains, the properties of private and shared representations are explicitly formulated. The learning target can be efficiently optimized through deriving variational bounds and auxiliary cost functions. We demonstrate the ability of DVIB in accurately decomposing the common or view-specific information from multiview data and improving the robustness in classification task on large-scale datasets.

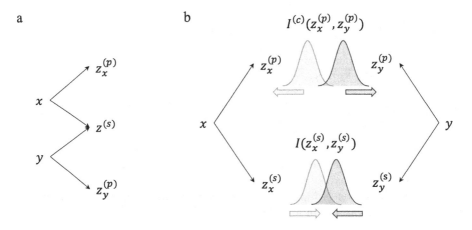

**Fig. 1.** Illustrations of (a) multiview disentangled representation learning concept and (b) proposed disentangle variational information bottleneck (DVIB) method.

## 2   Preliminaries and Existing Works

Representation learning is to extract effective, highly compact features from raw data containing various levels of noises and corruptions [4,8,12]. Efficient design of representation methods can greatly facilitate the down-streaming learning tasks. Learning highly compact representations from the view of information theory has attracted long-time attentions [13,14]. Pioneering work information bottleneck [15,16] aims to learn the representation from minimal input information but can predict the outcome well, by maximizing mutual information between the latent representations and output while minimizing mutual information between the input and latents.

The information bottleneck defines an elegant target for the optimization of compact representations from the view of information theory. However, efficiently calculation of the mutual information is challenging due to the intractable estimation of marginal distributions [17,18]. Recently, approximation of mutual information has been advanced greatly with the help of variational inference [18]. By replacing intractable margins with tractable approximators, we can alternatively seek to derive the variational upper or lower bound of mutual information. With the variational neural network and re-parameterization tricks [19], we can efficiently optimize the mutual information bounds.

The combination of information bottleneck and variational inference leads to the development of variational information bottleneck (VIB) [20]. VIB is able to learn maximally informative representations and shows robust performance with existence of perturbations. Following the idea of VIB, recently proposed methods enable more flexible representation learning for classification task [21] and information harmonization between multiview data (MVIB) [7].

Learning disentangled representation is an attracting task [22,23]. Some recent works, e.g. ($\beta$-VAE) [10] try to formulate general disentangled information learning from the view of variational inference and it is proved to be closely related to the information bottleneck [24]. However, explicitly modeling and quantifying the disentanglement is a challenge [25]. The recent development of mutual information and information bottleneck methods provides a theoretical fundamental of our method.

## 3   Disentangled Variational Information Bottleneck

To simplify the description, here we consider two views of features $x$ and $y$ collected from the same sample. Our goal is to decompose each single feature view, for example $x$, into two latent representations $z_x^{(s)}$ and $z_x^{(p)}$, where $z_x^{(s)}$ is the shared representation that preserves the common information from both views, and $z_x^{(p)}$ is the private representation that exhibits the view-specific property in $x$. Similarly, we can define the shared and private latents as $z_y^{(s)}$ and $z_y^{(p)}$ for view $y$.

### 3.1  Information Bottleneck for the Shared Representation

For the shared representations, we expect to capture the information shared by both $x$ and $y$ meanwhile neglecting the view-specific information. Following the definition of information bottleneck, we formulate the learning targets of $z_x^{(s)}$ and $z_y^{(s)}$ using mutual information,

$$\max\ I(x; z_x^{(s)}) + \lambda_x I(y; z_x^{(s)}) \tag{1}$$

$$\max\ I(y; z_y^{(s)}) + \lambda_y I(x; z_y^{(s)}) \tag{2}$$

where in Eqs. 1 and 2, first terms require the shared representation $z_*^{(s)}$ to have maximal mutual information with the view where it was generated from. The second term forces the shared representation, even it was learned from one view, can maintain high mutual information with the other view. Hyperparameters $\lambda_x \geq 0$ and $\lambda_y \geq 0$ balance the relative importance of two mutual informations. By maximizing two target functions, we can explicitly constrain shared latent representations to maintain maximal mutual information for both views simultaneously.

### 3.2  Information Bottleneck for the Private Representation

For the private representations, we restrict the learned $z_x^{(p)}$ and $z_y^{(p)}$ to only contain the unique information from the view it is generated from, but no information from the other view. Similarly, we define the learning targets as:

$$\max\ I(x; z_x^{(p)}) - \beta_x I(y; z_x^{(p)}) \tag{3}$$

$$\max\ I(y; z_y^{(p)}) - \beta_x I(x; z_y^{(p)}) \tag{4}$$

Similar as the information bottleneck [15], the maximization of first terms in Eqs. 3 and 4 require private latents and raw features ($z_x^{(p)}$ and $x$, or $z_y^{(p)}$ and $y$) to stay as similar as possible by mutual information metric. Meanwhile, the second terms require the $z_x^{(p)}$ (resp. $z_y^{(p)}$) to contain as less information as possible form the view $y$ (resp. $x$). Again, non-negative hyperparameters $\beta_x$ and $\beta_y$ define the trade-offs between the information to gain from the view where the private representation is learned and the information to suppress from the other view.

### 3.3  Variational Bounds

The optimization of introduced targets requires the calculation of a number of mutual information terms. However, it is known the mutual information is intractable for high dimensional variables [17,19,20]. We alternatively sort to derive the variational bounds of mutual information [18].

**Lower Bound of $I(x, z_x^{(s)})$.** We first consider the lower bound of mutual information between the latent representation (either shared or private) and the view it was generated from. Taking $I(x, z_x^{(s)})$ as an example, we have

$$I(x, z_x^{(s)}) = \mathbb{E}_{p(x,z_x^{(s)})} \log \frac{p(x|z_x^{(s)})}{p(x)} \tag{5}$$

$$= \mathbb{E}_{p(x,z_x^{(s)})} \log \frac{p(x|z_x^{(s)})}{q(x|z_x^{(s)})} \frac{q(x|z_x^{(s)})}{p(x)}$$

$$= \mathbb{E}_{p(x,z_x^{(s)})} \log \frac{q(x|z_x^{(s)})}{p(x)} + \mathbb{E}_{p(z_x^{(s)})} KL[p(x|z_x^{(s)})||q(x|z_x^{(s)})]$$

$$\geq \mathbb{E}_{p(x,z_x^{(s)})} \log q(x|z_x^{(s)}) + H(X)$$

where $KL[p||q] \geq 0$ represents the Kullback-Leibler divergence of two variables $p$ and $q$; $H(X)$ is the entropy of $x$; $q(x|z_x^{(s)})$ is the variational approximation of conditional distribution $p(x|z_x^{(s)})$. Entropy term $H(X)$ is determined by the dataset and is independent of the optimization process. Therefore, to maximize the $I(x, z_x^{(s)})$, we can alternatively maximize the $\mathbb{E}_{p(x,z_x^{(s)})} q(x|z_x^{(s)})$. Similarly, we can derive the lower bounds for $I(x, z_x^{(p)})$, $I(y, z_y^{(p)})$ and $I(y, z_y^{(s)})$. The full derivation of variational bounds can be found in Appendix A.

**Lower Bound of $I(y, z_x^{(s)})$.** Next we consider the mutual information between shared latent representations and features that are from different views. We take $I(y, z_x^{(s)})$ as an example. We follow the derivation in Ref. [7] and write the lower bound as

$$I(y, z_x^{(s)}) = I(z_x^{(s)}, z_y^{(s)}) + I(z_x^{(s)}; y|z_y^{(s)}) \tag{6}$$

$$\geq I(z_x^{(s)}, z_y^{(s)})$$

where the mutual information between a raw feature and a latent variable is approximated by the mutual information between two shared latent representations $z_x^{(s)}$ and $z_y^{(s)}$. We can derive the lower bound for $I(x, z_y^{(s)})$ symmetrically. Appendix B and Appendix Fig. 1 provide the derivation and illustration of the lower bound. Therefore, we can combine the optimization of $I(x, z_y^{(s)})$ and $I(y, z_x^{(s)})$ to the same target $I(z_y^{(s)}, z_x^{(s)})$.

**Upper Bound of $I(y; z_x^{(p)})$.** Finally we consider the view-private latent representations $z_x^{(p)}$ and $z_y^{(p)}$. Again, we take the mutual information $I(y; z_x^{(p)})$ as an example and derive the upper bound. We have:

$$I(y, z_x^{(p)}) = \mathbb{E}_{p(y,z_x^{(p)})} \log \frac{p_x^{(p)}(z_x^{(p)}|y)}{p(z_x^{(p)})} \tag{7}$$

$$= \mathbb{E}_{p(y,z_x^{(p)})} \log \frac{p_x^{(p)}(z_x^{(p)}|y)}{r(z_x^{(p)})} \frac{r(z_x^{(p)})}{p(z_x^{(p)})}$$

$$= \mathbb{E}_{p(y,z_x^{(p)})} \log \frac{p_x^{(p)}(z_x^{(p)}|y)}{r(z_x^{(p)})} - KL[r(z_x^{(p)})||p(z_x^{(p)})]$$

$$\leq \mathbb{E}_{p(y,z_x^{(p)})} \log \frac{p_x^{(p)}(z_x^{(p)}|y)}{r(z_x^{(p)})}$$

$$= \mathbb{E}_{p(y,z_x^{(p)})} \log \frac{p_x^{(p)}(z_x^{(p)}|y)}{p_y^{(p)}(z_y^{(p)}|y)} + \mathbb{E}_{p(y,z_x^{(p)})} \log \frac{p_y^{(p)}(z_y^{(p)}|y)}{r(z_x^{(p)})}$$

where $p_x^{(p)}$ and $p_y^{(p)}$ represent private encoders that learn private latent representations $z_x^{(p)}, z_y^{(p)}$ from raw features $x, y$. The upper bound tights on the approximation of marginal distribution $r(z_x^{(p)})$ to prior $p(z_x^{(p)})$. Two terms are in the upper bound: the first term is the encoding difference between two latent representations from $p_x^{(p)}$ and $p_y^{(p)}$ but with the same input $y$. It encourages two encoders to produce inconsistent encoding. The second term is the difference between encoder $p_y^{(p)}$ with the approximated margin $r(z_x^{(p)})$. Minimizing the upper bound requires the encoder from $y$ generates representations that are heterogeneous with both posterior and prior of $z_x^{(p)}$. Appendix C provides complete derivation.

Here, the estimation of the first term is not easy as it requires to input $y$ to encoder $p_x^{(p)}$. Because we use stochastic neural network mapping from raw data to the latent, the outputs from two view encoders can be greatly different. Besides, the second term enlarges the differences between $z_x^{(p)}$ prior and view-$y$ private encoder output and has the same optimization direction as the first term. Therefore, we simplify the upper bound to

$$\min I(y, z_x^{(p)}) \equiv \min \mathbb{E}_{p(y,z_x^{(p)})} \log \frac{p_y^{(p)}(z_y^{(p)}|y)}{r(z_x^{(p)})} \tag{8}$$

We note this formulation is the same as negative mutual information term in variational information bottleneck [20], which constrains information flow from raw features to the latent representations. Therefore, our optimization target has the same function to control the information flow from two views to latents.

## 3.4  Optimization

Finally, we combine the bounds and auxiliary targets introduced in the previous section together and derive the overall optimization function:

$$\max \ I_{total} = I_x + I_y \tag{9}$$
$$\geq I_{LB}(x; z_x^{(s)}; z_x^{(p)}) + I_{LB}(y; z_y^{(s)}; z_y^{(p)})$$
$$+ \lambda I(z_x^{(s)}, z_y^{(s)}) - \beta I^{(c)}(z_x^{(p)}, z_y^{(p)})$$

where the full formulation of each term is given in Appendix D. $I_{LB}(x; z_x^{(s)}; z_x^{(p)})$ and $I_{LB}(y; z_y^{(s)}; z_y^{(p)})$ are lower bounds of $I(x; z_x^{(s)}; z_x^{(p)})$ and $I(y; z_y^{(s)}; z_y^{(p)})$, respectively; $I^{(c)}(z_x^{(p)}, z_y^{(p)})$ represents the cross mutual information between private representations of two views (Appendix D).

To optimize the target function, we make use of the variational autoencoder structure and employ four encoders to output parameters that define posteriors of $z_x^{(s)}$, $z_y^{(s)}$, $z_x^{(p)}$ and $z_y^{(p)}$ while decoders are used to reconstruct raw features from latent representations. To maximize $I(z_x^{(s)}, z_y^{(s)})$, we use the neural mutual information estimators [16,18,26]. The exceptions over joint distributions are approximated by the empirically joint distributions [7,20].

## 4  Experiments

In this section, we calibrate the performance of DVIB and evaluate the quality of shared and private representations using various datasets. In the experiments, we focus on two questions: 1) the ability of private and shared latent representations to decompose entangled information and capture meaningful contents from each view; 2) how can the representations facilitate down-streaming analysis.

We implemented the DVIB in the framework of variational autoencoder where encoder and decoder networks (simple multi-layer neural network) were used to learn the representation distribution and reconstruct original signal. For prior distributions $r(z_x^{(p)})$ and $r(z_y^{(p)})$, we restrict them to follow $N(0, I)$ as described in [20]. To estimate the mutual information $I(z_x^{(s)}, z_y^{(s)})$, our implementation employed the Jensen-Shannon estimator [7,17].

### 4.1  Evaluation of Information Disentanglement on MNIST

We start from MNIST handwriting digit dataset [27], which involves simple sample categories and is easy to generate paired multiview data through image transformations. This can be an example to set up the baseline performance of DVIB. Here, for every sample in MNIST, we consider two transformation to simulate two-view data: 1) a rotation of the image in one of the following angle $[0, \pi/16, \pi/8, 3\pi/16, \pi/4]$ to generate view $x$; 2) a random flip from [None, horizontal, vertical, horizontal + vertical] to generate view $y$. We note each transformation choice is randomly performed on each digit so that the transformation is independent of the original MNIST digit labels. Because of the simple network used in DVIB cannot efficiently capture complicate image transformation, we firstly feed two view data into Inception-v3[1] that was pretrained on ImageNet [28] and the output of last fully connected layer (2,048 dimensions) is used.

To demonstrate the ability of DVIB latent representations in dissecting shared and latent representations, we evaluate the quality of shared representations ($z_x^{(s)}$ and $z_x^{(p)}$) by predicting the shared labels (digit identities) and the quality of private representations by predicting the view-specific labels (rotation angles for $z_x^{(p)}$ and flip type for $z_y^{(p)}$, respectively) using simple linear classifier.

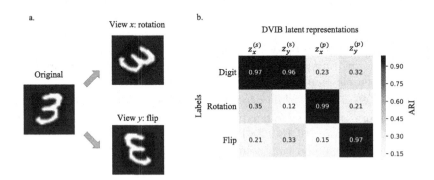

**Fig. 2.** (a) Example of simulated two-view data from original MNIST sample. (b) Performance of DVIB latent representations in predicting view-specific and -shared labels. Adjusted Rand index (ARI) is calculated between predicted labels and ground truths.

This experiment gives an initial demonstration how DVIB can capture the shared and unique information in views. The results (Fig. 2) demonstrate the shared representations ($z_x^{(s)}$ and $z_x^{(p)}$) can best capture the shared label information (digit) while each private representation ($z_x^{(p)}$ and $z_x^{(p)}$) has higher accuracy in predicting each view-specific labels. Meanwhile, the private representation from one view has poor performance to infer the private label in the other view or the shared digit labels, indicating view-specific information from raw feature is isolated into the private representations as expected.

## 4.2 Evaluation Using Corrupted Samples on ImageNet

One important application for multiview learning is to compensate from the additional view and recover the corrupted information in either view. Here, we consider the classification problem on ImageNet [28]. To simulate the image corruptions, we add Gaussian noise for the generation of view $x$ and use defocus blur for the view $y$ [29]. We use the existing image corruption implementation[2]

---

[1] Pretrained model provided by TensorFlow Hub (https://tfhub.dev/).
[2] https://github.com/bethgelab/imagecorruptions.

in the experiments. Again, to facilitate the efficient learning, we also employ the pretrained Inception-v3 framework to construct raw features. And multiview learning is built upon the deep features.

To calibrate the performance of our method, we consider general multiview deep neural network (DNN) [6]; multiview non-negative matrix factorization (M-NMF); deep canonical correlated autoencoder (DCCAE) [30] and its variational version (VCCA) [9]; variational autoencoder (VAE) [19]; information theory based method deep variational information bottleneck (VIB) [20], multiview information bottleneck (MIB) [7]. As we focus on the ability to remove the unwanted corruption information, we make use of the shared representations learned from two views. We note not all methods were designed for multiview study and we simply concatenate features from two views for these single-view methods. The classification is performed by softmax regression on the latent representations learned by each method.

**Table 1.** Classification accuracy of ImageNet using joint latent representations learned from different method.

| Method | Description | Accuracy |
|--------|-------------|----------|
| Baseline | Simple feature concatenations | 0.714 |
| DNN | Multiview neural network | 0.763 |
| M-NMF | Joint matrix factorization | 0.632 |
| DCCAE | A deep version of CCA | 0.716 |
| VCCA | A variational version of CCA | 0.758 |
| VAE | Variational autoencoder | 0.782 |
| VIB | Variational information bottleneck | 0.794 |
| MIB | Multiview information bottleneck | 0.821 |
| DVIB | The proposed method | **0.852** |

From the classification accuracies (Table 1), general multiview methods showed improved performance compared with feature concatenations. Variational methods (VCCA, VAE and VIB) have better robustness to the corruption than traditional methods DNN and NMF due to the design of variational inference. The recently proposed multiview information bottleneck method MIB, which aims to learn informative shared representations, further improved the accuracy. Our method, by explicitly formulating the property of shared latents from mutual information constrain, obtains the best accuracy.

## 4.3   Robustness to Image Corruption Levels

Based on the results on ImageNet, here we ask how robust is each method's performance to the corruption levels of inputs. To investigate this, we consider to increase the corruption level for each single modality or for both, and evaluate the accuracies correspondingly. As comparisons, we select top 5 methods from Table 1 and the baseline.

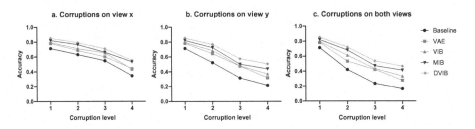

**Fig. 3.** Performance of classification on ImageNet with increasing corruption levels on (a) view $x$ alone, (b) view $y$ alone and (c) both.

From the overall results (Fig. 3), corruptions in image blur level (view $y$) have stronger effects to the accuracy than Gaussian noise (view $x$). All methods show decreasing trends when increasing the corruptions. MIB and DVIB maintain better performances compared with other method and DVIB has the best performance. It demonstrates the design of mutual information constrained structures improves the effective information extraction.

## 5    Discussion

In this work, we formulate the information disentanglement task from multiview data in the framework of information bottleneck. We explicitly define the desired information property of view-private and view-share latent representations with mutual information constrains and decompose each view. By deriving the variational bounds of mutual information, we effectively optimize the target function. We demonstrate the learned latent representations well preserve the view-shared and view-specific information and improve the classification robustness.

The DVIB model includes two hyperparameters $\lambda$ and $\beta$. How to simultaneously determine the appropriate values of these parameters is a challenge and requires further exploration. In our implementation, we consider same weights for two views as they are generated from the same source. However, for other types of multiview data which include different modalities (e.g. image and audio), this assumption might not stand. In the optimization of upper bound of mutual information (Eq. 7), we omit the first term in final loss function with the assumption that neural network encoders for two view will produce different encodings by default due to the random mapping property neural network. However, it requires abundant experiments study and rigorous mathematical proof. Taken together, with further extensive evaluation of the method, DVIB can be a potential powerful tool for the disentanglement of multiview data.

# References

1. Beyer, L., Hénaff, O.J., Kolesnikov, A., Zhai, X., van den Oord, A.: Are we done with ImageNet? arXiv preprint arXiv:2006.07159 (2020)
2. Tsipras, D., Santurkar, S., Engstrom, L., Ilyas, A., Madry, A.: From ImageNet to image classification: Contextualizing progress on benchmarks. In: International Conference on Machine Learning, pp. 9625–9635. PMLR (2020)
3. Yoshida, Y., Okada, M.: Data-dependence of plateau phenomenon in learning with neural network–statistical mechanical analysis. Adv. Neural. Inf. Process. Syst. **32**, 1722–1730 (2019)
4. Xu, C., Tao, D., Xu, C.: A survey on multi-view learning. arXiv preprint arXiv:1304.5634 (2013)
5. Wang, W., Arora, R., Livescu, K., Bilmes, J.: On deep multi-view representation learning. In: International Conference on Machine Learning, pp. 1083–1092. PMLR (2015)
6. Ngiam, J., Khosla, A., Kim, M., Nam, J., Lee, H., Ng, A.Y.: Multimodal deep learning. In: ICML, pp. 689–696 (2011)
7. Federici, M., Dutta, A., Forré, P., Kushman, N., Akata, Z.: Learning robust representations via multi-view information bottleneck. ICLR (2020)
8. Li, Y., Yang, M., Zhang, Z.: A survey of multi-view representation learning. IEEE Trans. Knowl. Data Eng. **31**(10), 1863–1883 (2018)
9. Wang, W., Yan, X., Lee, H., Livescu, K.: Deep variational canonical correlation analysis. arXiv preprint arXiv:1610.03454 (2016)
10. Irina Higgins, et al.: beta-VAE: learning basic visual concepts with a constrained variational framework. ICLR (2017)
11. Andrew, G., Arora, R., Bilmes, J., Livescu, K.: Deep canonical correlation analysis. In: International Conference on Machine Learning, pp. 1247–1255. PMLR (2013)
12. Bengio, Y., Courville, A., Vincent, P.: Representation learning: a review and new perspectives. IEEE Trans. Pattern Anal. Mach. Intell. **35**(8), 1798–1828 (2013)
13. Deng, Y., Bao, F., Deng, X., Wang, R., Kong, Y., Dai, Q.: Deep and structured robust information theoretic learning for image analysis. IEEE Trans. Image Process. **25**(9), 4209–4221 (2016)
14. Tishby, N., Zaslavsky, N.: Deep learning and the information bottleneck principle. In: 2015 IEEE Information Theory Workshop (ITW), pp. 1–5. IEEE (2015)
15. Tishby, N., Pereira, F.C., Bialek, W.: The information bottleneck method. arXiv preprint arXiv:physics/0004057 (2000)
16. Amjad, R.A., Geiger, B.C.: Learning representations for neural network-based classification using the information bottleneck principle. IEEE Trans. Pattern Anal. Mach. Intell. **42**(9), 2225–2239 (2019)
17. Belghazi, M.I., et al.: Mutual information neural estimation. In: International Conference on Machine Learning, pp. 531–540. PMLR (2018)
18. Poole, B., Ozair, S., Van Den Oord, A., Alemi, A., Tucker, G.: On variational bounds of mutual information. In: International Conference on Machine Learning, pp. 5171–5180. PMLR (2019)
19. Kingma, D.P., Welling, M.: Auto-encoding variational Bayes. ICLR (2014)
20. Alemi, A.A., Fischer, I., Dillon, J.V., Murphy, K.: Deep variational information bottleneck. ICLR (2017)
21. Hjelm, R.D., et al.: Learning deep representations by mutual information estimation and maximization. arXiv preprint arXiv:1808.06670 (2018)

22. Tran, L., Yin, X., Liu, X.: Disentangled representation learning GAN for pose-invariant face recognition. In: Proceedings of the IEEE Conference on Computer Vision and Pattern Recognition, pp. 1415–1424 (2017)

23. Zhang, Z., et al.: Gait recognition via disentangled representation learning. In: Proceedings of the IEEE/CVF Conference on Computer Vision and Pattern Recognition, pp. 4710–4719 (2019)

24. Burgess, C.P., et al.: Understanding disentangling in *beta*-VAE. arXiv preprint arXiv:1804.03599 (2018)

25. Locatello, F., et al.: Challenging common assumptions in the unsupervised learning of disentangled representations. In: International Conference on Machine Learning, pp. 4114–4124. PMLR (2019)

26. Belghazi, M.I., et al.: MINE: mutual information neural estimation. arXiv preprint arXiv:1801.04062 (2018)

27. Deng, L.: The MNIST database of handwritten digit images for machine learning research [best of the web]. IEEE Signal Process. Mag. **29**(6), 141–142 (2012)

28. Deng, J., Dong, W., Socher, R., Li, L.-J., Li, K., Fei-Fei, L.: ImageNet: a large-scale hierarchical image database. In: 2009 IEEE Conference on Computer Vision and Pattern Recognition, pp. 248–255. IEEE (2009)

29. Michaelis, C., et al.: Benchmarking robustness in object detection: autonomous driving when winter is coming. arXiv preprint arXiv:1907.07484 (2019)

30. Wang, W., Arora, R., Livescu, K., Bilmes, J.: On deep multi-view representation learning. In: Bach, F., Blei, D. (eds.) Proceedings of the 32nd International Conference on Machine Learning, volume 37 of Proceedings of Machine Learning Research, Lille, France, 07–09 July 2015, pp. 1083–1092. PMLR (2015)

# Real-Time Collision Warning and Status Classification Based Camera and Millimeter Wave-Radar Fusion

Lei Fan, Qi Yang, Yang Zeng[✉], Bin Deng, and Hongqiang Wang

National University of Defense Technology, Changsha, China

**Abstract.** Collision warning is the core content in the vehicle active safety system. However, relying on one sensor such as cameras and LIDARS will face some special detection difficulties. Cameras are limited to light and LIDARS are susceptible to rain or snow. Therefore, this paper proposes real-time collision warning and status classification based on camera and millimeter wave (mmw)-radar fusion. The proposed method can classify the object status automatically, which includes the danger, potential danger and safety. The networks for camera and radar are firstly constructed to detect and recognize targets respectively. Then the coordinates of cameras and radars are transformed into the same world coordinate by perspective transformation. Finally, both detection results of these two networks are fused at the decision level. Experimental results demonstrate that the proposed method provides an effective solution and owns robustness in detection and recognition.

**Keyword:** Collision warning · Camera and millimeter wave-radar · Deep learning · Status classification · Object detection

## 1 Introduction

Advanced automatic driving technology can solve the problem of urban traffic congestion and reduce traffic accidents [1, 2]. One of the most important parts for automatic driving technology is the vehicle active safety system, whose core content is the collision warning. Cameras, LIDARS and radars are the sensors used for collision warning widely. These sensors possess their advantages and disadvantages for detection [3]. The images captured by the camera have rich texture and shape information, which are intuitive and interpretable. But cameras are easily affected by light and weather. Moreover, the detection performance for targets located in the long range is poor. LIDARS can obtain abundant point clouds about scenes and targets. But LIDARS are susceptible to severe weather (e.g. heavy rain and snow) and their cost are expensive. Radars are sensitive to the range and velocity of targets and possess strong robustness to heavy weather and light. However, they are easy to produce some false alarm. Compared with the first two sensors, the interpretation of radar imaging results is obscure. Therefore, the current solutions are mostly considered to use multi-sensor detection.

© Springer Nature Switzerland AG 2021
L. Fang et al. (Eds.): CICAI 2021, LNAI 13070, pp. 103–114, 2021.
https://doi.org/10.1007/978-3-030-93049-3_9

From the perspective of which sensors are fused, it can be categorized into two classes generally: camera-LIDAR fusion and camera-radar fusion. As for camera-LIDAR fusion, the fusion usually contained two-stage detectors to extract the common feature from the bird-eye view (BEV) image recorded by LIDARs and images captured by the camera [4, 5]. But the fusion from the data source was difficult for realization in reality. As for camera-radar fusion, the fusion depended on the signal processing of the radar echo. Radar data can be divided into radar cube and point cloud further. Range-azimuth images are extracted from the radar cube to improve the capacity of range detection for optical images [6, 7]. The low resolution of azimuth was made up by the corresponding optical images. Interference of multi-channel radars is used to get point cloud about targets [8–10]. Combined with the vehicle kinematic model, collision warning was achieved well by the range and speed information. The influence of source-level, feature-level and decision-level feature fusion are analyzed [11]. Although the source-level and feature-level fusion brought the benefit for fusion detection, once one sensor was broken, the overall performance would sharply fall. Therefore, the sensors used by this paper were camera and radar. Decision-level fusion was chosen as the final fusion way.

With the development of deep learning, convolution neural network (CNN) has shown superior performance in image classification, object detection and super-resolution imaging [12, 13]. The deep learning for the radar field is faced with the dilemma of insufficient data and labels. Radar data own special properties. At present, commercially available radars can directly output radar cubes (called low-level features) by 3-D Fourier transform. They can output the point cloud about the target (called target-level features) by multi-channel interference. The specific content of point cloud incorporates the position of scattering centers, velocity and relative reflection intensity. These characteristics are contributive to design the network of radar.

In this paper, collision warning based camera and mmw-radar fusion is proposed. The images captured by camera and the data recorded by radar are first input into their corresponding networks to obtain the primary detection results. Especially, the radar data consider the low-level features and the target-level features, which can significantly reduce false alarm and computational burden. Then, the detection results of these two networks are transformed into the same world coordinate. Finally, according to the range and velocity of the detected target, the risk of objects status is predicted. The experimental results validate the superiority of the proposed method in object detection.

The remainder of this paper is organized as follows: In Sect. 2, the network architecture is described and the methods for fusion and status classification are presented in detail. The comprehensive experiment results demonstrate the effectiveness of the proposed method in Sect. 3. Finally, Sect. 4 summarizes the conclusion.

## 2 Method

### 2.1 FMCW Radar Signal Model

Frequency Modulated Continuous Wave (FMCW) radars own the ability to measure range (radial distance), velocity (Doppler), and azimuth angle [14, 15]. The frequency of FMCW radars is linearly modulated over the sweep period and can be expressed as

$$f_s = f_c + \frac{B}{T_s} t_s \tag{1}$$

where $f_s$ denotes frequency at the time $t_s$. $f_c$ denotes carrier frequency. $B$ denotes bandwidth. $T_s$ denotes sweep period. At the time $t$, its phase of transmit signal can be expressed as

$$\phi_T(t) = 2\pi f_s t \tag{2}$$

After reflecting on an object at the range $r(t)$, the phase of the received signal is

$$\phi_R(t) = 2\pi f_s(t - \tau) = \phi_T(t) - \phi(t) \tag{3}$$

where $c$ the light speed. $\tau = 2r(t)/c$ denotes the time delay of a round trip. $\phi(t)$ denotes the phase shift

$$\phi(t) = 2\pi f_s \tau = 2\pi f_s \frac{2r(t)}{c} \tag{4}$$

By measuring this phase shift, it can be deduced the range between the sensor and the reflected object. When the target is moving, its relative velocity can be accessed by the Doppler effect:

$$f_d = \frac{1}{2\pi} \frac{d\phi}{dt} = \frac{2v_R}{c} f_s \tag{5}$$

where $f_d$ denotes the Doppler frequency. $v_R$ denotes the radial velocity of the target. Hence measuring Doppler frequency can deduce the radial velocity.

Using the MIMO system with multi-Rx antennas, the azimuth angle $\theta$ of the target can be deduced from the phase shift $\Delta\phi_\theta$ of adjacent pairs Rx.

$$\Delta\phi_\theta = 2\pi f_s \frac{2h \sin\theta}{c} \tag{6}$$

where $h$ denotes the range separating the adjacent receivers.

## 2.2 Radar Network

Figure 1 shows the architecture of radar network. The overall pipeline is divided into three parts: pre-processing, radar network, post-processing. In the part of pre-progressing, a single frame of point cloud and the radar cube is first fetched. The absolute velocity can be compensated by ego-motion. In this paper, the speed of ego-car is zero. The point cloud and radar cube are considered to connect as follows. To filter the cluster and improve the correctness, the point cloud is executed by statistic judges. Then the information of point cloud is utilized to crop the radar cube to decrease the burden of calculation. In the radar network, the size of cropped radar cube is $L \times W \times H$, which represents range/azimuth/Doppler dimensions. Next, the $1 \times L \times W \times H$ sized cropped radar cube is transformed to $25 \times 1 \times 1 \times H$ by three 3-D convolutions (Conv.) and two max-pooling (MP.). This operation encodes the spatial feature of target and focuses on the speed distribution [16]. This encoded feature is then down-sampled by three Convs and MPs. The output of this module is a $64 \times 1 \times 1 \times H/8$. This feature is concatenated with the filtered point cloud. The total feature is fed into two fully connection layers (FC.), whose nodes are 256 and 128 respectively. Finally, the filter point cloud is clustered by the cluster method based density. The object of classification are persons, bicycles and car. Clustering results are denoted in box. The detail will be presented in Sect. 3.

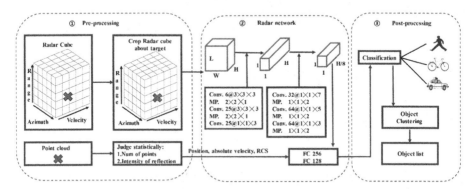

**Fig. 1.** The architecture of radar network

## 2.3 Radar and Camera Fusion

For the camera network, YOLO v4 is chosen as the baseline [12]. To decrease the cost of memory, the network is fine-tuned and the categories of output are reduced to three as above. Calibration between radars and cameras has been employed to the same image plane. Figure 2 shows the target point in the coordinate of the radar and camera. $(X_r, Y_r, Z_r)$ and $(X_c, Y_c, Z_c)$ denote the general radar coordinate and camera coordinate respectively. $r$ and $\alpha$ denotes the range and azimuth angle between the target and radar in the coordinate of radar. $(u, v)$ denotes the pixel coordinate in the coordinate of camera.

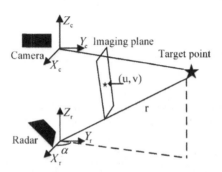

**Fig. 2.** Sketch of the target point in the radar and camera coordinate systems

Calibration of the camera involves the intrinsic and extrinsic parameters, which can be solved by Chessboard calibration [17]. The relation between pixel coordinates and real-world coordinate can be expressed as

$$sp = ABc \tag{7}$$

where $p = [p_x, p_y, 1]^T$ and $c = [c_x, c_y, c_z, 1]$ denote the pixel coordinate in the image and the real-world coordinates in reality. $s$ denotes a scaling factor. $A$ and $B$ denotes the intrinsic and extrinsic parameters and are defined as

$$A = \begin{bmatrix} f_x & 0 & a_x \\ 0 & f_y & a_y \\ 0 & 0 & 1 \end{bmatrix}, B = \begin{bmatrix} r_{11} & r_{12} & r_{13} & m_1 \\ r_{21} & r_{22} & r_{23} & m_2 \\ r_{31} & r_{32} & r_{33} & m_3 \end{bmatrix} \tag{8}$$

The transformation from the coordinate of radar to camera can be achieved by perspective transform [18, 19]. The transform matrix $T$ can be deduced by the position of radar and camera. The transform matrix can be express as

$$r_c = Tr_r = \begin{bmatrix} t_{11} & t_{12} & t_{13} \\ t_{21} & t_{22} & t_{23} \\ t_{31} & t_{32} & t_{33} \end{bmatrix} r_r \tag{9}$$

where $r_r$ and $r_c$ denote the target coordinate in the coordinate of radar and the coordinate transformed into the camera coordinate respectively.

## 2.4  Status Classification and Dataset

Figure 3 presents a sketch of status classification. Three risks of different areas are the danger, potential danger and safety respectively. It should be noted that the accurate collision warning can not only be determined by the range of the target but also influenced by speed [2, 20, 21]. According to the condition of collision occurrence, the collision time of different areas can be defined using the range and velocity. In this system, 0 s

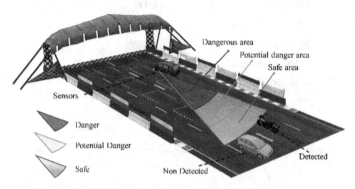

**Fig. 3.** Sketch of status classification

to 1.8 s is set as the time of dangerous range. 1.8 s ~ 3.6 s is set as the time of potential danger. Other time is the safe time.

Figure 4(a) shows the camera and radar used in the experiment. The FMCW radar chooses TI mmw-radar board AWR1642 with 2Tx and 4Rx producing a total of 8 virtual antennas [22, 23]. The ranging reliability for this radar has been carried out in the anechoic chamber. It proves that the range error is not more than 0.06 m. Since it is not our focus, we will no longer present the measuring process. The parameters of the radar are listed in Table 1. The image resolution captured by camera is 1280 × 720 pixels. The image and the recorded radar data are synchronized and have the same frame rate. Figure 4(b) shows the signal indicator light used for danger alarm. The indicator light can be bright as three colors: green, yellow and red, corresponding to the three status classifications of the target. The real-world dataset contains about 1 h of driving with the setups on campus and street. The target-level and low-level output of radar are recorded simultaneously. Figure 5 presents several typical scenarios for recording data.

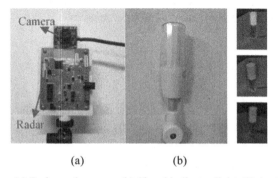

(a)                    (b)

**Fig. 4.** Setups. (a) Radar and camera. (b) Signal indicator light. (Color figure online)

**Table 1.** Parameters of the radar sensor

| Attribute | Value |
| --- | --- |
| Max. range | 80 m |
| Range resolution | 0.0375 m |
| Max. velocity | 20 m/s |
| Velocity resolution | 0.2 M/s |
| Max. azimuth | 120° |
| Azimuth resolution | 1.23° |
| Frame | 20 |

**Fig. 5.** Several typical scenarios for recording data

## 3    Experimental Results and Analysis

In this section, an ideal environment is firstly chosen to verify the proposed method. The detection results under different light conditions are compared to validate the advantages of multi-sensor detection. Then, two complex reality scenes are tested to validate the robustness against the general road condition. Finally, the quantitative analysis is given.

Experiments are carried out in the 64-bit Win7 system, and the software is mainly based on the deep learning architecture of Pytorch. The hardware is mainly based on Intel (R) Core (TM) i7-9700K @ 3.60 GHz CPU and one NVIDIA GTX 2080Ti GPU with CUDA 10.0 accessing computation.

### 3.1    Validation of Basic Functions

Figure 6 simulates the scenario that a stationary vehicle in front is detecting a driving bicycle behind. Figure 6(a–c) show three different status of the bicycle from far to near under the camera, which are safe, dangerous potentially and dangerous. Figure 6(d–f) and (g–i) present the range-Doppler results of radar cubes and the clustering results of point clouds in the horizontal plane. The red circle represents the position of the detected target.

It is not difficult to find that the ranges for these three statuses are 13.16 m, 10.86 m and 2.27 m and the mean velocities are about −3.5 m/s, −5 m/s and −3 m/s. Negative numbers of velocity indicate that the target is moving towards sensors. It can be deduced that the collision times for these three cases are 3.76 s, 2.17 s and 0.76 s respectively. According to the dangerous range time (1.8 s) and the safe range time (3.6 s), results of status classification for these three cases are affirmed. As the bicycle drives from far to near, the colors of the detection box varies from green to yellow, to red in turn, which are corresponding to its status classification: safety, potential danger, and danger. The total time is less than 0.05 s, which includes time for the radar signal processing and optical imaging processing.

Figure 7 is an example of the advantages of detection for multi-sensor. Due to the limitation of the environment, the comparative experiment takes only different lighting conditions as examples. Figure 7(a–b) are the images captured by camera under day and night conditions respectively. Figure 7(c–d) is the corresponding imaging results of the radar. It can be found that multi-sensor detection can overcome the shortcomings of a single sensor and realize robust detection intuitively.

**Fig. 6.** Detection results and status classification under the ideal condition. (a-c) The images captured by camera. (d-f) The range-Doppler images. (g-i) The clustering results of point cloud. (Color figure online)

## 3.2 Results in Real-World

Street and campus are chosen to validate the proposed method. To express the results more intuitively, the optical images, the clustering results of point clouds, and the range-Doppler results are drawn in one image, which are located in the left, northeast and southeast respectively. Figure 8 shows the detection results of one car or one bicycle at different ranges and velocities in the street. In the clustering results of point clouds of Fig. 8(a), the point cloud in the yellow circle is reflected by irony window roadside shown in the corresponding optical image, which is considered as static clutter. Observing the clustering results in Fig. 8(a), it can be seen that the range of the car is 8.1 m and the velocity is 4.8 m/s at this time. Assuming that the car is moving at this constant velocity, the collision time (1.68 s) is lower than the dangerous range time (1.8 s), so the result of status classification at this time should be danger. The color of the detected box is red. Similarly, Fig. 8(b) shows that the range and the velocity of the bicycle at this time is about 16.5 m and about 5 m/s. Assuming that the bicycle is moving at this constant

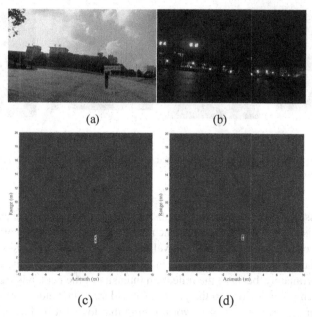

**Fig. 7.** Detection results of multi-sensor under different light conditions. (a-b) The images captured by camera. (d-f) The clustering results of point cloud. (Color figure online)

velocity, the collision time (3.3 s) is higher than the dangerous distance time and lower than the safe range time (3.6 s), so the status at this time is classified as potential danger and the color of the detected box is yellow.

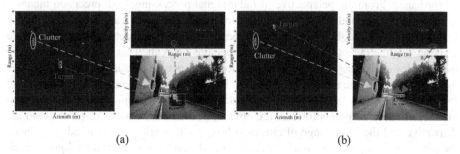

**Fig. 8.** Detection and classification results in the street. (a) One car. (b) One bicycle. (Color figure online)

To evaluate the robustness of the proposed method, the campus owning a more complex traffic environment is tested. Figure 9 shows the detection results of the multi cars and bicycles. The point cloud in the yellow circle is reflected by the irony fence roadside. The traffic volume of such road conditions is enormous and the velocities of targets are fast, which greatly increases the difficulty and requirements for radar hardware and signal processing. Figure 9(a) shows that the bicycle in front and the car

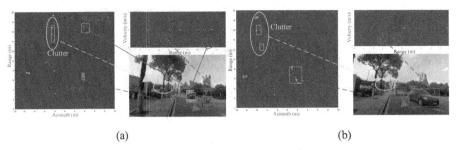

**Fig. 9.** Detection results on campus. (a) One car and one bicycle. (b) One car and one person. (Color figure online)

behind are classified corrected. It can be observed that the range and average velocity for the bicycle is about 5.4 m and 4 m/s. The range and average velocity are about 16 m and 5 m/s. According to both ranges and velocities, their status is classified as danger and potential danger as expected. The car in front and the person behind are shown in Fig. 9(b). Unfortunately, because the reflection intensity of the person is weak, it is hard to classify which points belong to the person. Considering the velocity of the person is slow and its danger level is low, it is worth noting that losing it in the cluster result of point cloud has little effect. In addition, the point cloud of the car can be clustered and its status is classified accurately, which verifies the effectiveness and robustness of the proposed method.

### 3.3 Qualitative Analysis

The precision, recall and F1 score are chosen as the metrics to evaluate the model performance. From the perspective of radar signal processing, high precision means a low false alarm rate and high recall means a low missed detection rate. Thus, the F1 score is an overall trade-off about precision and recall. From the perspective of F1 score, Table 2 shows that the detection performance of persons owns the best F1 score and the car owns the lowest F1 score. Because the location of the person is usually near, the pixels in optical images and the number of points in clustering results are relatively obvious. The location of the car is far. The pixels in optical images are occupied less. Although the number of reflected points for cars are obviously, their distributions of point cloud are dispersive and the error range of clustered boxes will magnify correspondingly. From the perspective of precision and recall, car owns the best precision and the lowest recall

**Table 2.** Performance metrics

|         | Precision | Recall | F1 Score |
|---------|-----------|--------|----------|
| Car     | 0.89      | 0.75   | 0.81     |
| Bicycle | 0.82      | 0.87   | 0.84     |
| Person  | 0.86      | 0.91   | 0.88     |

rare. It is because the cluttering box of car is susceptible to the clutter. It means that the size of clustering box will be magnified, which lead in more error. The recall rate of bicycle and person is higher than 0.85, which can guarantee the safety of detection system.

## 4  Conclusion

In this paper, a method of collision warning and status classification using camera and radar fusion is proposed. Especially, the radar network is designed by combining target-level features with low-level features. The performance for classification and detection is significantly improved. The final average F1 score 1 is 0.843, which realizes the high efficiency at the low-cost level. Sufficient experiment results prove the reliability and effectiveness of the proposed method in practical application.

**Acknowledgement.** Work supported by the National Natural Science Foundation of China (No. 61871386 and No. 61971427) and the Natural Science Fund for Distinguished Young Scholars of Hunan Province, China (No. 2019JJ20022).

## References

1. Schumann, O., Lombacher, J., Hahn, M., Wohler, C., Dickmann, J.: Scene understanding with automotive radar. IEEE Trans. Intell. Veh. **5**(2), 188–203 (2020). https://doi.org/10.1109/TIV. 2019.2955853
2. Wang, X., Xu, L., Sun, H., Xin, J., Zheng, N.: On-road vehicle detection and tracking using MMW radar and monovision fusion. IEEE Trans. Intell. Transp. Syst. **17**(7), 2075–2084 (2016). https://doi.org/10.1109/TITS.2016.2533542
3. Zhong, Z., Liu, S., Mathew, M., Dubey, A.: Camera radar fusion for increased reliability in ADAS applications. Electron. Imaging (2018). https://doi.org/10.2352/ISSN.2470-1173.201 8.17
4. Zhang, C., Luo, W., Urtasun, R.: Efficient convolutions for real-time semantic segmentation of 3D point clouds. In: International Conference on 3D Vision (2018)
5. Simony, M., Milzy, S., Amendey, K., Gross, H.: Complex-YOLO: an euler-region-proposal for real-time 3D object detection on point clouds. In: European Conference on Computer Vision (2018)
6. Daniel, L., et al.: Application of Doppler beam sharpening for azimuth refinement in prospective low-THz automotive radars. IET Radar Sonar Navig. **12**(10), 1121–1130 (2018). https://doi.org/10.1049/iet-rsn.2018.5024
7. Major, B., Fontijne, D., Ansari, A., Sukhavasi, R.T.: Vehicle detection with automotive radar using deep learning on range-azimuth-doppler tensors. arXiv: Computer Vision and Pattern Recognition (2019)
8. Song, W., Yang, Y., Fu, M., Qiu, F., Wang, M.: Real-time obstacles detection and status classification for collision warning in a vehicle active safety system. IEEE Trans. Intell. Transport. Syst. **19**(3), 758–773 (2018). https://doi.org/10.1109/TITS.2017.2700628
9. Palffy, A., Dong, J., Kooij, J.F.P., Gavrila, D.M.: CNN based road user detection using the 3D radar cube. IEEE Robot. Autom. Lett. (2020). https://doi.org/10.1109/LRA.2020.2967272
10. Dong, X., Wang, P., Zhang, P., Liu, L.: Probabilistic oriented object detection in automotive radar. In: Computer Vision and Pattern Recognition (CVPR) (2020)

11. Lim, T., Ansari, A., Major, B., Fontijne, D., Hamilton, M.: Radar and camera early fusion for vehicle detection in advanced driver assistance systems. In: Conference on Neural Information Processing Systems (NeurIPS) (2019)

12. Bochkovskiy, A., Wang, C., Liao, H.: YOLOv4: Optimal speed and accuracy of object detection. ArXiv, abs/2004.10934 (2020)

13. Bartsch, A., Fitzek, F., Rasshofer, R.H.: Pedestrian recognition using automotive radar sensors. Adv. Radio Sci. **10**, 45–55 (2012). https://doi.org/10.5194/ars-10-45-2012

14. Fuller, D.F., Saville, M.A.: A high-frequency multipeak model for wide-angle SAR imagery. IEEE Trans. Geosci. Remote Sens. **51**(7), 4279–4291 (2013). https://doi.org/10.1109/TGRS. 2012.2226732

15. Lekic, V., Babic, Z.: Automotive radar and camera fusion using generative adversarial networks. Comput. Vis. Image Underst. **184**, 1–8 (2019). https://doi.org/10.1016/j.cviu.2019. 04.002

16. Li, M., Stolz, M., Feng, Z., Kunert, M., Henze, R., Kucukay, F.: An adaptive 3D grid-based clustering algorithm for automotive high resolution radar sensor. In: IEEE International Conference on Vehicular Electronics and Safety (ICVES), pp. 1–7 (2018). https://doi.org/10.1109/ ICVES.2018.8519483

17. Wang, T., Zheng, N., Xin, J., Ma, Z.: integrating millimeter wave radar with a monocular vision sensor for on-road obstacle detection applications. Sensors **11**(9), 8992–9008 (2011). https://doi.org/10.3390/s110908992

18. Geyer, J., Kassahun, Y., Mahmudi, M.: A2D2: audi autonomous driving dataset. In: IEEE Conference on Computer Vision and Pattern Recognition (CVPR) (2020)

19. Ouaknine, A., Newson, A., Rebuty, J., Tupin, F., Perez, P.: CARRADA dataset: camera and automotive radar with range-angle-doppler annotations. arXiv: Computer Vision and Pattern Recognition (2020)

20. Lim, Q., He, Y., Tan, U.: Real-time forward collision warning system using nested Kalman filter for monocular camera. In: IEEE Intelligent Vehicles Symposium, pp. 868–873 (2018). https://doi.org/10.1109/ROBIO.2018.8665220

21. Perez, R., Schubert, F., Rasshofer, R., Biebl, E.: Single-frame vulnerable road users classification with a 77 GHz FMCW radar sensor and a convolutional neural network. In: International Radar Symposium (2018)

22. Schumann, O., Hahn, M., Dickmann, J., Wohler, C.: Supervised clustering for radar applications: on the way to radar instance segmentation. In: IEEE MTT International Conference on Microwaves for Intelligent Mobility (ICMIM), pp. 1–4 (2018). https://doi.org/10.1109/ ICMIM.2018.8443534

23. Scheiner, N., Appenrodt, N., Dickmann, J., Sick, B.: A multi-stage clustering framework for automotive radar data. In: International Conference on Intelligent Transportation Systems, pp. 2060–2067 (2019). https://doi.org/10.1109/ITSC.2019.8916873

# AD-DARTS: Adaptive Dropout for Differentiable Architecture Search

Ziwei Zheng[1], Le Yang[2], Liejun Wang[3], and Fan Li[1(✉)]

[1] School of Information and Communications Engineering,
Xi'an Jiaotong University, Xi'an, China
`nevermore@stu.xjtu.edu.cn, lifan@mail.xjtu.edu.cn`
[2] Department of Automation, Tsinghua University, Beijing, China
`yangle15@mails.tsinghua.edu.cn`
[3] College of Information Science and Engineering, Xinjiang University,
Urumqi, China
`wljxju@xju.edu.cn`

**Abstract.** Although Differentiable Architecture Search (DARTS) has achieved promising performance in many machine learning tasks, it still suffers from a problem during searching: due to those different operations in candidate set may need different levels of optimization, directly handling them with the same training scheme will make DARTS in favor of networks with fast convergence, resulting in a performance drop correspondingly. This problem will become more serious at the later searching stages. In this paper, we propose an adaptive dropout method for DARTS (AD-DARTS), which zeros the output of each operation with a probability according to structure parameters which can be considered as the variable representing the difficulty-level training such a candidate operation, thus serving to balance the training procedures for different operations. The operations with more parameters can be trained more adequately to strengthen the characterization ability of the network. Our analysis further shows that the proposed AD-DARTS are also with high search stability. The proposed method effectively solves the aforementioned problem and can achieve better performance compared with other baselines based on DRATS on CIFAR-10, CIFAR-100, and ImageNet.

**Keywords:** Differentiable neural architecture search · Adaptive dropout · Search stability

## 1 Introduction

Novel neural network architectures often lead to significant improvements on various machine learning tasks. However, these handcraft-designed architectures require a large amount of expert experience. Recently, the advent of neural architecture search (NAS) [1,8] has enabled machines to design the architectures of neural networks automatically, which achieve better performance than handcrafted deep models. The key to NAS is to find the optimal network architecture in a defined search space using heuristic search strategies [17,19]. However,

© Springer Nature Switzerland AG 2021
L. Fang et al. (Eds.): CICAI 2021, LNAI 13070, pp. 115–126, 2021.
https://doi.org/10.1007/978-3-030-93049-3_10

the searching stage of early NAS methods either relied on reinforcement learning [15,21] or genetic algorithms [12], which usually require plenty of resources with a long searching time, e.g., the NASNet [22] needs 2000 GPU days to find the optimal architectures, the AmoebaNet [16] even needs over 3000 GPU days.

Compared with the aforementioned methods, the DARTS based algorithms like [13,18], modeling the searching problem as a differentiable optimization problem, reduce the network search time from thousands of GPU days to just a few GPU days while achieve better performance by modeling the searching problem as a differentiable optimization problem. Specifically, different from the NAS methods such as [11], DARTS relaxes the decision variables and parameterizing the possible network structure, while these structure parameters are learnable. Moreover, similar to the idea of meta-learning, the search algorithm designed by DARTS can be viewed as a meta-learning machine [7,14] that teaches the machine to design the methodology of the network and uses a training set and a validation set for bi-level optimization. Due to the fast searching speed and its good performance, DARTS has attracted widespread interest in the NAS community for differentiable searches. GDAS [6] performs random sampling of an operation, which greatly reduces the search time; PC-DARTS [20] proposes channel sampling connection with edge normalization, which improves the network performance and also reduces the search time; P-DARTS [2] derives from the depth interval of search and validation network due to the different number of cell stacked, and established a depth progressive search algorithm.

However, classical DARTS treats all the candidate operations with the same training scheme, which can lead to architecture collapse during searching. Due to those different operations in the candidate set may need different levels of optimization, DARTS seems to be in favor of the operations without parameters and networks with fast convergence. This problem will become more serious at the end of searching, where the resulted architecture contains much more parameter-free operations. DARTS+ [10] proposed an early stop mechanism to alleviate this problem. The regularization of skip connection at the early stage of the search is widely adopted. NoisyDARTS [4] applies unbiased Gaussian noise at skip connection paths to suppress their frequent appearance. DARTS- [3] introduces additional hopping links and gives additional structural parameters to control the training, etc.

In this paper, we propose an adaptive dropout method for DARTS (AD-DARTS), which zeros the output of an operation with a probability according to the structure parameter during searching. Moreover, we provide an analysis of the optimality and stability of the obtained operations and architectures, which reveals the correlation between the corresponding weight of this operation and the final performance of the network. We further discuss whether the ranking of the weights corresponding to the operations is stable at the late stage of the search. Our contributions are listed as follows:

**Firstly**, we point out the unfair and inadequate training for each candidate operation of DARTS, the candidate operations without parameters are more likely to dominate, which has a significant impact on the network performance.

**Secondly**, we propose an adaptive dropout method at the operation level, i.e., the output of the operation is randomly set to zero according to the weight by probability, which plays a role in suppressing the overtraining operations which squeeze the training of the others.

**Thirdly**, we propose a new way to analyze the search stability. The operations in the discrete network obtained from a stable search process can be more adequately trained and make more contributions to performance improvement.

**Lastly**, the experimental results show that our proposed search method can alleviate the inequality in the search process while possessing better stability. We also achieve state-of-the-art networks on CIFAR-10/100 and ImageNet.

## 2    Existing Problems of DARTS

### 2.1    Differentiable Architecture Search

The network of DARTS consists of several normal cells and two reduction cells with shared weights. The cell can be seen as a directed acyclic graph. Each node inside is represented as a feature map. DARTS constructs a set of 8 different candidate operations {Zero, SepConv33, SepConv55, DilConv33, DilConv55, Maxpool, Avgpool, Identity} with a mixed operation between two nodes:

$$\overline{o}^{(i,j)}(x_i) = \sum_{o \in \Theta} \frac{\exp(\alpha_o^{(i,j)})}{\sum_{o' \in \Theta} \exp(\alpha_{o'}^{(i,j)})} \cdot o^{(i,j)}(x_i) \tag{1}$$

The above equation clarifies how the information of the ith node in the search process is passed to the jth node through the mixed operation, where $\alpha_o^{(i,j)}$ is a learnable network structure parameter whose result after *softmax* represents the weights of operations, and $o$ is a specific operation in the operation set $\Theta$. In this way, DARTS continuousizes the otherwise discrete operation selection and search process into a weighted form among the operations represented by the structure parameter, allowing the gradient to be updated by backpropagation. The network parameter $w$ is iteratively updated with the structure parameter $\alpha$ by the following bilevel optimization algorithm:

$$\min_{\alpha} \mathcal{L}_{val}(w^*(\alpha), \alpha)$$
$$\text{s.t. } w^*(\alpha) = \arg\min_{w} \mathcal{L}_{train}(w, \alpha) \tag{2}$$

That is, the network parameter corresponding to the structure parameter is optimized on the training set, while this optimized network is used to further optimize the structure parameter on the validation set. After the search, the construction of the discrete network is performed as follows:

$$x_j = \sum_{k < topk} \max_k(weight^{(i,j)}) \cdot o_{non-zero}^{(i,j)}(x_i), \ i < j \tag{3}$$

where $weight^{(i,j)}$ denotes the operational weights formed by the structure parameter $\alpha$ after $softmax$, and the largest $k$ of all edges connected to the forward node is selected to determine the discrete network structure, for CNN $topk = 2$.

## 2.2  Performance Collapse Caused by Parameter-Free Operations

Many studies have pointed out that DARTS has more skip connect operations in a single cell as the search process progresses, and this structure has a greater impact on the final network performance. Since skip connect is a parameter-free operation, this phenomenon leads to a shallow network and a significant decrease in network parameters.

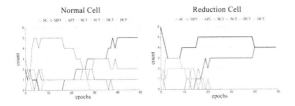

**Fig. 1.** Change of each operation in normal cell and reduction cell at different stages of the training process of DARTS on CIFAR-10.

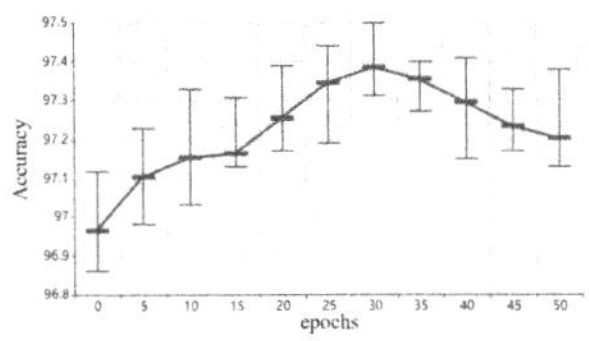

**Fig. 2.** The test accuracy of network at different stages of DARTS on CIFAR-10.

From Fig. 1, we can see that the number of skip connects in DARTS increases rapidly in the late stage of the search process and occupies most of the cell. Also, the number of maximum pooling in reduction cell shares the same trend which is also a parameter-free candidate operation.

The network obtained from DARTS does not show a trend of increasing performance as the search goes deeper and deeper from Fig. 2. We take 5 different random seeds for the test, and the red line indicates the average results. This reflects the performance collapse, which is not only caused by skip connect as it is reported in the existing paper, but also partly caused by maximum pooling, and we therefore extend this phenomenon to parameter-free operations.

## 2.3  Unfair and Inadequate Training Scheme

DARTS prefers to select parameter-free operations rather than others. It indicates the existence of unfair training, where the easy-to-fit and easy-to-converge operations significantly squeeze the training of the others, making them inadequately trained. From Fig. 3, the curve of skip connect is clearly different from the others. At the same time, the contrast between these four edges also shows that this dominance of skip connect falls as the depth increases, in other words, the dominance of skip connect in the normal cell is more obvious at shallow levels.

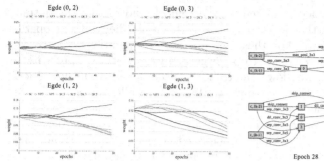

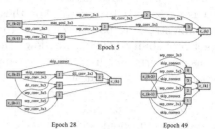

**Fig. 3.** The change curve of weights within the four edges in the normal cell of DARTS. Egde(i, j) represents the connection of the i-th node to the j-th node.

**Fig. 4.** Normal cell found at epoch 5/28/49 on CIFAR-10, representing the network structure at the beginning, middle and end of the search.

The unfair and inadequate training also appears between edge-to-edge at different positions in the cell. From Fig. 4, as the search progresses, because a single node has only two forward nodes connected to it, the interconnection between the four intermediate nodes in the normal cell is gradually replaced by the connection between the two input nodes to them. The four intermediate nodes tend to be equivalent and the network becomes progressively shallow, thus resulting in the performance drop.

## 3   The Adaptive Dropout Methodology

### 3.1   Motivation

Due to this unfair and inadequate training of DARTS, the network becomes shallower and shallower, and the overall number of parameters decreases dramatically. Therefore, this search method, which treats all operations in the same way, only results in a gradual decrease of the network representation capability. The search algorithm can also be regarded as an optimization algorithm, except that not only the optimized network parameters but also the hyperparameters that guide the design of the network structure are optimized, and it is obvious that the overfitting and the unfair competition problem in the structure parameter are not considered. Therefore, how to break this unfair competition and balance the training of different operations is the current concern of many studies.

A simple idea is regularization, i.e., regularizing parameter-free operations that are easy to train and converge, but many existing methods only consider representative skip connect and regularize it in different ways to limit the training. In this paper, we propose a more general approach to perform adaptive regularization for all operations determined by the probability according to the structure parameter obtained from the previous epoch and setting the output of all operations to 0 at the next epoch, so that the gradient of the operation is suspended to update. For those parameter-free operations, the corresponding structure parameter tends to be larger, so we zero the output with a larger probability, which slows down the training speed; for convolutional operations, the corresponding structure parameter becomes smaller due to the crush of parameter-free operations, so we zero the output with a smaller probability, and they will be trained more adequately. It is not absolutely possible to classify two categories as parameter-free or parameterized according to the trend, and this adaptive dropout approach is universally applicable to all operations.

## 3.2 Operation-Level Adaptive Dropout

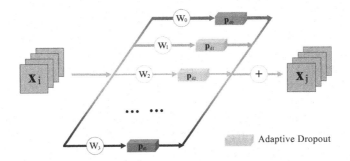

**Fig. 5.** Operation-level adaptive dropout.

The method is revealed by Fig. 5, in the process of connecting two nodes, the output of an operation is first weighted according to the structure parameter as below:

$$w^{(i,j)} = \frac{\exp(\alpha_o^{(i,j)})}{\sum_{o' \in \Theta} \exp(\alpha_{o'}^{(i,j)})} \tag{4}$$

Then it is set to 0 by the adaptive dropout module with probability $p_d$:

$$p_d^{(i,j)} = f_{AD}(w^{(i,j)}) \tag{5}$$

where $f_{AD}$ is a monotonic probability transformation function that describes the process of mapping the structure parameter to the probability of setting 0. Thus, the connection between two nodes can be described as:

$$o^{(i,j)}(x) = \sum_{o \in \Theta} o(x) \cdot m_o^{(i,j)} \tag{6}$$

$$Pr(m_o = 0) = p_d, \ Pr(m_o = 1) = 1 - p_d \tag{7}$$

The $m_o$ represents as the mask whether to retain the output of a certain operation, taking 0 or 1 accordingly. In our experiments, we found that the construction of the probabilistic transformation function $f_{AD}$ has an impact on the performance. If we adopt a plain idea, represented by an identity mapping, since the network structure parameter is the same for each operation at the initialization, the same dropout method is adopted in the pre-search stage to limit the update. This approach produces almost the same result as no restriction at all. Therefore, what should be done is to maximize the regularized variability between operations and operations in the early stage of the search, and we choose the following linear normalization method to map all probabilities to a fixed range while maintaining a positive correlation:

$$f_{AD}(w^{(i,j)}) = \frac{w^{(i,j)} - w_{min}^{(i,j)}}{w_{max}^{(i,j)} - w_{min}^{(i,j)}} \cdot s \tag{8}$$

where $s (s < 1)$ is the scaling factor to restrict the scaled probability in the range of $[0, s]$. $s = 0$ is the method of DARTS, and $s$ is used as the hyperparameter of AD-DARTS.

## 4 Experiments and Results

### 4.1 Implementation Details and Performance Comparation

**Search.** AD-DARTS searches on CIFAR10/100 following almost the same experimental settings as DARTS. We use a network with stacked 6 normal cells and 2 reduction cells, 16 channels, and the one-level approximation in DARTS for the experiments. The training and validation sets were randomly selected from 50,000 images with 50% probability each, and the batch size was set to 96 on a single Nvidia Geforce RTX 2080ti, and a total of 50 epochs are trained to the final discrete network. The network parameter and structure parameter are bilevel optimized according to different optimizers. The network parameter $w$ is set using the stochastic gradient descent (SGD) algorithm with an initial learning rate of 0.0375, a momentum of 0.9, and a weight decay of 0.0003; the structure parameter $\alpha$ is set using the Adam optimizer with an initial learning rate of 0.0003, a momentum of (0.5, 0.999), and a weight decay of 0.001. The scaling factor of the probability mapping function is set to $s = 0.20$.

**Evaluation.** The performance validation performs on CIFAR10/100 and ImageNet according to almost the same settings as DARTS. The overall network is stacked with 18 normal cells and 2 reduction cells, with an initial channel count of 36. The experiments are performed with a batch size of 192, an SGD optimizer, an initial learning rate of 0.05, a momentum of 0.9, and a weight decay of 0.0003, on a single Nvidia Geforce RTX 3090 with 600 epochs. The same cutout and network auxiliary tower as DARTS is also used, and no additional tricks are employed for data augmentation. We have obtained an average top-1 accuracy of 2.24% and 15.88% on CIFAR-10/100 respectively, and 23.4% on ImageNet in Table 1, which shows the best result compared to the current methods.

**Table 1.** The test error on CIFAR10/100 (left), we run 5 times with different random seeds. The test error on ImageNet (right), the method without mark indicates the structures searched on other datasets and migrated to ImageNet for retraining to verify the performance.

| Method | Error on C10(%) | Error on C100(%) | Params (M) | Cost (days) |
|---|---|---|---|---|
| DARTS [13] | 3 | 17.76 | 3.3 | 1.5 |
| SNAS [19] | 2.85 | - | 2.8 | 1.5 |
| GDAS [6] | 2.93 | 18.38 | 3.4 | 0.21 |
| P-DARTS [2] | 2.5 | 15.92 | 3.6 | 0.3 |
| PC-DARTS [20] | 2.50(2.57±0.07) | - | 3.6 | 0.1 |
| Fair-DARTS [5] | 2.49(2.54±0.05) | - | 3.32 | - |
| Noisy-DARTS [4] | 2.39 | - | 3.25 | - |
| DropNAS [9] | 2.26(2.58±0.14) | 16.39 | 4.1 | 0.7 |
| AD-DARTS(ours) | **2.24(2.32±0.08)** | **15.88** | **3.35** | 0.7 |

| Method | Top-1 Error(%) | Top-5 Error(%) | Params (M) | Cost (days) |
|---|---|---|---|---|
| DARTS [13] | 26.7 | 8.7 | 4.7 | 4 |
| SNAS [19] | 27.3 | 9.2 | 4.3 | 1.5 |
| ProxylessNAS [1][†] | 24.9 | 7.5 | 7.1 | 8.3 |
| P-DARTS [2] | 24.4 | 7.4 | 4.9 | 0.3 |
| PC-DARTS [20][†] | 24.2 | 7.3 | 5.3 | 3.8 |
| Fair-DARTS [5][†] | 24.4 | 7.4 | 4.3 | 3 |
| Noisy-DARTS [4][†] | 23.9 | 7 | 4.9 | - |
| DropNAS [9] | 23.5 | 6.7 | 5.7 | 0.7 |
| **AD-DARTS(C10)** | **23.4** | **6.6** | 5.6 | 0.7 |
| **AD-DARTS(C100)** | 23.5 | 6.7 | 6 | 0.7 |

[†] Directly searched on ImageNet.

## 4.2 Influence of Hyper-parameter

Table 2 reflects the effect of different probability transformation function scaling factor $s$ on the network performance. It can be seen that the best performance is achieved on both CIFAR-10/100 when $s = 0.2$, i.e., the probability of adaptive dropout is linearly normalized to the range of $[0, 0.2]$. Also, the results are improved considerably compared to the degradation of $s = 0$ to DARTS.

**Table 2.** The influence of different $s$ on CIFAR-10/100. We obtain 5 experiments for each value on CIFAR-10.

| Dataset | $s = 0$ | $s = 0.1$ | $s = 0.2$ | $s = 0.3$ | $s = 0.4$ |
|---|---|---|---|---|---|
| CIFAR-10 | $2.76 \pm 0.18$ | $2.50 \pm 0.12$ | $\mathbf{2.32 \pm 0.08}$ | $2.38 \pm 0.13$ | $2.82 \pm 0.10$ |
| CIFAR-100 | 17.76 | 16.05 | **15.88** | 15.93 | 17.87 |

## 4.3 Search Stability Study

**Search Consistency Factor.** We abstract a search consistency factor to measure the stability. The motivation is that if the variation of chosen operations is as small as possible during a given search process, the better. To facilitate quantification, we first encode all non-zero operations in the operation set $u$, e.g., skip connect as number 0, maximum pooling as number 1, etc., for a total of seven numbers. Define the maximum weight operation encoding vector within a cell at search moment $epoch = e$ as:

$$\boldsymbol{d}^e = [u_0, u_1, \ldots, u_{13}]^\top \tag{9}$$

To describe the variation, define the discrete difference mask vector $\boldsymbol{D}^e$ as:

$$D_k^e = \begin{cases} 1, & d_k^e - d_k^{e-1} \neq 0 \\ 0, & otherwise \end{cases}, k = 0, 1, \ldots, 13 \tag{10}$$

The changed location is set to 1. Since the value in weight vector $\boldsymbol{w}^e$ varies a lot from different locations (or different depths) in the cell, to measure search stability more accurately, we propose the change importance weighting vector $\boldsymbol{W}^e$ as:

$$W_k^e = \boldsymbol{w}^e[d_k^e], \; k = 0, 1, \ldots, 13 \tag{11}$$

If the edge corresponding to a larger weight operation has changed, then we define it as a more important change.

The discrete difference mask vector $\boldsymbol{D}^e$ and the change importance weighting vector $\boldsymbol{W}^e$ are with same length, representing the 14 edges in the cell. The final constructed search consistency factor $C_n$ is represented by a sliding average:

$$C_n = \frac{1}{n} \cdot \sum_{e=e_0}^{n-1} \boldsymbol{D}^{e\top} \cdot \boldsymbol{W}^e \tag{12}$$

$n$ described as the size of window of sliding average.

**The Stability of AD-DARTS.** Instead of making the whole search process unstable, the adaptive dropout method reinforces the stability of the search. The reasons are as follows: 1) even the largest dropout rate is limited to a small range due to the presence of the scaling factor $s$ of the probability transformation function; 2) the superiority of an operation is better reflected if the gradient backpropagation is prevented with a large probability throughout the search process but the weights still maintain large values.

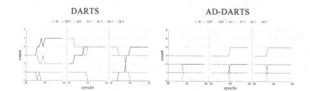

**Fig. 6.** Change of each operation in normal cell at late stage of the training process (epoch 30–49) of DARTS and AD-DARTS on CIFAR-10.

**Fig. 7.** The test accuracy of network at different stages of AD-DARTS on CIFAR-10.

The comparison in Fig. 6 shows that the ranking of operations in DARTS still changes a lot in the late search period, while AD-DARTS changes less, which further illustrates the improved stability. Figure 7 reflects the network performance of AD-DARTS at different stages of search on CIFAR-10. Unlike the performance degradation problem of DARTS, AD-DARTS can maintain a more stable performance improvement. Also, the impact of different random seeds on the performance at the late stage of search becomes relatively small. This reflects the stability of AD-DARTS in terms of performance growth, and the search process of the network always proceeds steadily in the direction of performance improvement.

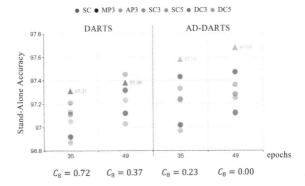

**Fig. 8.** The influence of stand-alone model on egde(1,2) of normal cell at two different stage of search on CIFAR-10.

**The Guiding Ability of Search Consistency Factor.** The network search consistency factor is proposed in AD-DARTS as a guide to the improvement of the structural performance. Figure 8 shows two different stages of the search, changing the selected operation to others, and testing its impact on network performance, and we call it the operation stand-alone accuracy. In DARTS, the operation selected by the discrete structure of this edge is the skip connect. But the network performance does not necessarily decrease when it is changed to others, and the final discrete structure even has a better performance for the $3 \times 3$ convolution; in AD-DARTS, this phenomenon does not exist, and the non-selected operations have a greater impact on the performance degradation, which in turn proves the superiority of the network selected ones. It can be seen that the value of the consistency factor $C_n$ with the sliding window of $n = 8$ corresponding to DARTS is significantly larger, which means that the discrete structure changes more during 8 epochs before the current epoch, while the value of the search consistency factor of AD-DARTS is smaller and even keeps the discrete structure without any change in the last 8 epochs, which illustrates that our method does substantially improve the search stability, especially in the late stage of the search.

## 5  Conclusion

We point out the performance collapse problem of DARTS, and the tendency of this search method to prefer parameter-free operations. We propose a novel differentiable search method for neural networks, AD-DARTS, which prevents structure parameter updates by probabilistically and adaptively setting the output of each operation to zero according to the weight, and substantially improves the performance of the discrete network. AD-DARTS achieves optimal performance compared to existing differentiable search algorithms. We also propose a network search consistency factor and analyze how AD-DARTS improves the network search stability.

# References

1. Cai, H., Zhu, L., Han, S.: Proxylessnas: direct neural architecture search on target task and hardware. In: International Conference on Learning Representations (ICLR) (2019)
2. Chen, X., Xie, L., Wu, J., Tian, Q.: Progressive differentiable architecture search: bridging the depth gap between search and evaluation. In: International Conference on Computer Vision (ICCV), pp. 1294–1303 (2019)
3. Chu, X., Wang, X., Zhang, B., Lu, S., Wei, X., Yan, J.: Darts-: robustly stepping out of performance collapse without indicators. arXiv preprint arXiv:2009.01027 (2020)
4. Chu, X., Zhang, B., Li, X.: Noisy differentiable architecture search. arXiv preprint arXiv:2005.03566 (2020)
5. Chu, X., Zhou, T., Zhang, B., Li, J.: Fair DARTS: eliminating unfair advantages in differentiable architecture search. In: Vedaldi, A., Bischof, H., Brox, T., Frahm, J.-M. (eds.) ECCV 2020. LNCS, vol. 12360, pp. 465–480. Springer, Cham (2020). https://doi.org/10.1007/978-3-030-58555-6_28
6. Dong, X., Yang, Y.: Searching for a robust neural architecture in four GPU hours. In: IEEE Conference on Computer Vision and Pattern Recognition (CVPR), pp. 1761–1770 (2019)
7. Franceschi, L., Frasconi, P., Salzo, S., Grazzi, R., Pontil, M.: Bilevel programming for hyperparameter optimization and meta-learning. In: International Conference on Machine Learning (ICML), pp. 1568–1577. PMLR (2018)
8. Gong, X., Chang, S., Jiang, Y., Wang, Z.: Autogan: neural architecture search for generative adversarial networks. In: International Conference on Computer Vision (ICCV), pp. 3224–3234 (2019)
9. Hong, W., Li, G., Zhang, W., Tang, R., Wang, Y., Li, Z., Yu, Y.: Dropnas: grouped operation dropout for differentiable architecture search. In: International Joint Conference on Artificial Intelligence (IJCAI) (2020)
10. Liang, H., et al.: Darts+: Improved differentiable architecture search with early stopping. arXiv preprint arXiv:1909.06035 (2019)
11. Liu, C., et al.: Progressive neural architecture search. In: European Conference on Computer Vision (ECCV), pp. 19–34 (2018)
12. Liu, H., Simonyan, K., Vinyals, O., Fernando, C., Kavukcuoglu, K.: Hierarchical representations for efficient architecture search. In: International Conference on Learning Representations (ICLR) (2018)
13. Liu, H., Simonyan, K., Yang, Y.: Darts: differentiable architecture search. In: International Conference on Learning Representations (ICLR) (2019)
14. Maclaurin, D., Duvenaud, D., Adams, R.: Gradient-based hyperparameter optimization through reversible learning. In: International Conference on Machine Learning (ICML), pp. 2113–2122. PMLR (2015)
15. Pham, H., Guan, M., Zoph, B., Le, Q., Dean, J.: Efficient neural architecture search via parameters sharing. In: International Conference on Machine Learning (ICML), pp. 4095–4104. PMLR (2018)
16. Real, E., Aggarwal, A., Huang, Y., Le, Q.V.: Regularized evolution for image classifier architecture search. In: AAAI conference on Artificial Intelligence (AAAI), pp. 4780–4789 (2019)

17. Stamoulis, D., Ding, R., Wang, D., Lymberopoulos, D., Priyantha, B., Liu, J., Marculescu, D.: Single-Path NAS: designing hardware-efficient ConvNets in less than 4 hours. In: Brefeld, U., Fromont, E., Hotho, A., Knobbe, A., Maathuis, M., Robardet, C. (eds.) ECML PKDD 2019. LNCS (LNAI), vol. 11907, pp. 481–497. Springer, Cham (2020). https://doi.org/10.1007/978-3-030-46147-8_29

18. Veniat, T., Denoyer, L.: Learning time/memory-efficient deep architectures with budgeted super networks. In: IEEE Conference on Computer Vision and Pattern Recognition (CVPR), pp. 3492–3500 (2018)

19. Xie, S., Zheng, H., Liu, C., Lin, L.: Snas: stochastic neural architecture search. In: International Conference on Learning Representations (ICLR) (2019)

20. Xu, Y., et al.: Pc-darts: partial channel connections for memory-efficient architecture search. In: International Conference on Learning Representations (ICLR) (2020)

21. Zhong, Z., Yan, J., Wu, W., Shao, J., Liu, C.L.: Practical block-wise neural network architecture generation. In: IEEE Conference on Computer Vision and Pattern Recognition (CVPR), pp. 2423–2432 (2018)

22. Zoph, B., Vasudevan, V., Shlens, J., Le, Q.V.: Learning transferable architectures for scalable image recognition. In: IEEE Conference on Computer Vision and Pattern Recognition (CVPR), pp. 8697–8710 (2018)

# An Improved DDPG Algorithm with Barrier Function for Lane-Change Decision-Making of Intelligent Vehicles

Tianshuo Feng[1($\boxtimes$)], Xin Xu[2], Xiaochuan Zhang[1], and Xinglong Zhang[2]

[1] Chongqing university of Technology, Chongqing, China
zxc@cqut.edu.cn
[2] National University of Defense Technology, Changsha, Hunan, China
zhangxinglong18@nudt.edu.cn

**Abstract.** As a decision-making problem with interaction between vehicles, it is difficult to describe intelligent vehicle lane change state space using a rule-based decision system. The deep deterministic policy gradient (DDPG) algorithm offers good performance for autonomous driving decision, but still has slow convergence and high collision probability in learning process when applied to lane change. Therefore, we propose an improved deep deterministic policy gradient algorithm with barrier function (DDPG-BF) algorithm to address these problems. The barrier function is constructed depending on the safety distance required for lane changes, and DDPG algorithm optimization is improved by guiding the vehicle to choose actions within safety constraints. Simulation results on TORCS confirmed that the proposed method converged in hundreds of training episodes, and reduced the unsafe behavior ratio to less than 0.05. Compared with DDPG and FEC-DDPG algorithm, the proposed method has the contribution to improve the convergence speed of learning and maintain the safe distance between vehicles in lane change.

**Keywords:** Lane change decision · Deep reinforcement learning · Deep deterministic policy gradient · Barrier function

## 1 Introduction

Lane change decision-making is a hot issue for autonomous driving, including intelligent vehicle multi module interaction, complex environment, and diverse traffic conditions [1]. Current autonomous driving decision-making modules are mostly based on artificial rules that can address most driving situations. However, it is difficult to enumerate all the situations and find optimal decision-making using artificially designed rules when driving environment become complex, involving unexpected situations.

© Springer Nature Switzerland AG 2021
L. Fang et al. (Eds.): CICAI 2021, LNAI 13070, pp. 127–139, 2021.
https://doi.org/10.1007/978-3-030-93049-3_11

Autonomous vehicles make optimal decisions in complex environments. This view is consistent with the one of reinforcement learning. Hence, many recent studies have applied deep reinforcement learning to autonomous driving decision-making, integrating decision-making and motion planning modules [2] to address problems in the driving environment. Wayve considered a smart car on a simple road without large prior dataset using the deep deterministic policy gradient (DDPG) algorithm [3], confirming that DDPG algorithm can effectively solve some autonomous driving decision-making problems with incomplete or difficult to obtain prior knowledge [4].

Although deep reinforcement learning is a potential policy generation method for autonomous driving, most previous deep reinforcement learning approaches fail to consider safety problems, which is likely to raise safety risk. However, some recent studies have considered reinforcement learning safety [5–8]. Bin et al. proposed FEC-DDPG [9] in 2019, which improves experience memory buffer sampling to enhance DDPG algorithm ability to avoid illegal actions (e.g., applying accelerator and brake simultaneously) during training. Cheng et al. added control barrier functions (CBF) to reinforcement learning [10] to extract actions from the safety set to participate in training, and verified the proposed method's effectiveness in terms of efficiency and safety assurance through vehicle following experiments.

Deep deterministic policy gradient has been applied to autonomous driving with some success, but most DDPG algorithm and the related improved algorithm application scenarios are relatively simple environments, such as vehicle following, lane keeping and so on. It is difficult to construct a deterministic safety state for decision scenarios with stochastic interactions, like the lane change shown in Fig. 1, due increased state space dimensionality, leading to problems such as the collision with interacting vehicles and slow convergence [11].

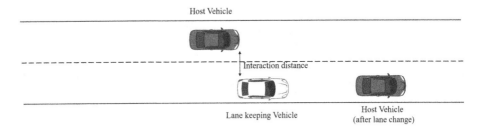

**Fig. 1.** Lane change scenario with surrounding vechicle.

Therefore, this paper proposes an improved DDPG algorithm with barrier function (DDPG-BF). Safety constraints were established from the safe distance model for lane change, and added to the loss function through the barrier function. Thus, the intelligent vehicle can be protected during reinforcement learning by the boundary limitation, and will tend to choose safe actions with high reward.

Our main contributions are as follows. (1) We construct a barrier function based on safety distances during lane changes. (2) We extend the DDPG algorithm by adding the barrier function term into the loss function. (3) We apply the proposed method to intelligent vehicle lane change scenario. It shows improved convergence speed and maintains the distance in the safety region for lane change interactions.

## 2    Background

### 2.1    A Safety Distance in Lane Change Decision

Vehicle lane change decision has always been a key issue in safe driving. Some scholars have constructed the dynamic model of safe driving from safety speed [12] and safety distance [13,14].

A critical distance constraint is required to prevent collision between the two vehicles at the next moment. Zang and his team put forward a calculation method of safety distance when changing the road by establishing the lateral safety distance model [13]. According to this distance model, we record the lane keeping vehicle as A, and the lane changing vehicle as B, assuming that there is no collision between the two vehicles after the lane change, the results can be divided into two cases as shown in Fig. 2:

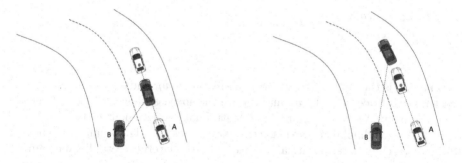

**Fig. 2.** The situations of vehicles in interaction after lane change.

Based on the relationship between the longitudinal speed and acceleration of the two vehicles, the classification analysis is carried out [13]. The corresponding time solutions of above lane change results are shown in Table 1 and Table 2.

**Table 1.** Time Solutions $(X_A > X_B)$

| Presupposition | Time solution |
|---|---|
| $\begin{cases} a_A - a_B cos\beta < 0 \\ v_A - v_B cos\beta > 0 \end{cases}$ | $t_1$ does not exist |
| $\begin{cases} a_A - a_B cos\beta < 0 \\ v_A - v_B cos\beta > 0 \end{cases}$ | $t_2 = \frac{(v_B cos\beta - v_A) + \sqrt{(v_A - v_B cos\beta)^2 + 2(a_A - a_B cos\beta)S_0}}{a_A - a_B cos\beta}$ |
| $\begin{cases} a_A - a_B cos\beta > 0 \\ v_A - v_B cos\beta < 0 \end{cases}$ | $t_3 = t_2$ |
| $\begin{cases} a_A - a_B cos\beta > 0 \\ v_A - v_B cos\beta > 0 \end{cases}$ | $t_4 = t_2$ |

**Table 2.** Time solutions $(X_A < X_B)$

| Presupposition | Time solution |
|---|---|
| $\begin{cases} a_A - a_B cos\beta < 0 \\ v_A - v_B cos\beta < 0 \end{cases}$ | $t_5 = \frac{(v_A - v_B cos\beta) + \sqrt{(v_A - v_B cos\beta)^2 + 2(a_B cos\beta - a_A)S_0}}{a_B cos\beta - a_A}$ |
| $\begin{cases} a_A - a_B cos\beta < 0 \\ v_A - v_B cos\beta > 0 \end{cases}$ | $t_6 = t_5$ |
| $\begin{cases} a_A - a_B cos\beta > 0 \\ v_A - v_B cos\beta < 0 \end{cases}$ | $t_7 = t_5$ |
| $\begin{cases} a_A - a_B cos\beta > 0 \\ v_A - v_B cos\beta > 0 \end{cases}$ | $t_8$ dose not exist |

where, $\beta$ is the angle between the two directions, $v_B$ and $a_B$ are the speed and acceleration of host vehicle, $v_A$ and $a_A$ are the speed and acceleration of vehicle A respectively, $X_B$ and $X_A$ are the longitudinal displacement of two cars, and $S_0$ is the longitudinal distance between two vehicles after lane change. According to the research on lateral critical distance [15], the lateral safety distance during lane change is expressed as:

$$D_0 = sin\beta(v_B t + \frac{a_B t^2}{2}) + 0.94 \quad \begin{cases} t = t_2, & X_B < X_A \\ t = t_5, & X_B > X_A \end{cases} \tag{1}$$

## 2.2 Autonomous Driving Based on DDPG Algorithm

DDPG algorithm has good performance in dealing with continuous action space problems. It improves the deterministic policy gradient (DPG) [16] method with technical points of deep Q network (DQN) [17], combined with actor-critic neural networks [3,18].

DDPG algorithm maps environment state $s$ to policy, state-action pair $(s, a)$ to value function, and uses deep neural network to calculate Q value instead of

Q table to store [16]. In the training process, the experience replay mechanism is used to store the quadruple $(s_t, a_t, r_t, s_{t+1})$ representing the state, action, reward and the next state into the memory buffer. When updating the network, multiple groups of experience samples are randomly selected from the memory pool to weaken the temporal correlation of the samples, which can more effectively optimize the network parameters.

At the same time, DDPG algorithm has target networks when using actor-critic method [3]. Actor-critic network is responsible for updating the parameters of policy network and value network. In light of the current state, it selects the corresponding actions to interact with the environment, and updates the parameters to the actor target network; The task of actor target network is to select the optimal action through the sampling state in the experience replay buffer; the critic target network $Q'$ will calculate a term of the target Q value, according to the sampling state action pair; the critic network calculates the current Q value according to the sampling state $s_j$ and action $a_j$, and constructs the mean square error loss function combined with the current target Q value $y_j$:

$$L = \frac{1}{m} \sum_{j}^{m} (y_j - Q(s_j, a_j|\omega))^2 \tag{2}$$

$$y_j = E\left[r_j + \gamma Q'\left(s_{j+1}, \pi_{\theta'}(s_{j+1})|\omega'\right)\right] \tag{3}$$

where $\theta'$, $\omega$ and $\omega'$ are the parameters of target actor network, critic network and target critic network, respectively. And $m$ is the number of samples, $\pi_{\theta'}$ is the policy from target actor network. For autonomous driving decision, DDPG algorithm obtains the state of the observation environment through the vehicle sensor interface:

$$state = \langle v, d, rpm, yaw, track\_pos, v_{wheels\_spin} \rangle \tag{4}$$

where $v$ is the velocity in three dimensions, $d$ is the actual lateral distance between two interacting vehicles. $rpm$, $yaw$, $track\_pos$ and $v_{wheels\_spin}$ are the engine speed, yaw, track position and velocity of wheels spin respectively. It gets the selected driving decision through the actor network, and controls the pedals of throttle and brake, steering and gear of the vehicle through controllers.

$$action = \langle p_{throttle}, p_{brake}, steering, gear \rangle \tag{5}$$

In order to solve the problem of exploration ability caused by deterministic policy, exploration noise is added, so that the action selection is determined by the output policy of actor network and exploration noise. After executing the action it receives the next state information, calculates the reward function:

$$r_{total} = r_{damage} + r_{yaw} + r_{track} + r_{progress} \tag{6}$$

The total reward can be formulated by four components: the reward for collision damage, the yaw angle limit, the lane tracking and the progress of vehicle. And then it stores the samples of state, action and reward value in the memory pool, and then randomly extracts groups of samples from the buffer to update the actor and critic network.

# 3   DDPG Algorithm with Barrier Function

## 3.1   The Barrier Function for Safe Lane Change

In order to improve the learning efficiency of DDPG algorithm and maintain vehicles at safe distance in lane change, we proposed an improved DDPG method with barrier function (DDPG-BF).

Barrier function [19] is a kind of constraint continuous function. As the point approaches the constraint boundary of optimization problem, the value of barrier function will increase. When it reaches the boundary of feasible region, the value of barrier function will increase to infinity. By designing an appropriate barrier function, the constrained optimization problem can be reconstructed into an unconstrained problem, and then the algorithm can converge through optimization iteration [20].

When the safety constraints of lane changing scenario are involved, the optimization problem of loss function becomes a kind of function minimum optimization problem under constraint conditions, which belongs to the solution scope of interior point method [21] in barrier function.

The key to deal with lane changing safety problem with interior point method is to express the safety constraints in the process through barrier function $B(s)$, add the safety boundary constraints in the DDPG training process, and select the optimal action in the feasible region. The feasible region $F$ within $B(s)$ boundary can be expressed as:

$$F : \{s \in R^n : 0 \leq B(s) \ll \infty\} \tag{7}$$

where $s$ is the state and $R^n$ is the state space. In the lane change scenario, the barrier function $B(s)$ is obtained through the transformation of safety distance constraints. The feasible region $F$ represents the set of states in which the distance between interacting vehicles is greater than the safety distance in the process of lane change.

## 3.2   Constructing Barrier Function by Interior Point Method

For DDPG algorithm, the objective function to be optimized is the loss function $L$, and there is an inequality constraint based on safety distance in the optimization process, which is denoted as $g(s)$. Then the optimization problem is expressed as:

$$min \left\{ L = \frac{1}{m} \sum_{j}^{m} (y_j - Q(s_j, a_j | \omega))^2 \right\} \text{ s.t. } g(s_j) > 0 \tag{8}$$

By defining function $B(s)$ as barrier function, the search point can be kept in the barrier function:

$$B(s_j) = \frac{1}{g(s_j)} \tag{9}$$

The solved function is transformed from $L$ to a new objective function $L_{BF}$. It is constructed by the original loss function and barrier function

$$L_{BF} = \frac{1}{m} \sum_{j}^{m} \left[ (y_j - Q(s_j, a_j | \omega))^2 + \mu B(s_j) \right] \tag{10}$$

The barrier factor $\mu$ is a very small number, which represents the influence of barrier function on loss function. The smaller the $\mu$ is, the smaller the influence of the barrier function on the loss function is, and the closer it is to the true solution; if $\mu$ is too large, the greater the influence of the barrier function is, and easier to get stuck at local optimum.

For the new loss function $L_{BF}$ with the structure of Eq. (10), when the state tends to the boundary of feasible region $F$, the function value of $L_{BF}$ tends to infinity. On the contrary, because $\mu$ is a very small number, the function value of $L_{BF}$ is approximately $L$, and the approximate solution of the original problem can be obtained by solving the new problem:

$$\begin{aligned} &min \ L_{BF} \\ &s.t. \ S \in F \end{aligned} \tag{11}$$

Where $S$ is the state set inside the safety constraint. According to the lateral safety distance in the process of lane changing, the corresponding barrier function can be constructed:

$$B(s_j) = \frac{1}{d_j - D_{0j}} \tag{12}$$

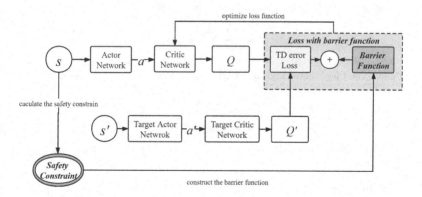

**Fig. 3.** Illustration of DDPG-BF training process.

## 3.3  Algorithm Process of DDPG-BF

The barrier function based on safety distance is introduced into the loss function optimization process of DDPG algorithm, and the loss function under safety constraints is used for the reinforcement learning training of intelligent vehicle lane change decision. The illustration and pseudo code of DDPG-BF algorithm are as follows (Fig. 3):

---

**Algorithm 1. DDPG-BF**

---

**Input:**
the parameters of networks $\theta$, $\theta'$, $\omega$, $\omega'$, discount factor $\gamma$, exploration noise $\mathcal{N}$, barrier factor $\mu$, longitudinal safety distance $S_0$, soft update coefficient $\tau$, sample numberm, maximum number of iterations $T$

**Output:**  optimal actor parameter $\theta$, and critic $\omega$
1: Randomly initialize the parameters $\theta$, $\omega$, $\omega' = \omega$, $\theta' = \theta$
2: Initialize replay memory buffer $D$
3: Initialize the random noise $\mathcal{N}$
4: **for** t from 1 to T **do**
5:     Initialize $s_t$ as the first state of the state sequence
6:     Select action $a = \pi_\theta(s_t) + \mathcal{N}$
7:     Execute action $a$, get the new state $s_{t+1}$ and observe reward $r$
8:     Compute barrier term $B(s_t)$ according to the safe constraint and the states of vehicles in interaction
9:     Store $(s_t, a_t, r_t, s_{t+1}, B(s_t))$ in memory buffer $D$
10:    $s_t = s_{t+1}$
11:    Sample a random minibatch of $m$ transitions$(s_j, a_j, r_j, s_{j+1}, B(s_j))$ from $D$
12:    Set $y_j = r_j + \gamma Q'(s_{j+1}, \pi_{\theta'}(s_{j+1})|\omega')$
13:    Update critic network via minimizing the loss:

$$L_{BF} = \frac{1}{m}\sum_{j=1}^{m}[(y_j - Q(s_j, a_j|\omega))^2 + \mu B(s_j)]$$

14:    Update actor network by the sampled policy gradient:

$$J = -\frac{1}{m}\sum_{j=1}^{m}Q(s_j, a_j|\omega)$$

15:    Update the target actor and target critic networks:

$$\omega' \leftarrow \tau\omega + (1-\tau)\omega'$$

$$\theta' \leftarrow \tau\theta + (1-\tau)\theta'$$

16:    **if** $s'$ is the ending state: **then**
17:        done
18:    **else**
19:        go to step 4
20: **return** return the output

---

# 4    Experiments

## 4.1    Experimental Preparations

This experiment is based on TORCS platform, setting lane change decision scene with vehicle interaction, and verifying the algorithm. The content of the comparative experiment includes the following two components:

(1) Through accumulated reward, the comparative experiments of DDPG [3], FEC-DDPG [9] and DDPG-BF are carried out. This work shows the improvement of the proposed algorithm in convergence speed.
(2) In order to have a comparison of above three methods in safety vehicle interaction, we record the behaviors out of safety distance boundary occurred at interaction each epoch and calculate the ratio of these epochs to the total training episodes.

In the experiment, the training map is selected as "Speedway No. 1" in TORCS, and the interactive vehicle driving model is selected as "Tita 4". The longitudinal safety distance $S_0$ is set as 28 m, the maximum lateral safety distance is 3.75 m as the world standard, so as to ensure that the initial action of the algorithm is selected in the feasible region. The value of experimental hyperparameters is shown in Table 3:

**Table 3.** Value of Hyperparameters

| Hyperparameter | Value |
| --- | --- |
| Discount rate $\gamma$ | 0.99 |
| Soft update coefficient $\tau$ | 0.001 |
| Number of extracting samples $m$ | 32 |
| Capacity of Memory Buffer $M$ | 100000 |
| Maximum step per epoch | 300 |

## 4.2    Comparative Analysis of Experimental Data

**Convergence Speed.** In Fig. 4, it displays the accumulated reward per episode for DDPG-BF. The experimental data comparison is shown in Fig. 5, abscissa for training episodes, ordinate for accumulated reward.

DDPG-BF not only sets the reward for collision damage, but also adds the barrier function constraint for the safety distance of vehicle interaction, and adds the safety distance constraint for the training of networks, so as to guide the agent to make action decisions and choices within the feasible region. In Fig. 5, After 1450 episodes of training, DDPG and FEC-DDPG algorithm can almost reach the accumulated reward value maintained by DDPG-BF, but the data is still in shock, not obtained convergence. From figures, we can see that accumulated reward of DDPG-BF has reached a relatively stable state after 700 episodes, and the convergence speed of DDPG-BF is obviously improved compared with the other two algorithms.

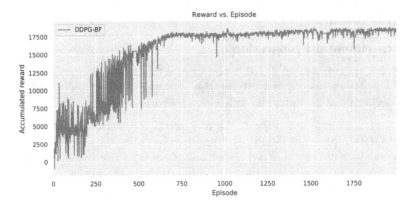

**Fig. 4.** The accumulated reward per episode for DDPG-BF.

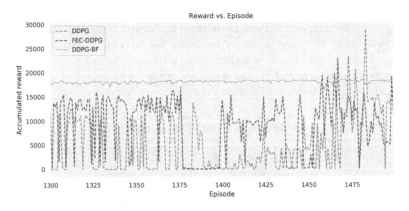

**Fig. 5.** The accumulated reward for DDPG, FEC-DDPG and DDPG-BF.

**Unsafe Behavior Ratio.** The safety of the three methods in lane change decision-making is compared by recording whether each episode has the behaviors made the gap less than safety distance at interaction, and calculating the ratio of the safety warning (the vehicles gap less than corresponding safety distance) episodes to the total training times.

From Fig. 6, we can see that the ratio of behaviors out of safety boundary is less than 0.05 in the whole training process of DDPG-BF. When using DDPG method alone and FEC-DDPG without barrier function, the ratios are almost above 0.15 and show the growth trend even in the later stages of training. Figure 7 illustrates the relationship between minimum lateral distance and the corresponding safety distance in the learning process of DDPG-BF. Values above the black line represent the safe set. And the safety distance boundary is calculating by lateral safety distance model [13]. In the later stages of DDPG-BF learning process, the lateral minimum distance is almostly maintained above the safety distance boundary.

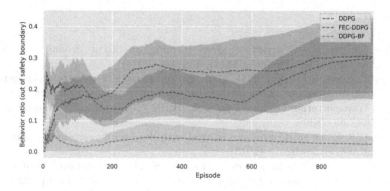

**Fig. 6.** The unsafe behavior ratio for DDPG, FEC-DDPG and DDPG-BF.

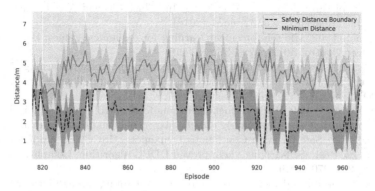

**Fig. 7.** The minimum lateral distance of vehicles for each training episode of DDPG-BF and the corresponding safety distance boundary.

## 5 Conclusion

The proposed DDPG-BF method was derived by analyzing DDPG autonomous driving policy framework and safe distance model during lane changes, particularly to address slow convergence speed and consequential high collision probability. Barrier function interference is added to loss function optimization of DDPG method, ensuring safety constraints act on driving action evaluation and selection. We verified that DDPG-BF improved convergence and significantly avoid collision by controlling the driving actions in the lane change safety region. However, the proposed algorithm only considered when the interaction target is other vehicles, and does not consider road emergency situations, pedestrians as interaction targets, and other complex road conditions. Future work will extend the approach to ensure safe driving considering multiple interaction objects.

**Acknowledgment.** This work was supported in part by the National Key Research and Development Program of China (Project No. 2018YFB1305105) and National Natural Science Foundation of China under Grant 62003361.

# References

1. Zhang, X., et al.: Overview of deep learning intelligent driving methods. J. Tsinghua Univ. (Sci. Technol.) **58**(4), 438–444 (2018)
2. Chae, H., Kang, C M., Kim, B D., et al.: Autonomous braking system via deep reinforcement learning. In: ITSC: 2017 IEEE 20th International Conference on Intelligent Transportation Systems, pp. 1–6. IEEE (2017). https://doi.org/10.1109/ITSC.2017.8317839
3. Lillicrap, T P., et al.: Continuous control with deep reinforcement learning. arXiv preprint arXiv:1509.02971 (2015)
4. Kendall, A., et al.: Learning to drive in a day. In: 2019 International Conference on Robotics and Automation (ICRA), Montreal, QC, Canada, 2019, pp. 8248–8254. https://doi.org/10.1109/ICRA.2019.8793742
5. Fulton, N., Platzer, A.: Safe reinforcement learning via formal methods: toward safe control through proof and learning. In: Thirty-Second AAAI Conference on Artificial Intelligence, vol. 32, no. 1 (2018)
6. Yang, Y., et al.: Safe reinforcement learning for dynamical games. Int. J. Robust Nonlinear Control **30**(9), 3706–3726 (2020)
7. Alshiekh, M., Bloem, R., Ehlers, R., et al.: Safe reinforcement learning via shielding. In: Thirty-Second AAAI Conference on Artificial Intelligence, vol. 32, no. 1 (2018)
8. Sibai, H., et al.: Safe Reinforcement Learning for Control Systems: A Hybrid Systems Perspective and Case Study (2019). http://publish.illinois.edu/husseinsibai/files/2019/10/Safe_RL_with_Continuous_Dynamics__HSCC2019-4.pdf
9. Zhang, B., et al.: Self-driving via improved DDPG algorithm. Comput. Eng. Appl. **55**(10), 264–270 (2019)
10. Cheng, R., et al.: End-to-end safe reinforcement learning through barrier functions for safety-critical continuous control tasks. In: Proceedings of the AAAI Conference on Artificial Intelligence, vol. 33(01), pp. 3387–3395 (2019). https://doi.org/10.1609/aaai.v33i01.33013387
11. Wang, P., Li, H., Chan, C.: Continuous control for automated lane change behavior based on deep deterministic policy gradient algorithm. In: IEEE Intelligent Vehicles Symposium (IV), Paris, France, 2019, pp. 1454–1460 (2019). https://doi.org/10.1109/IVS.2019.8813903
12. Zhang, L., Liu, J.: Calculation of safe speed of car in curve. Phys. Teacher **25**(7),62–63 (2004)
13. Zang, L., et al.: Modeling and simulation of lateral safety distance of automobile in the bend. J. Chongqing Jiaotong Univ. **219**(04), 15–20 (2020)
14. Bertolazzi, E., Biral, F., et al.: Supporting drivers in keeping safe speed and safe distance: the SASPENCE subproject within the European framework programme 6 integrating project PReVENT. IEEE Trans. Intell. Transp. Syst. **11**(3), 525–538 (2010). https://doi.org/10.1109/TITS.2009.2035925
15. Luo, Q., Xun, L., et al.: Simulation analysis and study on car-following safety distance model based on braking process of leading vehicle. In 2011 9th World Congress on Intelligent Control and Automation, pp. 740–743. https://doi.org/10.1109/WCICA.2011.5970612
16. Silver, D., Heess, N., et al.: Deterministic policy gradient algorithms. In: International Conference on Machine Learning, PMLR, pp. 387–395 (2014)
17. Barto, A.G., et al.: Neuron like elements that can solve difficult learning control problems. IEEE Trans. Syst. Man Cybern. **13**(5), 834–846 (1970)

18. Heess, N., Silver, D., et al.: Actor- critic reinforcement learning with energy-based policies, pp. 45–58. PMLR, In European Workshop on Reinforcement Learning (2013)
19. Nocedal, J., Wright, S.: Numerical optimization. Springer Science & Business Media (2006)
20. Guo, C., Li, D., Zhang, G., et al.: Dynamic interior point method for vehicular traffic optimization. IEEE Trans. Vehicular Technol. **69**(5), 4855–4868 (2020)
21. Boyd, S., Boyd, S.P., Vandenberghe, L.: Convex Optimization. Cambridge University Press, Cambridge (2004)

# Self-organized Hawkes Processes

Shen Yuan[1] and Hongteng Xu[2,3(✉)]

[1] University of Electronic Science and Technology of China, Chengdu, China
[2] Gaoling School of Artificial Intelligence, Renmin University of China,
Beijing, China
`hongtengxu@ruc.edu.cn`
[3] Beijing Key Laboratory of Big Data Management and Analysis Methods,
Beijing, China

**Abstract.** In this paper, we propose a novel self-organized Hawkes process (SOHP) to model complex event sequences based on extremely few observations. Motivated by the fact that the complicated global relations among events are often composed of simple local relations, we model the event sequences by a set of heterogeneous local Hawkes processes rather than a single Hawkes process. In the training phase, we learn the Hawkes processes with a self-organization mechanism, selecting training sequences adaptively for each Hawkes process by a bandit algorithm. The reward used in the algorithm is originally defined based on an optimal transport distance. Additionally, we leverage the superposition property of the Hawkes process to enhance the robustness of our algorithm to the data sparsity problem. We apply our SOHP method to sequential recommendation problems in the continuous-time domain and achieve encouraging performance in various datasets. The code is available at https://github.com/UESTC-DaShenZi/MHP.

**Keywords:** Hawkes process · Self-organization · Bandit algorithm · Optimal transport · Sequential recommendation

## 1 Introduction

Hawkes process (HP) is a powerful mathematical framework for modeling generative mechanisms of event sequences in the continuous-time domain. Since it was applied in modeling the patterns of earthquake [18], its ability to capture exogenous fluctuations of events and endogenous triggering patterns between different event types has made it a popular model in many application scenarios, *e.g.*, high frequency finance [3], social network [11], and recommendation systems [25], etc.

Although Hawkes processes provide competitive solutions to many important problems, their practical applications often suffer from some limitations:

---

H. Xu—Supported in part by Beijing Outstanding Young Scientist Program (No. BJJWZYJH012019100020098), National Natural Science Foundation of China (No. 61832017), and China Unicom Innovation Ecological Cooperation Plan.

L. Fang et al. (Eds.): CICAI 2021, LNAI 13070, pp. 140–151, 2021.
https://doi.org/10.1007/978-3-030-93049-3_12

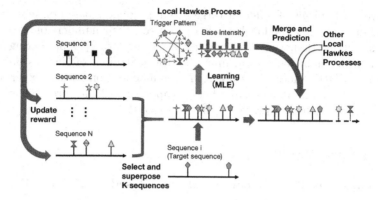

**Fig. 1.** An illustration of the proposed self-organized Hawkes process model.

(i) Real-world sequences may yield different models and the interrelation of the event types within each sequence can be complicated. Therefore, it is often difficult to model their generative mechanisms by a single Hawkes process [15,23]. (ii) Even in the scenario using a single Hawkes process, the number of event types is often huge while the observations can be extremely few in practice, making the learning task challenging [25,26]. Take sequential recommendation problem as an example. A sequential recommendation system needs to explore a huge set of items and recommend attractive ones to each user based on her/his sparse purchasing history. Moreover, the triggering patterns among the items for each user can be personalized, which reflects the diversity of the users' behaviors. Training a simple Hawkes process is often insufficient to handle such a complicated scenario.

To address the problems above, we propose a self-organized Hawkes process (SOHP) model, which learns heterogeneous local Hawkes processes based on subsets of observed event sequences. The proposed model is motivated by a fact that the complex global interrelation between events tend to consist of simple local structures [14]. Thus, it is feasible to capture the triggering patterns in the complicated event sequences by many local Hawkes processes. In particular, Fig. 1 illustrates the principle of our model. We select and superpose $K$ sequences for each target sequence and learn a Hawkes process by maximum likelihood estimation (MLE) in an iterative way. In each iteration, the $K$ sequences are selected based on a bandit algorithm, whose rewards are initialized by their optimal transport distance to the target sequence and updated according to the learned likelihood. Accordingly, the training sequences are organized adaptively as different subsets to support the learning of heterogeneous local Hawkes processes.[1] In this inference phase, we merge the local Hawkes processes and make prediction of future events for each sequence.

This self-organization mechanism adjusts event sequences belonging to different Hawkes processes with training progress, which helps to capture diverse

---

[1] Here, "heterogeneous" means that both the event types of different Hawkes processes and the triggering patterns among the event types are different.

generative mechanisms of different target event sequences based on few observations. Additionally, for the local Hawkes processes, their numbers of event types are much smaller than the total number of the event types appearing in the whole set, which is beneficial for improving the scalability of Hawkes processes. We test our SOHP model in sequential recommendation tasks. Experimental results show that our model owns better capacity and interpretability, which achieves higher prediction precision than state-of-the-art methods.

## 2    Proposed Model

### 2.1    Continuous-Time Recommendation Based on Hawkes Processes

Denote $\{(t_i, c_i)\}_{i=1}^I$ as an event sequence, where the timestamp $t \in [0, T]$, event type $c \in \mathcal{C}$, and $(t_i, c_i)$ represents the $i$-th event with type $c_i$ and at time $t_i$. A temporal point process models the event sequence as a counting process $N = \{N_c(t)|t \in [0, T], c \in \mathcal{C}\}$, where $N_c(t)$ is the number of type-$c$ events occurring till time $t$, and characterizes the expected instantaneous happening rate of the type-$c$ event at time $t$ by an intensity function:

$$\lambda_c(t) := \frac{\mathbb{E}[dN_c(t)|\mathcal{H}_t^{\mathcal{C}}]}{dt}, \ \forall \ c \in \mathcal{C} \text{ and } t \in [0, T]. \tag{1}$$

where $\mathcal{H}_t^{\mathcal{C}} = \{(t_i, c_i)|t_i < t, c_i \in \mathcal{C}\}$ represents historical observations till time $t$.

As a special kind of temporal point process, Hawkes process is able to capture the triggering patterns among the event types in an explicit way. It formulates the intensity function above as

$$\lambda_c(t) = \mu_c + \sum\nolimits_{t_i < t} \phi_{cc_i}(t, t_i) = \mu_c + \sum\nolimits_{t_i < t} a_{cc_i} \kappa(t - t_i), \ \forall \ c \in \mathcal{C}, \tag{2}$$

where $\mu_c$ represents the basic happening rate (a.k.a. base intensity) of the type-$c$ event. $\phi_{cc'}(t, t')$, $t' < t$ and $c, c' \in \mathcal{C}$, is called *impact function*, which represents the influence of the type-$c'$ event at time $t'$ on the type-$c$ event at time $t$. Generally, the impact function $\phi_{cc'}$ is assumed to be shift-invariant and parameterized by $a_{cc'} \kappa(t - t')$, where $a_{cc'} \kappa(t)$ is a weighted decay function. For convenience, we can represent a Hawkes process as $\mathrm{HP}(\boldsymbol{\mu}, \boldsymbol{A})$, where $\boldsymbol{\mu} = [\mu_c] \in \mathbb{R}^{|\mathcal{C}|}$ represents the base intensity of the event types and $\boldsymbol{A} = [a_{cc'}] \in \mathbb{R}^{|\mathcal{C}| \times |\mathcal{C}|}$ is the infectivity matrix capturing the triggering patterns among the event types.

Hawkes process owns many useful properties. Firstly, the infectivity matrix corresponds to the adjacency matrix of the *Granger causality graph* of event types [24], which represents the self- and mutually-triggering patterns hidden in event sequences explicitly. Secondly, the *superposition property* of Hawkes process [25] shows that when learning a Hawkes process from event sequences, we can superpose observed event sequences to obtain an event sequence with much denser events and learn the Hawkes process with a tighter bound of excess risk [25], which improves the robustness to the data sparsity problem.

Given a set of event sequences $\mathcal{N} = \{N^u\}_{u \in \mathcal{U}}$, we can learn the Hawkes process by the maximum likelihood estimation (MLE):

$$\min_{\theta} -\log \mathcal{L}(\mathcal{N}; \theta) + \gamma \mathcal{R}(\theta), \tag{3}$$

where $\theta = \{\mu, A\}$ represents the model parameter, $\mathcal{L}(\mathcal{N}; \theta)$ is the likelihood function of the event sequences [10]:

$$\mathcal{L}(\mathcal{N}; \lambda) = \prod_{u \in \mathcal{U}} \left( \prod_{(c_i^u, t_i^u) \in N^u} \lambda_{c_i^u}(t_i^u) \times \exp\left(-\sum_{c \in \mathcal{C}} \int_0^T \lambda_c^u(s) \mathrm{d}s\right) \right), \tag{4}$$

and $\mathcal{R}(\cdot)$ represents the regularization term imposed on the model parameters, such as the sparsity and the low-rank regularizers on $A$ [32]. This method applies the idea that the observed events are most probable and updates parameters to maximize the likelihood of the observed events.

After the intensity function is obtained, we could predict the type of next event in the future. In particular, given the history till time $t$ (i.e., $\mathcal{H}_t^{\mathcal{C}}$), the probability of the type-$c$ event at $t + \Delta t$ could be computed by [28]:

$$p(c|t + \Delta t, \mathcal{H}_t^{\mathcal{C}}) = \frac{\lambda_c(t + \Delta t)}{\sum_{c' \in \mathcal{C}} \lambda_{c'}(t + \Delta t)}. \tag{5}$$

The Hawkes process above has the potentials to capture purchasing behaviors of the users at the e-commercial platform and construct a sequential recommendation system in the continuous-time domain [25]. In such a situation, the event sequences correspond to the purchasing behaviors of different users and the event types correspond to different items. By learning a Hawkes process, we model the expected instantaneous purchasing rate of the users over time, where the infectivity matrix $A$ reflects the triggering patterns among the items and the base intensity $\mu$ reflects the intrinsic popularity of the items.

However, on one hand, the real-world behaviors of the users may yield different generative mechanisms, which correspond to different Hawkes processes. On the other hand, the purchasing behaviors of each individual are often very sparse, which are insufficient to learn a Hawkes process model. To overcome this conflict, we propose the self-organized Hawkes process model below, learning multiple local Hawkes processes robustly by organizing the event sequences of different users in an adaptive way.

## 2.2   Self-organized Hawkes Processes

As aforementioned, the proposed self-organized Hawkes process model is motivated by the work in [14], which captures the complicated global relations among all the items by learning and merging simple local relations among subsets of items. Therefore, in our SOHP model, each event sequence corresponds to a local Hawkes process.

The key challenge is how to solve the data sparsity problem—in the training phase, we need to suppress the risk of over-fitting caused by insufficient training events, while in the inference phase, we need to explore a sufficient large item space, predicting the items that may never appear in the sequence. Fortunately, the superposition property of Hawkes process helps us to build an algorithmic framework that is robust to this challenge above.

**Theorem 1 (Superposition Property** [25]). *For a set of independent Hawkes processes with a shared infectivity matrix, i.e., $\{N^u \sim HP(\boldsymbol{\mu}^u, \boldsymbol{A})\}_{u \in \mathcal{U}}$, the superposition of their sequences satisfies $\sum_{u \in \mathcal{U}} N^u \sim HP(\sum_{u \in \mathcal{U}} \boldsymbol{\mu}^u, \boldsymbol{A})$.*

With the help of this property, we can superpose sparse event sequences randomly and learn the infectivity matrix with much denser training events. The work in [25] demonstrates that this superposition-based strategy helps us to achieve a tighter bound of excess risk in the learning phase.

However, when learning multiple heterogeneous Hawkes processes in a complicated scenario, the event sequences belong to different Hawkes processes that have various infectivity matrices. In such a situation, superposing the event sequences randomly is likely to disobey the assumption imposed in Theorem 1 (*i.e.*, the Hawkes processes have the same infectivity matrix). **To overcome this problem, we design a new self-organization mechanism, selecting event sequences for each local Hawkes process and adjusting the selection with the training progress.** Mathematically, denote $\mathcal{N} = \{N^u\}_{u \in \mathcal{U}}$ as a set of real-world event sequences. Our learning task becomes

$$\min_{\{\theta^u\}_{u \in \mathcal{U}}} \min_{\mathcal{N}^u \subset \mathcal{N}} - \sum_{u \in \mathcal{U}} \frac{1}{|N^u \cup \mathcal{N}^u|} \log \mathcal{L}(N^u \cup \mathcal{N}^u; \theta^u), \qquad (6)$$

where $\mathcal{N}^u = \{N^{s_1}; \ldots; N^{s_K}\}$ represents the $K$ neighbors of the $u$-th sequence. $N^u \cup \mathcal{N}^u$ constructs a subset of event sequences for learning the $u$-th local Hawkes processes. Here, we aim at optimizing the parameters of the Hawkes processes $\{\theta^u\}_{u \in \mathcal{U}}$ and the selection of their training sets jointly.

## 3   Learning Algorithm

### 3.1   A Reward-Augmented Bandit Algorithm

The learning problem in (6) is NP-hard. Therefore, we propose a novel reward-augmented bandit algorithm to solve it heuristically. Intuitively, we hope that 1) the selected sequences are similar to the target sequence; 2) the selected sequences own some randomness to avoid the over-fitting problem. To achieve this aim, for each target sequence, we treat the selection of its neighbors (*i.e.*, the training set of a Hawkes process) as a multi-armed bandit problem [2], selecting their neighbors according to the potential *rewards*. The accumulation of the rewards formulates a benefit matrix $\boldsymbol{B} = [b_{u,k}] \in \mathbb{R}^{|\mathcal{U}| \times |\mathcal{U}|}$, where $b_{u,k}$ represents the benefit from selecting the $k$-th sequence for the $u$-th Hawkes process.

The key of our algorithm is the design and the update of the benefit matrix. Initially, we define the initial benefit based on the optimal transport distance between event sequences [27]. This distance is applicable to the sequences with heterogeneous event types. Specifically, for $N^u = \{N_i^u\}_{i \in \mathcal{C}_u}$ and $N^v = \{N_j^v\}_{j \in \mathcal{C}_v}$, where $\mathcal{C}_u$ and $\mathcal{C}_v$ are respectively the sets of event types appearing in $N^u$ and

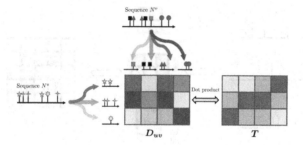

**Fig. 2.** The optimal transport distance between heterogeneous event sequences.

---

**Algorithm 1.** Learning SOHP via a reward-augmented bandit algorithm

---

**Input:** Event sequences $\{N^u\}_{u\in\mathcal{U}}$, distance matrix $\boldsymbol{D} = [d(N^u, N^v)] \in \mathbb{R}^{|\mathcal{U}|\times|\mathcal{U}|}$, maximum iterations $L$, the number of neighbors $K$, learning rate $\alpha$.
**Output:** benefit matrix $\boldsymbol{B} = [b_{u,k}] \in \mathbb{R}^{|\mathcal{U}|\times|\mathcal{U}|}$ and model parameters $\{\boldsymbol{\theta}^u\}_{u\in\mathcal{U}}$.

1: **for** $u = 1 : |\mathcal{U}|$ **do**
2:     Initialize $b_{u,k} = \max_v d(N^u, N^v) - d(N^u, N^k)$ for $k \in \mathcal{U}$.
3:     **for** $l = 1 : L$ **do**
4:         **if** $l < L$ **then**
5:             Set $\boldsymbol{p} = [\frac{b_{u,1}}{\sum_i b_{u,i}}, .., \frac{b_{u,|\mathcal{U}|}}{\sum_i b_{u,i}}]$, sample $\{N^{s_1}, .., N^{s_K}\}$ from $\mathcal{N}$ with $\boldsymbol{p}$.
6:         **else**
7:             Select $\{N^{s_1}, .., N^{s_K}\}$ with the $K$ highest benefits.
8:         **end if**
9:         Learn the model parameter $\boldsymbol{\theta}^u$ from $\{N^u\} \cup \{N^{s_1}, .., N^{s_K}\}$ by (3).
10:        **for** $k = s_1 : s_K$ **do**
11:           $b_{u,s_k} = b_{u,s_k} + \alpha\mathcal{L}(N^{s_k}; \boldsymbol{\theta}^u)$
12:        **end for**
13:     **end for**
14: **end for**

---

$N^v$, and $N_i^u$ is the counting process associated with the $i$-th event types in $N^u$. Then, the optimal transport distance between $N^u$ and $N^v$ is defined as

$$d(N^u, N^v) := \min_{\boldsymbol{T}\in\Pi(\frac{1}{|\mathcal{C}_u|}\mathbf{1}_{|\mathcal{C}_u|}, \frac{1}{|\mathcal{C}_v|}\mathbf{1}_{|\mathcal{C}_v|})} \sum_{i\in\mathcal{C}_u, j\in\mathcal{C}_v} T_{ij}d(N_i^u, N_j^v)$$
$$= \min_{\boldsymbol{T}\in\Pi(\frac{1}{|\mathcal{C}_u|}\mathbf{1}_{|\mathcal{C}_u|}, \frac{1}{|\mathcal{C}_v|}\mathbf{1}_{|\mathcal{C}_v|})} \langle \boldsymbol{D}_{uv}, \boldsymbol{T} \rangle \qquad (7)$$

where $\Pi(\frac{1}{|\mathcal{C}_u|}\mathbf{1}_{|\mathcal{C}_u|}, \frac{1}{|\mathcal{C}_v|}\mathbf{1}_{|\mathcal{C}_v|}) = \{\boldsymbol{T} \geq 0 | \boldsymbol{T}\mathbf{1} = \frac{1}{|\mathcal{C}_u|}\mathbf{1}_{|\mathcal{C}_u|}, \boldsymbol{T}^\top\mathbf{1} = \frac{1}{|\mathcal{C}_v|}\mathbf{1}_{|\mathcal{C}_v|}\}$ represents the set of joint distributions with marginals $\frac{1}{|\mathcal{C}_u|}\mathbf{1}_{|\mathcal{C}_u|}$ and $\frac{1}{|\mathcal{C}_v|}\mathbf{1}_{|\mathcal{C}_v|}$. $\boldsymbol{D}_{uv} = [d(N_i^u, N_j^v)] \in \mathbb{R}^{|\mathcal{C}_u|\times|\mathcal{C}_v|}$ is a distance matrix, where $d(N_i^u, N_j^v) = \frac{1}{T}\int_0^T |N_i^u(t) - N_j^v(t)|\mathrm{d}t$ represents the discrepancy between the sequence of the type-$i$ events and that of the type-$j$ events. The matrix $\boldsymbol{T} = [T_{ij}]$ that minimizes (7) is called the optimal transport matrix. Figure 2 illustrates the optimal transport distance. The optimal transport distance can be calculated efficiently by the Sinkhorn scaling algorithm [9].

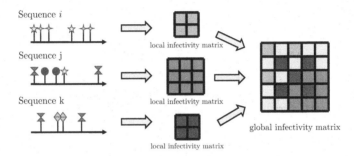

**Fig. 3.** An illustration of merging local infectivity matrices. The purple ones correspond to the overlapped event types, whose values are accumulated together. (Color figure online)

Accordingly, we initialize the benefit $b_{u,k}$ as $\max_v d(N^u, N^v) - d(N^u, N^k)$. In the training phase, we treat the normalized benefits as the probabilities of the sequences and select them accordingly. Then, we regard the likelihood of selected event sequences in each iteration as the intermediate reward and update the corresponding benefit by accumulating the reward accordingly. In the end of iteration, we select $K$ sequences with the highest benefits for each sequence $N^u$.[2] The steps of our reward-augmented bandit algorithm are shown in Algorithm 1. It should be noted that this algorithm reduces the computational complexity of learning Hawkes process: in each iteration, we only learn each Hawkes process based on the superposition of $K$ sequences, which greatly reduced the number of event types for each Hawkes process.

### 3.2   Merging Learned Hawkes Processes

In the inference phase, instead of leveraging the intensity function of each local Hawkes process to make predictions, we first merge the learned infectivity matrices by superposition operations, as shown in Fig. 3. Then, for each sequence, we predict its future events based on its local base intensity and the global infectivity matrix. Note that in the sequential recommendation scenario this inference strategy achieves an exploration-exploitation trade-off: the local base intensity reveals the preference of the user on purchased items while using the global infectivity matrix helps the system to explore the items that she never saw but might be interested in.

## 4   Related Work

**Sequential Recommendation.** By modeling the sequential behaviors from user historical records, sequential recommendation predicts future interests and

---

[2] We tried more sophisticated bandit algorithm like the Upper Confidence Bound (UCB) method [1] to select the sequences, but our experimental results show that the proposed greedy algorithm achieves the best performance.

recommend items. Besides the shopping recommendation, sequential recommendation has also been widely used in various application scenarios, such as web recommendation [31], music recommendation [6], and Point-of-Interest recommendation [7], etc. These years, many cutting-edge techniques have been applied into the sequential recommend, *e.g.*, FPMC [20] integrated matrix factorization and Markov chains, HRM [22] regarded representation learning as latent factors, they modeled the sequential behavior patterns via every two contiguous historical records. DREAM [29] is based on recurrent neural network (RNN) and learns the global sequential behavior patterns. These models all captured the contiguous behavior information without taking time interval between them into considering. Instead, we leverage the time stamps of every behavior to calculate efficiently the mutual effect of them, achieving continuous-time recommendation.

**Hawkes Process.** Because of its effectiveness on modeling the triggering patterns among real-world events, Hawkes process has been widely used in many scenarios, such as high frequency finance [3], and fake news mitigation [12]. Meanwhile, many variants of Hawkes process have been researched and developed, such as the Hawkes process with self-attention mechanisms [30,33]. Recently, the superposition of Hawkes process has been verified to be effective in both theory and experiments [25]. The model applied a random algorithm to select sequences and learned a single Hawkes process. Our work, however, demonstrate that making superposition with the help of a bandit algorithm is more valid for learning multiple Hawkes processes.

**Multi-armed Bandit Problem.** The multi-armed bandit problem denotes a problem where we need to make choices to maximize expected gain under some constraints [21]. These years, many bandit algorithms have been proposed, such as the Upper Confidence Bounds algorithm [5,8], adaptive epsilon-greedy strategy based on Bayesian ensembles [13], and behavior constrained Thompson Sampling [4], etc. In this paper, we make an attempt to apply the bandit algorithm to select training sequences for Hawkes processes.

## 5 Experiments

We experiment on the Amazon review dataset [16]. This dataset contains product reviews from Amazon spanning May 1996–July 2014. We select five product categories as our datasets to evaluate our model, including "Instant Video", "Musical Instruments", "Video Games", "Baby" and "Patio, Lawn and Garden". To be specific, we preprocess the datasets as follows. We select those items with more than 40 reviews. Then, their users need to satisfy three conditions: (i) the ratings they gave to these items are bigger than 4; (ii) there are at most 3 reviews spanning January 2014–April 2014; (iii) there are at least 1 review from April 2014 to July 2014. The statistics of the final datasets are shown in Table 1. After learning the behaviors of users from January 2014 to April 2014, our model predicts items for them from April 2014 to July 2014.

**Table 1.** Statistics of our datasets

| Categories | Musical instruments | Baby | Video games | Garden | Instant video |
|---|---|---|---|---|---|
| #Users | 471 | 1979 | 2142 | 1812 | 5948 |
| #Items | 678 | 2134 | 2104 | 2064 | 1344 |
| #Ratings | 1218 | 6070 | 6126 | 4976 | 15470 |

To demonstrate the superiority of our SOHP model, we adopt the following state-of-the-art methods as baselines for comparisons. **SVD**: Singular value decomposition, a classical method from linear algebra is getting popular in recommender systems; **kNN**: user-based K-nearest neighbors algorithm, a nonparametric classification and regression method; **BPR**: Bayesian personalized ranking [19], a popular method for top-N ranking recommendation; **SLIM**: Sparse linear method [17], a simple and effective method for top-N recommendation; **FPMC**: Factorized personalized Markov chains [20], one of the stat-of-the-art models for sequential recommendation based on matrix factorization and Markov chains. Each item is regarded as a basket in this model. **SHP**: The superposed Hawkes process [25] that learn a single Hawkes process by randomly superpose event sequences. For each model, denote the top-N recommended items for user $u$ as $R^u = \{r_1^u, \ldots, r_N^u\}$, where $r_i^u$ is ranked at the $i$-th position, and the set of real purchased items $T^u$, respectively. We use the top-N precision, recall and F1 score as the measurements.

When implementing our method, the number of iterations $L$ for each user is set as 20 and the learning rate $\alpha$ is set as 0.1. We primarily set the number of neighbor event sequences $K = 10$, and the effect of setting different $K$ is studied in the experiments. When learning the superposition of Hawkes processes, we empirically set the decay function $\kappa(t)$ as $\exp(-\beta t)$, where $\beta = 3e-4$, and use no regularization term. For each user, 10 items are recommended.

Table 2 summarizes the performance for the baselines and our method on the five datasets. For all the datasets except "Instant Video", our SOHP model achieves better performance than its competitors. Specifically, in the category "Musical instrument", the result of our model is 13% higher than that of the best baseline (*i.e.*, FPMC), and the difference would be up to 48% by tuning $K$ in the subsequent experiments. The results indicate the effectiveness of our model for sequential recommendation.

Essentially, for the users, the more neighbor event sequences are considered, the more intersections their training sets have. Accordingly the local Hawkes processes learned based on the sets may more similar. In such a situation, our SOHP model may tend to recommend different users similar items. We are curious about whether and how the number of neighbors affects the recommendation results. To do so, we study the performance of our model on F1@10 by tuning $K$ in the range of $\{2, 5, 10, 20, 50, 100\}$. Results are shown in Fig. 4.

For the categories "Musical Instruments" and "Baby", as the increase of $K$, the performance of them gets better. At this moment, the model learns more

**Table 2.** Summary of the performance for baselines and our model

| Datasets | Musical instruments | | | Baby | | | Video games | | | Garden | | | Instant video | | |
|---|---|---|---|---|---|---|---|---|---|---|---|---|---|---|---|
| Measures@10(%) | P | R | F1 | P | R | F1 | P | R | F1 | P | R | F1 | P | R | F1 |
| SVD | 0.106 | 1.061 | 0.193 | 0.121 | 0.735 | 0.207 | 0.065 | 0.498 | 0.113 | 0.055 | 0.395 | 0.094 | 0.126 | 1.052 | 0.221 |
| kNN | 0.382 | 2.671 | 0.649 | 0.389 | 2.513 | 0.638 | 0.661 | 4.915 | 1.163 | 0.237 | 1.751 | 0.405 | 0.817 | 6.901 | 1.435 |
| BPR | 0.467 | 3.750 | 0.811 | 0.389 | 2.469 | 0.635 | 0.658 | 4.864 | 1.112 | 0.110 | 0.762 | 0.185 | 0.859 | 7.049 | 1.503 |
| SLIM | 0.212 | 1.351 | 0.347 | 0.111 | 0.712 | 0.180 | 0.499 | 3.595 | 0.835 | 0.242 | 1.544 | 0.401 | **1.333** | **11.428** | **2.351** |
| FPMC | 0.594 | 4.193 | 1.006 | 0.283 | 1.912 | 0.470 | 0.556 | 3.799 | 0.927 | 0.171 | 1.117 | 0.285 | 0.931 | 7.413 | 1.622 |
| SHP | 0.361 | 2.406 | 0.604 | 0.258 | 1.734 | 0.432 | 0.317 | 2.037 | 0.525 | 0.199 | 1.350 | 0.331 | 0.933 | 7.406 | 1.623 |
| Our methods | **0.658** | **5.149** | **1.138** | **0.389** | **2.640** | **0.651** | **0.700** | **5.108** | **1.180** | **0.248** | **1.801** | **0.436** | 0.999 | 7.911 | 1.738 |

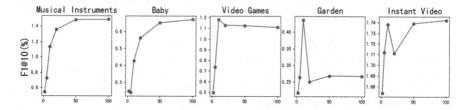

**Fig. 4.** Performance of our model under different number of neighbors $K$.

popular interests and will be more inclined to recommend the most popular items. For the category "Musical Instrument", the users tend to focus on a limited number of items and purchase similar ones. In the case of the category "Baby", babies cannot comment on the items, so parents often choose the most popular products.

For the categories "Video Games" and "Garden", the performance becomes the best at $K \approx 10$ and degrades a lot when further increasing neighbors. A potential reason for this phenomenon is that the purchasing behaviors in these two categories have diverse patterns. As aforementioned, when there are too many neighbors, our model will tend to recommend the most popular items to all the users, which is unsuitable for these two categories.

For the category "Instant Video", however, we find that our model is inferior to the SLIM method, and the performance is stable with respect to the number of neighbors. According to our analysis, a possible reason for this phenomenon is that for most users, their choice of instant video tend to be influenced by Amazon's recommendations or the most popular video list, which may not be suitable for sequential recommendation. Evidence supporting this explanation is that the FPMC, which is also a sequential recommendation model, has poor performance as well for this category.

## 6    Conclusion

In this paper, we proposed a framework combining bandit algorithm with Hawkes processes, which provides a new way to learn complicated sequential models robustly from insufficient observations. We designed the reward of the Bandit algorithm, and exerted the greedy strategy to update the reward iteratively. The proposed model achieved encouraging performance on continuous-time sequential recommendation. In the future, we plan to design new formulation of rewards and update rewards by more efficient strategies. Additionally, we will make attempts to set the number of neighbor event sequences adaptively to further improve our model.

# References

1. Audibert, J.Y., Munos, R., Szepesvári, C.: Exploration-exploitation tradeoff using variance estimates in multi-armed bandits. Theoret. Comput. Sci. **410**(19), 1876–1902 (2009)
2. Auer, P., Cesa-Bianchi, N., Freund, Y., Schapire, R.E.: The nonstochastic multi-armed bandit problem. SIAM J. Comput. **32**(1), 48–77 (2002)
3. Bacry, E., Mastromatteo, I., Muzy, J.F.: Hawkes processes in finance. Market Microstruct. Liquidity **1**(01), 1550005 (2015)
4. Balakrishnan, A., Bouneffouf, D., Mattei, N., Rossi, F.: Incorporating behavioral constraints in online AI systems. In: Proceedings of the AAAI Conference on Artificial Intelligence, vol. 33, pp. 3–11 (2019)
5. Bouneffouf, D., Parthasarathy, S., Samulowitz, H., Wistub, M.: Optimal exploitation of clustering and history information in multi-armed bandit. arXiv preprint arXiv:1906.03979 (2019)
6. Chen, S., Moore, J.L., Turnbull, D., Joachims, T.: Playlist prediction via metric embedding. In: Proceedings of the 18th ACM SIGKDD International Conference on Knowledge Discovery and Data Mining, pp. 714–722 (2012)
7. Cheng, C., Yang, H., Lyu, M.R., King, I.: Where you like to go next: successive point-of-interest recommendation. In: Twenty-Third International Joint Conference on Artificial Intelligence (2013)
8. Chu, W., Li, L., Reyzin, L., Schapire, R.: Contextual bandits with linear payoff functions. In: Proceedings of the Fourteenth International Conference on Artificial Intelligence and Statistics, pp. 208–214. JMLR Workshop and Conference Proceedings (2011)
9. Cuturi, M.: Sinkhorn distances: lightspeed computation of optimal transport. Adv. Neural. Inf. Process. Syst. **26**, 2292–2300 (2013)
10. Daley, D.J., Vere-Jones, D., et al.: An Introduction to the Theory of Point Processes: Volume II: General Theory and Structure. Springer, New York (2008)
11. Farajtabar, M., Rodriguez, M.G., Zamani, M., Du, N., Zha, H., Song, L.: Back to the past: source identification in diffusion networks from partially observed cascades. In: Artificial Intelligence and Statistics, pp. 232–240. PMLR (2015)
12. Farajtabar, M., et al.: Fake news mitigation via point process based intervention. In: International Conference on Machine Learning, pp. 1097–1106. PMLR (2017)
13. Gimelfarb, M., Sanner, S., Lee, C.G.: {\epsilon}-bmc: a bayesian ensemble approach to epsilon-greedy exploration in model-free reinforcement learning. arXiv preprint arXiv:2007.00869 (2020)
14. Lee, J., Bengio, S., Kim, S., Lebanon, G., Singer, Y.: Local collaborative ranking. In: Proceedings of the 23rd International Conference on World Wide Web, pp. 85–96 (2014)
15. Luo, D., Xu, H., Zhen, Y., Ning, X., Zha, H., Yang, X., Zhang, W.: Multi-task multi-dimensional hawkes processes for modeling event sequences. In: Proceedings of the 24th International Conference on Artificial Intelligence, pp. 3685–3691 (2015)

16. McAuley, J., Targett, C., Shi, Q., Van Den Hengel, A.: Image-based recommendations on styles and substitutes. In: Proceedings of the 38th International ACM SIGIR Conference on Research and Development in Information Retrieval, pp. 43–52 (2015)
17. Ning, X., Karypis, G.: Slim: sparse linear methods for top-n recommender systems. In: 2011 IEEE 11th International Conference on Data Mining, pp. 497–506. IEEE (2011)
18. Ogata, Y.: Statistical models for earthquake occurrences and residual analysis for point processes. J. Am. Stat. Assoc. **83**(401), 9–27 (1988)
19. Rendle, S., Freudenthaler, C., Gantner, Z., Schmidt-Thieme, L.: Bpr: Bayesian personalized ranking from implicit feedback. arXiv preprint arXiv:1205.2618 (2012)
20. Rendle, S., Freudenthaler, C., Schmidt-Thieme, L.: Factorizing personalized markov chains for next-basket recommendation. In: Proceedings of the 19th International Conference on World Wide Web, pp. 811–820 (2010)
21. Robbins, H.: Some aspects of the sequential design of experiments. Bull. Am. Math. Soc. **58**(5), 527–535 (1952)
22. Wang, P., Guo, J., Lan, Y., Xu, J., Wan, S., Cheng, X.: Learning hierarchical representation model for nextbasket recommendation. In: Proceedings of the 38th International ACM SIGIR Conference on Research and Development in Information Retrieval, pp. 403–412 (2015)
23. Xu, H., Carin, L., Zha, H.: Learning registered point processes from idiosyncratic observations. In: International Conference on Machine Learning, pp. 5443–5452. PMLR (2018)
24. Xu, H., Farajtabar, M., Zha, H.: Learning granger causality for hawkes processes. In: International Conference on Machine Learning, pp. 1717–1726. PMLR (2016)
25. Xu, H., Luo, D., Chen, X., Carin, L.: Benefits from superposed hawkes processes. In: International Conference on Artificial Intelligence and Statistics, pp. 623–631. PMLR (2018)
26. Xu, H., Luo, D., Zha, H.: Learning hawkes processes from short doubly-censored event sequences. In: International Conference on Machine Learning, pp. 3831–3840. PMLR (2017)
27. Xu, H., Luo, D., Zha, H.: Hawkes processes on graphons. arXiv preprint arXiv:2102.02741 (2021)
28. Xu, H., Wu, W., Nemati, S., Zha, H.: Patient flow prediction via discriminative learning of mutually-correcting processes. IEEE Trans. Knowl. Data Eng. **29**(1), 157–171 (2016)
29. Yu, F., Liu, Q., Wu, S., Wang, L., Tan, T.: A dynamic recurrent model for next basket recommendation. In: Proceedings of the 39th International ACM SIGIR Conference on Research and Development in Information Retrieval, pp. 729–732 (2016)
30. Zhang, Q., Lipani, A., Kirnap, O., Yilmaz, E.: Self-attentive hawkes processes. arXiv preprint arXiv:1907.07561 (2019)
31. Zhang, Y., Zhang, M., Liu, Y., Tat-Seng, C., Zhang, Y., Ma, S.: Task-based recommendation on a web-scale. In: 2015 IEEE International Conference on Big Data (Big Data), pp. 827–836. IEEE (2015)
32. Zhou, K., Zha, H., Song, L.: Learning social infectivity in sparse low-rank networks using multi-dimensional hawkes processes. In: Artificial Intelligence and Statistics, pp. 641–649. PMLR (2013)
33. Zuo, S., Jiang, H., Li, Z., Zhao, T., Zha, H.: Transformer hawkes process. In: International Conference on Machine Learning, pp. 11692–11702. PMLR (2020)

# Causal Inference with Heterogeneous Confounding Data: A Penalty Approach

Zhaofeng Lu[(✉)] and Bo Fu[(✉)]

Fudan University, Shanghai 200433, China
{18210980010,fu}@fudan.edu.cn

**Abstract.** Causal inference directly explores the causality among variables, in which average causal effect estimation is a fundamental task. But for heterogeneous confounding data, most previous methods fail to estimate causal effect accurately when confounders among heterogeneous subgroups are more complicated, as they may ignore local balance. Therefore, we propose a novel Heterogeneous subGroup Balance Adaptive Method (HGBAM), in which a penalty is elaborately proposed by employing the balance condition of covariates and heterogeneity among subgroups. The penalty constructs preferable balance constraints that facilitate better causal variable selection and de-confounding. In addition, a partially sharing parameters structure is designed, in which the confounders information among different subgroups could be investigated together. The structure helps to make full use of similarities and reserve heterogeneity among subgroups adaptively. Thus, our method will contribute to estimating multi-subgroups causal effect simultaneously meanwhile achieving the local and global balance. Besides, our theoretical analysis suggests that the method can make asymptotically unbiased estimation. The experimental results on both synthetic and real-world data demonstrate the efficacies of the proposed method for heterogeneous causal effect estimation.

**Keywords:** Causal inference · Heterogeneous causal effect · Penalty approach · Covariates balance

## 1 Introduction

Most machine learning (ML) algorithms employing the associations between variables to make predictions, in contrast, causal inference aims to explore causality between variables and assesses the causal effect of a treatment. It is well-known that "association doesn't imply causality". So some ML algorithms may have shortcomings in stability and explainability [9,24]. Learning causality from observational data enable us to answer "counterfactual" question, such as "would the patient recover had he/she received another treatment?" in medical data mining. Besides, causal inference could improve the performance of ML and deep learning methods in domain adaptation [9,27], removing confounding

© Springer Nature Switzerland AG 2021
L. Fang et al. (Eds.): CICAI 2021, LNAI 13070, pp. 152–163, 2021.
https://doi.org/10.1007/978-3-030-93049-3_13

bias in pre-training models [25,26]. And causal inference has many important applications in statistic, econometrics, computer science, etc. [24].

A fundamental task in causal inference is to estimate causal effect from observational data. As the distribution of covariates between treat units and control units is not balanced due to some confounders available, it is hard to obtain average causal/treatment effect estimation or counterfactual outcome prediction directly. There are mainly two frameworks, the potential outcome framework [19] and structure causal model (SCM) [16]. As the two frameworks could make full use of observational data to explore causality, motivated by them, many methods have been proposed.

Many related literatures typically focus on estimating average causal effect, however, there are also some cases where heterogeneous causal effects are present, and different subgroups may possess different causal effects. For example, the same treatment may have different effects on subtypes of a complex disease. When heterogeneous data is present, confounding among subgroups is more complicated. Subgroups may have their unique confounders and share similar confounders. So most previous methods fail to estimate subgroup causal effect accurately [3,12] and they could not achieve subgroup balance and global balance simultaneously.

Therefore, in this paper, we focus on heterogeneous causal effect estimation and propose a novel Heterogeneous subGroup Balance Adaptive Metho (HGBAM). For purpose of employing similarities among different subgroups meanwhile reserving heterogeneity, we partially pools subgroups together to share information meanwhile conducts elaborately designed penalty to impose balance constraints and select causal variables. A special parameters structure is established, and it can explore heterogeneity adaptively. Such pooling may result in some weakly associated covariates and more complicated confounding effect, which makes heterogeneous causal inference more challenging. Thus, in the penalty, heterogeneity and balance criterions are taken into consideration, varying with different subgroups and covariates. In this way, the shared confounders information and designed penalty enable the method to eliminate confounding in heterogeneous subgroups and achieve subgroup balance. By theoretical analysis and empirical experiments, we establish that the proposed method can achieve better performance in the heterogeneous data.

The contributions of this paper are summarized as following:

- For heterogeneous data, we propose a novel Heterogeneous subGroup Balance Adaptive Method (HGBAM) to estimate multi-subgroup causal effects simultaneously, where the connections and differences among subgroups are fully employed by the parameters sharing structure.
- Bringing in balance criterions and heterogeneity, an adaptive penalty term is specially designed for heterogeneous causal inference. It contributes to eliminating complicated confounding and achieving distribution balance in both local and global data.
- We give theoretical analysis to prove its good property for addressing the problem. We conduct various experiments in synthetic and real-world data,

and the results demonstrate the advantages of our method. Especially, the performance in complicated simulation scenarios validates its advantages.

## 2  Related Work

Two main challenges of estimating causal effect are unbalanced distribution and selection bias. Under assumption of ignorability, some methods have been proposed to achieve balance by propensity score re-weighting or matching [19,21]. There are some cases where models may be slightly misspecified. So, in order to improve robustness, augmented estimator [17] and weighted regression [5] are proposed. In addition, there are some literatures directly focusing on optimizing balance criterions. Imai et al. [7] proposed covariates balance method by optimizing balance function induced from log-likelihood. By analyzing theoretical bias and efficiency, Fan et al. [4] provided the optimal choice of the covariates balance function and improved the method. Also, some papers investigated direct estimation of sample weights by entropy balance [6], approximate residual balance [1]. The above-mentioned methods provide effective approaches to estimate average causal effect, but for heterogeneous data, they could not estimate heterogeneous causal effect accurately.

As not all covariates are confounders, Brookhart et al. [2] proposed variable selection criterion for causal inference. Unlike the selection procedure in prediction tasks of ML, we should consider the causalities of covariates, treatment and potential outcome. So Shortreed et al. [20] introduced outcome adaptive method to select appropriate covariates by adaptive penalty. In addition, group lasso was applied to conduct simultaneously modelling of outcome and treatment [8].

Recently, there is a trend to combine causal inference and ML. For multiple domains, causal inference was employed to seek invariant subset [18] or explore invariant conditional distribution [13] to improve performance on new test task. Kuang et al. [9] applied direct balance scheme to make stable prediction across unknown domains. Besides, eliminating confounders caused by background knowledge could make improvements without extra complexity [26,27].

## 3  Heterogeneous subGroup Balance Adaptive Method

### 3.1  Problem Formulation

For heterogeneous data, assuming it could be divided into $K$ subgroups based on an indicator $\mathcal{G}$. Denote $\mathcal{G}_i \in \{1, 2, \cdots, K\}$ the subgroup label of sample $i$. $X^{(k)} \in \mathbb{R}^{N^{(k)} \times P}$, $Y^{(k)} \in \mathbb{R}^{N^{(k)}}$, $T^{(k)} \in \{0, 1\}$ are respectively pre-treatment covariates, potential outcome and treatment of subgroup $k$, where $N^{(k)}$ is the sample size of subgroup $k$ and $P$ is the number of covariates, $\sum_k N^{(k)} = N$. We are interested in heterogeneous causal effect $\tau^{(k)} = \mathbb{E}\{Y(1) - Y(0) \mid X, \mathcal{G} = k\}$, for different subgroups $k, h$, satisfying $\tau^{(k)} \neq \tau^{(h)}$.

Following work [3], we assume basic assumptions (SUTVA, ignorability, overlap) of causal inference still hold for heterogeneous data. For purpose of balancing distribution of confounders, we prefer to employ propensity score $\pi(X)$, which is defined as the probability of receiving treatment given covariates, $\pi(X) = \Pr(T = 1 \mid X)$. And we have the following proposition.

**Proposition 1.** *The propensity score balances the distribution of $X$ in heterogeneous subgroups.*

$$X \perp T \mid \pi(X, \mathcal{G}), \quad \mathcal{G} = 1, 2, \cdots K.$$

### 3.2   Heterogeneous subGroup Balance Adaptive Method

In order to detect similarities and reserve heterogeneity adaptively, we apply a partially sharing structure $\beta^{(k)} = \mu + \gamma^{(k)}$ to formulate the involved parameters of subgroup $k$, where $\mu$ represents the similarities and $\gamma^{(k)}$ represents heterogeneity.

The confounders in different subgroups may be related, so confounding effects are more complicated. It is quite hard to achieve subgroup and global balance. To address the challenges, we partially pool different subgroups together and conduct elaborately designed penalty constraints to guarantee that the model can explore the balance and heterogeneity of each subgroup adaptively. Therefore, we propose Heterogeneous subGroup Balance Adaptive Method (HGBAM) as:

$$\underset{\mu,\ \gamma}{\arg\min} \sum_{k=1}^{K} \mathcal{L}\left(T^{(k)}, X^{(k)}\right) + \lambda \left[\|\mu\|_1 + \sum_{k=1}^{K}\sum_{j=1}^{P} g^{(k)} w_j^{(k)} \left|\gamma_j^{(k)}\right|\right] \tag{1}$$

where $\mathcal{L}()$ is the logistic regression loss, and note that it can also take other appropriate forms, such as calibration loss $\mathcal{L}() = \mathbb{E}(T\exp(-X^\top\beta) + (1-T)X^\top\beta)$. The term $w_j^{(k)}$ would conduct penalty to $X_{\cdot j}^{(k)}$ adaptively based on the covariates balance, meanwhile, $g^{(k)}$ conducting adaptive penalty to subgroup $k$ determined by its heterogeneity. And $\lambda$ would impose an appropriate penalty constraint based on global balance. Unlike Lasso, note that $w_j^{(k)}, g^{(k)}, \lambda$ are specially designed for motivations of heterogeneous causal inference, and we will make more illustrations about it in later section.

By splitting parameters into two parts $\mu, \gamma^{(k)}$ with adaptive penalty as described in the Eq. 1, HGBAM can control model heterogeneity depending on distributions of different subgroups. Working with a large enough $g^{(k)}$, $\gamma^{(k)}$ will be forced to go to zero and $\beta^{(k)} \approx \mu$, *i.e.* "one model for all". Working with large enough $\lambda$, the approach will be reduced to $K$ different models.

Based on work [15], we see that the penalty in the Eq. 1 encourages that $\beta^{(k)}$ possesses a sparse $\mu$ and sparse heterogeneity $\gamma^{(k)}$. Thus HGBAM can identify causal variables and eliminate complicated confounding to achieve subgroup and global balance. We can get new transformation $\tilde{T} = \left(T^{(1)\top}, \cdots, T^{(K)\top}\right)^\top$, similarly getting $Z$ by rearranging $X^{(1)}, \cdots, X^{(K)}$, getting $\beta$ by rearranging $\beta^{(1)}, \cdots, \beta^{(K)}$, and employ a convenient optimizing method for Eq. 1 by coordinate descent or subgradient descent. Meanwhile, after obtaining propensity score from Eq. 1, inverse propensity score weighting (IPW) is applied to estimate heterogeneous causal effect $\tau^{(k)}$.

$$\hat{\tau}^{(k)} = \frac{1}{N^{(k)}} \sum_{i=1}^{N^{(k)}} \frac{T_i^{(k)} Y_i^{(k)}}{\hat{\pi}\left(X_{i\cdot}^{(k)}\right)} - \frac{1}{N^{(k)}} \sum_{i=1}^{N^{(k)}} \frac{\left(1 - T_i^{(k)}\right) Y_i^{(k)}}{1 - \hat{\pi}\left(X_{i\cdot}^{(k)}\right)} \tag{2}$$

### 3.3   Theoretical Analysis

Introduce $\beta^*$ the true solution of Eq. 1, support set $M(\beta^*) = \{j \mid \beta_j^* \neq 0\}$, cardinality $J = |M(\beta^*)|$. We assume the noise $\epsilon_i^{(k)}$ is $i.i.d.$ and obey sub-Gaussian distribution with parameter $\sigma > 0$. Assuming there is a constant $C > 0$, such that the minimal eigenvalue $\Lambda_{\min}(Z_M^\top Z_M) \geq C$, $\|X_{\cdot j}^{(k)}\|/\sqrt{N^{(k)}} \leq 1$. Then we can state Theorem 1.

**Theorem 1.** *The Eq. 1 has a unique solution and the bias upper bound with high probability grater than $1 - 4\exp(c_1 N\lambda^2) \to 1$ satisfies:*

$$\left\|\hat{\beta}_M - \beta_M^*\right\|_\infty \leq \lambda \left[\left\|(Z_M^\top Z_M/N)^{-1}\right\|_\infty + \frac{4\sigma}{\sqrt{C}}\right] \leq \lambda \left(\frac{\sqrt{J}}{C} + \frac{4\sigma}{\sqrt{C}}\right) \tag{3}$$

Further, without losing generality, assuming $\|Z_M^\top Z_M/N\|_\infty = \mathcal{O}(1)$, setting $\lambda = \mathcal{O}(\sqrt{\log(P)/N})$, the bias upper bound will satisfy:

$$\left\|\hat{\beta} - \beta^*\right\|_2 = \mathcal{O}\left(\lambda\sqrt{J}\right) = \mathcal{O}\left(\sqrt{\frac{J\log(P)}{N}}\right) \tag{4}$$

When $N \to \infty$, the bias $\|\hat{\beta} - \beta^*\|_2$ will go to zero. And note that inverse propensity score weighting (IPW) is an unbiased estimator. Thus the theorem suggests that the proposed HGBAM could provide a reliable solution and make asymptotically unbiased estimation for the heterogeneous problem. In addition, the theorem is an extended result of literatures [15,22].

### 3.4   Choice of Hyperparameters

The penalty is constructed to balance confounders and capture heterogeneity in different subgroups. Therefore, we elaborately proposed hyperparameters setting approach for heterogeneous causal inference.

**Choice of $g^{(k)}$.** Following group lasso, a sensible choice is $g^{(k)} = 1/\sqrt{|V^{(k)}|}$, where $|V^{(k)}|$ is the cardinality of subgroup $k$. However, it can not guarantee consistent selection and oracle property. Meanwhile, we need to reserve the heterogeneity among different subgroups, and $g^{(k)}$ must entail differences among heterogeneous subgroups. Therefore, we specially set:

$$g^{(k)} = \left\|\beta^{(k)\mathrm{MLE}} - \bar{\beta}^{\mathrm{MLE}}\right\|^{-c} \tag{5}$$

where $\beta^{(k)\text{MLE}}$ is the maximum likelihood estimation (MLE) of $\beta^{(k)}$ with samples in subgroup $k$, $\bar{\beta}^{\text{MLE}}$ the average of all $\beta^{(k)\text{MLE}}$, $c$ a positive constant.

The Eq. 5 suggests that the more heterogeneous a subgroup is, the less penalty will be conducted. So that the heterogeneity would be reserved.

**Choice of $w_j^{(k)}$.** In prediction tasks, a popular choice is to set $w_j^{(k)}$ based on the corresponding parameter. But here we need to consider distribution balance. By investigating balance property of different covariates, we give the penalty as:

$$w_j^{(k)} = \left| \frac{1}{N^{(k)}} \sum_{i=1}^{N^{(k)}} \left( \frac{T_i}{\hat{\pi}(X_{i\cdot}^{(k)})^{\text{MLE}}} - 1 \right) x_{ij} \right|^c \tag{6}$$

where $\hat{\pi}(X_{i\cdot}^{(k)})^{\text{MLE}}$ is MLE of propensity score with samples in subgroup $k$.

Following works [4,14], a smaller value of Eq. 6 means that $X_{\cdot j}^{(k)}$ is more balanced. It indicates that IPW is hard to balance its distribution if the value is too larger. Therefore, we impose a larger penalty on the more unbalanced covariate, as we prefer to decrease the dependence of unbalanced covariates [23] and achieve approximate balance when the dimension is high [14]. In this way, it will make a smaller impact to subgroup and global balance. And we can obtain more accurate heterogeneous causal effect estimation.

Meanwhile, there is another approach to determine $w_j^{(k)}$ based on associations of $Y$ and $X$, called outcome adaptive lasso (OAL) [20]. We also apply the setting approach into our method, called Heterogeneous subGroup Outcome Adaptive Method (HGOAM). And we will compare their performances in experiments.

**Choice of $\lambda$.** In most ML algorithms, $\lambda$ is determined by errors on the validation data. However, causal inference is a counterfactual problem and the true causal effect is not available. Moreover, tuning $\lambda$ based on AUC of the treatment assignments is also not preferred, as the motivation is to get accurate causal effect estimation. For better balance, we use exact balance criterion [11] to tune $\lambda$. It is an attractive property we would like to achieve.

$$h_j^{(k)}(\lambda) = \frac{\sum_{i=1}^{N^{(k)}} x_{ij}^{(k)} T_i^{(k)} \left( 1 - \hat{\pi}_i^{(k)} \right)}{\sum_{i=1}^{N^{(k)}} T_i^{(k)} \left( 1 - \hat{\pi}_i^{(k)} \right)} - \frac{\sum_{i=1}^{N^{(k)}} x_{ij}^{(k)} \left( 1 - T_i^{(k)} \right) \hat{\pi}_i^{(k)}}{\sum_{i=1}^{N^{(k)}} \left( 1 - T_i^{(k)} \right) \hat{\pi}_i^{(k)}} \tag{7}$$

The bets situation is $h_j^{(k)} = 0$, which indicates the method has achieved balance between control units and treat units. So the optimal $\lambda^* = \arg\min \sum_k \sum_j h_j^{(k)}(\lambda)$.

From another perspective, dual form of the $L_1$ penalty Eq. 1 is similar to sample weights $\zeta$ direct optimizing methods with constraints, such as entropy balance [6], residual balance [1]. To some extent, the adaptive penalty serves as the constraint, satisfying $\|X_{\cdot j}^{\mathsf{T}(k)}(\zeta \odot T) - \bar{X}_{\cdot j}^{(k)}\| \leq w_j^{(k)}$. Thus our method can impose preferable balance constraints based on the inner property of covariates.

# 4    Experiments

In this section, we illustrate the performance of our method on both synthetic and real-world datasets. For purpose of evaluating the performance comprehensively, various simulation scenarios are designed.

## 4.1    Datasets and Settings

**Synthetic Data.** Assuming subgroup number $K = 4$, for each subgroup, $X_i^{(k)}$ is generated from multivariate standard Gaussian distribution, $X_i^{(k)} \sim N(0, \Sigma^{(k)})$. With logistic regression, propensity score is generated as $\pi_i^{(k)} = \text{logit}(X_{i.}^{(k)\top}\beta^{(k)})$, and binary treatment satisfies $T_i^{(k)} \sim \text{Bernoulli}(\pi_i^{(k)})$. Then we generate potential outcome $Y$ with respect to $X$, $T$, and heterogeneous causal effect $\tau^{(k)}$ as $Y_i^{(k)} = \tau^{(k)}T_i^{(k)} + X_{i.}^{(k)\top}\alpha^{(k)} + \epsilon_i^{(k)}$. Depending on subgroup heterogeneity, sparsity, and model generation mechanisms, there are five important scenarios.

- Scenario1. For each subgroup, the coefficient $\beta^{(k)}$ varies with $k$, but $\alpha^{(k)}$ keeps same. Meanwhile, the sparsity is relatively smaller.
- Scenario2. For each subgroup, the coefficients $\beta^{(k)}$, $\alpha^{(k)}$ both vary with $k$. And the sparsity of them becomes larger.
- Scenario3. The coefficients settings are the same as scenario2, the treatment model keeping the same. But the potential outcome model becomes different.

$$Y_i^{(k)} = \tau^{(k)}T_i^{(k)} + X_{i.}^{(k)\top}\alpha^{(k)} + S_{i.}^{(k)\top}\alpha + \epsilon_i^{(k)} \tag{8}$$

- Scenario4. The potential outcome keeps the same with scenario2, but treatment model becomes different.

$$\pi_i^{(k)} = \text{Pr}(T_i^{(k)} = 1) = \text{logit}\left(X_{i.}^{(k)\top}\beta^{(k)} + S_{i.}^{(k)\top}\beta\right) \tag{9}$$

**Fig. 1.** Model generation in scenario3        **Fig. 2.** Model generation in scenario4

The motivation of scenario3,4 (Figs. 1 and 2) is that we may not observe all related covariates in the real-world data. Some unobserved variables may be available, like $S_i^{(k)}$ in scenario3,4. In addition, for heterogeneous subgroups, there could be similar inducing mechanisms, such as the example given in section1. So scenario3,4 are more close to the real applications.

- Scenario5. We also test the performance of our model in high-dimensional situations when $P = 100, 200$. The coefficient is an extension of scenario2.

$$\beta_{\text{high}}^{(k)} = \left(\beta^{(k)}, 0, 0, \cdots, 0\right), \ \alpha_{\text{high}}^{(k)} = \left(\alpha^{(k)}, 0, 0, \cdots, 0\right) \tag{10}$$

**Lalonde Data.** Lalonde (Jobs) [10] is a famous dataset, which is employed to assess the effect of a job training program on income. By analysis, taking 9 and 12 years of education as partition nodes, we divide it into three heterogeneous subgroups. The quite different income also verifies the rationale of the division.

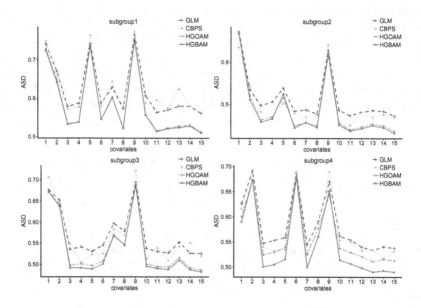

**Fig. 3.** Covariates ASD of different methods in scenario2

**Baselines and Metrics.** As our method is based on logistic regression, we consider generalized linear model (GLM) a baseline. In terms of covariates balance, we implement covariate balance propensity score (CBPS) [7]. When covariates number is increasing in scenario3,4,5, we also compare it with OAL [20]. Besides, HGOAM is also implemented, as described in Sect. 3.4.

Mean Square Error (MSE) is adopted as a performance metric. Moreover, the challenge of causal inference is to balance the distribution of confounders, so absolute standardized difference (ASD) is used to assess covariates balance. The smaller value of ASD means better performance.

$$\text{ASD}(X_{\cdot j}^{(k)}) = \left| \frac{X_{\cdot j}^{(k)\top}(w_1^{(k)} \odot T^{(k)})}{w_1^{(k)\top} T^{(k)}} - \frac{X_{\cdot j}^{(k)\top}\left(w_0^{(k)} \odot \left(1 - T^{(k)}\right)\right)}{w_0^{(k)\top}\left(1 - T^{(k)}\right)} \right| \Bigg/ \sqrt{\frac{\hat{\sigma}_{j,t}^{(k)2}}{N_t^{(k)}} + \frac{\hat{\sigma}_{j,c}^{(k)2}}{N_c^{(k)}}}$$

Where $\hat{\sigma}_{j,t}^{(k)2}, \hat{\sigma}_{j,c}^{(k)2}$ are the sample variances of $X_{\cdot j}^{(k)}$ in treat and control units separately, and $w_{1i}^{(k)} = 1/\hat{\pi}_i^{(k)}$, $w_{0i}^{(k)} = 1/\left(1 - \hat{\pi}_i^{(k)}\right)$.

## 4.2   Results and Analysis

**Experiments on Synthetic Data.** In synthetic data, we test the method over different scenarios and calculate the MSE as well as ASD of each subgroup.

We report the results of different methods in Table 1, and plot the ASD of each covariate in Fig. 3 for scenario2. From Table 1, we have the following observational results. (1). In both subgroup and global situations, comparing with other methods, HGBAM possesses the best performance under each scenario. It indicates that our method can achieve subgroup and global balance simultaneously for heterogeneous causal inference, and estimate heterogeneous causal effect accurately. (2). Another adaptive penalty applied in HGOAM, comparing with it, the better performance of HGBAM demonstrates the effectiveness of the proposed hyperparameters setting approach. So the adaptive penalty of covariate balance is beneficial for de-confounding. (3). Besides, our method can also improve the results in complicated scenario3,4, demonstrating its efficacies.

**Table 1.** Relative performance of different methods in synthetic data

| Method | Subgroup1 | | Subgroup2 | | Subgroup3 | | Subgroup4 | | ASD sum |
|---|---|---|---|---|---|---|---|---|---|
| | MSE | ASD | MSE | ASD | MSE | ASD | MSE | ASD | |
| Scenario1 | | | | | | | | | |
| GLM | 0.421 | 5.783 | 0.491 | 4.473 | 0.692 | 4.966 | 1.005 | 5.102 | 20.324 |
| CBPS | 0.454 | 5.860 | 0.392 | 4.390 | 0.683 | 4.980 | 1.030 | 5.174 | 20.404 |
| HGOAM | 0.367 | 5.652 | 0.451 | 4.420 | 0.640 | 4.883 | 0.985 | 5.123 | 20.078 |
| HGBAM | **0.357** | **5.618** | **0.445** | 4.430 | **0.633** | **4.836** | **0.927** | **5.002** | **19.886** |
| Scenario2 | | | | | | | | | |
| GLM | 0.379 | 8.763 | 0.388 | 6.734 | 0.626 | 7.320 | 0.922 | 7.701 | 30.519 |
| CBPS | 0.363 | 8.797 | 0.356 | 6.466 | 0.605 | 7.129 | 0.918 | 7.682 | 30.074 |
| HGOAM | 0.301 | 8.397 | 0.356 | 6.562 | 0.574 | 7.106 | 0.883 | 7.620 | 29.685 |
| HGBAM | **0.293** | **8.362** | **0.351** | **6.514** | **0.562** | **7.034** | **0.822** | **7.416** | **29.325** |
| Scenario3 | | | | | | | | | |
| GLM | 0.418 | 8.860 | 0.506 | 6.619 | 0.722 | 7.329 | 1.043 | 7.658 | 30.466 |
| CBPS | 0.406 | 8.737 | 0.492 | 6.529 | 0.685 | 7.337 | 1.012 | 7.586 | 30.188 |
| OAL | 0.447 | 8.624 | 0.571 | 6.834 | 0.756 | 7.460 | 1.095 | 7.672 | 30.590 |
| HGOAM | 0.359 | 8.519 | 0.462 | 6.450 | 0.666 | 7.137 | 0.997 | 7.630 | 29.736 |
| HGBAM | **0.356** | 8.535 | **0.458** | **6.432** | **0.657** | **7.084** | **0.930** | **7.473** | **29.525** |
| Scenario4 | | | | | | | | | |
| GLM | 0.710 | 12.766 | 0.657 | 11.539 | 0.976 | 11.522 | 1.554 | 11.659 | 47.485 |
| CBPS | 0.655 | 12.786 | 0.687 | 11.422 | 1.020 | 11.216 | 1.338 | 11.342 | 46.766 |
| OAL | 0.693 | 13.092 | 0.702 | 11.662 | 0.964 | 11.848 | 1.515 | 11.968 | 48.570 |
| HGOAM | **0.517** | 11.969 | **0.534** | 10.910 | 0.831 | 10.942 | 1.389 | 11.518 | 45.339 |
| HGBAM | 0.519 | **11.939** | 0.535 | **10.929** | **0.821** | **10.919** | **1.183** | **10.978** | **44.764** |

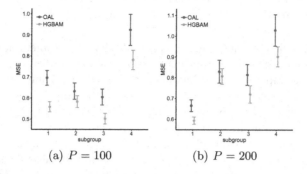

(a) $P = 100$                              (b) $P = 200$

**Fig. 4.** MSE of different subgroups in scenario5 when $P = 100$ and $P = 200$

From Fig. 3, we can clearly see that HGBAM can achieve better balance for most covariates as it can efficiently employ all confounding information among different subgroups. Especially, at the end of the plot curve, our method has much smaller ASD. It shows that our method is able to conduct appropriate causal variables selection, as we set the coefficients of these variables nearly to zero in synthetic data. In addition, for high-dimensional situations, varying $P$ from 100 to 200, we plot MSE in Fig. 4. Comparing with OAL, which is a strong baseline for high-dimensional causal inference, the better results also verify the advantages of our method.

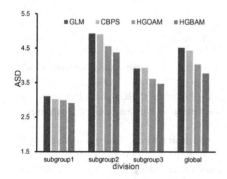

**Fig. 5.** Subgroup balance and global balance condition in Lalonde data

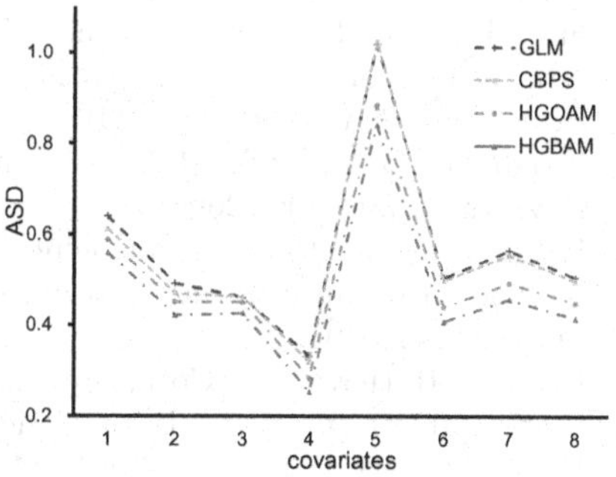

**Fig. 6.** Global ASD of different covariates in Lalonde data

**Experiments on Lalonde.** As the true causal effect is not available in real-world data, for heterogeneous subgroups, we plot subgroups ASD and global ASD in Fig. 5. Due to space limitation, we only display global ASD of each covariate in Fig. 6. Similar to results in synthetic data, our method consistently outperforms other methods in both subgroup and global data.

# 5   Conclusion

In this paper, we focus on heterogeneous causal inference problem. Most previous methods have deficiencies in estimating heterogeneous causal effect, as they could not achieve subgroup and global balance simultaneously. Therefore, we develop heterogeneous subgroup balance adaptive method by employing connections of different subgroups and imposing elaborately designed penalty constraints. The penalty guarantees that the method can explore heterogeneity adaptively, meanwhile eliminate confounding in subgroup and global data. By theoretical analysis, we prove the method could make accurate estimation for heterogeneous causal effect. The extensive experimental results show that our method has improved performance and validate advantages of the proposed penalty constraints.

**Acknowledgements.** This work was supported by the National Natural Science Foundation of China (Grant No. 12071089 and Grant No. 71991471).

# References

1. Athey, S., Imbens, G.W., Wager, S.: Approximate residual balancing: debiased inference of average treatment effects in high dimensions. J. Roy. Stat. Soc. (2018)
2. Brookhart, M.A., Schneeweiss, S., Rothman, K.J., Glynn, R.J., Avorn, J., Stürmer, T.: Variable selection for propensity score models. Am. J. Epidemiol. **163**(12), 1149–1156 (2006)
3. Dong, J., Zhang, J.L., Zeng, S., Li, F.: Subgroup balancing propensity score. Stat. Methods Med. Res. **29**(3), 659–676 (2020)
4. Fan, J., Imai, K., Liu, H., Ning, Y., Yang, X.: Improving covariate balancing propensity score: A doubly robust and efficient approach. Princeton University, Princeton, USA, Tech. rep. (2016)
5. Freedman, D.A., Berk, R.A.: Weighting regressions by propensity scores. Eval. Rev. **32**(4), 392–409 (2008)
6. Hainmueller, J.: Entropy balancing for causal effects: a multivariate reweighting method to produce balanced samples in observational studies. Political analysis, pp. 25–46 (2012)
7. Imai, K., Ratkovic, M.: Covariate balancing propensity score. Journal of the Royal Statistical Society: Series B: Statistical Methodology, pp. 243–263 (2014)
8. Koch, B., Vock, D.M., Wolfson, J.: Covariate selection with group lasso and doubly robust estimation of causal effects. Biometrics **74**(1), 8–17 (2018)
9. Kuang, K., Cui, P., Athey, S., Xiong, R., Li, B.: Stable prediction across unknown environments. In: Proceedings of the 24th ACM SIGKDD International Conference on Knowledge Discovery & Data Mining, pp. 1617–1626 (2018)
10. LaLonde, R.J.: Evaluating the econometric evaluations of training programs with experimental data. The American economic review, pp. 604–620 (1986)
11. Li, F., Morgan, K.L., Zaslavsky, A.M.: Balancing covariates via propensity score weighting. J. Am. Stat. Assoc. **113**(521), 390–400 (2018)
12. Li, F., Zaslavsky, A.M., Landrum, M.B.: Propensity score weighting with multilevel data. Stat. Med. **32**(19), 3373–3387 (2013)

13. Magliacane, S., van Ommen, T., Claassen, T., Bongers, S., Versteeg, P., Mooij, J.M.: Domain adaptation by using causal inference to predict invariant conditional distributions. In: Proceedings of the 32nd International Conference on Neural Information Processing Systems, pp. 10869–10879 (2018)
14. Ning, Y., Sida, P., Imai, K.: Robust estimation of causal effects via a high-dimensional covariate balancing propensity score. Biometrika **107**(3), 533–554 (2020)
15. Ollier, E., Viallon, V.: Regression modelling on stratified data with the lasso. Biometrika **104**(1), 83–96 (2017)
16. Pearl, J.: Causality. Cambridge University Press, Cambridge (2009)
17. Robins, J.M., Rotnitzky, A., Zhao, L.P.: Analysis of semiparametric regression models for repeated outcomes in the presence of missing data. J. Am. Stat. Assoc. **90**(429), 106–121 (1995)
18. Rojas-Carulla, M., Schölkopf, B., Turner, R., Peters, J.: Invariant models for causal transfer learning. J. Mach. Learn. Res. **19**(1), 1309–1342 (2018)
19. Rosenbaum, P.R., Rubin, D.B.: The central role of the propensity score in observational studies for causal effects. Biometrika **70**(1), 41–55 (1983)
20. Shortreed, S.M., Ertefaie, A.: Outcome-adaptive lasso: Variable selection for causal inference. Biometrics **73**(4), 1111–1122 (2017)
21. Stuart, E.A.: Matching methods for causal inference: a review and a look forward. Stat. Sci. **25**(1), 1–21 (2010)
22. Wainwright, M.J.: Sharp thresholds for high-dimensional and noisy sparsity recovery using l1-constrained quadratic programming (lasso). IEEE Trans. Inf. Theory **55**(5), 2183–2202 (2009)
23. Wang, Y., Zubizarreta, J.R.: Minimal dispersion approximately balancing weights: asymptotic properties and practical considerations. Biometrika **107**(1), 93–105 (2020)
24. Yao, L., Chu, Z., Li, S., Li, Y., Gao, J., Zhang, A.: A survey on causal inference. ArXiv abs/2002.02770 (2020)
25. Yue, Z., Zhang, H., Sun, Q., Hua, X.S.: Interventional few-shot learning. In: Advances in Neural Information Processing Systems, pp. 2734–2746 (2020)
26. Zhang, D., Zhang, H., Tang, J., Hua, X.S., Sun, Q.: Causal intervention for weakly-supervised semantic segmentation. In: Advances in Neural Information Processing Systems, pp. 655–666 (2020)
27. Zhang, K., Gong, M., Stojanov, P., Huang, B., LIU, Q., Glymour, C.: Domain adaptation as a problem of inference on graphical models. In: Advances in Neural Information Processing Systems, pp. 4965–4976 (2020)

# Optimizing Federated Learning on Non-IID Data Using Local Shapley Value

Zuoqi Tang[1,2], Feifei Shao[1], Long Chen[1,3], Yunan Ye[1], Chao Wu[1(✉)], and Jun Xiao[1]

[1] Zhejiang University, Hangzhou, China
{tangzq,sff,longc,chryleo,chao.wu,junx}@zju.edu.cn
[2] Guizhou University, Guiyang, China
[3] Columbia University, New York, USA

**Abstract.** Federated learning (FL) was originally proposed as a new distributed machine learning paradigm that addresses the data security and privacy protection issues with a global model trained by ubiquitous local data. Currently, *FL* techniques have been applied in some data-sensitive areas such as finance, insurance, and healthcare. Although *FL* has broad application scenarios, there are still some significant and fundamental challenges, one of which is the training on *Not independently and identically distributed (Non-IID)* data. More concretely, the global model aggregation and collaboration of a massive number of participants on the *Non-IID* data remain an unsolved problem. We find that most of the model aggregation optimization algorithms in the literature suffer from significant accuracy loss in the *Non-IID* setting for *FL*. To this end, in this paper, we propose a novel model aggregation algorithm terms *FedSV*, which dynamically updates global model aggregation weights according to each local participant's contribution in each training round. Furthermore, to evaluate the participants' contribution, we propose a quantization algorithm based on *Local Federated Shapley Value*, which dynamically computes the contribution by the properties of the participant. Extensive experiments on *Non-IID* data partition, such as CIFAR-10 and MNIST, demonstrate that our approach can improve accuracy during training compared with existing methods.

**Keywords:** Federated learning · Model aggregation · Shapley value

## 1 Introduction

Federated Learning (FL) conception was initially proposed by Google [16] in 2016 as a decentralized model aggregation approach while keeping data localized. Now it is gradually used to solve the problem of data island, privacy protection, and

---

L. Chen—This work was done when Long Chen was a Ph.D. student at Zhejiang University.

L. Fang et al. (Eds.): CICAI 2021, LNAI 13070, pp. 164–175, 2021.
https://doi.org/10.1007/978-3-030-93049-3_14

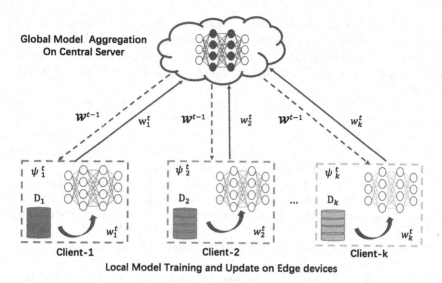

**Fig. 1.** The Overview of federated learning and model aggregation process. FL mainly consists of central aggregation server and a large number of edge devices (e.g.,smart devices, IoT devices, and organizations). The lower three dotted boxes (local models) to the upper dotted box (global model) depict the model aggregation process. The different color network connecting lines indicate different weights of the model. The upper dotted box represents the weights of the central server global model. The lower dotted box represents the weights of edge clients' local model. The middle parts (including dotted lines and solid lines) represents the process model aggregation. The same structure of the global model and local model, $\mathcal{W}$ is global model weights, $w$ is the local model weights, $D_k$ is the local datasets, and $\psi_k$ is the weights for aggregation. The different colors dataset icons represent the Non-IID datasets. (Color figure online)

security [24]. As can be observed from Fig. 1, the classical federated learning process consists of two main steps. The central server randomly chooses some clients to run local stochastic gradient descent (SGD) to update the local model parameters, and then the central server performs model aggregation according to receive local model parameters. Currently, we face two core problems that federated learning has to solve. The first critical issue is how to evaluate the value of clients' data, and the second fundamental problem is how to solve the problem of model aggregation according to clients' contributions in system heterogeneity and data heterogeneity setting [9,14].

The most primitive solution is the FedAvg [16] method. Intrinsically, the FedAvg algorithm takes a simple weighted average according to the size of local datasets. Currently, the state-of-the-art methods, for example, the FedPox [14] method adds a proximal term that restricts the local updates in order to close to the global aggregation model by automatic set number of local epochs. The FedMA [22] method takes advantage of leveraging layer-wise and extracting signature by similar features to design the shared global model through, which adapts to most neural network architectures. The FedAtt [6] method uses the attention mechanism as a model aggregation optimization algorithm, which only adapts to the neural language modeling. However, these methods ignore the real contributions of a massive amount of clients' data during federated learning training.

We find that most of the model optimization algorithms in the previous work still suffer significant accuracy loss in the Non-IID setting for federated learning, in particular under a massive number of edge devices [26], i.e., a large number of participants. To address this problem, from a cooperative game theory viewpoint, we present a novel model aggregation algorithm, which dynamic updates global model weights according to the contribution of every device. However, in practice, directly calculating exact Shapley value requires exponential time, as the number of participants increases. So we focus on finding an efficient algorithm to compute Shapley Value. We present *local Shapley Value*, which is a modified form of exact Shapley value and computed more efficiently than the exact Shapley Value.

Federated learning has two fundamental characteristics. First, the client's data is distributed over a vast number of devices. Second, training data is Non-IID. We argue that the optimization model aggregation approach for federated learning is still a research hot spot in research communities [8,13].

**Contributions.** In this paper, our primary contributions include three parts:

- We propose a novel global model aggregation strategy on the central server, which is especially suitable for the situation where many Non-IID federated learning nodes participate in the collaborative training of a global model.
- In order to address the core issues of computing the weights of clients according to clients' private data, we present an evaluation algorithm based on local Shapley Value, which can dynamically quantify the contribution of participants on the entire federated learning process.
- We perform extensive experiments on different settings to verify the efficiency of our approach on the MNIST and CIFAR-10 datasets, in particular, in Non-IID data and a massive number of client scenarios.

## 2    Related Work

### 2.1    Federated Learning Settings

In heterogeneous system setting [1], for example, federated learning is currently deployed in a large-scale distributed heterogeneous network [3], where including most of the edge computing devices, such as mobile phones, IoT devices, autonomous vehicles, various sensors, etc. [17]. In these settings, federated learning is unreliable and low effectiveness due to some of the massive numbers of devices are unreliable in the heterogeneous system [20]. In addition, for heterogeneous data, the most significant difference between federated learning and traditional large-scale distributed machine learning is that one is IID data, and the other is Non-IID data.

## 2.2    Model Optimization Algorithm

McMahan et al. introduced the Federated Averaging algorithm (FedAvg) [16], one of the most common methods for optimizing federated learning settings. Indeed, facing statistical heterogeneity challenges, the FedAvg method shows the significant divergence in the training process, even when jointly learning a single shared global model. From the perspective of optimization theory, Federated Averaging is a leading optimization algorithm which both simplicity and effectiveness. However, due to lack of theoretical basis for Non-IID data, in order to provide insight for a conceptual understanding of FedAvg, Li et al. formulated strongly convex and smooth problems, establish a convergence rate $\mathcal{O}(\frac{1}{T})$ by analyzing the convergence of FedAvg [15].

Currently, some approaches proposed aiming to address critical issues of the modeling for Non-IID data. For example, to solve system heterogeneity and statistical heterogeneity challenges, Li et al. presented a federated optimization algorithm framework, name FedProx [14], which is a generalization and re-parametrization of FedAvg. The core of this method is to add a proximal term that restricts the local updates, then to close to the global model by automatically set the number of local epochs.

Federated learning optimization algorithm facing another crucial challenge is communication-efficient, aim to deal with communication bottlenecks and convergence oscillation. The fundamental method of reducing communication costs is structured updates or sketched updates approach [10]. Rothchild et al. proposed a novel optimization algorithm (FetchSGD) [18], which compresses model parameter updates employing Count Sketch. Some novel optimization algorithms from other areas are used in federated learning. For instance, Wang et al. developed a novel optimzing federated learning mehtod using reinforcement learning [21], which is an experience driven control framework.

## 3    Preliminary

### 3.1    Federated Learning

For a federated learning process, we typically define the empirical risk over local data as local objective function $f(w) = \frac{1}{n}\sum_{i=1}^{n} f_i(w)$, where $f_i(w) = \ell(x_i, y_i; w)$, $\ell(.;.)$ is a user-defined loss function, and $n$ represents the number of local available samples. By above analysis, we focus on the minimize finite-sum objective function of the form [13, 16]:

$$\underset{w\in\mathbb{R}^d}{\arg\min}\, F(w)\,, \quad \text{where } F(w) = \sum_{k=1}^{K} p_k F_k(w)\,. \tag{1}$$

Where, $K$ denotes the total number of devices, $p_k$ indicates the weight of the $k$-th device ($p_k \geq 0$ and $\sum_k p_k = 1$), with two generally settings being $p_k = \frac{1}{n}$ or $p_k = \frac{n_k}{n}$, where $n = \sum_k n_k$ denote the total amount of samples, and $F_k$ denote the local objective function for the $k$-th device.

## 3.2  Shapley Value

The concept of *Shapley Value (SV)* was first presented as a theoretical way to solve the cooperative game problem [19], and it has been widely used in computer science, in particular, use the SV method to evaluate data value for machine learning.

The *SV* takes as input a set function $v : 2^N \to \mathbb{R}$, $N$ is a set ($n$ players). The *SV* produces attributions $\phi(i)$ for each player $i \in N$ that add up to $v(N)$. The *SV* of a player $i$ is given by:

$$\phi_i(N,v) = \sum_{S \subseteq N \setminus i} \frac{|S|! * (|N| - |S| - 1)!}{N!} \left( v(S \cup i) - v(S) \right). \tag{2}$$

where, $v(S \cup i) - v(S)$ denotes each player $i$ expected marginal contribution, $v$ indicates utility function, $S$ is the subset of all players. The Shapley value method satisfies the following properties [2,4,7]:

1. **Additivity:** The total sum of the SV of each participant is equal to the SV of the union of participants' dataset, denoted by $\sum_i^m \phi(\mathcal{D}_i) = \phi(N)$, also known as *Group Rationality*.
2. **Symmetry:** If any two participants $P_i$ and $P_j$, for every subsets $S$ of $N$, if $v_x(S \cup \{i\}) = v_x(S \cup \{j\})$, $\forall S \subseteq N \setminus \{i,j\}$, then $\phi_x(i) = \phi_x(j)$, also known as *Fairness*.
3. **Monotonicity:** Given two participants $P_i$ and $P_j$ , let $m_x$ and $m_x'$ represent the associated utility functions, and let $\phi(x)$ and $\phi_x'$ represent the associated Shapley values, if $m_x(S,i) \geq m_x'(S,i)$ for all subsets $S$, so we are assure that $\phi_x(i) \geq \phi_x'(i)$.
4. **Dummy:** Given a participant $P_i$, if $v_x(S \cup \{i\}) = v_x(S))$, then $\phi_x(i) = 0$, so $P_i$ is null player.

## 4  Proposed Method

### 4.1  Problem Formulation

As shown in Fig. 1, the classical federated learning paradigm involves learning a global shared model from data stored in a massive number of remote edge devices. It aims to collaboratively train a global model while decoupling the model training and device-generated data, with only periodically updates model parameters with a central server. In this work, we assume assumption is *unbalanced and Non-IID* dataset. We denote $K$ as the total number of clients (devices or organizations), which hold a fixed local dataset. At the beginning of training, a coordinate server randomly chooses a fraction (denoted by $C$) of clients, and the server broadcasts the current global initial parameters to each of the participants selected. Each participant selected conduct training by their local dataset, and upload local model parameters to the central server. Finally, the server aggregates the model by different strategies, and repeat this process until subject to requirements (e.g., accuracy, communication rounds).

## 4.2 Local Federated Shapley Value

We observe that classical federated learning is essentially a collaborative training process in which participants randomly selected by the coordinator join in each training process in a two-stage sampling order manner. Therefore, a variant of FedSV [23] for federated learning called *local FedSV* was defined.

**Definition 1 (Local FedSV).** *Let* $\mathcal{C} = \{1, \cdots, n\}$ *represents the subset of participants selected by the coordinator server during entire $T$-round fedrated learning process. Let $\mathcal{C}_t$ be the coalition of clients selected in round $t$ and $\mathcal{C}_t \subseteq \mathcal{C}$. Then, the* local FedSV *of participant $\mathcal{P}_i$ at round $t$ is defined as:*

$$\phi_i^t(v) := \frac{1}{|\mathcal{C}_t|} \sum_{S \subseteq \mathcal{C}_t \setminus \{i\}} \frac{1}{\binom{|\mathcal{C}_t|-1}{|S|}} \mathcal{V}(S) \tag{3}$$

*where* $\mathcal{V}(S) = [v(\mathcal{C}_{1:t-1} + (S \cup \{i\})) - v(\mathcal{C}_{1:t-1} + S)]$, $i \in \mathcal{C}_t$, *and $S$ indicates the number of chosen participants $t$-th round.*

**Definition 2 (Global FedSV).** *The ultimate weighed FedSV of participant $\mathcal{P}_i$ takes the sum of the values of all rounds, $W_i$ is the total number of training rounds for participants selected during the entire federated learning process. Then, the* global FedSV *of participant $\mathcal{P}_i$ is defined as:*

$$\Psi_i^t = \frac{1}{W_i} \sum_{t=1}^{T} \psi_i^t \tag{4}$$

## 4.3 Dynamical Weights Update

The goal of a federated learning optimization approach is to learn an optimal global model that can share with all clients by using private data training. In our proposed model aggregation optimization method, we consider it as searching and matching an optimal global model that is close to the client local models in parameter space concerning the contribution of selected client local models while aggregating on a central server. So we define the optimization objective as follows [6]:

$$\underset{\mathcal{W}^{t+1}}{\arg\min} f(\mathcal{W}^{t+1}), where f(\mathcal{W}) = \sum_{k=1}^{K} \left[ \frac{1}{2} \psi_k L \left( \mathcal{W}, w_k^{t+1} \right)^2 \right]. \tag{5}$$

where, $\mathcal{W}^t$ is the parameters of the global model at communication round $t$, $w_k^{t+1}$ indicates the parameters of the $k$-th client local model at communication round $t + 1$, $L(\cdot, \cdot)$ denotes the distance between local model parameters and global model parameters, and $\psi_k$ represents the shapley weight to measure the importance of weights for the client models.

Then, we apply softmax on the similarity to calculate the weights of the $k$-th client in Eq. 6.

$$\psi_k^t = softmax(\phi_k^t) = \frac{e^{\phi_k^t}}{\sum_{i=1}^{m} e^{\phi_i^t}} \tag{6}$$

---

**Algorithm 1.** Dynamic Federated Averaging *(Dynamic FedAvg)*.

The $K$ clients are indexed by $k$, $C$ is the ratio of selected clients, $B$ is the local minibatch size, $E$ is the number of local epochs, $\eta$ is the learning rate, $\beta$ is momentum on server, $\mathcal{P}_k$ is the index of datasets on participant $k$, $\mathcal{W}$ is global model weights, $w$ is the local model weights, and $\mathbb{W}$ is global model weights with momentum.

---

    **Server executes:**    // *Run on server*
1: **Initialize:** $\mathcal{W}^0$
2: **Input:** server parameters $\mathcal{W}^t$ at round $t$, client parameters $w_1^{t+1}, ..., w_m^{t+1}$ at round $t+1$.
3: **Output:** aggregated server parameters $\mathcal{W}^{t+1}, \mathbb{W}^{t+1}$.
4: **for** each round $t = 1, 2, ...$ **do**
5:      $m \leftarrow \max(C \cdot K, 1)$
6:      $S_t \leftarrow$ (random set of $m$ clients)
7:      **for** each client $k \in S_t$ **in parallel do**
8:          $w_k^{t+1} \leftarrow$ ClientUpdate$(k, \mathcal{W}^t)$
9:      **end for**
10:     $\mathcal{W}^{t+1} \leftarrow$ **ModelAggregation**$(w_k^{t+1})$ // *call Algorithm 2*
11:     $\mathbb{W}^{t+1} \leftarrow \beta \mathcal{W}^t + (1-\beta)\mathcal{W}^{t+1}$
12: **end for**
    **ClientUpdate**$(k, \mathcal{W}^t)$**:**     // *Run on client k*
13: $\mathcal{B} \leftarrow$ (split $\mathcal{P}_k$ into batches of size $B$)
14: **for** each local epoch $i$ from 1 to $E$ **do**
15:      **for** batch $b \in \mathcal{B}$ **do**
16:          $w_k^{t+1} \leftarrow w_k^t - \eta \triangledown \ell(\mathcal{W}^t : b)$
17:      **end for**
18: **end for**
19: **return** $w_k^{t+1}$ to central server

---

For the active online selected set of $m$ clients, we perform stochastic gradient descent (SGD) to update the parameters of the global model in Eq. 7 as:

$$\mathbb{W}^{t+1} \leftarrow \beta \mathcal{W}^t + (1-\beta)\mathcal{W}^{t+1}. \tag{7}$$

where $\beta$ is the momentum. The detail procedure of our presented model optimization algorithm is depicted in **Algorithm 1**. It takes the parameters of the serve global model $\mathcal{W}^t$ at round $t$ and the parameters of the clients local model $w_1^{t+1}, ..., w_m^{t+1}$ at round $t+1$, and returns the updated parameters of the global model on central server.

### 4.4  Algorithm

**Global Model Aggregation on Central Server.** We describe the main steps of the global model aggregation in detail. The central server firstly initializes the global model parameters, and broadcasts these parameters to all clients who maybe will federated learning participants, or randomly choose a number of

**Algorithm 2.** Model Aggregation Optimization using Shapley Value Weights. $k$ is the ordinal of clients, $t$ is communication rounds, $\phi_k$ is the Shapley value of $k$-th client, $\psi_k$ is the Shapley weights of the $k$-th client, and $w^{t+1}$ is local model weights of the $k$-th client at $t+1$ round.

---

$\text{ModelAggregation}(w_k^{t+1})$
1: **Input**: client parameters $w_1^{t+1}, ..., w_m^{t+1}$ at round $t+1$ .
2: **Output**: aggregated server parameters $\mathcal{W}^{t+1}$.
3: **Initialize** $\phi_i = \{\phi_1, \dots, \phi_k, \dots, \phi_m\}$
4: **for** each clients $l = 1, 2, ... m$ **do**
5:     $\mathcal{C}_t \leftarrow$ (Random choose $\tau$ tuple from $m$ clients)
6:     **for** each user k  **do**
7:         $\phi_k^t(v) := \frac{1}{|\mathcal{C}_t|} \sum_{S \subseteq \mathcal{C}_t \setminus \{k\}} \frac{1}{\binom{|\mathcal{C}_t|-1}{|S|}} \mathcal{V}(S)$ // *calculated by Eq. (3)*
8:     **end for**
9:     $\psi_k^t = softmax(\phi_k^t) = \frac{e^{\phi_k^t}}{\sum_{i=1}^m e^{\phi_i^t}}$
10: **end for**
11: $\mathcal{W}^{t+1} \leftarrow \sum_{k=1}^m \psi_k^t w_k^{t+1}$
12: return $\mathcal{W}^{t+1}$

---

participants to join federated learning. Then, this server waits for active online participants for local model training. During a federated learning task, the central server periodically receives the model updated parameters and performs the server model aggregation optimization after the selected number of participants finish the local model train and update. Generally, one communication round includes the global model parameter sending and local model parameter receiving. Our present optimization algorithm conducts in line 10 to line 11 in the **Algorithm** 1 and **Algorithm** 2.

**Local Model Training and Update on Clients.** We introduce the main steps of the local model training and update. Each participant (or online active client) receives the global model parameters and conducts standalone local training using their devices-generated data. For neural network modeling (e.g., CNN, ResNet), clients commonly perform local Stochastic Gradient Descent (SGD) to update local client models. After several local epochs of training, the participants send the parameters of the global shared models to the central server by secure communication channels or encrypt the model parameters with differential private. The details are elaborated in line 13 to 19 in the **Algorithm** 1.

## 5    Experiments

We evaluate Dynamic FedAvg for image classification on two datasets (MNIST [12] and CIFAR-10 [11]). Because there are not the benchmarks and libraries to adequately support diverse algorithmic comparisons for federated learning [5], in our experiments, we compare Dynamic FedSV with FedAvg, to keep the fairness of comparison, The models and data partition are the same as the FedAvg experimental setup.

## 5.1   Experimental Setup

**Model and Datasets Partition.** In our experiment, we use a standard neural network architecture, which is sufficient for our experiments, as our goal is to evaluate our model optimization algorithm and data evaluation method [25], not achieve the best possible accuracy on this task. More concretely, the 2NN [16] model consists of two $5 \times 5$ convolution layers (the first with 32 channels, the second with 64, each followed with $2 \times 2$ max pooling), a fully connected layer with 512 units and ReLu activation, and a final softmax output layer. Also, for the original MNIST and CIFAR-10 dataset, we do not use data augmentation and normalize each image, as we hope to evaluate the actual value of clients' local data.

To simulate the experimental settings of the real-world image classification task, we conduct data partition on popular image modeling datasets. We use two data partition methods to mimic a federated learning scenario: For MNIST dataset, (1)*IID data partition*, where the data is shuffled, and then divide into 100 clients each receiving 600 samples, and (2)*Non-IID data partition*, where we first sort the data by digit label, split it into 200 shards of size 300, and assign every of 100 clients two shards. Besides, we also perform experiments on the CIFAR-10 dataset, which contains ten classes of $32 \times 32$ images with three RGB channels, 50,000 training examples, and 10,000 testing examples. We split it into 100 clients, each containing 500 training samples. In particular, in the Non-IID data experiment, we use a 30% label distribution deviation as data partition methods to evaluate our algorithm.

**Baselines.** We perform several group experiments for comparison. There are two baselines totally in these comparisons. The basic settings of baselines and our presented approach are as follows.

- *FedSGD:* Federated stochastic gradient descent takes all the clients for federated aggregation and every client conducts one epoch of gradient descent.
- *FedAvg:* Federated averaging random chooses a fraction of clients for each iteration and online participants can take several steps of gradient descent.
- *FedAvg⁺:* In the process of model aggregation, our metod adds global momentum as the weights of the previous global model and current global model.
- *FedSV:* Our proposed approach takes a similar setting as FedAvg, but uses an improved dynamical model aggregation algorithm, details are described in **Algorithm2**

**Federated Learning Setting.** We deploy our experiments under a simulated federated learning environment where we set a centralized node as central server and 100 distributed nodes as clients. The number of local epochs $E$ is 10, local batch size $B$ is 50, local learning rate $lr$ is 0.01, local SGD momentum $\alpha$ is 0.5, the local optimizer is *sgd*, the fraction of clients $C$ is 0.1 and the global momentum $\beta$ is 0.5.

## 5.2 Experimental Results

In this section, we employ the MNIST and CIFAR-10 datasets to investigate the properties of our presented algorithm to improve the federated averaging algorithm. We present an empirical study of our methods with performance under two data partitions.

**Table 1.** Trained models summary for 2NN trained on CIFAR-10 and MNIST. (1) ($*$) denotes each client contains the same size training samples with two labels. (2) (†) indicates Non-IID partition with 30% label skew. (3) Total 100 clients. (4) To achieve final accuracy, we run 100 rounds federated learning on MNIST and 500 rounds federated learning on CIFAR-10, respectively.

| Settings | Dataset | FedSGD | FedAvg | FedAvg$^+$(ours) | FedSV(ours) |
|---|---|---|---|---|---|
| Non-IID | MNIST($*$) | 99.03 | 97.61 | 96.93 | 97.52 |
| | MNIST(†) | 99.03 | 90.01 | **93.78** | 92.64 |
| | CIFAR-10($*$) | 60.27 | 39.22 | 43.16 | **46.02** |
| | CIFAR-10(†) | 60.27 | 49.49 | 51.19 | **52.43** |
| IID | MNIST | 99.03 | 97.92 | 96.89 | 97.86 |
| | CIFAR-10 | 60.27 | 52.15 | 53.87 | **54.52** |

As shown in Table 1, dynamic federated Shapley value algorithm is superior to the original federated averaging algorithm, especially on non-IID Settings under a massive number of clients. Besides, FedAvg$^+$ and dynamic FedSV can reduce the oscillation by adding global momentum during the process of model aggregation on the central server compared to the *FedAvg* algorithm.

## 6    Conclusion

In this work, we presented a novel dynamic model aggregation strategy based on the Local Shapley value for federated learning. For IID and Non-IID data partition, we conduct extensive experiments and show our approaches are available. Moving forward, we plan to quantify participants' contributions through statistical metrics such as local data characteristics such as quality, quantity, dissimilarity, etc. This research results provide a new idea for solving optimization problems and data valuation for federated learning.

**Acknowledgments.** This work was supported by the National Key Research and Development Project of China (No. 2018AAA0101900), the National Natural Science Foundation of China (U19B2042, U19B2043, 61976185), Zhejiang Natural Science Foundation (LR19F020002), Zhejiang Innovation Foundation (2019R52002), and the Fundamental Research Funds for the Central Universities.

# References

1. Bonawitz, K., et al.: Towards federated learning at scale: system design. In: Conference on Systems and Machine Learning (2019)
2. Chen, J., Song, L., Wainwright, M.J., Jordan, M.I.: L-shapley and c-shapley: efficient model interpretation for structured data. In: International Conference on Learning Representations (2018)
3. Dean, J., et al.: Large scale distributed deep networks. In: Advances in Neural Information Processing Systems, pp. 1223–1231 (2012)
4. Ghorbani, A., Zou, J.: Data shapley: equitable valuation of data for machine learning. In: International Conference on Machine Learning, pp. 2242–2251 (2019)
5. He, C., et al.: Fedml: a research library and benchmark for federated machine learning. In: Advances in Neural Information Processing Systems (2020)
6. Ji, S., Pan, S., Long, G., Li, X., Jiang, J., Huang, Z.: Learning private neural language modeling with attentive aggregation. In: 2019 International Joint Conference on Neural Networks (IJCNN), pp. 1–8. IEEE (2019)
7. Jia, R., et al.: Towards efficient data valuation based on the shapley value. In: The 22nd International Conference on Artificial Intelligence and Statistics, pp. 1167–1176. PMLR (2019)
8. Kang, J., Xiong, Z., Niyato, D., Yu, H., Liang, Y.C., Kim, D.I.: Incentive design for efficient federated learning in mobile networks: a contract theory approach. In: 2019 IEEE VTS Asia Pacific Wireless Communications Symposium (APWCS), pp. 1–5. IEEE (2019)
9. Karimireddy, S.P., Kale, S., Mohri, M., Reddi, S., Stich, S., Suresh, A.T.: Scaffold: Stochastic controlled averaging for federated learning. In: International Conference on Machine Learning, pp. 5132–5143. PMLR (2020)
10. Konečný, J., McMahan, H.B., Yu, F.X., Richtárik, P., Suresh, A.T., Bacon, D.: Federated learning: strategies for improving communication efficiency (2016). arXiv preprint arXiv:1610.05492
11. Krizhevsky, A., Hinton, G., et al.: Learning multiple layers of features from tiny images (2009)
12. LeCun, Y., Bottou, L., Bengio, Y., Haffner, P.: Gradient-based learning applied to document recognition. Proc. IEEE **86**(11), 2278–2324 (1998)
13. Li, T., Sahu, A.K., Talwalkar, A., Smith, V.: Federated learning: challenges, methods, and future directions. IEEE Signal Process. Mag. **37**(3), 50–60 (2020)
14. Li, T., Sahu, A.K., Zaheer, M., Sanjabi, M., Talwalkar, A., Smith, V.: Federated optimization in heterogeneous networks (2018). arXiv preprint arXiv:1812.06127
15. Li, X., Huang, K., Yang, W., Wang, S., Zhang, Z.: On the convergence of fedavg on non-iid data. In: International Conference on Learning Representations (2019)
16. McMahan, B., Moore, E., Ramage, D., Hampson, S., y Arcas, B.A.: Communication-efficient learning of deep networks from decentralized data. In: Artificial Intelligence and Statistics, pp. 1273–1282 (2017)
17. Nishio, T., Yonetani, R.: Client selection for federated learning with heterogeneous resources in mobile edge. In: International Conference on Communications (2019)

18. Rothchild, D., et al.: Fetchsgd: communication-efficient federated learning with sketching. In: International Conference on Machine Learning (2020)
19. Shapley, L.S.: A value of n-person games. In: Contributions to the Theory of Games, pp. 307–317 (1953)
20. Tang, H., et al.: Distributed learning over unreliable networks. In: International Conference on Machine Learning (2019)
21. Wang, H., Kaplan, Z., Niu, D., Li, B.: Optimizing federated learning on non-iid data with reinforcement learning. In: IEEE INFOCOM 2020-IEEE Conference on Computer Communications, pp. 1698–1707. IEEE (2020)
22. Wang, H., Yurochkin, M., Sun, Y., Papailiopoulos, D., Khazaeni, Y.: Federated learning with matched averaging. In: International Conference on Learning Representations (2020)
23. Wang, T., Rausch, J., Zhang, C., Jia, R., Song, D.: A principled approach to data valuation for federated learning. In: Yang, Q., Fan, L., Yu, H. (eds.) Federated Learning. LNCS (LNAI), vol. 12500, pp. 153–167. Springer, Cham (2020). https://doi.org/10.1007/978-3-030-63076-8_11
24. Yang, Q., Liu, Y., Chen, T., Tong, Y.: Federated machine learning: concept and applications. ACM Trans. Intell. Syst. Technol. (TIST) 10(2), 1–19 (2019)
25. Yoon, J., Arik, S.O., Pfister, T.: Data valuation using reinforcement learning. In: International Conference on Machine Learning (2020)
26. Zhao, Y., Li, M., Lai, L., Suda, N., Civin, D., Chandra, V.: Federated learning with non-iid data (2018). arXiv preprint arXiv:1806.00582

# Boosting Few-Shot Learning with Task-Adaptive Multi-level Mixed Supervision

Duo Wang[1], Qianxia Ma[1], Ming Zhang[2], and Tao Zhang[1,3(✉)]

[1] Department of Automation, Tsinghua University, Beijing, China
mqx15@mails.tsinghua.edu.cn
[2] College of Engineering and Physical Sciences, Aston University, Birmingham, UK
m.zhang21@aston.ac.uk
[3] Beijing National Research Center for Information Science and Technology,
Tsinghua University, Beijing, China
taozhang@tsinghua.edu.cn

**Abstract.** In this paper, we propose a novel task-adaptive few-shot learning (FSL) method called Multi-Level Mixed Supervision (MLMS), which adapts a classifier specifically for each task by mixed supervision. Our method complements the supervised training with a multi-level unsupervised loss including the instance-level certainty term, set-level divergence term, and group-level consistency term. We further modify the set-level divergence term under the unbalanced prior situation where different classes of the unlabeled set contain different numbers of samples. Besides, we propose an approximate solution of minimizing our MLMS loss which is faster than the gradient-based method. Extensive experiments on multiple FSL datasets demonstrate that our method outperforms several recent models by an obvious margin on both transductive FSL and semi-supervised FSL tasks. Codes and trained models are available at https://github.com/Wangduo428/few-shot-learning-mlms.

**Keywords:** Few-shot learning · Task-adaptive · Multi-level mixed supervision · Transductive FSL · Semi-supervised FSL · Unbalanced prior

## 1 Introduction

Deep learning has achieved tremendous success, even outperforms human in various Artificial Intelligence (AI) tasks. However, its performance will degenerate severely when the amount of training data is limited, making Few-Shot Learning (FSL) a very active research topic recently [14,18]. Recent FSL works follow the similar idea that extracting task-agnostic prior knowledge from a large annotated dataset of some base categories to assist the learning of novel categories. Such a large base dataset is exploited by either training a generic feature extractor

© Springer Nature Switzerland AG 2021
L. Fang et al. (Eds.): CICAI 2021, LNAI 13070, pp. 176–187, 2021.
https://doi.org/10.1007/978-3-030-93049-3_15

through standard supervised learning, or a meta-model that can generate good classifiers in an episodic manner, or both.

To tackle the low-data regime of novel tasks, some works use unlabeled auxiliary samples for feature matching or model adapting, which can be the entire query set (transductive FSL [1,6,9,11–13,21,24]) or an additional unlabeled set (semi-supervised FSL [8,15,23]). In this paper, we follow this setting and propose a novel and effective method called Multi-Level Mixed Supervision (MLMS). For each FSL task from a novel set of categories, we train a classifier adaptively with mixed supervision integrating 4 loss terms: (1) supervised cross-entropy term from the labeled support set, (2) instance-level certainty term encouraging confident predictions of unlabeled samples, (3) set-level divergence term avoiding identical predictions of the whole unlabeled set and (4) group-level consistency term favoring consistent predictions of similar samples. We further consider a more practical unbalanced situation where different classes of unlabeled set contain different numbers of samples and propose a weighted et-level divergence term as modification towards the unbalanced prior. Besides, inspired by ADMM, we propose an approximate solution of minimizing our MLMS loss which is faster than gradient-based optimization without losing too much accuracy. We conduct comprehensive experiments on multiple popular FSL datasets and achieve a series of SOTA results on both transductive and semi-supervised FSL tasks.

**Related Works.** Recently, FSL with additional unlabeled data (including transductive FSL and semi-supervised) has attracted lots of attention and produced great performance improvement over supervised inductive counterparts. Related works can be generally categorized into: (1) label propagation or assigning [4,6,11,12,21,24], (2) feature space learning or inducing [9,13,16], (3) adaptive model fine-tuning [1,3,5,8,15,23]. Our method belongs to (3) and is closely related to [3] and [1], with the common ground of fine-tuning the model adaptively based on the information entropy of unlabeled predictions. The difference is two-fold. First, we additionally design a group-level consistency loss term to further regularize the model fine-tuning. Second, we modify the set-level divergence loss term by introducing weights over categories towards the unbalanced prior of unlabeled samples.

## 2    Proposed Method

### 2.1    Problem Definition

In this paper, we focus on the 'C-way, K-shot, M-test' few-shot learning (FSL) problem. Typically, a FSL task $\mathcal{T}$ defined on some novel set of categories $\mathcal{C}_n$ consists of a small labeled support set $\mathcal{S}$ and a large unlabeled query set $\mathcal{Q}$. $\mathcal{S} = \{(\boldsymbol{x}_i, \boldsymbol{y}_i)|, i = 1, 2, ...C \times K\}$ and $\mathcal{Q} = \{(\boldsymbol{x}_i, \boldsymbol{y}_i)|, i = 1, 2, ...C \times M\}$ contain $K$ and $M$ samples of each of $C$ different classes, respectively. $\boldsymbol{x}_i$ denotes the raw image data and $\boldsymbol{y}_i$ denotes its one-hot label. To tackle data scarcity, we consider to augment the small support set $\mathcal{S}$ by an unlabeled auxiliary set $\mathcal{U}$, which can be either the query set $\mathcal{Q}$ of the given FSL task following transductive FSL setting or an additional unlabeled set denoted by $\mathcal{U} = \{(\boldsymbol{x}_i, \boldsymbol{y}_i)|, i = 1, 2, ...C \times U\}$ following semi-supervised FSL. Let $g_\theta$ denote the feature extractor parameterized by $\theta$ that maps the input image to a $d$-dimensional feature embedding. A large labeled base dataset $\mathcal{D}_b = \{(\boldsymbol{x}_i, \boldsymbol{y}_i)|, i = 1, 2, ...N_b\}$ from a disjoint class set $\mathcal{C}_b$ (i.e. $\mathcal{C}_b \cap \mathcal{C}_n = \phi$) is also available to facilitate FSL tasks from novel classes. In this paper, we exploit $\mathcal{D}_b$ to first pre-train $g_\theta$ following standard supervised learning then following the typical episodic training method [17] to produce class-agnostic metric space for better generalization to different novel FSL tasks. The influence of episodic pre-training will be evaluated in Sect. 3.

## 2.2   Task-Adaptive FSL with Multi-level Mixed Supervision

For a given '$C$-way, $K$-shot, $M$-test' FSL task $\mathcal{T} = (\mathcal{S}, \mathcal{Q}, \mathcal{U})$ from some novel set of classes, we adaptively define a classifier with parameters $\boldsymbol{W} = [\boldsymbol{w}_1, \boldsymbol{w}_2, ...\boldsymbol{w}_C] \in \mathbb{R}^{d \times C}$ that produces posterior distributions over categories of input features based on the distance to each class weight:

$$p_{ia} = \frac{\exp(-\frac{\tau}{2}||\boldsymbol{w}_a - \boldsymbol{e}_i||^2)}{\sum_{b=1}^{C} \exp(-\frac{\tau}{2}||\boldsymbol{w}_b - \boldsymbol{e}_i||^2)} \tag{1}$$

where $\boldsymbol{e}_i$ denote the $L2$-normalized featrue embedding of sample $\boldsymbol{x}_i$ from pre-trained backbone $g_\theta$. $p_{ia}$ denotes the probability that the sample $\boldsymbol{x}_i$ belongs to the $a$th category.

We propose Multi-Level Mixed Supervision to optimize the parameters of task-adaptive classifier $\boldsymbol{W}$ with both the labeled support set $\mathcal{S}$ and unlabeled set $\mathcal{U}$ (or $\mathcal{Q}$ for transductive FSL). For labeled $\mathcal{S}$, the supervised loss with standard cross-entropy between the prediction and true label is calculated by:

$$L_{sup} = -\frac{1}{|\mathcal{S}|} \sum_{i \in \mathcal{S}} \sum_{a=1}^{C} y_{ia} \log(p_{ia}) \tag{2}$$

For unlabeled $\mathcal{U}$, we design a multi-level unsupervised loss which contains the **instance-level certainty** term, **set-level divergence** term, and **group-level consistency**. The **instance-level certainty** term forces the classifier to produce highly certain predictions for the unlabeled samples since they come from the same set of categories as the labeled ones. One common way to measure the certainty (or uncertainty) of random variables is the Shannon Entropy, which has been exploited by many semi-supervised learning works, given by:

$$L_{cer} = -\frac{1}{|\mathcal{U}|} \sum_{i \in \mathcal{U}} \sum_{a=1}^{C} p_{ia} \log(p_{ia}) \tag{3}$$

Optimizing $\boldsymbol{W}$ merely with the certainty loss may lead to a degenerate situation where all unlabeled samples are classified into a single category. Therefore, the **set-level divergence** term is exploited to increase the overall diversity of predicted labels of the unlabeled set. High diversity of prediction means that each category should contain substantial samples. In other words, the marginal distribution over categories estimated by the whole unlabeled set should be uncertain, yielding large Shannon Entropy, given by:

$$L_{div} = -\sum_{a=1}^{C} \hat{p}_a \log(\hat{p}_a) \tag{4}$$

where the estimated marginal distribution $\hat{p}_a$ is calculated by the average of posterior distributions of all unlabeled samples, i.e. $\hat{p}_a = \frac{1}{|\mathcal{U}|}\sum_{i\in\mathcal{U}} p_{ia}$.

Besides, we propose the **group-level consistency** term to encourage the classifier $W$ to make consistent predictions among similar feature samples. Concretely, for each unlabeled feature $e_i$, we find its $N$-nearest-neighbors from the unlabeled set let $\mathcal{N}(i)$ denote the set of their indices. Then the cross-entropy between the prediction of $e_i$ and those of its neighbors are minimized, given by:

$$L_{con} = -\frac{1}{N|\mathcal{U}|}\sum_{i\in\mathcal{U}}\sum_{j\in\mathcal{N}(i)}\sum_{a=1}^{C} p_{ia}\log(p_{ja}) \tag{5}$$

The final loss to adapt $W$ towards the given FSL task is the weighted sum of the terms introduced above:

$$L = \lambda L_{sup} - \delta L_{div} + \alpha L_{cer} + \beta L_{con} \tag{6}$$

### 2.3   Multi-level Mixed Supervision with Unbalanced Prior

The **set-level divergence** term imposes the estimated marginal distribution to be close to uniform, following an implicit assumption that the unlabeled set $\mathcal{U}$ contains an identical number of samples for all categories, which may not hold in reality. When the category prior of unlabeled set is unbalanced, maximizing the plain entropy of the estimated marginal distribution may produce degraded performance. In this paper, we propose a simple yet effective weighted entropy of category marginal distribution as a replacement to optimize $W$:

$$L_{div-un} = -\sum_{a=1}^{C} r_a * \hat{p}_a \log(\hat{p}_a) \tag{7}$$

where the weight $r_a$ control the bias of entropy towards category $a$, which is proportional to the current estimated marginal distribution:

$$r_a = \hat{p}_a * C \tag{8}$$

## 2.4   FSL Model Solution

We initialize the classifier parameters $\boldsymbol{w}_a$ by the mixure of support features of class $a$ and the weighted aggregation of unlabeled features:

$$\boldsymbol{w}_a^0 = \frac{\sum_{i\in\mathcal{U}} s_{ia} * \boldsymbol{e}_i + \sum_{i\in\mathcal{S}} y_{ia}\boldsymbol{e}_i}{\sum_{i\in\mathcal{U}} s_{ia} + \sum_{i\in\mathcal{S}} y_{ia}} \tag{9}$$

where $s_{ia}$ is defined as the similarity between the $i$th unlabeled feature and the support feature of class $a$ normalized by softmax. The classifier can be optimized by minimizing the loss given by Eq. 6 with the common gradient-descent-based (GD) method. Additionally, motivated by the Alternating Direction Method of Multipliers and a recent work [1], we derive an approximate solution of the optimal parameters $\boldsymbol{W}^*$ which is of higher computing efficiency without losing much performance. We present the fast approximate solution with the set-level divergence term of unbalanced prior. Solving with the balanced prior is a special case when $r_a$ is set to 1. Plug the loss terms Eq. 2 3 7 5 into Eq. 6 and introduce auxiliary variables $\boldsymbol{q}$ to replace the posterior distributions of unlabeled samples, we have a multi-variable objective function:

$$L(\boldsymbol{W}, \boldsymbol{q}) = -\frac{\lambda}{|\mathcal{S}|}\sum_{i\in\mathcal{S}}\sum_{a=1}^{C} y_{ia}\log(p_{ia}) + \delta\sum_{a=1}^{C} r_a\hat{q}_a \log(\hat{q}_a) - \frac{\alpha}{|\mathcal{U}|}\sum_{i\in\mathcal{U}}\sum_{a=1}^{C} q_{ia}\log(p_{ia})$$

$$-\frac{\beta}{N|\mathcal{U}|}\sum_{i\in\mathcal{U}}\sum_{j\in\mathcal{N}(i)}\sum_{a=1}^{C} q_{ia}\log(p_{ja}) + \frac{1}{|\mathcal{U}|}\sum_{i\in\mathcal{U}}\sum_{a=1}^{C} q_{ia}\log(\frac{q_{ia}}{p_{ia}})$$

$$\text{s.t.} \sum_{a=1}^{C} q_{ja} = 1, \quad q_{ja} \geq 0, \quad \forall i \in \mathcal{U} \tag{10}$$

where the last term is the KL-divergence between $\boldsymbol{q}$ and $\boldsymbol{p}$ to guarantee the equivalence to the original objective. Eq. 10 can be solved by alternately optimizing $\boldsymbol{q}$ while fixing $\boldsymbol{W}$ ($\boldsymbol{q}$-step) and then vice versa ($\boldsymbol{W}$-step). For $\boldsymbol{q}$-step, the KL-divergence implicitly eliminates the inequality constraint and yields a strictly convex problem. By introducing Lagrange variables and solving KKT conditions, the optimal $\boldsymbol{q}^*$ should satisfy:

$$q_{ia} = \frac{p_{ia}^{1+\alpha}(\prod_{j\in\mathcal{N}(i)} p_{ja})^{\beta/N} e^{-r_a\delta}}{(\frac{1}{|\mathcal{U}|}\sum_{i\in\mathcal{U}} q_{ia})^{r_a\delta}} \frac{1}{\sum_{a=1}^{C}\frac{p_{ia}^{1+\alpha}(\prod_{j\in\mathcal{N}(i)} p_{ja})^{\beta/N} e^{-r_a\delta}}{(\frac{1}{|\mathcal{U}|}\sum_{i\in\mathcal{U}} q_{ia})^{r_a\delta}}} \tag{11}$$

The optimal $\boldsymbol{q}^*$ can not be directly solved from Eq. 11. Alternatively, we set $\boldsymbol{q}$ in the right side of Eq. 11 to the value from the previous step and calculate $\boldsymbol{q}^*$ in an iterative way.

For $\boldsymbol{W}$-step, substitute $\boldsymbol{p}$ in Eq. 10 by Eq. 1, we have:

$$L(\boldsymbol{W}) = \frac{\tau\lambda}{2|S|} \sum_{i\in S}\sum_{a=1}^{C} y_{ia}||\boldsymbol{w}_a - \boldsymbol{e}_i||^2 + \frac{\lambda}{|S|}\sum_{i\in S}\sum_{a=1}^{C} y_{ia}\log(\sum_{b=1}^{C}\exp(-\frac{\tau}{2}||\boldsymbol{w}_b - \boldsymbol{e}_i||^2))$$

$$+ \frac{\tau(\alpha+1)}{2|\mathcal{U}|}\sum_{i\in\mathcal{U}}\sum_{a=1}^{C} q_{ia}||\boldsymbol{w}_a - \boldsymbol{e}_i||^2 + \frac{\alpha+1}{|\mathcal{U}|}\sum_{i\in\mathcal{U}}\sum_{a=1}^{C} q_{ia}\log(\sum_{b=1}^{C}\exp(-\frac{\tau}{2}||\boldsymbol{w}_b - \boldsymbol{e}_i||^2))$$

$$+ \frac{\tau\beta}{2|\mathcal{U}|}\sum_{i\in\mathcal{U}}\sum_{j\in\mathcal{N}(i)}\sum_{a=1}^{C} q_{ia}||\boldsymbol{w}_a - \boldsymbol{e}_j||^2 + \frac{\beta}{N|\mathcal{U}|}\sum_{i\in\mathcal{U}}\sum_{j\in\mathcal{N}(i)}\sum_{a=1}^{C} q_{ia}\log(\sum_{b=1}^{C}\exp(-\frac{\tau}{2}||\boldsymbol{w}_b - \boldsymbol{e}_j||^2))$$

$$+ \delta\sum_{a=1}^{C} r_a\hat{q}_a\log(\hat{q}_a) + \frac{1}{|\mathcal{U}|}\sum_{i\in\mathcal{U}}\sum_{a=1}^{C} q_{ia}\log(q_{ia}) \qquad (12)$$

The last line of Eq. 12 can be considered as constant w.r.t $\boldsymbol{W}$. For each of the other lines, the first term is quadratic and the second term can be approximated linearly at the current point by first-order Taylor expansion, making Eq. 12 also a strictly convex problem without constraint. Thus, the optimal $\boldsymbol{W}^*$ can be obtained by setting its gradient to 0, which is:

$$\boldsymbol{w}_a^{n+1} = \frac{1}{\frac{\lambda}{|S|}\sum_{i\in S} y_{ia} + \frac{\alpha+\beta+1}{|\mathcal{U}|}\sum_{i\in\mathcal{U}} q_{ia}}(\frac{\lambda}{|S|}\sum_{i\in S}(p_{ia}^n(\boldsymbol{w}_a^n - \boldsymbol{e}_i) + y_{ia}\boldsymbol{e}_i)$$

$$+ \frac{\alpha+1}{|\mathcal{U}|}\sum_{i\in\mathcal{U}}(p_{ia}^n(\boldsymbol{w}_a^n - \boldsymbol{e}_i) + q_{ia}\boldsymbol{e}_i) + \frac{\beta}{N|\mathcal{U}|}\sum_{i\in\mathcal{U}}\sum_{j\in\mathcal{N}(i)}(p_{ja}^n(\boldsymbol{w}_a^n - \boldsymbol{e}_j) + q_{ia}\boldsymbol{e}_j)) \qquad (13)$$

## 3   Experiments

### 3.1   Setup

**Datasets.** We conduct experiments on 3 common few-shot learning datasets: **mini-ImageNet** [18], **tiered-ImageNet** [15], and **CUB-200** [19], which are split into 64-16-20, 351-97-160, and 100-50-50 in classes as the base, validation, and novel subsets respectively, following general protocol. The base and validation subsets are used to pre-train and validate the backbone and the novel subsets provide FSL tasks for model evaluation.

**Implementation Details.** We exploit 2 backbones in this paper which are **ResNet-12** and **ResNet-18**. We first follow [22] to pre-train ResNet-12 and follow [20] to pre-train ResNet-18. Then we follow [17] to conduct metric meta-training to both backbones. The number of meta-training epochs is 200 and each epoch contains 100 synthesized FSL tasks from the base split. SGD optimizer is exploited with the initial learning rate of 0.001, Nesterov momentum 0.9, and weight decay 0.0005. The learning rate is reduced by half every 40 epochs. Images are resized to 84×84 and augmented by random cropping and horizontal flipping. For updating the task-adaptive classifier $\boldsymbol{W}$, the gradient-descent-based method uses ADAM optimizer with the learning rate of 0.001. The updating steps $n_{out}$ is set to 80 in both methods and the steps of iterating $\boldsymbol{q}$ $n_{\mathcal{U}}$ is set to 10. Other hyper-parameters including the weights of different loss terms, $\tau$, and $N$ are determined through grid search and will be introduced in Sect. 3.3.

**Table 1.** Comparative results of transductive single-domain FSL with mini-ImageNet and tiered-ImageNet datasets. Statistical over 2000 FSL tasks. "–" means not given. The best and second-best results are marked in bold and underline.

| Method | Backbone | Mini-ImageNet | | Tiered-ImageNet | |
|---|---|---|---|---|---|
| | | 1-shot | 5-shot | 1-shot | 5-shot |
| CAN-T [4] | R-12 | $67.19 \pm 0.55$ | $80.64 \pm 0.35$ | $73.21 \pm 0.58$ | $84.93 \pm 0.38$ |
| DPGN [21] | R-12 | $67.77 \pm 0.32$ | $84.60 \pm 0.43$ | $72.45 \pm 0.51$ | $87.24 \pm 0.39$ |
| TAFSSL [9] | R-18 | $73.73 \pm 0.27$ | – | $80.60 \pm 0.27$ | – |
| TIM [1] | R-18 | 73.9 | 85.0 | 80.0 | 88.5 |
| MCT [7] | R-12 | $\underline{78.30 \pm 0.81}$ | $\underline{86.48 \pm 0.42}$ | $80.89 \pm 0.84$ | $87.30 \pm 0.49$ |
| Trans-Fine [3] | W-28 | $65.73 \pm 0.68$ | $78.40 \pm 0.52$ | $73.34 \pm 0.71$ | $85.50 \pm 0.50$ |
| TRPN [12] | W-28 | $68.25 \pm 0.50$ | $85.40 \pm 0.39$ | $70.25 \pm 0.50$ | $85.21 \pm 0.37$ |
| SIB [5] | W-28 | $70.0 \pm 0.6$ | $79.2 \pm 0.4$ | – | – |
| BD-CSPN [10] | W-28 | $70.31 \pm 0.93$ | $81.89 \pm 0.60$ | $78.74 \pm 0.95$ | $86.92 \pm 0.63$ |
| LaplacianShot [24] | W-28 | $74.86 \pm 0.19$ | $84.13 \pm 0.14$ | $80.18 \pm 0.21$ | $87.56 \pm 0.15$ |
| TAFSSL [9] | Dense | $77.06 \pm 0.26$ | $84.99 \pm 0.14$ | $\mathbf{84.29 \pm 0.25}$ | $\underline{89.31 \pm 0.15}$ |
| MLMS-gd | R-12 | $\mathbf{79.23 \pm 0.58}$ | $\mathbf{87.73 \pm 0.29}$ | $80.58 \pm 0.59$ | $88.88 \pm 0.32$ |
| MLMS-fast | R-12 | $79.06 \pm 0.58$ | $87.66 \pm 0.29$ | $80.39 \pm 0.63$ | $88.81 \pm 0.32$ |
| MLMS-gd | R-18 | $77.19 \pm 0.61$ | $85.38 \pm 0.33$ | $\underline{83.01 \pm 0.56}$ | $\mathbf{89.41 \pm 0.32}$ |
| MLMS-fast | R-18 | $76.98 \pm 0.62$ | $85.55 \pm 0.32$ | $82.77 \pm 0.57$ | $89.31 \pm 0.33$ |

**Table 2.** Comparative results of the CUB-200 dataset on transductive single-domain and cross-domain FSL. Statistical over 2000 FSL tasks. "–" means not given. The best and second-best results are marked in bold and underline.

| Method | Backbone | CUB-200 | | Mini-ImageNet→CUB-200 | |
|---|---|---|---|---|---|
| | | 1-shot | 5-shot | 1-shot | 5-shot |
| DPGN[21] | R-12 | $75.71 \pm 0.47$ | $91.48 \pm 0.33$ | – | – |
| TEAM[13] | R-18 | 80.16 | 87.17 | – | – |
| LaplacianShot[24] | R-18 | 80.96 | 88.68 | $\underline{55.46}$ | 66.33 |
| TIM[1] | R-18 | 82.2 | 90.8 | – | $\underline{71.10}$ |
| BD-CSPN | W-28 | $\underline{87.45}$ | $\underline{91.74}$ | – | – |
| MLMS-gd | R-12 | $\mathbf{91.14 \pm 0.42}$ | $\mathbf{94.08 \pm 0.20}$ | $\mathbf{58.68 \pm 0.67}$ | $\mathbf{75.63 \pm 0.46}$ |
| MLMS-fast | R-12 | $91.00 \pm 0.41$ | $93.90 \pm 0.19$ | $57.32 \pm 0.66$ | $74.56 \pm 0.46$ |

**Table 3.** Comparative results of semi-supervised FSL with mini-ImageNet and tiered-ImageNet datasets. Statistical over 2000 FSL tasks. "–" means not given. The best and second-best results are marked in bold and underline.

| Method | Backbone | Mini-ImageNet | | Tiered-ImageNet | |
|---|---|---|---|---|---|
| | | 1-shot | 5-shot | 1-shot | 5-shot |
| SKM[15] | C-4 | 62.10 | 73.60 | 68.60 | 81.00 |
| TPN[11] | C-4 | 62.70 | 74.20 | 72.10 | 83.30 |
| TEAM[13] | R-18 | $54.81 \pm 0.59$ | $68.92 \pm 0.38$ | – | – |
| LST[8] | R-12 | $70.10 \pm 1.90$ | $78.70 \pm 0.80$ | $77.70 \pm 1.60$ | $85.20 \pm 0.80$ |
| TransMatch[23] | W-28 | $63.02 \pm 1.07$ | $82.24 \pm 0.59$ | – | – |
| TAFSSL[9] | Dense | $\underline{80.11 \pm 0.25}$ | $\underline{85.78 \pm 0.13}$ | $\mathbf{86.00 \pm 0.23}$ | $\mathbf{89.39 \pm 0.15}$ |
| MLMS-gd | R-12 | $\mathbf{81.33 \pm 0.57}$ | $\mathbf{88.23 \pm 0.29}$ | $\underline{83.25 \pm 0.55}$ | $\underline{89.20 \pm 0.33}$ |
| MLMS-fast | R-12 | $80.62 \pm 0.57$ | $87.76 \pm 0.26$ | $82.44 \pm 0.55$ | $88.69 \pm 0.34$ |

## 3.2   Comparison with SOTA Works

We compare our proposed method with extensive state-of-the-art transductive FSL and semi-supervised FSL works. Transductive FSL results on mini-ImageNet and tiered-ImageNet are listed in Table 1, and results on CUB-200 and a more challenging cross-domain setting 'mini-ImageNet→CUB-200' [2] are listed in Table 2. Results of semi-supervised FSL are shown in Table 3. All the results are presented in the form of average classification accuracy and 95% confidence interval over 2000 randomly-sampled '5-way, 1-shot/5-shot, 15-test' FSL tasks of the novel subsets. For semi-supervised FSL, we exploit 50 unlabeled samples per class. Results show that our proposed methods outperform a series of SOTA works by an obvious margin almost under all FSL settings and datasets except some cases on tiered-ImageNet. However, our methods still rank in top-2 and the best competitor TAFSSL [9] exploits DenseNet as backbone which costs more computational resources than ResNet. The results of TAFSSL with ResNet backbone are inferior to ours (see Table 1). Besides, the performance of the fast solution is very similar to that of the gradient-based method, demonstrating the effectiveness of the approximation.

## 3.3   Ablation Study

**Hyper-parameters.** We conduct the grid search to determine the hyper-parameters including weights of different loss terms, $\tau$, and $N$. Limited by space, we only present the results. $\delta$ and $\alpha$ are set to 1.0 and 0.1 for all experiments. $\tau$ is set to 10 for 1-shot mini-ImageNet and 15 for the rest. $\beta$ and $N$ are set to 0.1 and 4 for 1-shot setting and 0.02 and 2 for 5-shot setting. $\lambda$ is set to 0.1 for mini-ImageNet and 0.5 for 1-shot tiered-ImageNet, and 0.2 for the rest.

**Loss Terms and Training Options.** We evaluate the influence of all loss terms and training options including meta pre-training and initialization with provided samples. The results are listed in Table 4. We can see that all the unsupervised terms consistently enhance the accuracy over the plain supervised training. However, the improvement of $L_{cer}$ is very limited and the possible reason is that $L_{cer}$ will uncontrollably lead to a degenerate situation where all samples are classified into the same category. Regularizing it further with the divergence loss and consistency loss will bring remarkable improvement. The meta pre-training also plays an important role in the final performance. In a word, the model equipped with all terms and options outperform other counterparts across all 3 datasets.

184    D. Wang et al.

**Table 4.** Ablation study of loss terms and training options over 600 FSL tasks.

| $L_{sup}$ | $L_{div}$ | $L_{cer}$ | $L_{con}$ | MetaPre | Init | 1-shot/5-shot Mini | Tiered | Cub |
|---|---|---|---|---|---|---|---|---|
| ✓ | | | ✓ | ✓ | ✓ | 69.39/83.26 | 71.25/85.70 | 82.40/90.72 |
| ✓ | ✓ | | ✓ | ✓ | ✓ | 73.20/85.63 | 75.26/87.46 | 86.77/91.83 |
| ✓ | | ✓ | ✓ | ✓ | ✓ | 69.47/84.15 | 71.51/86.19 | 85.52/91.48 |
| ✓ | ✓ | ✓ | | ✓ | ✓ | 78.58/87.71 | 79.78/88.71 | 87.37/93.38 |
| ✓ | ✓ | | ✓ | ✓ | ✓ | 79.35/87.07 | 80.48/88.38 | 90.76/92.70 |
| ✓ | ✓ | ✓ | ✓ | | ✓ | 75.43/85.01 | 77.25/86.13 | 83.75/90.06 |
| ✓ | ✓ | ✓ | ✓ | ✓ | | 79.24/87.92 | 80.84/88.84 | 90.81/93.49 |
| | ✓ | ✓ | ✓ | ✓ | ✓ | 79.30/87.36 | 80.45/88.28 | 90.57/93.16 |
| ✓ | ✓ | ✓ | ✓ | ✓ | ✓ | **79.66/88.03** | **80.89/88.89** | **90.89/93.56** |

**Numbers of Unlabeled Samples.** We vary the number of unlabeled samples per class from 5 to 50 and test our method on mini-ImageNet and tiered-ImageNet datasets. The results are illustrated in Fig. 1(a) (transductive FSL) and (b) (semi-supervised FSL). As is shown in the figure, the accuracy generally increases as more unlabeled samples are included.

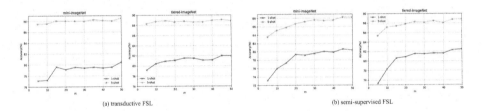

(a) transductive FSL          (b) semi-supervised FSL

**Fig. 1.** Different numbers of unlabeled samples and accuracies of mini-ImageNet and tiered-ImageNet. (a) Transductive FSL; (b) Semi-supervised FSL.

**Unbalanced Prior.** For the FSL with unbalanced prior, different classes in a task contain different numbers of unlabeled samples. Here we set the median number to 15 and a 5-way FSL task contains $15-2*un\_b$, $15-un\_b$, 15, $15+un\_b$, $15+2*un\_b$ unlabeled samples for the 5 categories, where $un\_b$ is the unbalanced factor and varies from 0 to 7 in our experiments. The true prior distribution over classes can be calculated by the number of samples each class contains divided by the total unsupervised number. Results of transductive FSL and semi-supervised FSL with mini-ImageNet and tiered-ImageNet are shown in Fig. 2. "TP" means True Prior, where the $L_{div}$ is substituted by the cross-entropy between the true prior distribution and the prediction of the model. "NP" means None Prior, where the unbalanced prior is not taken into consideration, and

$L_{div}$ remains unchanged as Eq. 4. "EP" denotes Estimated Prior that calculates $L_{div}$ by Eq. 7, and "EP-F" means its fast approximation solution. As is shown in Fig. 2, introducing prior in the training loss will keep the accuracy much more stable. As the unbalance becomes large, performance without true prior will drop, and the divergence loss term with estimated prior outperforms that without prior by an obvious margin. However, under small unbalance, the model without prior performs better. This is probably because the bias between the estimated prior and small unbalanced prior is larger than that between balance prior and small unbalanced prior.

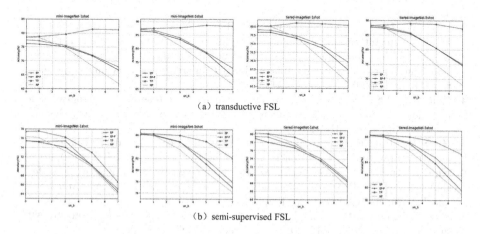

(a) transductive FSL

(b) semi-supervised FSL

**Fig. 2.** Different unbalanced factors and accuracies of mini-ImageNet and tiered-ImageNet. (a) Transductive FSL; (b) Semi-supervised FSL.

**Running Times.** We report running times of 600 "5-way, 1-shot/5-shot, 15-test" FSL tasks from mini-ImageNet with ResNet-12 backbone. All the experiments are run on two GTX 1080 Ti GPUs. For 1-shot setting, MLMS-fast spends 55.7 s totally, i.e. 0.093 s/task, while MLMS-gd spends 78.5 s/600 = 0.131 s/task. For 5-shot setting, MLMS-fast spends 67.1 s/600 = 0.112 s/task and MLMS-gd spends 81.2 s/600 = 0.135 s/task. MLMS-fast behaves more efficient than the gradient-based solution.

## 4  Conclusion

In this paper, we propose a simple yet effective FSL method called Multi-Level Mixed Supervision (MLMS) that adapts the classifier specifically for a given FSL task based on the mixed supervision. In conjunction with the episodic pre-training of backbone with base classes, our method achieves a series of state-of-the-art results on both transductive and semi-supervised FSL benchmarks.

Incorporating weights over categories induced by the estimated prior into set-level divergence term yields great improvement in the unbalanced situation. The fast solution based on ADMM and first-order approximation will speed up the inference with little performance gap. We believe our work in this paper could provide substaintial benefit to the FSL research in the future.

## References

1. Boudiaf, M., Masud, Z.I., Rony, J., Dolz, J., Piantanida, P., Ayed, I.B.: Transductive information maximization for few-shot learning (2020). arXiv preprint arXiv:2008.11297
2. Chen, W., Liu, Y., Kira, Z., Wang, Y.F., Huang, J.: A closer look at few-shot classification. In: International Conference on Learning Representations, ICLR (2019)
3. Dhillon, G.S., Chaudhari, P., Ravichandran, A., Soatto, S.: A baseline for few-shot image classification. In: International Conference on Learning Representations, ICLR (2020)
4. Hou, R., Chang, H., Ma, B., Shan, S., Chen, X.: Cross attention network for few-shot classification. In: Advances in Neural Information Processing Systems, NeurIPS, pp. 4005–4016 (2019)
5. Hu, S.X., et al.: Empirical bayes transductive meta-learning with synthetic gradients. In: International Conference on Learning Representations, ICLR (2020)
6. Kim, J., Kim, T., Kim, S., Yoo, C.D.: Edge-labeling graph neural network for few-shot learning. In: Proceedings of the IEEE/CVF Conference on Computer Vision and Pattern Recognition, pp. 11–20 (2019)
7. Kye, S.M., Lee, H.B., Kim, H., Hwang, S.J.: Transductive few-shot learning with meta-learned confidence (2020). arXiv preprint arXiv:2002.12017
8. Li, X., et al.: Learning to self-train for semi-supervised few-shot classification. In: Advances in Neural Information Processing Systems: 33rd Conference on Neural Information Processing Systems (NeurIPS 2019), Vancouver, Canada, December, vol. 8, pp. 1–11 (1906)
9. Lichtenstein, M., Sattigeri, P., Feris, R., Giryes, R., Karlinsky, L.: TAFSSL: task-adaptive feature sub-space learning for few-shot classification. In: Vedaldi, A., Bischof, H., Brox, T., Frahm, J.-M. (eds.) ECCV 2020. LNCS, vol. 12352, pp. 522–539. Springer, Cham (2020). https://doi.org/10.1007/978-3-030-58571-6_31
10. Liu, J., Song, L., Qin, Y.: Prototype rectification for few-shot learning. In: Vedaldi, A., Bischof, H., Brox, T., Frahm, J.-M. (eds.) ECCV 2020. LNCS, vol. 12346, pp. 741–756. Springer, Cham (2020). https://doi.org/10.1007/978-3-030-58452-8_43
11. Liu, Y., et al.: Learning to propagate labels: transductive propagation network for few-shot learning. In: International Conference on Learning Representations, ICLR (2019)
12. Ma, Y., et al.: Transductive relation-propagation network for few-shot learning. In: Proceedings of the Twenty-Ninth International Joint Conference on Artificial Intelligence, IJCAI-20, pp. 804–810 (2020). https://doi.org/10.24963/ijcai.2020/112
13. Qiao, L., Shi, Y., Li, J., Wang, Y., Huang, T., Tian, Y.: Transductive episodic-wise adaptive metric for few-shot learning. In: Proceedings of the IEEE/CVF International Conference on Computer Vision, pp. 3603–3612 (2019)
14. Ravi, S., Larochelle, H.: Optimization as a model for few-shot learning. In: International Conference on Learning Representations, ICLR (2017)

15. Ren, M., et al.: Meta-learning for semi-supervised few-shot classification. In: International Conference on Learning Representations, ICLR (2018)

16. Simon, C., Koniusz, P., Nock, R., Harandi, M.: Adaptive subspaces for few-shot learning. In: IEEE/CVF Conference on Computer Vision and Pattern Recognition (CVPR) (2020)

17. Snell, J., Swersky, K., Zemel, R.: Prototypical networks for few-shot learning. In: Advances in Neural Information Processing Systems, NeurIPS, pp. 4077–4087 (2017)

18. Vinyals, O., Blundell, C., Lillicrap, T., Wierstra, D., et al.: Matching networks for one shot learning. In: Advances in Neural Information Processing Systems, NeurIPS, pp. 3630–3638 (2016)

19. Wah, C., Branson, S., Welinder, P., Perona, P., Belongie, S.: The caltech-ucsd birds-200-2011 dataset (2011)

20. Wang, Y., Chao, W., Weinberger, K.Q., van der Maaten, L.: Simpleshot: revisiting nearest-neighbor classification for few-shot learning. arXiv e-prints (2019)

21. Yang, L., Li, L., Zhang, Z., Zhou, X., Zhou, E., Liu, Y.: DPGN: distribution propagation graph network for few-shot learning. In: IEEE/CVF Conference on Computer Vision and Pattern Recognition (CVPR) (2020)

22. Ye, H.J., Hu, H., Zhan, D.C., Sha, F.: Few-shot learning via embedding adaptation with set-to-set functions. In: IEEE Conference on Computer Vision and Pattern Recognition, CVPR, pp. 8808–8817 (2020)

23. Yu, Z., Chen, L., Cheng, Z., Luo, J.: Transmatch: a transfer-learning scheme for semi-supervised few-shot learning. In: Proceedings of the IEEE/CVF Conference on Computer Vision and Pattern Recognition, pp. 12856–12864 (2020)

24. Ziko, I., Dolz, J., Granger, E., Ayed, I.B.: Laplacian regularized few-shot learning. In: International Conference on Machine Learning, pp. 11660–11670. PMLR (2020)

# Learning Bilevel Sparse Regularized Neural Network

Xin Xu[1(✉)], Liangliang Zhang[2], and Qi Kong[2]

[1] Autonomous Driving Division, JD.com, Beijing, China
`xuxin178@jd.com`
[2] Autonomous Driving Division, JD.com American Technologies Corporation,
Mountain View, CA 94043, USA
{`liangliang.zhang,qi.kong`}`@jd.com`

**Abstract.** Sparse regularization has attracted considerable attention in machine learning community these years, which is a quite powerful and widely used strategy for high dimensional learning problems. However, when applied in deep neural networks (DNNs), sparse regularizers have a lot of redundant weights and unnecessary connections, and little work has been devoted to regularizer-based method for DNNs sparsification. Therefore, we aim to develop a proper sparse regularizer that can avoid augmenting excessive computation complexity in DNNs. In this paper, we find that the sparse regularizer learning corresponds to learning a activation function. Further, the regularizer is learned by the bilevel optimization method for smaller number of function evaluations. Moreover, we design a novel learning method, named bilevel sparse regularized neural network (BSRL) to learn the regularization parameters based on the prior knowledge of the system. Experimental results on standard benchmark datasets show that the proposed BSRL framework outperforms other models with state-of-the-art sparse regularizers.

**Keywords:** Sparse regularizer · Neural networks · Bilevel optimization

## 1 Introduction

Recent progress in deep learning [8] has improved the state-of-the-art performance in a range of applications. The multiple layers of non-linear transformations in a deep neural network (DNN), or related network variations, allow complex and difficult data to be well modelled. However, its high-level abstraction and representation of input features make it difficult to interpret the DNN parameters. This can cause various issues for improving parameter estimation, and network generalisation.

To reduce over-fitting, regularisation techniques are commonly used in DNN training. Weight decay adds a squared $L_2$-norm term of the DNN parameters to the cost function. This penalises large weights during parameter optimisation.

---

This work was supported by JD.com, Beijing, China.

L. Fang et al. (Eds.): CICAI 2021, LNAI 13070, pp. 188–199, 2021.
https://doi.org/10.1007/978-3-030-93049-3_16

Rather than modifying the criterion, dropout [10] randomly turns off, drops, a set of nodes during the training procedure; as a result, the final DNN can be viewed as an ensemble model of many small DNNs. This averaging helps reduce over-fitting to the training data.

The sparse optimization method can be utilized to regularizers to produce sparse solutions. The $\ell_0$ norm, which counts the number of non-zero elements, is the most intuitive form of sparse regularizers and can promote the sparsest solution. However, minimizing the $\ell_0$ problem is combinatory and usually NP-hard [3,11,21]. The $\ell_1$ norm is the most commonly used surrogate [6,25], which is convex and can be solved easily. Although $\ell_1$ enjoys several good properties, it is sensitive to outliers and may cause serious bias in estimation [9,13]. To overcome this defect, many norms are proposed and analyzed, including smoothly clipped absolute deviation (SCAD) [9,17], log penalty [6,32], capped $\ell_1$ [7,31], minimax concave penalty (MCP) [14,31], $\ell_p$ penalty with $p \in (0,1)$ [5,30], the difference of $\ell_1$ and $\ell_2$ norms [18,26].

However, as far as we know, existing regularizers on these efficient algorithms either have relatively poor ability in promoting the sparseness of the solution, or lack applicability due to the tough choices of hyper-parameters. And we also realize that existing works rarely discuss the regularization based on bilevel optimization method, which has the potential to reduce over-fitting and improve the capability to generalise DNNs.

In this paper, we propose a novel network dubbed bilevel sparse regularized neural network using activation function (BSRL), whose generalization is improved utilizing bilevel sparse regularization. Bilevel optimization problem contains two levels of optimization tasks, whose optimal solutions to the lower level problem become possible feasible candidates to the upper level problem. The main contributions of this paper can be summarized into the following aspects:

- We find a strong connection between sparse regularizers and activation functions, which converts the problem of learning the sparse regularizer into that of learning a activation function. (see Subsect. 2.1).
- We propose a novel bilevel optimization method to learn a sparse regularizer based on the idea of meta learning, which is a nested optimization problem that involves two levels of optimization tasks. The approach is capable of handling complicated bilevel optimization problems and learning the optimal regularizer in relatively smaller number of function evaluations (see Subsect. 2.2).
- We set up a novel learning method, named bilevel sparse regularized neural network (BSRL), where the outer level objective measures an expectation of the error over the training data and the inner level problem measures the regularized data misfit (see Subsect. 2.3). Experimental results show that BSRL outperforms the networks with existing sparse regularizers, both in terms of classification accuracy and regularization effectiveness.

## 2  Proposed Approaches

### 2.1  A Strong Connection Between Regularization and Activation Function

Proximal algorithms are very generally applicable. Their base operation is evaluating the proximal operator of a function, which involves solving a small convex optimization problem that often admits a closed-form solution. In particular, the proximal operator often admits a closed-form solution. In particular, the proximal operator $\text{prox}_f(x) : \mathbb{R} \to \mathbb{R}$ of a function is defined as

$$\text{prox}_f(x) = \arg\min_y \frac{1}{2}\|y - x\|^2 + f(y), \tag{1}$$

where $f(\cdot)$ can be a sparse regularizer. If $f$ is convex, the fixed points of the proximal operator of $f$ are precisely the minimizers of $f$. In other words, $\text{prox}_f(x^*) = x^*$ iff $x^*$ minimizes $f$. This fix-point property motivates the simplest proximal method called the proximal point algorithm which iterates $x^{(n+1)} = \text{prox}_f(x^{(n)})$. All the proximal algorithms used here are based on this fix-point property. Note that even if the function $f(\cdot)$ is not differentiable (e.g., $\ell_1$ norm) there might exist a closed-form or easy-to-compute proximal operator.

Then we define a regularizer $\mathcal{F}(x)$ :

$$\mathcal{F}(x) = \sum_{i=1}^n \int_0^{x_i} \left( \xi^{-1}(y) - y \right) dy = \sum_{i=1}^n \int_0^{x_i} \xi^{-1}(y) dy - \frac{1}{2}\|x\|^2, \tag{2}$$

where $\xi(\cdot) : \mathbb{R} \to \mathbb{R}$ is a activation function, $\xi^{-1}(y)$ is the inverse function of $\xi(\cdot)$. Then we can verify that the solution to the proximal operator:

$$\text{prox}_{\mathcal{F}}(x) = \underset{y}{\text{argmin}} \frac{1}{2}\|y - x\|^2 + \mathcal{F}(y) \tag{3}$$

is exactly $y = \text{prox}_{\mathcal{F}}(x) = \xi(x)$ [19], where $\xi(x)$ is applied to $x$ entrywise and a non-decreasing Lipschitz continuous function of $x$, which can be saturating (e.g., sigmoid and tanh) and non-differentiable (e.g., ReLU and leaky ReLU).

We notice a strong connection between the regularizer $\mathcal{F}(x)$ and the activation function $\xi(x)$ by the solution to proximal operator. For instance, once we decide upon a choice of regularizer, the activation function is dictated by that choice. On the other hand, if we choose a activation function first, then the regularization is dictated by that choice. With the above analysis, regularization in a model can be transformed into learning a activation function which is non-decreasing.

### 2.2  Learnable Bilevel Sparse Regularizer

Due to the computational complexity of learning activation functions, traditional optimization algorithms suffer from low efficiency since they usually require a huge number of function evaluations. In order to solve this problem, we introduce the idea of bilevel optimization which is widely used in meta learning [23]. Bilevel optimization is a class of problems which exhibit a two-level structure, and its goal is to minimize an outer objective function (4) with variables which are constrained to be the optimal solution to an inner optimization problem (5). In this section, we consider learning a regularizer as a bilevel optimization problem:

$$\min_{y \in \mathcal{Y}_a} \mathcal{L}(y) = \frac{1}{2m} \sum_{i=1}^{m} \left\| \xi^{(i)}(y) - x_{true}^{(i)} \right\|^2, \tag{4}$$

s.t. $\xi^{(i)}(y) = \arg\min_{x \in X_a} \mathcal{J}^{(i)}(x) = \arg\min_{x \in X_a} \frac{1}{2} \|x - y\|^2 + \mathcal{F}(x) = \phi(x) + \mathcal{F}(x),$

$$\tag{5}$$

where $x = \xi(y)$ based on the analysis of Subsect. 2.1, $\mathcal{Y}_a$ is a closed convex and nonempty admissible set for $y$ and $X_a$ is a closed, convex, nonempty admissible set which is contained in the solution space $X$ ($x \in X_a \subseteq X$).

Then, we can write our inner minimization problem (5) with a generalized regularizer as an average:

$$\xi^{(i)}(y) = \arg\min_{x \in X_a} \mathcal{J}^{(i)}(x) = \arg\min_{x \in X_a} \frac{1}{2m} \sum_{i=1}^{m} \left[ \left\| x^{(i)} - y^{(i)} \right\|^2 + \mathcal{F}(x^{(i)}) \right], y \in \mathcal{Y}_a.$$

$$\tag{6}$$

To solve this problem in (6), we will employ derivative based methods [2] such as projected gradient descent. The directional derivative of $\mathcal{J}$ in a direction $\vec{h}$ in (6) w.r.t $y$ in its variational form is, for $i = 1, ..., m$,

$$\mathbf{D}\mathcal{J}^{(i)} \left( x^{(i)} \right) [\vec{h}] = \frac{1}{m} \left[ ((x^{(i)} - y^{(i)}), \vec{h}) + \left( (\partial_{x^{(i)}} \xi)^* (\partial_\xi \mathcal{F}) \mathcal{F}, \vec{h} \right) \right], \tag{7}$$

where $(\partial_{x^{(i)}} \xi)^*$ is the adjoint of $\partial_{x^{(i)}} \xi$.

The iteration rule of generalized proximal gradient (GPG) method [19] for solving the minimization problem (5) is as follows:

$$x^{(k+1)} = \arg\min_x \phi \left( x^{(k)} \right) + \left\langle \nabla\phi \left( x^{(k)} \right), x - x^{(k)} \right\rangle + \frac{L}{2} \left\| x - x^{(k)} \right\|_F^2 + \mathcal{F}(x)$$

$$= \arg\min_x \frac{L}{2} \left\| x - x^{(k)} + \frac{1}{L} \nabla\phi \left( x^{(k)} \right) \right\|_F^2 + \mathcal{F}(x),$$

$$\tag{8}$$

where $L$ is the Lipschitz constant of $\nabla\phi(\cdot)$, which guarantees the convergence of generalized proximal gradient (GPG) method, satisfying

$$\|\nabla\phi(x) - \nabla\phi(y)\|_F \leq L\|x - y\|_F. \tag{9}$$

We denote $\mathcal{R} = x^{(k)} - \frac{1}{L} \nabla\phi \left( x^{(k)} \right)$, and solving (8) requires solving the following optimization problem

$$\text{Prox}_{\mathcal{F}}(\mathcal{R}) = \arg\min_x \frac{1}{2} \|x - \mathcal{R}\|_F^2 + \mathcal{F}(x), \tag{10}$$

where $\text{Prox}_{\mathcal{F}}(\cdot)$ is the proximal operator that is associated with the regularizer $\mathcal{F}(\cdot)$. We are now able to get the optimal solution to (10) by the updating $x^{(i)}$:

$$x^{(i)} = \text{Prox}_{\mathcal{F}}\left(x^{(i-1)} - \frac{1}{L}\nabla\phi\left(x^{(i-1)}\right)\right). \tag{11}$$

Then, the projected gradient descent scheme for solving (6), for the network layers (optimization iteration) $j = 1, ..., n$, is given by

$$x_j^{(i)} = \text{Prox}_{x \in X_a}\left(x_{j-1}^{(i)} - \frac{1}{L}\nabla\mathcal{J}\left(x_{j-1}^{(i)}\right)\right). \tag{12}$$

(12) is also known as the forward propagation. We are using $\nabla$ to denote the gradient and $\mathbf{D}$ to denote the directional derivative in (7). Now substitute the gradient from (7) in (12) to arrive at

$$x_j^{(i)} = \text{Prox}_{x \in X_a}\left(x_{j-1}^{(i)} - \frac{1}{mL}\left[\left(x_{j-1}^{(i)} - y^{(i)}\right) + \left(\partial_{x_{j-1}^{(i)}}\xi\right)^*(\partial_\xi\mathcal{F})\mathcal{F}\right]\right). \tag{13}$$

## 2.3   Learning Bilevel Sparse Regularized Neural Network (BSRL)

Putting it all together, we now describe our proposed bilevel sparse regularized neural network learning architecture (BSRL). Suppose we have $m$ distinct samples, and $n$ layers in our network. Let $x_{true}^{(i)}$ and $y$ be the known true solution and its corresponding experimental data for the $i$-th sample, with $i = 1, ..., m$. Then, we formulate our learning problem as, for $j = 1, ..., n$,

$$\min_{y \in \mathcal{Y}_a} \mathcal{L}(y) = \frac{1}{2m}\sum_{i=1}^{m}\left\|\xi^{(i)}(y) - x_{true}^{(i)}\right\|^2 = \frac{1}{2m}\sum_{i=1}^{m}\left\|x^{(i)} - x_{true}^{(i)}\right\|^2, \tag{14}$$

$$\text{s.t. } x_j^{(i)} = \text{Prox}_{x \in X_a}\left(x_{j-1}^{(i)} - \frac{1}{L}\nabla\mathcal{J}\left(x_{j-1}^{(i)}\right)\right)$$

$$= \text{Prox}_{x \in X_a}\left(x_{j-1}^{(i)} - \frac{1}{mL}\left[\left(x_{j-1}^{(i)} - y^{(i)}\right) + \left(\partial_{x_{j-1}^{(i)}}\xi\right)^*(\partial_\xi\mathcal{F})\mathcal{F}\right]\right). \tag{15}$$

To solve the outer level problem for $y \in \mathcal{Y}_a$, we again use the projected gradient descent method, as described above, with learning rate $\beta$ and $z$ iterations,

$$y_{l+1} = \text{Prox}_{y \in \mathcal{Y}_a}\left(y_l - \beta\nabla_{y_l}\mathcal{L}(y_l)\right), \quad l = 0, ..., z - 1, \tag{16}$$

where $\text{Prox}_{y \in \mathcal{Y}_a}(\cdot)$ is the projection onto the admissible set. It then remains to evaluate $\nabla_{y_l}\mathcal{L}(y_l)$. After applying the chain rule, we obtain that

$$\nabla_{y_l}\mathcal{L}(y_l) = \frac{1}{m}\sum_{i=1}^{m}\int_\Omega\left(x_n^{(i)} - x_{true}^{(i)}\right)\frac{dx_n^{(i)}}{dy}\bigg|_{y=y_l} d\Omega, \tag{17}$$

where $\Omega \subset \mathbb{R}^n$ with $n \geq 1$ is a bounded domain.

The computation of sensitivity of $x$ w.r.t. $y$ is challenging, because at each network layer, $y$ depends on the previous iterate, as well as $y$. We evaluate $\left. \frac{dx_n^{(i)}}{dy} \right|_{y=y_l}$ in (17) by implicit differentiation. This results in an iterative system of equation that we need to solve. For each sample index $i$, it is explicitly derived as follows, for $j = 1, \dots, n$,

$$\left. \frac{dx_j}{dy} \right|_{y=y_l} = \frac{\partial x_j}{\partial x_{j-1}} \cdot \left. \frac{dx_{j-1}}{dy} \right|_{y=y_l} + \left. \frac{\partial x_j}{\partial y} \cdot \frac{dy}{dy} \right|_{y=y_l}, \tag{18}$$

$$
\begin{aligned}
\frac{\partial x_j}{\partial x_{j-1}} =& I - \frac{1}{mL} \Bigg[ I + \frac{\partial}{\partial x_{j-1}} \left( \frac{\partial \xi}{\partial x_{j-1}} \right)^* \left( \frac{\partial \mathcal{F}}{\partial \xi} \right)^* \mathcal{F} \\
&+ \left( \frac{\partial \xi}{\partial x_{j-1}} \right)^* \frac{\partial}{\partial x_{j-1}} \left( \frac{\partial \mathcal{F}}{\partial \xi} \right)^* \mathcal{F} \\
&+ \left( \frac{\partial \xi}{\partial x_{j-1}} \right)^* \left( \frac{\partial \mathcal{F}}{\partial \xi} \right)^* \left( \frac{\partial \mathcal{F}}{\partial \xi} \cdot \frac{\partial \xi}{\partial x_{j-1}} \right) \Bigg],
\end{aligned}
\tag{19}
$$

and,

$$
\begin{aligned}
\frac{\partial x_j}{\partial y} =& -\frac{1}{mL} \Bigg[ -I + \left( \frac{\partial}{\partial y} \left( \frac{\partial \xi}{\partial x_{j-1}} \right)^* \right) \left( \frac{\partial \mathcal{F}}{\partial \xi} \right)^* \mathcal{F} \\
&+ \left( \frac{\partial \xi}{\partial x_{j-1}} \right)^* \left( \frac{\partial}{\partial y} \left( \frac{\partial \mathcal{F}}{\partial \xi} \right)^* \right) \cdot \mathcal{F} + \left( \frac{\partial \xi}{\partial x_{j-1}} \right)^* \left( \frac{\partial \mathcal{F}}{\partial \xi} \right)^* \cdot \frac{\partial \mathcal{F}}{\partial \xi} \frac{\partial \xi}{\partial y} \Bigg].
\end{aligned}
\tag{20}
$$

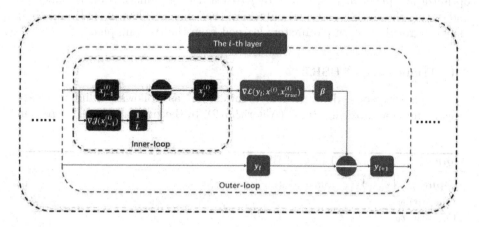

**Fig. 1.** The key architecture of the deep BSRL.

Substituting (19) and (20) in (18) yields the sensitivity of $x$ w.r.t. $y$. Now we have the key architecture of the deep BSRL, as shown in Fig. 1. We divide our network into a training phase and a testing phase, as is common in a standard

---

**Algorithm 1.** Training Phase of BSRL

---

**Input:** $\left\{x_{true}^{(i)}, y^{(i)}\right\}_{i=1}^{m}$, $m$ training samples

**Output:** $y^*$

**repeat**

    Initialize $x_0, \frac{dx_0}{dy}$ and $y_0$.

    **for** $l = 0$ to $z - 1$ **do**

        **for** $j = 1$ to $n$ **do**

            Compute $x^{(i)}$ and $\frac{dx_n^{(i)}}{dy}$ for all $i = 1, \ldots, m$ :

$$x_j^{(i)} = \text{Prox}_{x \in X_a} \left( x_{j-1}^{(i)} - \frac{1}{L} \nabla \mathcal{J} \left( x_{j-1}^{(i)} \right) \right)$$

$$= \text{Prox}_{x \in X_a} \left( x_{j-1}^{(i)} - \frac{1}{mL} \left[ \left( x_{j-1}^{(i)} - y^{(i)} \right) + \left( \partial_{x_{j-1}^{(i)}} \xi \right)^* (\partial_\xi \mathcal{F}) \mathcal{F} \right] \right),$$

$$\left. \frac{dx_j^{(i)}}{dy} \right|_{y=y_l} = \left. \frac{\partial x_j^{(i)}}{\partial x_{j-1}^{(i)}} \cdot \frac{dx_{j-1}^{(i)}}{dy} \right|_{y=y_l} + \left. \frac{\partial x_j^{(i)}}{\partial y} \cdot \frac{dy}{dy} \right|_{y=y_l}.$$

        **end for**

        Compute the gradient of $\mathcal{L}(y)$ :

$$\nabla_{y_l} \mathcal{L}(y_l) = \frac{1}{m} \sum_{i=1}^{m} \int_\Omega \left( x_n^{(i)} - x_{true}^{(i)} \right) \left. \frac{dx_n^{(i)}}{dy} \right|_{y=y_l} d\Omega,$$

        Update $y : y_{l+1} = \text{Prox}_{y \in \mathcal{Y}_a} \left( y_l - \beta \nabla_{y_l} \mathcal{L}(y_l) \right).$

    **end for**

**until** BSRL is convergent

---

machine learning framework. During the training phase, we solve the bilevel optimization problem ((14) and (15)) to learn the regularization parameters, and during the testing phase we only solve the inner problem in ((14) and (15)) using the regularization parameters learned from the training phase.

## 2.4   Framework of BSRL

We summarize our deep BSRL architecture as follows: training Phase (Algorithm 1) and testing Phase (Algorithm 2). In the training phase, we pass in

---

**Algorithm 2.** Testing Phase of BSRL

---

**Input:** $y^*, \left\{y_{test}^{(i)}\right\}_{i=1}^{m_{test}}$, $m_{test}$ testing samples

**Output:** $x$

Initialize $x_0$.

**for** $j = 1$ to $n_{test}$ **do**

    Compute $x$ for all $i = 1, \ldots, m_{test}$ :

$$x_j^{(i)} = \text{Prox}_{x \in X_a} \left( x_{j-1}^{(i)} - \frac{1}{mL} \left[ \left( x_{j-1}^{(i)} - y_{test}^{(i)} \right) + \left( \partial_{x_{j-1}^{(i)}} \xi \right)^* (\partial_\xi \mathcal{F}) \mathcal{F} \right] \right).$$

**end for**

---

**Table 1.** Performance of different methods on the datasets.

| Dataset | Measure | $\ell_1$ | $\ell_2$ | $\ell_{1-2}$ | SGL | CGES | SCAD | capped-$\ell_1$ | LSP | MCP | BSRL (ours) |
|---|---|---|---|---|---|---|---|---|---|---|---|
| Fashion-MNIST | Accuracy | 0.9012 | 0.9124 | 0.9281 | 0.8924 | 0.8873 | 0.8671 | 0.8982 | 0.9031 | 0.9127 | **0.9389** |
| | Parameter | 1 | 0.2398 | 0.4363 | 0.4218 | 0.2819 | 0.5728 | 0.6629 | 0.2763 | 0.3397 | **0.1784** |
| MNIST | Accuracy | 0.9752 | 0.9642 | 0.9538 | 0.9863 | 0.9837 | 0.9824 | 0.9563 | 0.9563 | 0.9623 | **0.9887** |
| | Parameter | 1 | 0.1727 | 0.2735 | 0.1029 | 0.2013 | 0.1197 | 0.1126 | 0.0928 | 0.3328 | **0.0753** |
| DIGITS | Accuracy | 0.8563 | 0.8638 | 0.8837 | 0.8542 | 0.8837 | 0.8682 | 0.8538 | 0.8772 | 0.8831 | **0.8993** |
| | Parameter | 1 | 0.3387 | 0.2928 | 0.2901 | 0.4283 | 0.4419 | 0.2765 | 0.5319 | 0.4019 | **0.1972** |
| CIFAR-10 | Accuracy | 0.8103 | 0.8238 | 0.8188 | 0.8092 | 0.8542 | 0.8452 | 0.8562 | 0.8458 | 0.8229 | **0.8642** |
| | Parameter | 1 | 0.6784 | 0.5829 | 0.5429 | 0.4492 | 0.5186 | 0.6294 | 0.5529 | 0.3165 | **0.2754** |
| CIFAR-100 | Accuracy | 0.7029 | 0.7329 | 0.7219 | 0.6872 | 0.7239 | 0.6549 | 0.7129 | 0.7278 | 0.7362 | **0.7641** |
| | Parameter | 1 | 0.5587 | 0.4982 | 0.8829 | 0.7623 | 0.4927 | 0.6549 | 0.5498 | 0.4892 | **0.3361** |
| SDD | Accuracy | 0.9658 | 0.9829 | 0.9669 | 0.9539 | 0.9827 | 0.9567 | 0.9632 | 0.9862 | 0.9685 | **0.9937** |
| | Parameter | 1 | 0.3092 | 0.4294 | 0.2397 | 0.4962 | 0.2981 | 0.3982 | 0.5729 | 0.4839 | **0.1993** |
| PENDIGITS | Accuracy | 0.9828 | 0.9852 | 0.9902 | 0.9762 | 0.9683 | 0.9719 | 0.9629 | 0.9739 | 0.9827 | **0.9938** |
| | Parameter | 1 | 0.6931 | 0.3397 | 0.6791 | 0.3018 | 0.2973 | 0.7538 | 0.5392 | 0.4492 | **0.1833** |

**Table 2.** Training and test time of networks with each regularizer.

| Dataset | Measure | $\ell_1$ | $\ell_2$ | $\ell_{1-2}$ | SGL | CGES | SCAD | capped-$\ell_1$ | LSP | MCP | BSRL (ours) |
|---|---|---|---|---|---|---|---|---|---|---|---|
| CIFAR-10 | Seconds per batch | 0.0293 | 0.0198 | 0.0231 | 0.0231 | 0.0178 | 0.0155 | 0.0239 | 0.0301 | 0.2931 | **0.0129** |
| | Seconds per test dataset | 0.3129 | 0.2783 | 0.2873 | 0.2183 | 0.2933 | 0.2319 | 0.2754 | 0.2583 | 0.2307 | **0.2018** |
| CIFAR-100 | Seconds per batch | 0.3984 | 0.4185 | 0.4029 | 0.3938 | 0.4294 | 0.4036 | 0.4173 | 0.4219 | 0.3992 | **0.3889** |
| | Seconds per test dataset | 14.6723 | 14.9029 | 13.9987 | 14.9824 | 14.8528 | 13.8237 | 13.4986 | 14.8399 | 13.9643 | **13.0829** |

$m$ training samples $\left\{ x_{true}^{(i)}, y^{(i)} \right\}_{i=1}^{m}$ to learn the optimal $y$ which we denote by $y^*$. The depth of the BSRL at the training phase is $z$ sets of $n$ layers.

In the testing phase, we use the $y^*$ learned from the training phase and testing data $\left\{ y_{\text{test}}^{(i)} \right\}_{i=1}^{m_{\text{test}}}$ to Algorithm 2. The depth of the network at the testing phase is $n_{\text{test}}$ layers.

## 3    Experiments

In this section, we perform experiments on several real-world public classification datasets. We adopt accuracy as the evaluation metric.

We use Tensorflow framework to implement the models. Except for our proposed BSRL, in other cases, we employ the ReLU function $f(x) = \max(0, x)$ as the activation function. As for the output layer, we apply the softmax activation function.

Besides, one-hot encoding is used to encode different classes. We initialize the weights of the network by random initialization according to a normal distribution. The size of minibatch is varied depending on the scale of the datasets. We choose the standard cross-entropy loss as the loss function. On one specific dataset, we use the same network architecture for various penalties to keep the comparison fair. To obtain more reliable results, we repeatedly run the training process five times in each experiment. Experiments are repeated 20 times, and the averaged performance are reported.

### 3.1 Baselines

To demonstrate the superiority of our proposed BSRL solved by proximal algorithm, we compare it with several representative state-of-the-art baselines:

- **Network with $\ell_1$** [6]. $\ell_1$ is a biased and commonly used convex regularizer. It can only achieve sparsity in connection level.
- **Network with $\ell_2$** [20]. It is a fully connected network and is utilized as a reference to illustrate the sparsity-promoting ability of the competitors. $\ell_2$ cannot promote sparsity and it is only used to improve the generalization ability and the performance of the model.
- **Network with $\ell_{1-2}$** [28]. This regularization term is used to illustrate the superiority of BSRL among other non-convex regularizers. We choose $\ell_{1-2}$ as the non-convex competitor.
- **Network with sparse group lasso (SGL)** [24]. The SGL is a regularizer that combines group sparsity and $\ell_1$ regularizer. The group sparsity is used to introduce neuronlevel sparsity and $\ell_1$ regularizer is still utilized to promote sparsity among connections.
- **Network with combined group and exclusive sparsity (CGES)** [29]. The CGES combines group sparsity and exclusive sparsity. It differs from the SGL in that the CGES uses exclusive sparsity instead of $\ell_1$ to promote connection-level sparsity.
- **Network with SCAD, capped-$\ell_1$,LSP, and MCP** [12]. They are a general class of nonconvex penalties and adaptive nonconvex lowrank regularizers.

### 3.2 Datasets

- **Fashion-MNIST** [27]. This dataset consists of a training set with 60,000 instances and a test set with 10,000 examples. Each example is a $28 \times 28$ grayscale image, associated with a label from 10 classes.
- **MNIST** [16]. This dataset consists of 70,000 $28 \times 28$ grayscale images of handwritten digits, which can be classified into 10 classes. The number of training instances and test samples is 60,000 and 10,000, respectively.
- **DIGITS** [22]. This is a toy dataset of handwritten digits, composed of 1,797 $8 \times 8$ grayscale images.
- **CIFAR-10** [15]. This dataset consists of 60,000 $32 \times 32$ color images in 10 classes, with 6,000 images per class.
- **CIFAR-100** [15]. This dataset comprises 60,000 $32 \times 32$ pixels color images as in CIFAR-10. However, these images can be divided into 100 categories instead of 10 classes and each class has 600 images.

- **Sensorless Drive Diagnosis (SDD)** [4]. This dataset is downloaded from the UCI repository. It contains 58,508 examples obtained under 11 different operating conditions.
- **PENDIGITS** [1]. This dataset is composed of 10,992 4×4 grayscale images of handwritten digits 0–9, where there are 7,494 training instances and 3,498 test samples.

### 3.3   Experimental Results and Analysis

In this subsection, we compare our proposed BSRL with several baselines to verify the superiority of our model. To quantitatively measure the performance of various models, two metrics are utilized, including the prediction accuracy and the corresponding number of parameters used in the network. A higher accuracy means that the model can train a better network to implement classification tasks. The smaller the number of parameters used is, the better the regularizer is. The experimental result about number of parameters in Table 1 is expressed as a percentage of the used parameters relative to $\ell_2$ norm.

We list the results in Table 1. The best results are highlighted in bold face. As seen from Table 1, our model has the best performance when compared with other baselines. The reason why our network can achieve best performance might be that such network is dense and can learn more information from the input data. Generally speaking, the performances of our proposed BSRL are better than that of other regularizers. In detail, our model achieves the best results in terms of all two indicators (the prediction accuracy and the corresponding number of parameters) among all regularizers.

Next, we verify the acceleration effect of the introduction of sparse regularization terms on the network. We list the time which takes for the model to train a data batch (seconds per batch) and to test the whole test dataset (seconds per test dataset). Since the models constructed on FashionMNIST, MNIST, DIGITS, SDD and PENDIGITS are all simple and with small size, the training and test time are short and the acceleration effect of the sparse regularizers is not obvious. Therefore, we implement this experiment on more complex dataset, including CIFAR-10 and CIFAR-100. We list the results in Table 2. As can be seen in Table 2, the training process and the test process are both accelerated due to the introduction of sparse regularization terms and our proposed BSRL has best acceleration.

## 4   Conclusion

In this paper, we proposed a new neural network model BSRL for learning sparse regularizer. In this model, we find a correspondence between sparse regularizers and activation functions via proximal operators. BSRL, which divides the network into a training phase and a testing phase, can be learned by solving bilevel optimization problems with sparse regularizers. Experiments have demonstrated that our proposed BSRL framework achieves better results than other regularizer-based ones.

# References

1. Alimoglu, F., Alpaydin, E.: Combining multiple representations and classifiers for pen-based handwritten digit recognition. In: Proceedings of the Fourth International Conference on Document Analysis and Recognition, vol. 2, pp. 637–640 (1997)
2. Arzeno, N.M., Deng, Z.D., Poon, C.S.: Analysis of first-derivative based QRS detection algorithms. IEEE Trans. Biomed. Eng. **55**(2), 478–484 (2008)
3. Atserias, A., Müller, M.: Automating resolution is np-hard. J. ACM (JACM) **67**(5), 1–17 (2020)
4. Bayer, C., Enge-Rosenblatt, O., Bator, M., Mönks, U.: Sensorless drive diagnosis using automated feature extraction, significance ranking and reduction. In: 2013 IEEE 18th Conference on Emerging Technologies Factory Automation (ETFA), pp. 1–4 (2013)
5. Bore, J.C., Ayedh, W.M.A., Li, P., Yao, D., Xu, P.: Sparse autoregressive modeling via the least absolute lp-norm penalized solution. IEEE Access **7**, 40959–40968 (2019)
6. Candes, E.J., Wakin, M.B., Boyd, S.P.: Enhancing sparsity by reweighted L1 minimization. J. Fourier Anal. Appl. **14**(5–6), 877–905 (2008)
7. Chen, M., Wang, Q., Chen, S., Li, X.: Capped $l$-1-norm sparse representation method for graph clustering. IEEE Access **7**, 54464–54471 (2019)
8. Deng, L., Yu, D.: Deep learning: methods and applications. Found. Trends Signal Process. **7**(3–4), 197–387 (2014)
9. Fan, J., Li, R.: Variable selection via nonconcave penalized likelihood and its oracle properties. J. Am. Stat. Assoc. **96**(456), 1348–1360 (2001)
10. Gal, Y., Hron, J., Kendall, A.: Concrete dropout. In: Advances in Neural Information Processing Systems, pp. 3581–3590 (2017)
11. Hillar, C.J., Lim, L.H.: Most tensor problems are np-hard. J. ACM (JACM) **60**(6), 1–39 (2013)
12. Hu, E.L., Kwok, J.T.: Low-rank matrix learning using biconvex surrogate minimization. IEEE Trans. Neural Netw. Learn. Syst. **30**(11), 3517–3527 (2019)
13. Issa, I., Gastpar, M.: Computable bounds on the exploration bias. In: 2018 IEEE International Symposium on Information Theory (ISIT), pp. 576–580. IEEE (2018)
14. Jiang, H., Zheng, W., Luo, L., Dong, Y.: A two-stage minimax concave penalty based method in pruned adaboost ensemble. Appl. Soft Comput. **83**, 105674 (2019)
15. Krizhevsky, A., Hinton, G., et al.: Learning multiple layers of features from tiny images (2009)
16. LeCun, Y., Bottou, L., Bengio, Y., Haffner, P.: Gradient-based learning applied to document recognition. Proc. IEEE **86**(11), 2278–2324 (1998)
17. Li, Z., Wan, C., Tan, B., Yang, Z., Xie, S.: A fast dc-based dictionary learning algorithm with the scad penalty. Neurocomputing **429**, 89–100 (2020)
18. Lou, Y., Yin, P., He, Q., Xin, J.: Computing sparse representation in a highly coherent dictionary based on difference of L1 and L2. J. Sci. Comput. **64**(1), 178–196 (2015)
19. Lu, C., Zhu, C., Xu, C., Yan, S., Lin, Z.: Generalized singular value thresholding. In: Proceedings of the AAAI Conference on Artificial Intelligence, vol. 29 (2015)
20. Luo, X., Chang, X., Ban, X.: Regression and classification using extreme learning machine based on l1-norm and l2-norm. Neurocomputing **174**, 179–186 (2016)
21. Natarajan, B.K.: Sparse approximate solutions to linear systems. SIAM J. Comput. **24**(2), 227–234 (1995)

22. Netzer, Y., Wang, T., Coates, A., Bissacco, A., Wu, B., Ng, A.Y.: Reading digits in natural images with unsupervised feature learning (2011)

23. Rajeswaran, A., Finn, C., Kakade, S.M., Levine, S.: Meta-learning with implicit gradients. In: Wallach, H., Larochelle, H., Beygelzimer, A., d'Alché-Buc, F., Fox, E., Garnett, R. (eds.) Advances in Neural Information Processing Systems, vol. 32, pp. 113–124 (2019)

24. Simon, N., Friedman, J., Hastie, T., Tibshirani, R.: A sparse-group lasso. J. Comput. Graph. Stat. **22**(2), 231–245 (2013)

25. Tsagkarakis, N., Markopoulos, P.P., Sklivanitis, G., Pados, D.A.: L1-norm principal-component analysis of complex data. IEEE Trans. Signal Process. **66**(12), 3256–3267 (2018)

26. Wu, S., et al.: $l$1-norm batch normalization for efficient training of deep neural networks. IEEE Trans. Neural Netw. Learn. Syst. **30**(7), 2043–2051 (2018)

27. Xiao, H., Rasul, K., Vollgraf, R.: Fashion-mnist: a novel image dataset for benchmarking machine learning algorithms (2017). arXiv preprint arXiv:1708.07747

28. Yin, P., Lou, Y., He, Q., Xin, J.: Minimization of 1–2 for compressed sensing. SIAM J. Sci. Comput. **37**(1), A536–A563 (2015)

29. Yoon, J., Hwang, S.J.: Combined group and exclusive sparsity for deep neural networks. In: International Conference on Machine Learning, pp. 3958–3966 (2017)

30. Zhang, M., Ding, C., Zhang, Y., Nie, F.: Feature selection at the discrete limit. In: Proceedings of the AAAI Conference on Artificial Intelligence, vol. 28 (2014)

31. Zhang, T.: Analysis of multi-stage convex relaxation for sparse regularization. J. Mach. Learn. Res. **11**(3), 1081–1107 (2010)

32. Zhang, Y., Zhang, H., Tian, Y.: Sparse multiple instance learning with non-convex penalty. Neurocomputing **391**, 142–156 (2020)

# Natural Language Processing

# DGA-Net: Dynamic Gaussian Attention Network for Sentence Semantic Matching

Kun Zhang[1,2(✉)], Guangyi Lv[3], Meng Wang[1,2], and Enhong Chen[3]

[1] Key Laboratory of Knowledge Engineering with Big Data, Hefei University of Technology,
Hefei, China
zhkun@hfut.edu.cn
[2] School of Computer Science and Information Engineering, Hefei University of Technology,
Hefei, China
[3] School of Computer Science and Technology, University of Science and Technology
of China, Hefei, China
{gylv,cheneh}@mail.ustc.edu.cn

**Abstract.** Sentence semantic matching requires an agent to determine the semantic relation between two sentences, where much recent progress has been made by advancement of representation learning techniques and inspiration of human behaviors. Among all these methods, attention mechanism plays an essential role by selecting important parts effectively. However, current attention methods either focus on all the important parts in a static way or only select one important part at one attention step dynamically, which leaves a large space for further improvement. To this end, in this paper, we design a novel *Dynamic Gaussian Attention Network (DGA-Net)* to combine the advantages of current static and dynamic attention methods. More specifically, we first leverage pre-trained language model to encode the input sentences and construct semantic representations from a global perspective. Then, we develop a Dynamic Gaussian Attention (DGA) to dynamically capture the important parts and corresponding local contexts from a detailed perspective. Finally, we combine the global information and detailed local information together to decide the semantic relation of sentences comprehensively and precisely. Extensive experiments on two popular sentence semantic matching tasks demonstrate that our proposed *DGA-Net* is effective in improving the ability of attention mechanism.

## 1 Introduction

Sentence semantic matching is a long-lasting theme of Natural Language Processing (NLP), which requires an agent to determine the semantic relations between two sentences. For example, in Natural Language Inference (NLI), it is used to determine whether a hypothesis can be inferred reasonably from a given premise [15]. In Paraphrase Identification (PI), it is utilized to identify whether two sentences express the same meaning or not [9]. Figure 1 gives us two representative examples of NLI and PI.

As a fundamental technology, sentence semantic matching has been applied successfully in many NLP fields, e.g., information retrieval [7,26], question answering [18], and dialog system [24]. With advanced representation learning techniques

L. Fang et al. (Eds.): CICAI 2021, LNAI 13070, pp. 203–214, 2021.
https://doi.org/10.1007/978-3-030-93049-3_17

**Fig. 1.** Two example from different sentence semantic matching datasets (colored words are the important parts that need attention). (Color figure online)

[8,11,31], numerous efforts have been dedicated to this task, where the dominant trend is to build complex structures with attention. For example, self-attention [28] can generate better representations by relating elements at different positions in a single sentence. Co-attention [15,34] focuses on sentence interaction from a detailed perspective. Dynamic re-read attention [35] is able to select the important parts in a dynamic way based on learned information. They all help to achieve impressive performance.

However, most work either focuses on all the important parts in a static way [3] or only selects one important part at each selection in a dynamic way [35]. They either are incapable of adapting to dynamic changes during the sentence understanding process or ignore the importance of local structures. For example, in Fig. 1, colored words illustrate the focus points. When selecting the important parts as the static attention methods do, the representations of two sentences may be similar since many of the important words are the same (e.g.,*woman, shirt*). When employing the dynamic attention methods [35], the attributes of the selected parts may be missed since dynamic methods only select one important word at each step and ignore the local contexts (e.g.,*woman with purple shirt, woman with blue shirt*). All these will lead to a wrong decision. Therefore, how to leverage attention mechanism to select proper information for precise sentence semantic understanding and matching is the main challenge that we need to consider.

To this end, in this paper, we propose an effective *Dynamic Gaussian Attention Network (DGA-Net)* approach to combine the advantages of current static and dynamic attention methods. In concerned details, we first utilize pre-trained BERT to model the semantic meanings of input words and sentences globally. Based on the dynamic attention mechanism and Gaussian distribution, we develop a novel Dynamic Gaussian Attention (DGA) to pay close attention to one important part and corresponding local contexts among sentences at each attention step simultaneously. Along this line, we can not only focus on the most important part of sentences dynamically, but also use the local context to support the understanding of these selected parts precisely. Extensive evaluations on two popular sentence semantic matching tasks (i.e., NLI and PI) demonstrate the effectiveness of our proposed *DGA-Net* method and its advantages over state-of-the-art sentence encoding-based baselines.

## 2    Related Work

With the available large annotated datasets, such as SNLI [1], SCITAIL [14], and Quora Question Pair [13], as well as various neural networks, such as LSTM [4], GRU [6], and attention mechanism [22,28,35–37], plenty of methods have been developed to represent and evaluate sentence semantic meanings. Among all methods, attention mechanism has become the essential module, which helps models capture semantic relations and properly align the elements of sentences. For example, Liu et al. [19] proposed inner-attention to pay more attention to the important words among sentences. In order to better capture the interaction of sentences, Kim et al. [15] utilized co-attention network to model the interaction among sentence pairs. Moreover, Cho [12] and Shen [25] proposed to utilize multi-head attention to model sentence semantics and interactions from multiple aspects without RNN/CNN structure. They took full advantage of attention mechanism for better sentence semantic modeling and achieved impressive performance on sentence semantic matching task.

Despite the success of using attention mechanism in a static way, researchers also learn from human behaviors and propose dynamic attention methods. By conducting a lab study, Zheng et al. [38] observed that users generally read the document from top to bottom with the reading attention decays monotonically. Moreover, in a specific scenario (e.g., Answer Selection), users tend to pay more attention to the possible segments that are relevant to what they want. They will reread more snippets of candidate answers with more *skip* and *up* transition behaviors, and ignore the irrelevant parts [17]. Furthermore, Zhang et al. [35] designed a novel dynamic re-read attention to further improve model performance. They tried to select one important word at each attention calculation and repeated this operation for precise sentence semantic understanding.

However, static attention methods select all the important parts at one time, which may lead to a misunderstanding of sentence semantics since there are too many similar but semantically different important parts. Dynamic methods only select one important part at each operation, which may lose some important attributes of the important parts. Thus, we propose a *DGA-Net* to select the important parts and corresponding local context in sentences for better sentence semantic understanding and matching.

## 3    Problem Statement and Model Structure

In this section, we formulate the NLI task as a supervised classification problem and introduce the structure and technical details of our proposed *DGA-Net*.

### 3.1    Problem Statement

First, we define our task in a formal way. Given two sentences $s^a = \{w_1^a, w_2^a, ..., w_{l_a}^a\}$ and $s^b = \{w_1^b, w_2^b, ..., w_{l_b}^b\}$. Our goal is to learn a classifier $\xi$ which is able to precisely predict the relation $y = \xi(s^a, s^b)$ between $s^a$ and $s^b$. Here, $w_i^a$ and $w_j^b$ are one-hot vectors which represent the $i^{th}$ and $j^{th}$ word in the sentences. $l_a$ and $l_b$ indicate the total number of words in $s^a$ and $s^b$, respectively.

In order to model sentence semantic meanings more precisely and comprehensively, the following important challenge should be considered:

– How to overcome the shortcomings of static and dynamic attention methods, and leverage attention operation to select proper information for precise sentence semantic understanding and matching?

To this end, we propose a novel *Dynamic Gaussian Attention Network (DGA-Net)* to tackle the above issue and doing better sentence semantic matching.

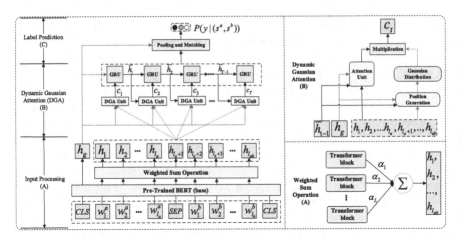

**Fig. 2.** Architecture of *Dynamic Gaussian Attention Network (DGA-Net)*.

## 3.2 Dynamic Gaussian Attention Network

The overall architecture of *DGA-Net* is shown in Fig. 2, which consists of three main components: 1) *Input Processing*: utilizing pre-train BERT to generate the extravagant representation of input words; 2) *Dynamic Gaussian Attention*: selecting one important part and proper local structure at each step and dynamically reading these contextual parts with all learned information; 3) *Label Prediction*: predicting the final results based on the expanded semantic representations.

**Input Processing.** By making full use of large corpus and multi-layer transformers, BERT [8] has accomplished much progress in many natural language tasks and become a powerful tool to process the raw input sentences. Therefore, we also employ BERT to encode the input sentences. In order to make full use of BERT and encode sentence comprehensively, we use the weighted sum of all the hidden states from different transformer layers of BERT as the final contextual representations of input sentences. Specifically, the input sentence $s^a = \{w_1^a, w_2^a, ..., w_{l_a}^a\}$ and $s^b = \{w_1^b, w_2^b, ..., w_{l_b}^b\}$ will be split into BPE tokens [23]. Then, we leverage a special token *"[SEP]"* to concatenate two sentences and add *"[CLS]"* token at the beginning and the end of concatenated sentences. As illustrated in Fig. 2(A), suppose the final number of tokens in the sentence pair is $l_{ab}$, and BERT generates $L$ hidden states for each BPE token $BERT_t^l, (1 \leq l \leq L, 1 \leq t \leq l_{ab})$. The contextual representation for $t^{th}$ token in input sentence pair at token level is then a per-layer weighted sum of transformer block output, with weights $\alpha_1, \alpha_2, ..., \alpha_L$.

$$h_t = \sum_{l=1}^{L} \alpha_l BERT_t^l, \quad 1 \leq t \leq l_{ab}, \tag{1}$$

where $\alpha_l$ is the weight for the $l^{th}$ layer in BERT and will be learned during the training. $h_t$ is the representation for the $t^{th}$ token. Moreover, we treat the output $BERT_0^L$ of the first special token "*[CLS]*" in the last block as the contextual representation $h_g$ for input sentences globally. Along this line, we can model the semantic meanings of words and sentences comprehensively, which lays a good foundation for subsequent study.

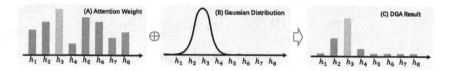

**Fig. 3.** The processing of Dynamic Gaussian Attention (DGA) calculation.

**Dynamic Gaussian Attention.** As introduced in Sect. 1, static attention methods select all the important parts at one time, which may lead to a misunderstanding of sentence semantic meanings since there are too many similar but semantically different important parts. Meanwhile, dynamic attention methods [35] try to select one important part at step, which can alleviate the problem that static attention methods suffer from. However, it still causes the model to lose some important attributes of important parts and lead to an incorrect result. Therefore, it is crucial to employ attention mechanism in a proper way for better sentence semantic understanding and matching.

Inspired by previous work [35,38] and Gaussian distribution, we design a novel *Dynamic Gaussian Attention (DGA)* unit to select the important part and proper local context simultaneously. Figure 2(B) and Fig. 3 illustrate the entire processing of DGA calculation. During each DGA operation, we first calculate the attention weight among input sequence. Meanwhile, we leverage a position generation method $G(\cdot)$ to predict the focus point, which can be visualized as the yellow bar in Fig. 3. Then, we generate a Gaussian distribution with the focus point as the center. Next, we multiply the attention weight and Gaussian distribution to get the DGA result. Along this line, the attention weights of the words that are close to the important part will be preserved, and the rest will be discarded. In other words, we can focus on the important part and corresponding local context for better semantic understanding. Inspired by DRr-Net [35], we also repeat DGA operation many times for the final decision.

Specifically, DGA unit treats $\{h_i | i = 1, 2, ..., l_{ab}\}$ as the inputs, and produces an important position $p_t$ at attention step $t$. The representation $c_t$ for this position is derived as a weighted summation over the inputs within the window $[p_t - \frac{D}{2}, p_t + \frac{D}{2}]$; $D$ is the window size. Since we select these important parts in a sequential manner, GRU is adopted to encoder these important parts. This process can be formulated as follows:

$$\begin{aligned} H = [h_1, h_2, ..., h_{l_{ab}}], \quad p_t &= G(H, \bar{h}_{t-1}, h_g), \\ c_t = F(p_t, H, \bar{h}_{t-1}, h_g), \quad \bar{h}_t &= GRU(\bar{h}_{t-1}, c_t), \quad t = 1, 2, ..., T, \end{aligned} \tag{2}$$

where $G(\cdot)$ is position generation function. $F(\cdot)$ denotes DGA function. $T$ is the dynamic attention length. In order to take global information into consideration, we also treat the global representation $h_g$ as an additional context in $G(\cdot)$ and $F(\cdot)$. $\bar{h}_T$ can be regarded as the dynamic locally-aware representation for the input sentence pair.

Different from DRr-Net [35] that treats the word that has biggest weight as the current selection, we intend to use MLP to predict the focus point at current step. More specifically, we utilize position generation function $G(\cdot)$ to generate the important position $p_t$ at attention step $t$ as follows:

$$m_t = \sum_{i=1}^{l_{ab}}(W_1^p h_i) + W_2^p \bar{h}_{t-1} + W_3^p h_g,$$

$$p_t = l_{ab} \cdot \text{sigmoid}(v_p^T \tanh(U_p m_t)), \tag{3}$$

where $\{W_1^p, W_2^p, W_3^p, v_p, U_p\}$ are trainable parameters. T is transposition operation. As the result of $\text{sigmoid}(\cdot)$ function, $p_t \in [0, l_{ab}]$. Along this line, we are able to use all the learned information to generate the important position at each attention step.

After getting the important position $p_t$, it is urgent to ensure its exact meaning in the sentence, which is in favor of overcoming the issue in Sect. 3.1. Inspired by the observation that adjacent words contribute more for understanding current phrase than distant ones, we develop a novel DGA method by placing a Gaussian distribution centered around $p_t$ to further process the attention weights. The implementation function $F(\cdot)$ can be formulated as follows:

$$g_t = exp(-\frac{(s - p_t)^2}{2\sigma^2}),$$

$$\alpha^a = \omega_d^T \tanh(W_d H + (U_d \bar{h}_{t-1} + M_d h_g) \otimes e_{l_{ab}}), \tag{4}$$

$$\bar{\alpha}^a = \alpha^a \cdot g_t, \quad c_t = \sum_{i=1}^{l_{ab}} \frac{exp(\bar{\alpha}_i^a)}{\sum_{k=1}^{l_{ab}} exp(\bar{\alpha}_k^a)} h_i,$$

where $\{\omega_d, W_d, U_d, M_d\}$ are trainable parameters. $g_t$ is Gaussian distribution centered around $p_t$, $\sigma = \frac{D}{2}$, and $e_{l_{ab}} \in \mathbb{R}^{l_{ab}}$ is a row vector of 1. In this operation, we utilize the Gaussian distribution to optimize the original attention value $\alpha^a$ so that the model can focus on the important position and its corresponding context, capture the local structure of sentences, and represent the sentence semantic more precisely.

**Label Prediction.** After finishing the dynamic selections, we first adopt attention pooling to fuse all the selected important parts to generate a locally-aware representation $\bar{h}$ from a detailed perspective as follows:

$$\bar{H} = [\bar{h}_1, \bar{h}_2, ..., \bar{h}_{l_{ab}}], \quad \alpha^b = \omega^T \tanh(W\bar{H} + b),$$

$$\bar{h} = \sum_{i=1}^{l_{ab}} \frac{\exp(\alpha_i^b)}{\sum_{k=1}^{l_{ab}} \exp(\alpha_k^b)} \bar{h}_i. \tag{5}$$

After getting the locally-aware representation $\bar{h}$, we leverage heuristic matching [3] between $h$ generated from a global aspect and $\bar{h}$ generated from a detailed aspect. Then we send the result $u$ to a two-layer MLP for final classification. This process is formulated as follows:

$$u = [h_g, \bar{h}, h_g \odot \bar{h}, \bar{h} - h_g], \quad P(y|(s^a, s^b)) = \text{MLP}(u), \tag{6}$$

**Table 1.** Performance (accuracy) of models on different SNLI test sets and SICK test set.

| Model | Full test | Hard test | SICK test |
|---|---|---|---|
| (1) CENN [33] | 82.1% | 60.4% | 81.8% |
| (2) BiLSTM with Inner-Attention [19] | 84.5% | 62.7% | 85.2% |
| (3) Gated-Att BiLSTM [2] | 85.5% | 65.5% | 85.7% |
| (4) CAFE [27] | 85.9% | 66.1% | 86.1% |
| (5) Gumbel TreeLSTM [5] | 86.0% | 66.7% | 85.8% |
| (6) Distance-based SAN [12] | 86.3% | 67.4% | 86.7% |
| (7) DRCN [15] | 86.5% | 68.3% | 87.4% |
| (8) DSA [32] | 87.4% | 71.5% | 87.7% |
| (9) DRr-Net [35] | 87.5% | 71.2% | 87.8% |
| (10) BERT-base [8] | 90.3% | 80.8% | **88.5%** |
| (11) *DGA-Net* | **90.72%** | **81.44%** | 88.36% |

where concatenation can retain all the information [33]. The element-wise product is a certain measure of "similarity" of two sentences [21]. Their difference can capture the degree of distributional inclusion in each dimension [30].

## 4  Experiment

In this section, we first present the details about the model implementation. Then, we introduce the datasets that we will evaluate our model on, including four benchmark datasets for two sentence semantic matching tasks, which cover different domains and exhibit different characteristics. Next, we will make a detailed analysis about the model and experimental results.

### 4.1  Experimental Setup

**Loss Function.** Since sentence semantic matching task can be formulated as classification task, we employ *cross-entropy* as the loss function:

$$L = -\frac{1}{N} \sum_{i=1}^{N} \boldsymbol{y}_i \log P(y_i|(\boldsymbol{s}_i^a, \boldsymbol{s}_i^b)) + \epsilon \|\boldsymbol{\theta}\|_2 , \tag{7}$$

where $\boldsymbol{y}_i$ is the one-hot representation for the true class of the $i^{th}$ instance. $N$ represents the number of training instances. $\epsilon$ is the weight decay. $\theta$ denotes the trainable parameters in the model and $\|\boldsymbol{\theta}\|_2$ is l2-norm for these parameters.

**Model Initialization.** We have tuned the hyper-parameters on validation set for best performance. An *Early-Stop* operation is employed to select the best model. Some common hyper-parameters are listed as follows:

The vocabulary is the same as the vocabulary of BERT-base. The window size in Gaussian distribution is $D = 4$. The dynamic attention length in DGA is $T = 4$. The

210    K. Zhang et al.

**Table 2.** Experimental results on Quora and MSRP datasets.

| Model | Quora | MSRP |
|---|---|---|
| (1) CENN [33] | 80.7% | 76.4% |
| (2) MP-CNN [10] | – | 78.6% |
| (3) BiMPM [29] | 88.2% | – |
| (4) DRCN [15] | 90.2% | 82.5% |
| (5) DRr-Net [35] | 89.7% | 82.5% |
| (6) BERT-base [8] | 91.1% | 84.3% |
| (7) *DGA-Net* | **91.7%** | **84.5%** |

**Table 3.** Ablation performance (accuracy) of *DGA-Net*.

| Model | SNLI test | SICK test |
|---|---|---|
| (1) BERT-base | 90.3% | 88.5% |
| (2) *DGA-Net* (w/o vector $h_g$) | 85.3% | 83.2% |
| (3) *DGA-Net* (w/o vector $\bar{h}$) | 89.4% | 87.5% |
| (4) DRr-Net | 87.5% | 87.8% |
| (5) Multi-GRU + DGA | 88.4% | 88.1% |
| (6) *DGA-Net* (w/o local context) | 90.5% | 88.5% |
| (7) *DGA-Net* | **90.72%** | **88.36%** |

attention size in DGA is set to 200. The hidden state size of GRU is set to 768. The initial learning rate is set to $10^{-4}$. An Adam optimizer with $\beta_1 = 0.9$ and $\beta_2 = 0.999$ is adopted to optimize all trainable parameters.

**Dataset.** In order to evaluate the model performance comprehensively, we employ two sentence semantic matching tasks: *Natural Language Inference (NLI)* and *Paraphrase Identification (PI)* to conduct the experiments. NLI task requires an agent to predict the semantic relation from premise sentence to hypothesis sentence among "*Entailment, Contradiction, Neutral*". We select two well-studied and public available datasets: SNLI [1] and SICK [20]. Meanwhile, PI task requires an agent to identify whether two sentences express the same semantic meaning or not. For this task, we select Quora [13] and MSRP [9] to evaluate the model performance.

### 4.2 Experiment Results

In this section, we will give a detailed analysis about the models and experimental results. Here, we use *Accuracy* on different test sets to evaluate the model performance.

**Performance on SNLI and SICK.** Table 1 reports the results of *DGA-Net* compared with other published baselines. We can observe that *DGA-Net* achieves highly comparable performance on different NLI test set. Specifically, we make full use of pre-trained language model to get the comprehensive understanding about the semantic meanings. This is one of the important reasons that *DGA-Net* is capable of outperforming other BERT-free models by a large margin. Furthermore, we develop a novel DGA unit to further improve the capability of dynamic attention mechanism. Instead of only selecting one important part at each attention operation, DGA can select the important part and proper local context simultaneously at each step. Therefore, the local context of the sentence can be fully explored, and sentence semantics can be represented more precisely. This is another reason that *DGA-Net* achieves better performance than all baselines, including the BERT-base model.

Among all baselines, DRr-Net [35] and DSA [32] are current state-of-the-art methods without BERT. DSA [32] modifies the dynamic routing in capsule network and develops a DSA to model sentences. It utilizes CNN to capture the local context information and encodes each word into a meaningful representation space. DRr-Net adopts multi-layer GRU to encode the sentence semantic meanings from a global perspective and designs a dynamic re-read attention to select one important part at each attention step for detailed sentence semantic modeling. They all achieved impressive performances. However, both RNN and CNN structures have some weaknesses in extracting features or generating semantic representations compared with BERT. We can observe from Table 1 that the BERT-base model outperforms them by a large margin. Meanwhile, their attention operations either select too many important parts at one time or only focus on one important part at each operation, which may lead to a misunderstanding of the sentence semantic meanings. Thus, their performance is not as good as *DGA-Net* reaches. On the other hand, apart from the powerful encoding ability, BERT still focuses on the importance of words to the sequence and has some weaknesses in distinguishing the exact meanings of sentences. By taking the local context into consideration and leveraging DGA to get the precise meanings of sentences, *DGA-Net* is able to achieve better performance than BERT.

**Performance on Quora and MSRP.** Besides NLI task, we also select PI task to better evaluate the model performance on sentence semantic similarity identification. Table 2 illustrates the experimental results on Quora and MSRP datasets. Different from the results on NLI datasets, our proposed *DGA-Net* achieves the best performance compared with other baselines on both test sets, revealing the superiority of our proposed *DGA-Net*. Besides, we can obtain that almost all the methods have better performance on Quora dataset and the improvement of our proposed *DGA-Net* on Quora dataset is also larger than the improvement on MSRP dataset. Quora dataset [13] has more than 400k sentence pairs, which is much larger than MSRP dataset. Large data is capable of helping to model to better analyze the data and get close to the upper bound of the performance. Meanwhile, we also speculate that the inter-sentence interactions is probably another possible reason. Quora dataset contains many sentence pairs with less complicated interactions (e.g., many identical words in two sentences) [16].

### 4.3  Ablation Performance

The overall performance has proven the superiority of *DGA-Net*. However, which part is more important for performance improvement is still unclear. Thus, we conduct an ablation study on two NLI test sets to examine the effectiveness of each component. Recall the model structure, two important semantic representations are $h_g$ from BERT output and $\bar{h}$ from DGA output. As illustrated in Table 3(2)–(3), when we remove the global representation $h$, we can observe that the model performance has a big drop. This result is in line with our intuitive. We should have a comprehensive understanding about the sentence before making a decision. Only the important parts are insufficient for the decision making.

Meanwhile, when removing the detailed representation $\bar{h}$, model performance is worse than BERT-base model. we speculate that DGA is in the training process but not in the predicting process, which decreases the model performance. Besides, we

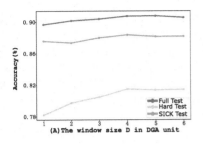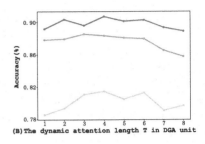

**Fig. 4.** Performance of *DGA-Net* with different window sizes (1–6), and attention lengths (1–8).

investigate the effectiveness of BERT encoder and local context. When replacing BERT with multi-layer GRUs, we can observe that its performance is still better than DRr-Net, suggesting the importance of local context utilization. Meanwhile, its performance is not comparable with BERT-base, let alone the entire *DGA-Net*, proving the importance of BERT. When removing the local context, the performance of *DGA-Net* is capable of optimizing the BERT-base model, proving the effectiveness of local context utilization. In other words, both BERT encoder and local context utilization are indispensable for *DGA-Net* to achieve better performance.

### 4.4  Sensitivity of Parameters

There are two hyper-parameters that affect the model performance: 1) The window size $D$ in DGA unit; 2) The dynamic attention length $T$ in DGA unit. Therefore, we evaluate *DGA-Net* performance on two NLI test sets with different hyper-parameter settings. The results are summarized in Fig. 4.

When talking about the window size in DGA unit, we can observe that the model performance first increases and then becomes smooth with the increase of window size. We speculate that a too small or too big window cannot help to capture the local structure for precisely semantic understanding. When the window size is $D = 4$, DGA will consider two words on each side of the center word, which is suitable for leveraging local context to enhance the semantic understanding of sentences.

As for the dynamic attention length, Bowman et al. [1] has conducted that the average length is 14.1 for premise and 8.3 for hypothesis in SNLI. From Fig. 4(B), 4 is suitable for dynamic attention length. Too short reading length may cause the model to ignore some important parts. Meanwhile, too long reading length may weaken the ability of precisely local structure capturing and semantic understanding.

## 5  Conclusion and Future Work

In this paper, we proposed an effective *Dynamic Gaussian Attention Network (DGA-Net)* approach for sentence semantic matching, a novel architecture that not only models sentence semantics in a global perspective, but also utilizes local structure to support the analysis of the important parts step by step. To be specific, we first make full use of pre-trained language model to evaluate semantic meanings of words and sentences from a global perspective. Then, we design a novel Dynamic Gaussian Attention (DGA) to pay close attention to one important part and corresponding local context among sentences simultaneously at each attention operation. By taking the local information into consideration, *DGA-Net* is capable of measuring the sentence semantics more comprehensively. Finally, we integrate the global semantic representation from Bert and detailed semantic representation from DGA to further improve the model performance on sentence semantic matching. Extensive evaluations on two sentence semantic matching tasks (i.e., NLI and PI) demonstrate the superiority of our proposed *DGA-Net*. In the future, we will focus on providing more information for dynamic attention to better local important parts selecting and sentence semantic understanding.

**Acknowledgements.** This work was supported in part by grants from the National Natural Science Foundation of China (Grant No. 62006066), the Open Project Program of the National Laboratory of Pattern Recognition (NLPR), and the Fundamental Research Funds for the Central Universities, HFUT.

# References

1. Bowman, S.R., Angeli, G., Potts, C., Manning, C.D.: A large annotated corpus for learning natural language inference. In: EMNLP (2015)
2. Chen, Q., Zhu, X.D., Ling, Z.H., Wei, S., Jiang, H., Inkpen, D.: Recurrent neural network-based sentence encoder with gated attention for natural language inference. In: RepEval@EMNLP, pp. 36–40 (2017)
3. Chen, Q., Zhu, X., Ling, Z., Wei, S., Jiang, H., Inkpen, D.: Enhanced LSTM for natural language inference. In: ACL, pp. 1657–1668 (2017)
4. Cheng, J., Dong, L., Lapata, M.: Long short-term memory-networks for machine reading. In: EMNLP, pp. 551–561 (2016)
5. Choi, J., Yoo, K.M., Lee, S.G.: Learning to compose task-specific tree structures. In: AAAI (2018)
6. Chung, J., Gulcehre, C., Cho, K., Bengio, Y.: Empirical evaluation of gated recurrent neural networks on sequence modeling (2014). CoRR abs/1412.3555
7. Clark, P., et al.: Combining retrieval, statistics, and inference to answer elementary science questions. In: AAAI (2016)
8. Devlin, J., Chang, M.W., Lee, K., Toutanova, K.: Bert: pre-training of deep bidirectional transformers for language understanding (2018). arXiv preprint arXiv:1810.04805
9. Dolan, W.B., Brockett, C.: Automatically constructing a corpus of sentential paraphrases. In: IWP (2005)
10. He, H., Gimpel, K., Lin, J.: Multi-perspective sentence similarity modeling with convolutional neural networks. In: EMNLP, pp. 1576–1586 (2015)
11. Hong, R., He, Y., Wu, L., Ge, Y., Wu, X.: Deep attributed network embedding by preserving structure and attribute information. IEEE TSMC: S **51**, 1434–1445 (2019)
12. Im, J., Cho, S.: Distance-based self-attention network for natural language inference (2017). CoRR abs/1712.02047
13. Iyer, S., Dandekar, N., Csernai, K.: First quora dataset release: question pairs (2017)
14. Khot, T., Sabharwal, A., Clark, P.: Scitail: a textual entailment dataset from science question answering. In: AAAI (2018)
15. Kim, S., Hong, J.H., Kang, I., Kwak, N.: Semantic sentence matching with densely-connected recurrent and co-attentive information (2018). CoRR abs/1805.11360

16. Lan, W., Xu, W.: Neural network models for paraphrase identification, semantic textual similarity, natural language inference, and question answering. In: COLING, pp. 3890–3902 (2018)
17. Li, X., Mao, J., Wang, C., Liu, Y., Zhang, M., Ma, S.: Teach machine how to read: reading behavior inspired relevance estimation. In: Proceedings of the 42nd International ACM SIGIR Conference on Research and Development in Information Retrieval, pp. 795–804 (2019)
18. Liu, Q., et al.: Finding similar exercises in online education systems. In: SIGKDD, pp. 1821–1830. ACM (2018)
19. Liu, Y., Sun, C., Lin, L., Wang, X.: Learning natural language inference using bidirectional LSTM model and inner-attention (2016). CoRR abs/1605.09090
20. Marelli, M., Bentivogli, L., Baroni, M., Bernardi, R., Menini, S., Zamparelli, R.: Semeval-2014 task 1: evaluation of compositional distributional semantic models on full sentences through semantic relatedness and textual entailment. In: SemEval, pp. 1–8 (2014)
21. Mou, L., et al.: Natural language inference by tree-based convolution and heuristic matching. In: ACL, pp. 130–136 (2016)
22. Parikh, A.P., Täckström, O., Das, D., Uszkoreit, J.: A decomposable attention model for natural language inference. In: EMNLP, pp. 2249–2255 (2016)
23. Sennrich, R., Haddow, B., Birch, A.: Neural machine translation of rare words with subword units (2015). arXiv preprint arXiv:1508.07909
24. Serban, I.V., Sordoni, A., Bengio, Y., Courville, A.C., Pineau, J.: Building end-to-end dialogue systems using generative hierarchical neural network models. In: AAAI, vol. 16 (2016)
25. Shen, T., Zhou, T., Long, G., Jiang, J., Pan, S., Zhang, C.: Disan: directional self-attention network for rnn/cnn-free language understanding (2017). CoRR abs/1709.04696
26. Sun, P., Wu, L., Zhang, K., Fu, Y., Hong, R., Wang, M.: Dual learning for explainable recommendation: towards unifying user preference prediction and review generation. In: WWW, pp. 837–847 (2020)
27. Tay, Y., Tuan, L.A., Hui, S.C.: A compare-propagate architecture with alignment factorization for natural language inference (2017). CoRR abs/1801.00102
28. Vaswani, A., et al.: Attention is all you need. In: NIPS, pp. 5998–6008 (2017)
29. Wang, Z., Hamza, W., Florian, R.: Bilateral multi-perspective matching for natural language sentences (2017). CoRR abs/1702.03814
30. Weeds, J., Clarke, D., Reffin, J., Weir, D., Keller, B.: Learning to distinguish hypernyms and co-hyponyms. In: COLING, pp. 2249–2259 (2014)
31. Wu, L., Yang, Y., Zhang, K., Hong, R., Fu, Y., Wang, M.: Joint item recommendation and attribute inference: an adaptive graph convolutional network approach. In: SIGIR, pp. 679–688 (2020)
32. Yoon, D., Lee, D., Lee, S.: Dynamic self-attention: computing attention over words dynamically for sentence embedding (2018). arXiv preprint arXiv:1808.07383
33. Zhang, K., Chen, E., Liu, Q., Liu, C., Lv, G.: A context-enriched neural network method for recognizing lexical entailment. In: AAAI, pp. 3127–3133 (2017)
34. Zhang, K., Lv, G., Chen, E., Wu, L., Liu, Q., Chen, C.P.: Context-aware dual-attention network for natural language inference. In: PAKDD, pp. 185–198 (2019)
35. Zhang, K., et al.: Drr-net: dynamic re-read network for sentence semantic matching. In: Proceedings of the AAAI Conference on Artificial Intelligence, vol. 33, pp. 7442–7449 (2019)
36. Zhang, K., et al.: Multilevel image-enhanced sentence representation net for natural language inference. IEEE TSMC: S **51**, 3781–3795 (2019)
37. Zhang, K., Wu, L., Lv, G., Wang, M., Chen, E., Ruan, S.: Making the relation matters: Relation of relation learning network for sentence semantic matching. In: AAAI (2021)
38. Zheng, Y., Mao, J., Liu, Y., Ye, Z., Zhang, M., Ma, S.: Human behavior inspired machine reading comprehension. In: SIGIR, pp. 425–434 (2019)

# Disentangled Contrastive Learning for Learning Robust Textual Representations

Xiang Chen[1,2], Xin Xie[1,2], Zhen Bi[1,2], Hongbin Ye[1,2], Shumin Deng[1,2],
Ningyu Zhang[1,2(✉)], and Huajun Chen[1,2(✉)]

[1] Zhejiang University & AZFT Joint Lab for Knowledge Engine, Hangzhou, China
{xiang_chen,xx2020,bi_zhen,yehongbin,231sm,
zhangningyu,huajunsir}@zju.edu.cn
[2] Hangzhou Innovation Center, Zhejiang University, Hangzhou, China

**Abstract.** Although the self-supervised pre-training of transformer models has resulted in the revolutionizing of natural language processing (NLP) applications and the achievement of state-of-the-art results with regard to various benchmarks, this process is still vulnerable to small and imperceptible permutations originating from legitimate inputs. Intuitively, the representations should be similar in the feature space with subtle input permutations, while large variations occur with different meanings. This motivates us to investigate the learning of robust textual representation in a contrastive manner. However, it is non-trivial to obtain opposing semantic instances for textual samples. In this study, we propose a disentangled contrastive learning method that separately optimizes the uniformity and alignment of representations without negative sampling. Specifically, we introduce the concept of momentum representation consistency to align features and leverage power normalization while conforming the uniformity. Our experimental results for the NLP benchmarks demonstrate that our approach can obtain better results compared with the baselines, as well as achieve promising improvements with invariance tests and adversarial attacks. The code is available in https://github.com/zxlzr/DCL.

**Keywords:** Natural language processing · Contrastive learning · Adversarial attack

## 1 Introduction

The self-supervised pre-training of transformer models has revolutionized natural language processing (NLP) applications. Such pre-training with language modeling objectives provides a useful initial point for parameters that generalize well to new tasks with fine-tuning. However, there is a significant gap between task performance and model generalizability. Previous approaches have indicated

---

X. Chen and X. Xie—Equal contribution and shared co-first authorship.

© Springer Nature Switzerland AG 2021
L. Fang et al. (Eds.): CICAI 2021, LNAI 13070, pp. 215–226, 2021.
https://doi.org/10.1007/978-3-030-93049-3_18

that neural models suffer from poor **robustness** when encountering *randomly permuted contexts* [21] and *adversarial examples* [11,13].

To address this issue, several studies have attempted to leverage data augmentation or adversarial training into pre-trained language models (LMs) [11], which has indicated promising directions for the improvement of robust textual representation learning. Such methods generally augment textual samples with synonym permutations or back translation and fine-tune downstream tasks on those augmented datasets. Representations learned from instance augmentation approaches have demonstrated expressive power and contributed to the performance improvement of downstream tasks in robust settings. However, the previous augmentation approaches mainly focus on the supervised setting and neglect large amounts of unlabeled data. Moreover, it is still not well understood whether a robust representation has been achieved or if the leveraging of more training samples have contributed to the model robustness.

Specifically, a robust representation should be similar in the feature space with subtle permutations, while large variations occur with different semantic meanings. This motivates us to investigate robust textual representation in a contrastive manner. It is intuitive to utilize data augmentation to generate positive and negative instances for learning robust textual representation via auxiliary contrastive objects. However, it is non-trivial to obtain opposite semantic instances for textual samples. For example, given the sentence, "Obama was born in Honululu," we are able to retrieve a sentence such as, "Obama was living in Honululu," or, "Obama was born in Hawaii." There is no guarantee that these randomly retrieved sentences will have negative semantic meanings that contradict the original sample.

In this study, we propose a novel disentangled contrastive learning (DCL) method for learning robust textual representations. Specifically, we disentangle the contrastive object using two subtasks: feature alignment and feature uniformity [27]. We introduce a unified model architecture to optimize these two sub-tasks jointly. As one component of this system, we introduce momentum representation consistency to align augmented and original representations, which explicitly shortens the distance between similar semantic features that contribute to feature alignment. As another component of this system, we leverage power normalization to enforce the unit quadratic mean for the activations, by which the scattering features within the same batch implicitly contribute to the feature uniformity. Our DCL approach is a unified, unsupervised, and model-agnostic approach, and therefore it is orthogonal to existing approaches. The contributions of this study can be summarized as follows:

- We investigate robust textual representation learning problems and introduce a disentangled contrastive learning approach.
- We introduce a unified model architecture to optimize the sub-tasks of feature alignment and uniformity, as well as providing theoretical intuitions.
- Extensive experimental results related to NLP benchmarks demonstrate the effectiveness of our method in the robust setting; we performed invariance tests and adversarial attacks and verified that our approach could enhance state-of-the-art pre-trained language model methods.

## 2    Related Work

Recently, studies have shown that pre-trained models (PTMs) [5] on the large corpus are beneficial for downstream NLP tasks, such as in GLUE, SQuAD, and SNLI. The application scheme of these systems is to fine-tune the pre-trained model using the limited labeled data of specific target tasks. Since training distributions often do not cover all of the test distributions, we would like a supervised classifier or model to perform well on. Therefore, a key challenge in NLP is learning robust textual representations. Previous studies have explored the use of data augmentation and adversarial training to improve the robustness of pre-trained language models. [12] introduced a novel text adversarial training with token-level perturbation to improve the robustness of pre-trained language models. However, supervised instance-level augmentation approaches ignore those unlabeled data and do not guarantee the occurrence of real robustness in the feature space. Our work is motivated by contrastive learning [23], which aims at maximizing the similarity between the encoded query $q$ and matched key $k^+$, while distancing randomly sampled keys $\{k_0^-, k_1^-, k_2^-, ...\}$. By measuring similarity with a score function $s(q, k)$, a form of contrastive loss function is considered as:

$$\mathcal{L}_{contrast} = -\log \frac{\exp(s(q, k^+))}{\exp(s(q, k^+)) + \sum_i \exp(s(q, k_i^-))}, \tag{1}$$

where $k^+$ and $k^-$ are positive and negative instances, respectively. The score function $s(q, k)$ is usually implemented with the cosine similarity $\frac{q^T k}{\|q\| \cdot \|k\|}$. $q$ and $k$ are often encoded by a learnable neural encoder (e.g., BERT [5]). Contrastive learning have increasingly attracted attention, which is beneficial for unsupervised or self-supervised learning from computer vision [3,10,25,30,34] to natural language processing [9,17,18,20,31,33].

## 3    Preliminaries on Learning Robust Textual Representations

**Definition 1. Robust textual representation** indicates that the representation is vulnerable to small and imperceptible permutations originating from legitimate inputs. Formally, we have the following:

$$g(X + z) = g(X), \text{ and } \text{Sim}(f(X + z), f(X)) \geq \epsilon, \tag{2}$$

where $z$ refers to the random or adversarial permutation of the input text and $g(.)$ takes input from $x$ and outputs a valid probability distribution for tasks. $f(.)$ is the feature encoder, such as BERT. We are interested in deriving methods for pre-training representations that provide guarantees for the movement of inputs such that they are robust to permutations. Therefore, a robust representation should be similar in the feature space with subtle permutations, while large variations are observed for different semantic meanings. Such constraints are related to the well-known contrastive learning [2] schema as follows:

218   X. Chen et al.

**Remark.** Robust representation is closely related to regularizing the feature space with the following constraints:

$$L_{contrast} = \sum(\sum_1^m |f(X) - f(X+z)| - \sum_1^n |f(X) - f(X')|) \tag{3}$$

where $m$ and $n$ are the number of positive and negative instances, respectively, regarding the original input, $X$, $X+z$ and $X'$ are the positive and negative instances, respectively. Note that we can obtain $X+z$ via off-the-shelf tools such as data augmentation or back-translation. However, it is non-trivial to obtain negative instances for textual samples. Previous approaches [4,6,7,28] regard random sampling of the remaining instances from the corpus as negative instances; however, there is no guarantee that those random instances are semantically irrelevant. Recent semantic-based information retrieval approaches [32] can obtain numerous similar semantic sentences via an approximate nearest neighbor [14], which further indicates that negative sampling for sentences may result in noise. In this study, inspired by the approach utilized by [27], we disentangle the contrast loss with the two following properties:

- *Alignment*: two samples forming a positive pair should be mapped to nearby features and therefore be (mostly) invariant to unneeded noise factors.
- *Uniformity*: feature vectors should have an approximately uniform distribution on the unit hypersphere.

$$\begin{aligned} L_{contrast} &= \mathbb{E}\left[-\log \frac{e^{f_x^T f_y/\tau}}{e^{f_x^T f_y/\tau} + \sum_i e^{f_x^T f_{y_i^-}/\tau}}\right] \\ &= \mathbb{E}\left[-f_x^T f_y/\tau\right] + \mathbb{E}\left[\log\left(e^{f_x^T f_y/\tau} + \sum_i e^{f_x^T f_{y_i^-}/\tau}\right)\right] \end{aligned} \tag{4}$$

$$\overset{P[f,v=f_y)]=1}{=} \underbrace{\mathbb{E}\left[-f_x^T f_y/\tau\right]}_{positive\ alignment} + \underbrace{\mathbb{E}\left[\log\left(e^{1/\tau} + \sum_i e^{f_x^T f_{y_i^-}/\tau}\right)\right]}_{uniformity}$$

The alignment loss can be defined straightforwardly as follows:

$$\mathcal{L}_{align}(f;\alpha) \triangleq - \mathop{\mathbb{E}}_{(x,y)\sim p_{pos}} [\|f(x) - f(y)\|_2^\alpha], \quad \alpha > 0 \tag{5}$$

where $f(.)$ is the feature encoder and $x,y$ are positive instance pairs. The uniformity metric refers to optimizing this metric should converge to a uniform distribution. The loss can be defined with the radial basis function (RBF) kernel $G_t : \mathcal{S}^d \times \mathcal{S}^d \to \mathbb{R}_+$ [27]. Formally, we have:

$$\begin{aligned} L_{uniform}(f;t) &\triangleq \log \mathop{\mathbb{E}}_{x,y \overset{ii.d.}{\mathbb{E}}} [G_t(u,v)] \\ &= \log \mathop{\mathbb{E}}_{x,y \overset{id.d.}{\sim} p_{data}} \left[e^{-t\|f(x)-f(y)\|_2^2}\right], \quad t > 0 \end{aligned} \tag{6}$$

where $t$ is a fixed parameter.

# 4    Disentangled Contrastive Learning

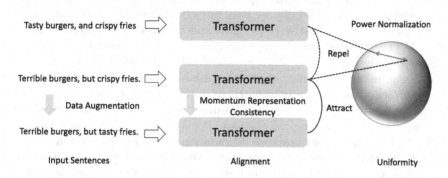

**Fig. 1.** Disentangled contrastive learning for robust textual representations.

In this section, we present a preliminary study on how to learn robust textual representation via disentangled contrastive learning, as represented in Fig. 1.

## 4.1    Feature Alignment with Momentum Representation Consistency

There are multiple ways to align a textual representation. We utilize two transformers with a consistent momentum representation to explicitly guarantee feature alignment [8]. The two networks are defined by a set of weights $\theta$ and $\xi$. We use the exponential moving average of the parameters $\theta$ to get $\xi$. Formally, we have:

$$\xi \leftarrow \tau\xi + (1 - \tau)\theta \tag{7}$$

Given a sentence $X$ and its augmentation $X'$ (e.g., via data augmentation) from the first original network, we may obtain output representations $q \triangleq f_\theta(X)$ and $p \triangleq f_\theta(X')$. Note that previous works [3,8] indicates that an projection $p$ in feature space improve the performance. We then leverage a projection function $g(p_\theta)$ and $\ell_2$-normalize both $g(p_\theta)$ and $q_\xi$ to $\bar{g}(p_\theta) \triangleq g/\|g(p_\theta)\|_2$ and $\bar{q}_\xi \triangleq q_\xi/\|q_\xi\|_2$, respectively. We leverage the mean squared loss as follows:

$$\mathcal{L}_{\text{align}} \triangleq \|\bar{g}(q) - \bar{p}_\xi\|_2^2 = 2 - 2 \cdot \frac{\langle g(q_\theta), p_\xi \rangle}{\|g(q_\theta)\|_2 \cdot \|p_\xi\|_2} \tag{8}$$

Additionally, we make the losses symmetrical $\mathcal{L}_{\text{align}}$ by feeding $X$ to the augmented network and $X'$, separately. We optimize $\mathcal{L}_{\text{align}} + \widetilde{\mathcal{L}}_{\text{align}}$ with respect to $\theta$ only, but *not* $\xi$, via the stop-gradient.

## 4.2  Feature Uniformity with Power Normalization

To ensure that feature vectors should have an approximately uniform distribution, we can directly optimize the Eq. 6. However, different from computer vision, in the original loss of BRET [5], we have already utilized the next sentence prediction loss. Such a contrastive object has explicitly made the sentence representation $f(.)$ scattered in the feature space; thus, the model may quickly collapse without learning. Inspired by [22], we argue that batch normalization can identify the common-mode between examples of a mini-batch and removes it using the other representations in the mini-batch as implicit negative examples. We can, therefore, view batch normalization as a novel method of implementing feature uniformity on embedded representations. Because vanilla batch normalization will lead to significant performance degradation when naively used in NLP, we leverage an enhanced power normalization [24] to guarantee feature uniformity. Specifically, we leverage the unit quadratic mean rather than the mean/variance of running statistics with an approximate backpropagation method to compute the corresponding gradient. Formally, we have the following:

$$\widehat{X}^{(t)} = \frac{X^{(t)}}{\psi^{(t-1)}}$$
$$Y^{(t)} = \gamma \odot \widehat{X}^{(t)} + \beta \qquad (9)$$
$$\left(\psi^{(t)}\right)^2 = \left(\psi^{(t-1)}\right)^2 + (1-\alpha)\left(\psi_B^2 - \left(\psi^{(t-1)}\right)^2\right)$$

Note that we compute the gradient of the loss regarding the quadratic mean of the batch. In other words, we utilize the running statistics to conduct backpropagation, thus, resulting in bounded gradients, which is necessary for convergence in NLP (see proofs in [24]).

## 4.3  Implementation Details

We leverage synonyms from WordNet categories to conduct data augmentation for computation efficiency. We combine all the momentum representation consistency and power normalization results in a unified architecture with the mask language model object. We leverage the same architecture of the BERT-base [5]. We first pre-train the model in a large-scale corpus unsupervisedly (e.g., the same corpus and training steps with BERT) and then fine-tune the model using task datasets.

# 5  Experiment

## 5.1  Datasets and Setting

We conducted experiments on three benchmarks: GLUE, SQuAD, SNLI, and DialogRE.

**Table 1.** Summary of results on GLUE.

| Model | | CoLA | SST-2 | MRPC | QQP | MNLI (M/MM) | QNLI | RTE | GLUE Avg |
|---|---|---|---|---|---|---|---|---|---|
| NORMAL | BERT | 56.8 | 92.3 | 89.7 | 89.6 | 84.6/85.2 | 91.5 | 69.3 | 82.3 |
| | BERT+DA | 58.6 | 93.2 | 86.5 | 86.7 | 84.2/84.4 | 91.1 | 68.9 | 81.7 |
| | DCL | **60.9** | 93.0 | **89.7** | **90.0** | 84.7/84.6 | **91.7** | **69.7** | **83.0** |
| ROBUST | BERT | 46.4 | 91.8 | 88.1 | 84.9 | 81.6/82.2 | 89.2 | 67.1 | 78.9 |
| | BERT+DA | 53.8 | 92.9 | 85.6 | 85.5 | 83.1/83.4 | 90.7 | 66.3 | 80.1 |
| | DCL | 48.4 | **92.4** | 86.0 | **85.5** | 82.5/82.7 | 89.7 | **68.8** | 79.5 |

**GLUE** [26] is an NLP benchmark aimed at evaluating the performance of downstream tasks of the pre-trained models. Notably, we leverage nine tasks in GLUE, including CoLA, RTE, MRPC, STS, SST, QNLI, QQP, and MNLI-m/mm. We follow the same setup as the original BERT for single sentence and sentence pair classification tasks. We leverage a multi-layer perception with a softmax layer to obtain the predictions.

**SQuAD** is a reading comprehension dataset constructed from Wikipedia articles. We report results on SQuAD 1.1. Here also, we follow the same setup as the original BERT model and predict an answer span—the start and end indices of the correct answer in the correct context.

**SNLI** is a collection of 570k human-written English sentence pairs that have been manually labeled for balanced classification with entailment, contradiction, and neutral labels, thereby supporting the task of natural language inference (NLI). We add a linear transformation and a softmax layer to predict the correct label of NLI.

**DialogRE** is a dialogue-based relation extraction dataset, which contains 1,788 dialogues from a famous American television situation comedy Friends.

To evaluate the robustness of our approach, we also conduct invariance testing with CheckList[1] [21] and adversarial attacks[2]. To generate label-preserving perturbations, we used WordNet categories (e.g., synonyms and antonyms). We selected context-appropriate synonyms as permutation candidates. To generate adversarial samples, we leverage a probability-weighted word saliency (PWWS) [19] method based on synonym replacement. We manually evaluate the quality of the generated instances. We also conduct experiments that apply data augmentation and adversarial training to the BERT model. We utilize PyTorch to implement our model. We use Adam optimizer with a cosine decay learning rate schedule. We set the initial learning rate as 1e−5. We use a batch size of 32 over eight Nvidia 1080Ti GPUs.

---

[1] https://github.com/marcotcr/checklist.git.
[2] https://github.com/thunlp/OpenAttack.

## 5.2  Results and Analysis

### Main Results

From Table 1 and 2, we can observe the following: 1) Vanilla BERT achieves poor performance in the robust set on both GLUE and SQUAD, which indicates that the previous fine-tuning approach cannot obtain a robust textual representation. This will lead to performance decay with permutations.

**Table 2.** Summary of results on SQuAD.

| Model | | F1 | EM |
|---|---|---|---|
| NORMAL | BERT | 88.5 | 80.8 |
| | BERT+DA | 88.2 | 80.4 |
| | DCL | **88.4** | **81.0** |
| ROBUST | BERT | 86.7 | 77.8 |
| | BERT+DA | 87.8 | 79.9 |
| | DCL | 86.8 | 78.1 |

2) With data augmentation, BERT can obtain improved performance in the robust set; however, a slight performance decay is observed in the original test set. We argue that data augmentation can obtain better performance by fitting to task-specific data distribution; there is no guarantee that more data will result in robust textual representations.

3) Our DCL approach achieves improved performance in both the original test set and robust set compared with vanilla BERT. Note that our DCL is an unsupervised approach, and we leverage the same training instances with BERT. The performance improvements indicate that our approach can obtain more robust textual representations that enhance the performance of the system.

### Adversarial Attack Results

From Table 3, we can observe the following: 1) Vanilla BERT achieves a poor performance with adversarial attacks; BERT with adversarial training can obtain a good performance. However, we notice that there exists a performance decay for adversarial training in the original test set. Note that adversar-

**Table 3.** Summary of results on CoLA, SNLI, DialogRE.

| Model | | CoLA | SNLI | DialogRE |
|---|---|---|---|---|
| NORMAL | BERT | 56.8 | 91.0 | 63.0 |
| | BERT+Adv | 55.0 | 90.9 | 64.3 |
| | DCL | **58.8** | 91.0 | 64.2 |
| ADVERSARIAL | BERT | 47.0 | 87.4 | 59.0 |
| | BERT+Adv | 55.1 | 90.3 | 62.9 |
| | DCL | 48.2 | 90.5 | 63.2 |

ial training methods would lead to standard performance degradation [29], i.e., the degradation of natural examples. 2) Our DCL approach achieves improved performance in the test set with and without an adversarial attack, which further demonstrates that our approach can obtain robust textual representations that are stable for different types of permutations.

### Quantitative Analysis of Textual Representation

As we hypothesize that power normalization can implicitly contribute to feature uniformity, we conduct further experiments to analyze the effects of normalization [1]. Specifically, we random sample instances and leverage the cosine similarity of the original input projection vectors and the augmented projection vectors. We calculate the average cosine similarity between positive instances (in

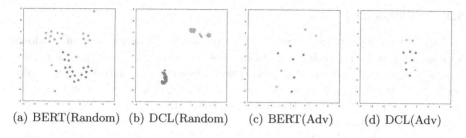

(a) BERT(Random)    (b) DCL(Random)    (c) BERT(Adv)    (d) DCL(Adv)

**Fig. 2.** T-SNE visualizations of sentence embeddings.

blue) and random instances (in red) with different strategies, including without normalization (No Norm), batch normalization (BN), and power normalization.

From Fig. 3, we observe that with no normalization in $p$ or $q$, the feature space is aligned for both positive and negative instances, which shows that there exists a feature collapse for textual representation learning. Considering DCL training (i.e., with power normalization), we notice that the textual representations are relatively more similar between the positive instances (0.9842) than random (negative) ones (0.7904); thus, we can obtain different vectors.

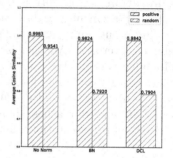

Next, we give an intuitive explanation of preventing feature collapse for textual representation learning. Given an input instance without negative examples, the model may always output the projection vector $z$ with $[0, 1, 0, 0, \ldots]$. Thus, the model can achieve a perfect prediction through

**Fig. 3.** Cosine similarity of the original input projection vectors with the augmented input projection vectors. (Color figure online)

learning a simple identity function, which in other words, collapse in the feature space. With normalization, the output vector $z$ cannot obtain such singular values. Since the outputs will be redistributed regarding the learned mean and standard deviation, we can implicitly learn robust representations.

### Qualitative Analysis of Textual Representation

We randomly selected instances to visualize a sentence with T-SNE [15] to better understand the behaviors of textual representations. The different color refers to the different sentence pairs for both random permutation and adversarial attack settings. From Fig. 2, it may be observed that our approach can obtain a relatively similar semantic representation with permutations in both invariant tests and adversarial attack settings. Note that we explicitly align the projection of the textual representation with a random permutation, thereby encouraging similar semantic instances to have relatively similar representations.

### 5.3   Discussion

**Robust Representation with Contrastive Learning.** Conventional approaches usually try to leverage instance-level augmentation aimed at achieving good performance on a robust set. However, there is no guarantee that robust textual representations will be obtained. Intuitively, directly aligning the representation of input tokens with slight permutations may contribute to robust representations. However, without any negative constraints, the model will easily collapse with a sub-optimal solution. In this study, we observe that power normalization identifies this common mode between examples. In other words, it can remove those trivial samples by using the other representations in the batch as implicit negative instances. We can, therefore, view normalization as an implicitly contrastive learning method.

**Limitations.** This work is not without limitations. We only consider the synonym replacement as a data augmentation strategy due to the efficiency of processing a huge amount of data. Other strong data augmentation methods can also be leveraged. Another issue is representation alignment, as there are lots of augmentations. We cannot enumerate all positive pairs for alignments; thus, there is still some room for designing more efficient feature-aligning algorithms. Moreover, as we utilize the square root loss, which is absolutely a Euclidean distance. Recent approaches [16] indicates that Euclidean space may be sub-optimal for textual representations, and we leave this for future works.

## 6   Conclusion

We investigated robust textual representation learning and proposed a disentangled contrastive learning approach. We introduced feature alignment with a momentum representation consistency and feature uniformity with power normalization. We empirically observed that our approach could obtain an improved performance compared with baselines in NLP benchmarks and achieve a robust performance with invariant tests and adversarial attacks.

**Acknowledgments.** We want to express gratitude to the anonymous reviewers for their hard work and kind comments. This work is funded by NSFC91846204/NSFCU19B2027.

## References

1. Abe, F., Josh, A.: Understanding self-supervised and contrastive learning with bootstrap your own latent (BYOL). https://untitled-ai.github.io/understanding-self-supervised-contrastive-learning.html
2. Arora, S., Khandeparkar, H., Khodak, M., Plevrakis, O., Saunshi, N.: A theoretical analysis of contrastive unsupervised representation learning. arXiv preprint arXiv:1902.09229 (2019)

3. Chen, T., Kornblith, S., Norouzi, M., Hinton, G.: A simple framework for contrastive learning of visual representations. In: Proceedings of Machine Learning and Systems 2020, pp. 10719–10729 (2020)
4. Chi, Z., et al.: InfoXLM: an information-theoretic framework for cross-lingual language model pre-training. arXiv preprint arXiv:2007.07834 (2020)
5. Devlin, J., Chang, M.W., Lee, K., Toutanova, K.: BERT: pre-training of deep bidirectional transformers for language understanding. In: Proceedings of NAACL, pp. 4171–4186. Minneapolis, Minnesota, June 2019. https://doi.org/10.18653/v1/N19-1423
6. Fang, H., Xie, P.: CERT: contrastive self-supervised learning for language understanding. arXiv preprint arXiv:2005.12766 (2020)
7. Giorgi, J.M., Nitski, O., Bader, G.D., Wang, B.: DeCLUTR: deep contrastive learning for unsupervised textual representations. arXiv preprint arXiv:2006.03659 (2020)
8. Grill, J.B., et al.: Bootstrap your own latent: a new approach to self-supervised learning. arXiv preprint arXiv:2006.07733 (2020)
9. Gunel, B., Du, J., Conneau, A., Stoyanov, V.: Supervised contrastive learning for pre-trained language model fine-tuning. arXiv preprint arXiv:2011.01403 (2020)
10. He, K., Fan, H., Wu, Y., Xie, S., Girshick, R.B.: Momentum contrast for unsupervised visual representation learning. CoRR abs/1911.05722 (2019)
11. Jin, D., Jin, Z., Zhou, J.T., Szolovits, P.: Is BERT really robust? A strong baseline for natural language attack on text classification and entailment. arXiv:1907 (2019)
12. Li, L., Qiu, X.: TextAT: adversarial training for natural language understanding with token-level perturbation. arXiv preprint arXiv:2004.14543 (2020)
13. Li, L., et al.: Normal vs. adversarial: salience-based analysis of adversarial samples for relation extraction. arXiv preprint arXiv:2104.00312 (2021)
14. Liu, T., Moore, A.W., Yang, K., Gray, A.G.: An investigation of practical approximate nearest neighbor algorithms. In: Advances in Neural Information Processing Systems, pp. 825–832 (2005)
15. Maaten, L.V.D., Hinton, G.: Visualizing data using t-SNE. J. Mach. Learn. Res. 9(Nov), 2579–2605 (2008)
16. Meng, Y., Zhang, Y., Huang, J., Zhang, Y., Zhang, C., Han, J.: Hierarchical topic mining via joint spherical tree and text embedding. In: Proceedings of the 26th ACM SIGKDD International Conference on Knowledge Discovery & Data Mining, pp. 1908–1917 (2020)
17. Mikolov, T., Sutskever, I., Chen, K., Corrado, G.S., Dean, J.: Distributed representations of words and phrases and their compositionality. In: Advances in Neural Information Processing Systems 26, pp. 3111–3119 (2013)
18. Mnih, A., Kavukcuoglu, K.: Learning word embeddings efficiently with noise-contrastive estimation. In: Advances in Neural Information Processing Systems 26, pp. 2265–2273 (2013)
19. Ren, S., Deng, Y., He, K., Che, W.: Generating natural language adversarial examples through probability weighted word saliency. In: Proceedings of the 57th Annual Meeting of the Association for Computational Linguistics, pp. 1085–1097 (2019)
20. Rethmeier, N., Augenstein, I.: A primer on contrastive pretraining in language processing: methods, lessons learned and perspectives. arXiv preprint arXiv:2102.12982 (2021)

21. Ribeiro, M.T., Wu, T., Guestrin, C., Singh, S.: Beyond accuracy: behavioral testing of NLP models with checklist. In: Jurafsky, D., Chai, J., Schluter, N., Tetreault, J.R. (eds.) Proceedings of the 58th Annual Meeting of the Association for Computational Linguistics, ACL 2020, Online, 5–10 July 2020, pp. 4902–4912. Association for Computational Linguistics (2020). https://www.aclweb.org/anthology/2020.acl-main.442/

22. Santurkar, S., Tsipras, D., Ilyas, A., Madry, A.: How does batch normalization help optimization? In: Advances in Neural Information Processing Systems, pp. 2483–2493 (2018)

23. Saunshi, N., Plevrakis, O., Arora, S., Khodak, M., Khandeparkar, H.: A theoretical analysis of contrastive unsupervised representation learning. In: Proceedings of the 36th International Conference on Machine Learning, vol. 97, pp. 5628–5637. PMLR, 9–15 June 2019, Long Beach, California, USA. http://proceedings.mlr.press/v97/saunshi19a.html

24. Shen, S., Yao, Z., Gholami, A., Mahoney, M.W., Keutzer, K.: PowerNorm: rethinking batch normalization in transformers. In: The proceedings of the International Conference on Machine Learning (ICML) (2020)

25. Tian, Y., Krishnan, D., Isola, P.: Contrastive multiview coding. CoRR abs/1906.05849 (2019). arxiv:1906.05849

26. Wang, A., Singh, A., Michael, J., Hill, F., Levy, O., Bowman, S.R.: GLUE: a multi-task benchmark and analysis platform for natural language understanding. In: 7th International Conference on Learning Representations, ICLR 2019. OpenReview.net (2019). https://openreview.net/forum?id=rJ4km2R5t7

27. Wang, T., Isola, P.: Understanding contrastive representation learning through alignment and uniformity on the hypersphere. arXiv preprint arXiv:2005.10242 (2020)

28. Wei, X., Hu, Y., Weng, R., Xing, L., Yu, H., Luo, W.: On learning universal representations across languages. arXiv preprint arXiv:2007.15960 (2020)

29. Wen, Y., Li, S., Jia, K.: Towards understanding the regularization of adversarial robustness on neural networks (2019)

30. Wu, Z., Xiong, Y., Yu, S.X., Lin, D.: Unsupervised feature learning via nonparametric instance discrimination. In: 2018 IEEE Conference on Computer Vision and Pattern Recognition, CVPR 2018, pp. 3733–3742. IEEE Computer Society (2018)

31. Wu, Z., Wang, S., Gu, J., Khabsa, M., Sun, F., Ma, H.: CLEAR: contrastive learning for sentence representation. arXiv preprint arXiv:2012.15466 (2020)

32. Xiong, L., et al.: Approximate nearest neighbor negative contrastive learning for dense text retrieval. arXiv preprint arXiv:2007.00808 (2020)

33. Ye, H., et al.: Contrastive triple extraction with generative transformer. arXiv preprint arXiv:2009.06207 (2020)

34. Ye, M., Zhang, X., Yuen, P.C., Chang, S.: Unsupervised embedding learning via invariant and spreading instance feature. In: IEEE Conference on Computer Vision and Pattern Recognition, CVPR 2019, pp. 6210–6219. Computer Vision Foundation/IEEE (2019)

# History-Aware Expansion and Fuzzy for Query Reformulation

Wei Pang[1(✉)] and Ruixue Duan[2]

[1] School of Artificial Intelligence, Beijing University of Posts
and Telecommunications, Beijing 100876, China
`pangweitf@bupt.edu.cn`
[2] School of Computer Science, Beijing Information Science and Technology
University, Beijing 100101, China

**Abstract.** Query reformulation is the task of rewriting users' query to predict their information need. A user often struggles to modify a query by adding or removing terms when interacting with search engines. To address this issue, we propose a history-aware expansion and fuzzy model for query reformulation that improves follow-up queries based on successful history click-through logs. A probabilistic model is thus presented to calculate term weight in history and expand meaningful terms or fuzz trivial terms to follow-up query. Experimental results show that reformulated query can improve search engine results on low-frequency and long-tailed queries.

**Keywords:** Query reformulation · Query expansion · Query fuzzy

## 1 Introduction

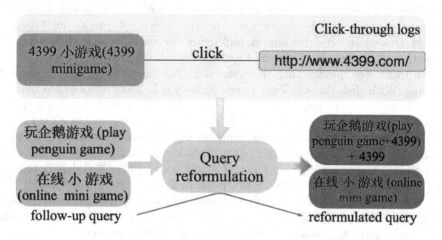

**Fig. 1.** An illustration of history-aware expansion and fuzzy for query reformulation.

Query reformulation is a proven approach to deal with a mismatch between query and document that is a challenging problem for search engines [1]. On the

ⓒ Springer Nature Switzerland AG 2021
L. Fang et al. (Eds.): CICAI 2021, LNAI 13070, pp. 227–238, 2021.
https://doi.org/10.1007/978-3-030-93049-3_19

one hand, users and documents often use different words leading to a mismatch between query terms and document terms [1]. On the other hand, many people usually try to modify queries during the search to make their real needs more accurate, especially when seeking unclear requirements. For microblog search, because tweets only contain a dozen words, the mismatch problem is extensive. This paper focuses on the above problem with the two reformulation methods of query expansion and fuzzy.

Most prior work focuses on expanding query by adding new terms into the original query. Especially, the new terms mainly come from thesaurus [2] and external resources [3,4], such as WordNet [2,3,5,6] and Freebase [7]. Pseudo-relevance feedback [1,8,9] method selects expanded words from the initially returned documents and then searches again, achieving state-of-the-art results. However, this method highly relies on the initial search result. Word embedding [4,10–12] is also used for query expansion and chooses similar words for query terms as expansion. Although these types of query expansion can augment some of the original query term's synonyms, they cannot help understand users' hidden needs. Yiqun Liu et al. [13] proposed a snippet click model, mining keywords from snippets clicked by users.

Query deduction aims to drop query terms with little information in a long query [14]. However, removing some words from the user's query is not good. In some cases, dropping terms could hide the user's practical search intention.

A good query matches well and retrieves many relevant documents ranked at the top position, e.g., at the top three positions on the head page. However, a poor query might get bad search results because it is not adequate to understand the user's need, or the satisfying results might rank in later positions since the score is lower due to the term mismatch problem. Although the search result is irrelevant in the returned search result, some users often browse the later result until finding a relevant result, click the URL, and complete this search action. In comparison, some users may struggle to rewrite a series of queries until success [15]. Motivated by this, we aim at improving the users' search experience by extracting the knowledge from the successful search experience.

We think two queries that share co-click documents might have a similar meaning. With click-through logs, we can actively help users rewrite their queries through the related co-clicked queries. We calculate the weight of terms with co-clicked queries, where the meaningful terms are used for expanding a query and the trivial terms for fuzzing. For example, in Fig. 1, there are three queries that co-click the same document http://www.4399.com/, where "4399 小游戏 (4399 mini-game)" is a hot query that has real search need, while "在线小游戏 (online mini-games)" and "玩企鹅游戏 (play penguin game)" are low-frequency and retrieves poor results. With query reformulation, we expand the word "4399" to "玩企鹅游戏 (play penguin game)", and fuzzy "小 (mini)" in "在线小游戏 (online mini-games)", and thus satisfying users' information needs.

This paper proposes a novel history-aware expansion and fuzzy framework to implement the above ideas. It consists of four modules: History-aware Term Weight (HaTwei), Index, Matching, and Fine-tune (Ftune). HaTwei computes weight distributions of terms based on co-click behavior in click-through logs. Index stores the results of HaTwei as an index database, which is used for the Matching module to find expanded or fuzzy terms for follow-up queries. Ftune module fine-tunes the documents by expanded or fuzzy terms. Experimental results on two different datasets show that our model achieves better results compared with three strong baselines.

To summarize, our contributions are mainly three-fold:

- We propose a novel history-aware expansion and fuzzy framework for query reformulation. Based on click-through logs, we combine query expansion and query reduction through a probabilistic model.
- We present query fuzzy strategy that reduces trivial terms in the weight calculating, instead of query reduction to drop them.
- We evaluate our method on two different datasets and explicitly improve search engines' quality compared with the three baselines.

## 2    Related Work

Query reformulation plays an essential role in query understanding in search engines [1,16,17]. It helps solve the mismatch problem [1] between query and document, rewriting a query to make their meaning more effective. Generally, query reformulation consists of spelling error correction, stemming, query segmentation, query expansion, and query deduction [1]. This paper focuses on two basic types, query expansion with adding meaningful terms and query deduction via fuzzy trivial terms.

Query expansion is a well-known method to deal with the term mismatch problem [1,3,6,10,18]. One typical way is to use external resources [1,2] to select additional terms. A common practice is to derive from knowledge bases, such as Freebase [7], Probase [18], WordNet [5], ConceptNet, and Wikipedia.

Another method is pseudo-relevance feedback [1,8], which assumes the top-retrieved document is relevant, selects the expansion terms from the initially returned documents and then prompts a second search with the original query plus expansion terms. Recently, query expansion based on word embedding [4,10–12] is widely applied. They train the Word2Vec model over the entire corpus and select semantically related terms to the raw query in the word2vec space [4,10–12]. These expansion methods tend to find synonyms or related words of a query word [6] and solve mismatch problems. In [4], Qian Liu et al. also use word embedding to select top similar words with the initial query via cosine similarity.

Reinforcement Learning is also used for query reformulation tasks [19,20]. Nogueira et al. [19] propose a neural network with reinforcement learning to model relationships of expansion terms and document recall, selecting some terms to maximize the recall rate of relevant documents returned.

Qian Liu et al. [4] proposed four fuzzy rules to re-weight the expansion words, which differs from our fuzzy method. We focus on fuzzing uninformative terms in the raw query.

In contrast to query expansion, query deduction is another query reformulation technique via the removal of trivial terms from the query [1,21]. However, instead of removing terms in our task, we focus on fuzzing them in the original query. We refer to this type of reformulation as query fuzzy. One advantage of fuzzy reformulation is that we can preserve as much search need of users' original intention as possible.

## 3   Models

Figure 1 shows the overview of our History-aware Expansion and Fuzzy model for Query Reformulation (HaEFQR). There are four modules: History-aware Term •Weight (HaTwei), Index, Matching, and Fine-tune (Ftune). HaTwei calculates term weight based on history click-through logs. The three modules of Index, Matching, and Fine-tune are applied to a microblog retrieval task. Details of each module are given as follows.

### 3.1   History-Aware Term Weight (HaTwei)

The goal of HaTwei is to find essential or trivial terms according to history click-through logs. For example, if many users issue the query "Alipay fast payment", they mostly click an official website at the top position. A few people submit a query "Alipay" and click the same official website on the fourth page, and they might share the same need because they have the same co-click behavior. This co-click information can help users understand their intention actively, e.g., extracting the meaningful phrase "fast payment". If a user searches for "Alipay", the search engine might automatically expand it with "fast payment", which performs retrieving documents with the reformulated query "Alipay fast payment".

On the contrary, a query "natural logarithm transformation" [1], "natural" is invaluable in understanding user need; we fuzz the trivial "natural" to obtain more relevant results.

We group all the queries by co-click, $Q = \{q^{(i)}\}_{i=1}^{N}$ denotes a set of $N$ queries that share the same clicked documents, $U = \{u^{(i)}\}_{i=1}^{M}$ denotes the co-clicked documents given $Q$, $u$ indicates a document and M is the total number of documents. Let $t$ indicates a query term, $q$ indicates a query.

To simplify the discussion, we initially consider two co-clicked queries $q^{(i)} = \{t^{(i)}\}_{i=1}^{n}$ and $q^{(j)} = \{t^{(i)}\}_{i=1}^{m}$, each query is represented as a bag of words of size $n$ and $m$, respectively. If we want to choose some terms from $q^{(j)}$ to reformulate $q^{(i)}$, how to measure the weight of each query term of $q^{(j)}$. In this work, our main objective is to calculate conditional probability, i.e. $p(t \in q^{(j)}|q^{(i)})$.

A click graph is seen in Fig. 2. We start at a query $q^{(i)}$ and allow two random walks via a document to $q^{(j)}$. There exist multiple paths between the two queries. Along each path, we define the conditional probability $p(t \in q^{(j)}|q^{(i)})$ using the multiplication rule. The term weight can be expressed as the sum conditional probabilities of multiple paths, as written in Eq. 1:

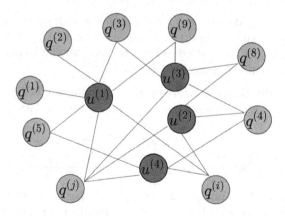

**Fig. 2.** A co-click graph among queries and their clicked documents, red nodes denote documents, the yellow mean queries, and the edge represents click behavior between query and document. (Color figure online)

$$p(t \in q^{(j)}|q^{(i)}) = \int_{u \in U} p(t|u)p(u)du, \qquad (1)$$

where $p(u)$ denotes the marginal distribution of document $u$, $p(t|u)$ is the term weight of $t$ under the document $u$. Notes that in this work $p(u)$ can be extended to include $q^{(i)}$ and $q^{(j)}$, then it is defined as Eq. 2:

$$p(u) = \prod_{q \in \{q^{(i)}, q^{(j)}\}} p(u|q)p(q). \qquad (2)$$

Based on the analysis of the relationship between $q^{(i)}$ and $q^{(j)}$, we present a novel probabilistic model from the viewpoint of random walk. The more important the walking path is, the larger conditional probability will get. The term might play a crucial role in representing the user's needs in $q^{(i)}$. As an approximation, we calculate $p(t|u)$ by using its source query $q(j)$, and $p(t|u)$ is given by $p(t|u) \approx p(q^{(j)}|u)$. The new model is then written as Eq. 3:

$$p(t \in q^{(j)}|q^{(i)}) = \sum_{k=1}^{m} p(u^{(k)}|q^{(j)})p(q^{(j)}|u^{(k)})p(u^{(k)}|q^{(i)}). \qquad (3)$$

Let $p(q^{(j)}|u^{(k)})$ denote click weight of a special query $q^{(j)}$ under given clicked document $u^{(k)}$. Similarly, $p(u^{(k)}|q^{(i)})$ can be represented as clicked weight of a special document $u^{(k)}$ when given a query $q^{(i)}$. Formally, we give a detailed calculation as following subsections.

**Query Weight Score (qwei)** is a distribution of queries co-clicking the same document, abbreviated as qwei, its empiric formula is defined as Eq. 4:

$$p(q|u) = \frac{\frac{click\_count^{1.75}(<q,u>)}{search\_count(q)}}{0.01 \sum_{e \in \{e|e\ co-clicks\ u\}} \frac{click\_count^{1.75}(<e,u>)}{search\_count(e)}}, \qquad (4)$$

where $search\_count(q)$ denotes the total search count of query $q$, $click\_count(< q, u >)$ is the total clicked times between the pair of $< q, u >$. The symbol clicked-count squared 1.75 we used in Eq. 4 because we think that users' clicking behavior plays an important role in relevance.

**Document Weight Score (dwei)** means a distribution of documents clicked by the same query, short for dwei, and also has an empiric formula as written in Eq. 5:

$$p(u|q) = \frac{\frac{click\_count^{1.75}(<u,q>)}{search\_count(q)}}{0.01 \sum_{d \in \{d | d \ co-clicks \ by \ q\}} \frac{click\_count^{1.75}(<d,q>)}{search\_count(q)}}. \tag{5}$$

In our work, we find many pairs of $\{q^{(i)}, q^{(j)}\}_{j \neq i}$ that co-clicks the same documents as shown in Fig. 2. Such crowd-sourced user experience is valuable in inferencing what a user wants to know and yield insights into how people generate a query using keywords step by step. More generally, integrating all co-clicked pairs $< q, u >$ that are associated with $q^{(i)}$, we total probabilities of the term as Eq. 6:

$$p(t|q^{(i)}) = \sum_{j \neq i} p(t \in q^{(j)} | q^{(i)}). \tag{6}$$

Terms (words or phrases) are ranked in descending order according to Eq. 6. In the experiment, we choose expanded terms with the top $k$ for query expansion, the tail $l$ as fuzz terms for fuzzing. We observe that the reformulated query enables us to grasp users' needs. As an illustration detail, Table 2 shows some examples for query reformulation.

**Merger Query Expansion and Fuzzy.** We merged query expansion and query fuzzy strategy into the same framework. However, query reformulation may bring several problems, such as ambiguous terms, irrelevant statistically correlated terms [2], and the complicated problem is topic drift [1,21], indicating a change in query intent. To avoid these issues, we adopt three strategies that perform well together in the experiment.

First, we reduce the coverage ratio of query expansion and fuzzy algorithm to hold down the number of impacted queries. Furthermore, to deal with noise in the click-through data [1,13], we use some smoothing methods.

Second, we utilize the knowledge extracted from click-through logs, representing historical relevance knowledge of crowd-sourced feedback by many users.

Finally, information retrieval mainly includes three steps: inverted index, matching, and ranking [2]. In our work, the expanded term or fuzz term plays an auxiliary role in retrieving. Firstly, for intersecting, the expanded or fuzz term is transparent to the engine. The fuzz term does not participate in intersecting. The recall rate will be increased significantly in the query fuzzy task. Secondly, based on the initial documents retrieved from the index, we are interested in the documents containing the expanded terms. We tune the relevance score via the proximity-weighted scoring function sum over the expanded terms. Similarly, if some of the matched words are fuzz terms, the score will be discounted.

**Parameter Setting.** Here we provide two methods to estimate the model parameters, dwei, and qwei.

First, we only consider a document, which means group all the queries that co-click a specific document, called a local query model. Then we pair a query with each other and calculate the conditional probability in each pair according to Eq. 1–5.

Second, a query might click many documents with other queries, as seen in Fig. 2. We sum the probability along other click paths according to Eq. 6. We refer to a global query model. Specially, we use a hash lookup table to store the expanded terms and fuzz terms for each query. To avoid repeated terms and reduce memory space, we decompose model parameters into two tables. One table is HashLookUp Table, storing hash value for query with additional information, including the hash of terms. The other one is the terms table, only memory unique terms, consisting of expansion and fuzz terms. During looking up, the key is the hash value of the query and then perform the hash search as quickly as possible.

**Estimate Score with Smoothing Methods.** Click-through logs with much noise and sparseness [1,13,22] are challenging. For example, if we have one click and two impressions, then the CTR would be 50%, while if we have 50 clicks and 100 impressions, the CTR (Clickthrough rate) is the same. However, the two cases have different meanings; the latter one is more reliable. However, the Wilson score of 1 click in 2 impressions is significantly lowered to 21.13%. Therefore, we use two tricks to alleviate the issues as follows:

(1) We exploit the click count to the power of 1.75 for reducing sparseness of click information in the click-through logs.
(2) We use the Wilson score confidence interval [23,24], abbreviated as Wilson score, to compute the Click-Through Rate avoiding biased ratio caused by sample size [23]. Similarly, we also adopt another weighting function as in Eq. 7, which is introduced by Pennington et al. [25] for discounting the search count and click count.

$$f(x) = \begin{cases} (x/x_{max})^{3/4} & if \ x \ < \ x_{max} \\ 1.0 & otherwise \end{cases} \tag{7}$$

Compared with the traditional CTR method, it is clear that the Wilson score method helps limit noises in click-through data. In search engine, 50 clicks occur in 100 impressions might represent more robust and more relevant feedback than one click out of 2 impressions. In the SougouQ dataset, the number of impressions is no more than 20 possess about 73.76%, the Wilson score of CTR can balance the impact of small sample size caused by random noises [1].

## 4    Experiment

**Datasets.** We conduct an extensive experiment on two datasets. The SogouQ dataset consists of search and click-through logs released by the Sogou Labs

[26]. The SogouQ dataset contains queries and its clicked document per line, including four columns: query, rank position, click order, and clicked document, where click order means which ranked position the click behavior occurs. In total, there are 20.43M queries. We obtain 3.02M of distinct query collection after data preprocessing, 8.2M separate documents. With word segmentation, we get a total number of unique words is 695,148, the average length of queries is 2.488. We collected MicroBlog data from Sina, containing 74.6K tweets. We use MicroBlog to evaluate our model and baseline methods.

**Search: Matching and Ranking.** We apply the proposed model to a microblog retrieval task for practical use. It involves three parts: query understanding, matching, and ranking. In query understanding phrase, we design a simple strategy to expand short query or fuzzy long query for a new query as [14] does.

**Expanding Short Query.** We consider a short query in our work as the size of terms is less than 5, which requires expanding for improving the retrieved accuracy.

How many distinct terms k we chose to expand? With more terms for query expansion, it might cause topic drift problems [1,27]. While fewer terms expand to a query would not benefit. Our focus is the relationship between the CTR and the number of query terms, as heuristics information to help us select a fair number of expanded terms. Finally, we empirically set up the number of expanded terms k to 2.

**Fuzzing Long Query.** We consider a long query as the size of query terms is more than 7. We restrict fuzzy query strategy to long query and fuzz trivial query terms for increasing document recall rate.

Query fuzzy strategy, a type of query reformulation, is defined as the reverse formation of query expansion through fuzzing query terms with little meaning. We consider trivial terms as the minimum weight of conditional probability $p(t|q^{(i)})$ and set $l$ to 2, as shown in Table 2.

**Query Understanding.** Query understanding plays a crucial role in search engines [1]. Coming to a new query, we first check whether or not the query is long or short. If so, then compute the hash value of the query and finally conduct a binary search on the hash lookup table. Then the strategy of query reformulation is triggered. The original query, plus expanded or fuzz terms and their weights, are used to retrieve documents.

**Matching Between Query and Document.** In matching, based on the inverted index, we cannot take the expansion terms or fuzz terms into account. The hidden relevant documents can be matched by intersecting the posting list associated with the other query terms. Therefore, the document recall rate might not be decreased by expansion or fuzzy operation.

**Ranking of Relevance Score.** With initially matched documents, we boost the documents that contain expansion terms with a boost weight, which is defined as Eq. 8:

$$boost = \alpha \times \frac{\sum_{ex} wei(ex)idf(ex)}{\sum_{ori} idf(ori)} \times Jaccard(ex, ori), \tag{8}$$

where ex denotes the expanded terms, ori denotes the original query terms. $Jaccard(ex, ori)$ is a function that denotes the Jaccard similarity coefficient between hit sentences of expansion terms and original query terms, is given by Eq. 9:

$$Jaccard = \frac{|HitSents(ex) \cap HitSents(ori)|}{|HitSents(ex) \cup HitSents(ori)|} \tag{9}$$

$idf(t) = 1.0 + log(df/(df(t) + 0.5))$ is the inverse document frequency of term $t$, $df(t)$ is the number of documents in which t occurs, $df$ is the total number of documents in the corpus, $wei()$ stands for the expansion weight.

We access the IDF value for every term for ranking, pre-calculated, and stored in indexing. This boost weight is added to relevance scores between query and document, and then the score of most relevant documents may be ahead during ranking. We achieve the aim of query reformulation in helping users to get a better result by one-time search.

In terms of hitting fuzzy terms containing little click information in the original query, the document may be demoted to avoid matching irrelevant words. We set demote weight as Eq. 10 to fuzz terms in the matching of query and document,

$$demote = \begin{cases} \alpha \dfrac{|BM25(fuzz\ terms)|}{|BM25(all\ terms)|} & if\ fuzz\ terms\ hit \\ 1.0 & otherwise, \end{cases} \tag{10}$$

where $\alpha$ is a damping coefficient, we set it to 0.85 by default. $BM25$ is a widely used method for computing term scores [1,4]. Both Demote and Boost weight are used to weight factors that adjust the final ranked score of the document.

**Baseline Models.** We compare with three baseline methods. The first method is pseudo-relevance feedback-based query expansion [1,8,9], the expanded terms obtained from initially retrieved top-ranked documents. Here, the title of the top 10 documents is used as resources. We find expanded terms using the TF-IDF algorithm. The second one is based on word embeddings for query expansion [4,9,11,12], which selects expanded terms with cosine similarity in word vector space. We train word vectors on the MicroBlog data using the word2vec tool [28,29] and use the top 2 most similar words related to query terms for expanding. The third one chose terms from Freebase [7].

**Comparisons on NDCG.** Table 1 shows the result compared with strong baselines, 'Good' denotes the result of the reformulated query is better than the original query's, 'Bad' is the opposite, 'TheSame' means that both 'Good' and 'Bad' are as good as each other. Our model is better than compared methods and shows that the terms come from historical queries containing the user's real search need. However, the number of queries impacted is much lower than the baselines since we try to reduce topic drift [1]. We also note that there is a

**Table 1.** Performance of NDCG on the WeiBo dataset

| Method | Measures | | | Query impacted(%) | User review(%) | | |
|---|---|---|---|---|---|---|---|
| | MAP | NDCG@5 | P@5 | | Good | Bad | TheSame |
| Embedding | 0.2490 | 0.4785 | 0.4401 | 53.60 | 11.5 | 6.5 | 75.50 |
| Pseudo-relevance | 0.2460 | 0.5372 | 0.4943 | 48.50 | 16.28 | 10.23 | 73.49 |
| Freebase | 0.2166 | 0.3672 | 0.3741 | 34.43 | 10.87 | 9.39 | 79.74 |
| HaEFQR (Ours) | **0.2826** | **0.5443** | **0.5060** | 17.75% | **21.5** | **5.7** | **72.8** |

**Table 2.** Query reformulation examples for the proposed model.

| Original query | Candidate terms | Reformulation | New query |
|---|---|---|---|
| Sina | Weibo, video | Expansion | Sina expand (weibo, video) |
| Sohu | Video, news | Expansion | Sohu expand (video, news) |
| Trade war | ZTE, negotiate | Expansion | Trade war expand (ZTE, negotiate) |
| Amazing ! this logo | ! this | Fuzzy | Amazing logo Fuzzy (this, !) |
| The wife wants to go home in half an hour | The, wants, to, in | Fuzzy | Wife go home half an hour Expand (The, wants, to, in) |

**Table 3.** Comparison performance of parameter k on NDCG@k.

| Method | NDCG@k | | | |
|---|---|---|---|---|
| | k = 1 | k = 2 | k = 5 | k = 10 |
| Embedding | 0.1207 | 0.1821 | 0.4647 | 0.5244 |
| Pseudo-relevance | 0.1164 | 0.2420 | 0.5272 | 0.5263 |
| HaEFQR (Ours) | **0.1691** | **0.3114** | **0.5391** | **0.5718** |

tradeoff between query drift and query reformulation. Given a query, we add more terms that usually diversify the search results, while it is easy to fall into the topic drift.

Table 2 gives some examples. We find that the expanded terms might grasp the user's need. Furthermore, expanded terms have two types: the categorical term representing the query topic and the functional term that often makes the user's intention more concrete and accurate.

Table 3 shows the result of 150 queries in the experiment, and it demonstrates that our model gains good results. However, the limitation of the proposed model is that the affected query's coverage ratio is lower than the baseline methods.

# 5   Conclusion

With this paper, we propose a History-aware Expansion and Fuzzy model for Query Reformulation (HaEFQR), integrating query expansion and query fuzzy into a probabilistic framework. The basic idea is that successful historical experiences from previous users might help follow-up users. Especially, we develop a fuzzy strategy for trivial terms to reformulate long query, rather than removing them directly. Experiments show that the expanded terms are usually closely related to the original query. The reformulated query provides a more enriched description than the original query and naturally diversifies the search results.

**Acknowledgements.** We thank the reviewers for their comments and suggestions. This paper is supported by NSFC (No. 61906018), Huawei Noah's Ark Lab, MoE-CMCC "Artificial Intelligence" Project (No. MCM20190701), Beijing Natural Science Foundation (Grant No. 4204100), and BUPT Excellent Ph.D. Students Foundation (No. CX2020309).

# References

1. Li, H., Jun, X.: Semantic matching in search. In: Foundations and Trends in Information Retrieval **7**, pp. 343–469 (2014)
2. Manning, C.D., Raghavan, P., SchAijtze, H.: Introduction to Information Retrieval. Cambridge University Press, Cambridge (2008)
3. Carpineto, C., Romano, G.: A Survey of Automatic Query Expansion in Information Retrieval. ACM Computing Surveys, volume 44 of CSUR, pp. 1–50, January 2012
4. Liu, Q., Huang, H., Lu, J., Gao, Y., Zhang, G.: Enhanced word embedding similarity measures using fuzzy rules for query expansion. In: Fuzzy Systems, pp. 571–578 (2017)
5. Meili, L., Sun, X., Wang, S., Lo, D., Duan, Y.: Query expansion via wordnet for effective code search. In: Software Analysis. Evolution and Reengineering, SANER, pp. 545–549, Montreal, QC, Canada (2015)
6. Ooi, J., Ma, X., Qin, H., Liew, S.C.: A survey of query expansion, query suggestion and query refinement techniques. In: 4th International Conference on Software Engineering and Computer Systems. ICSECS, pp. 545–549, Kuantan, Pahang, Malaysia (2015)
7. Xiong, C., Callan, J.: Query expansion with freebase. In: 2015 International Conference on The Theory of Information Retrieval. ICTIR, pp. 111–120. Northampton, Massachusetts, USA (2015)
8. Singh, J., Sharan, A.: A new fuzzy logic-based query expansion model for efficient information retrieval using relevance feedback approach. In: Neural Computing and Applications, vol. 28, pp. 2557–2580 (2017)
9. Almasri, M., Berrut, C., Chevallet, J.-P.: A comparison of deep learning based query expansion with pseudo-relevance feedback and mutual information. In: European Conference on Information Retrieval, pp. 709–715, March 2016
10. Kuzi, S., Shtok, A., Kurland, O.: Query expansion using word embeddings. In: CIKM, pp. 1929–1932, Indianapolis, IN, USA (2016)
11. Diaz, F., Mitra, B., Craswell, N.: Query expansion with locally trained word embeddings. In: ACL, vol. 1, pp. 367–377 (2016)

12. Roy, D., Paul, D., Mitra, M., Garain, U.: Using word embeddings for automatic query expansion. In: SIGIR, vol. 1, July 2016
13. Liu, Y., Miao, J., Zhang, M., Ma, S., Ru, L.: How do users describe their information need: query recommendation based on snippet click model. In: Expert Systems with Applications, vol. 38, pp. 13847–13856 (2011)
14. Balasubramanian, N., Kumaran, G., Carvalho, V.R.: Exploring reductions for long web queries. In: SIGIR, pp. 571–578, Geneva, Switzerland, July 2010
15. Pramanik, R., Pal, S., Chakraborty, M.: What the user does not want ?: Query reformulation through term inclusion-exclusion. In: ACM Conference on Data Sciences, pp. 116–117, Bangalore, India (2015)
16. Sloan, M., Yang, H., Wang, J.: A term-based methodology for query reformulation understanding. Inf. Retrieval J. 18(2), 145–165 (2015). https://doi.org/10.1007/s10791-015-9251-5
17. Mottin, M., Bonchi, F., Gullo, F.: Graph query reformulation with diversity. In: SIGKDD, Sydney, NSW, Australia (2015)
18. Wang, Y., Huang, H., Feng, C.: Query expansion based on a feedback concept model for microblog retrieval. In: WWW, pp. 559–568, Perth, Australia (2017)
19. Nogueira, R., Cho, K.: Task-oriented query reformulation with reinforcement learning. In: EMNLP, pp. 585–594, Copenhagen, Denmark, September 2017
20. Buck, C., et al.: Ask the right questions: active question reformulation with reinforcement learning. In: ICLR, Vancouver, Canada (2018)
21. Odijk, D., White, R.W., Awadallah, A.H., Dumais, S.T.: Struggling and success in web search. In: CIKM, Melbourne, Australia (2015)
22. Wang, X., Zhai, C.: Mining term association patterns from search logs for effective query reformulation. In: CIKM, pp. 479–488, Napa Valley, California, USA, October 2008
23. Thulin, M.: The cost of using exact confidence intervals for a binomial proportion. Electron. J. Stat. 8, 817–840 (2014)
24. Wallis, S.: A binomial confidence intervals and contingency tests: mathematical fundamentals and the evaluation of alternative methods. J. Quant. Linguist. 20, 178–208 (2013)
25. Pennington, J., Socher, R., Manning, C.D.: GloVe: global vectors for word representation. In: EMNLP, pp. 1532–1543 (2014)
26. Liu, Y., Miao, J., Zhang, M., Ma, S., Liyun, R.: How do users describe their information need: query recommendation based on snippet click model. Expert Syst. Appl. 38(11), 13847–13856 (2011)
27. Zhou, X.W.D., Zhao, W.: Query expansion with enriched user profiles for personalized search utilizing folksonomy data. In: EMNLP, vol. 29, pp. 1536–1548, July 2017
28. Mikolov, T., Sutskever, I., Chen, K., Corrado, G., Dean, J.: Distributed representations of words and phrases and their compositionality. In: NeurIPS (2013)
29. Mikolov, T., tau Yih, W., Zweig, G.: Linguistic regularities in continuous space word representations. In: NAACL (2013)

# Stance Detection with Knowledge Enhanced BERT

Yuqing Sun and Yang Li[✉]

College of Information and Computer Engineering, Northeast Forestry University,
Harbin, Heilongjiang, China
yli@nefu.edu.cn

**Abstract.** Microblog stance detection aims to determine an author's stance (for or against) towards a specific topic or claim in a post. It has become a key component in applications like truth finding, intention mining and rumor detection. Recently, researchers have becoming more and more interested in using neural models to detect user's stance. Most of the work directly models the word sequence in the text and learns its text representation. However, few researches have explored the integration of external knowledge into stance detection to enrich the learned text representation. In this paper, a knowledge-enhanced BERT model for Microblog stance detection is proposed. In this model, the triples in knowledge graphs are used as domain knowledge injected into the sentences. We conduct experiments and test the proposed method on a public Chinese Microblog stance detection dataset. Experimental results show that our model significantly outperforms the competitive baseline methods. Furthermore, the incorporation of knowledge graph gives more than 11.3% improvement in F1 score compared with state-of-the-art method.

**Keywords:** Stance detection · Knowledge · Pre-trained language model

## 1 Introduction

With the increasing popularity of social media, people express their attitude towards almost everything at any time through online websites. Recently, much attention has been paid to automatic stance classification (detection) because of its wide range of applications [1,2], especially in the field of social media analysis, opinion mining, and rumor detection. The early work of stance detection focused on argumentative debates in online-forums [11,18,22]. Gradually, stance detection began to be studied on online social media such as Twitter and Microblog.

Microblog Stance detection represents a well-established task in Natural Language Processing and is often described by having two inputs: (1) a target and (2) a post or comment made by an author. In detail, given these two inputs, the purpose is to automatically determine the author's stance (*Favor*, *Against* or *None*) of the post towards the target. The target here may be a product, an event, a government policy or even a social phenomenon. Users may not explicitly mention the target or express their stance in microblog posts, which brings challenges to the task of microblog stance detection [28].

© Springer Nature Switzerland AG 2021
L. Fang et al. (Eds.): CICAI 2021, LNAI 13070, pp. 239–250, 2021.
https://doi.org/10.1007/978-3-030-93049-3_20

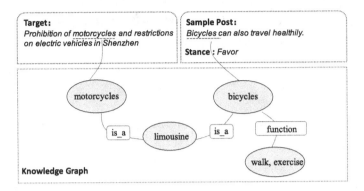

**Fig. 1.** Example of stance detection with knowledge graph.

So far, a considerable amount of literatures have been published on text stance detection [4,7,25,31]. Most work on stance detection regards it as a text classification task, in which the text of post and target is used as the features in traditional machine learning methods, such as Logistic Regression, Naive Bayes, Decision Tree and Support Vector Machine [7,10,17,21]. With the wide application of deep learning models, some researchers use deep neural networks like RNN, CNN and LSTM to learn the representation of a microblog post and then perform text classification based on the learned representation. Furthermore, pre-trained language models such as BERT [6], GPT [19] and XLNet [30] have shown great potential in learning effective representations recently, and have achieved state-of-the-art performance on various natural language processing tasks.

It has been demonstrated that the combination of knowledge graph (KG for short) and language representation model can improve the performance in specific domain tasks [14,23,32]. Therefore, we use the pre-trained model BERT in large-scale open domain corpora to obtain universal language representations, and then fine-tuned it in stance detection task. In Fig. 1, given the target *"Prohibition of motorcycles and restrictions on electric vehicles in Shenzhen"*, it is difficult to determine the stance of the post *"Bicycles can also travel healthily"*. Nevertheless, if we have the knowledge that bicycling is a green transport that can replace motorcycles or electric vehicles, it may be inferred that the author's stance towards the target is "**Favor**".

In order to overcome the above challenges, we propose a microblog stance detection framework based on pre-trained language model and external knowledge. Inspired by previous studies [14,23], our model extracts triples from a knowledge graph CN-DBpedia [26], and integrates the informative entities in KGs into the stance detection model based on BERT. Experimental results on a Chinese microblog benchmark dataset demonstrate that our proposed method outperforms state-of-the-art methods for stance detection. The main contributions of this paper can be summarized as follows:

- We propose a microblog stance detection framework based on a knowledge enhanced BERT model. As far as we know, this is the first work to combine knowledge graph and pre-trained language models into an integrated framework for this task.

– We thoroughly investigate several baseline methods including recent neural attention-based models and pre-trained language model BERT for comparison. Experiment results show that by taking full advantage of lexical, syntactic, and knowledge information, our model outperforms various state-of-the-art methods.

## 2   Method

In this section, we present a Microblog Stance Detection framework with Knowledge enhanced BERT model (**K-BERT-MSD** for short). As shown in Fig. 2, the K-BERT-MSD model consists of two parts: external knowledge integration and stance detection. In the rest of this section, we will present each of these two parts in detail.

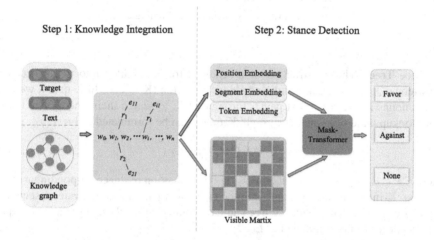

**Fig. 2.** The model structure of K-BERT-MSD.

### 2.1   Task Definition

Let $D = \{x_i = (s_i, t_i, y_i)_{i=1}^{N}$ SPSVERBc2be a dataset with $N$ examples, each consisting of a sentence $s_i$ (a text), a target $t_i$, and a stance label $y_i$. In addition, the sentence is denoted as a sequence $s = \{w_0, w_1, ..., w_n\}$, and the target is expressed as $t = \{t_0, t_1, ..., t_m\}$, where $n$, $m$ is the length of the sentence and the target, respectively. Each token $w_i$ or $t_i$ is included in the vocabulary $\mathbb{V}$, $w_i \bigcap t_i \in \mathbb{V}$. The knowledge graph $\mathbb{K}$ used by the model is CN-DBpedia, which represents knowledge as triples, $\epsilon = (w_i, r_j, w_k)$, where $w_i$ and $w_k$ are entities and the relationship is represented by $r_j$.

### 2.2   Knowledge Integration

**Target-Text Pair.** Given a set of microblog data and five targets respectively, we first segment the text data using the BERT preprocess tokenizer. In detail, the [CLS] and [SEP] identifiers are added to the sentence for indicating the begin of a sentence or separating two sentence.

**Knowledge Graph.** KG is a domain related conceptual model. The common knowledge graphs include WiKiData [24], ConceptNet [13], CN-DBpedia [26] and so on. We use the commonsense knowledge base CN-DBpedia as knowledge sources in our model. CN-DBpedia is a large open field encyclopedia KG developed by the knowledge work laboratory of Fudan University, which classifies knowledge into triples (as shown in the Table 1 below). We use the refined CN-DBpedia by deleting triples whose entity names are less than 2 or contain special characters [14].

Table 1. Example of knowledge triples from CN-DBpedia.

| Entity | Relation | Entity |
|---|---|---|
| Jay Chou | Date of birth | 1979-01-18 |
| Real Madrid | Fans' nicknames | Meilinger |

**Sentence Tree.** The construction of a sentence tree is divided into two processes: Knowledge Query (K-query) and Knowledge Injection (K-inject). In K-query, we find out the nouns in the target-text pair *ts* through NER, to search them one by one in the entity lookup table of knowledge graph. K-query can be formulated as Eq. 1:

$$E = K\_Query\left(ts, \mathbb{K}\right), \tag{1}$$

where $E = \{(w_i, r_{i0}, e_{i0}), ..., (w_i, r_{ik}, e_{ik})\}$ is a collection of the corresponding triples.

Next, K-inject injects the triple into the target-text pair *ts* and generates a sentence tree T. We specify that a sentence tree can have multiple branches, but at a fixed depth of 1. K-inject can be formulated as Eq. 2,

$$SenT = KInject\left(ts, E\right), \tag{2}$$

Specifically, given an input target $t = \{t_0, t_1, ..., t_m\}$, an input sentence $s = \{w_0, w_1, ..., w_n\}$, and a knowledge graph $\mathbb{K}$, then the sentence tree $SenT = \{w_0, w_1, ..., w_i\{(r_{i0}, e_{i0}), ..., (r_{ik}, e_{ik})\}, ..., w_{n+m}\}$.

### 2.3 Stance Detection

**Embedding Layer.** The function of the Embedding Layer(EL) is to transform the sentence tree into an embedding representation, and the embedding vector is similar to BERT. The input vector of BERT consists of three parts. While for the input vector, the difference between our model and BERT is that we input a sentence tree rather than a token sequence. Next, we introduce the embedding methods in our model detailedly.

- *Token Embedding*: The part converts words into limited common subword units through WordPiece (word segmentation). Tokens in the sentence tree need to be rearranged before they are embedded, and the tokens in the branch are inserted following the correspond node, then the subsequent tokens are moved backwards.

- *Segment Embedding*: It is used to distinguish two input sentences.
- *Soft-Position Embedding*: For encoding the position of words as eigenvectors. Here, soft-position embedding and visible matrix can well solve the problem of lost structured information of sentence tree if it is forced to be tilted into a sequence input model. In the sentence tree shown in Fig. 3, the Numbers in Step 2 are hard-position indexes and the Numbers in Step 3 are soft-position indexes. For token embedding, tile the token in the sentence tree into a sequence of token embedding according to its hard position index, and soft position index is embedded as position token together with token embedding.

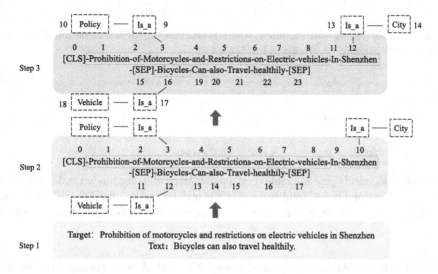

**Fig. 3.** The formation process of soft-position embedding.

**Visible Matrix.** In the process of knowledge integration, it is inevitable to introduce knowledge noise (KN), that is, too much knowledge will make the sentence deviate from the correct meaning. The introduction of visible matrix limits the visible area of each token so that the problem of KN is alleviated. For a visible matrix $M$, the blue points are visible to each other, while the beige points are invisible. The visual matrix is defined as follows:

$$M_{ij} = \begin{cases} 0 & w_i \ominus w_j \\ -\infty & w_i \oslash w_j \end{cases} \qquad (3)$$

where $w_i \ominus w_j$ refers $w_i$ and $w_j$ are in the same branch, vice versa.

**Mask-Transformer.** We add a visible matrix to obtain some structural information. However, since the encoder of Transformer cannot receive the input of visible matrix, we need present a Mask-Transformer to process. As BERT, we denote the number of layers as $L$, the hidden size as $H$, and the number of mask-self-attention heads as $A$.

Transformer layer includes two main sub-layers: MHA (multi-head attention) and FFN (fully connected feed-forward network). While MHA is actually the combination of multiple self-attention structures. Each head learns the features in different representation spaces. The Q (query), K (key) and V (value) matrices in the self-attention mechanism are all from the same input. First, we need to calculate the dot product between Q and K, and then divide it by a scale $\sqrt{d_k}$ to prevent the result from being too large, where $\sqrt{d_k}$ is the dimension of Q and K vectors. Next, the softmax operation is used to normalize the result to probability distribution, and then multiply by matrix V to get the representation of weight sum. The operation can be expressed as Eq. 4:

$$Attention\,(Q, K, V) = softmax\left(\frac{QK^{\top}}{\sqrt{d_k}}\right)V, \tag{4}$$

We propose a mask-self-attention, which is an extension of self-attention. Formally, the mask-self-attention is shown in Eq. 5, 6 and 7:

$$Q^{i+1}, K^{i+1}, V^{i+1} = h^i W_q, h^i W_k, h^i W_v, \tag{5}$$

$$S^{i+1} = softmax\left(\frac{Q^{i+1}K^{i+1^{\top}} + M}{\sqrt{d_k}}\right), \tag{6}$$

$$h^{i+1} = S^{i+1}V^{i+1}, \tag{7}$$

where $W_q$, $W_k$ and $W_v$ are trainable model parameters. $h^i$ is the hidden state of the $i$-th mask-self-attention blocks. Intuitively, if $W_k$ is invisible to $W_j$, the $M_{jk}$ will mask the attention score $S^{i+1}_{jk}$ to 0, which means $W_k$ make no contribution to the hidden state of $W_j$.

**Model Training.** We use cross-entropy and SGD with Adam optimizer to train the model parameters. The loss function is defined as follows:

$$loss = -\sum_{i=1}^{N}\sum_{j=1}^{C} y_j^i \log p_j^i \tag{8}$$

where $y_j^i$ represents the true label of the $i$-th instance in the dataset, and the $p_j^i$ is the probability value of the stance prediction. $C$ is the number of categories of stance labels, and $N$ is the total number of instances.

## 3 Experiment

The method is applied to Microblog stance detection task, and its performance is evaluated. In this section, we design experiments to answer the following research questions: (1) Compared with other baseline methods, how helpful is the pre-trained language model BERT for stance detection? (2) Is the external knowledge useful to this task? If so, how much improvement can be made to the task of stance detection by using the knowledge enhanced BERT model in this paper?

### 3.1   Dataset

Our dataset was released by NLPCC-ICCPOL 2016 task 4 *"Chinese Microblog Stance Detection Task"* [1]. The dataset contains 4000 Chinese microblogs with stance labels, among which 3000 microblogs are for training and 1000 microblogs for testing.

The data is represented by the format of (*"stance"*,*"target"*,*"text"*), where *"stance"* is a stance label, divided into three categories: **Favor**, **Against** and **None**. The five targets are *"IPhone SE"*, *"Set off firecrackers in the Spring Festival (SF for short)"*, *"Russian anti-terrorism operations in Syria (RA for short)"*, *"Two child policy (TP for short)"* and *"Prohibition of motorcycles and restrictions on electric vehicles in Shenzhen (PM for short)"*. For the five different targets, the distribution of instances in the training data is shown in Table 2 [28].

**Table 2.** Distribution of instances for five different targets in training data.

| Targets | Training dataset | | | Testing dataset | | |
|---|---|---|---|---|---|---|
| | Favor | Against | None | Favor | Against | None |
| iPhone SE | 245 | 209 | 146 | 75 | 104 | 21 |
| SF | 250 | 250 | 100 | 88 | 94 | 18 |
| RA | 250 | 250 | 100 | 94 | 86 | 20 |
| TP | 260 | 200 | 140 | 99 | 95 | 6 |
| PM | 160 | 300 | 140 | 63 | 110 | 27 |

### 3.2   Experimental Settings

**Baseline Methods.**  For comparison, we consider the following baseline methods:

- *RUC MMC*: Dian et al. [7] used five manually selected features as input of Random Forest and SVM model. They achieve the best results in task 4 of NLPCC-ICCPOL 2016.
- *ATA*: A two-stage attention model proposed by Yue et al. [31]. Firstly, the attention mechanism is applies to model target, then the context is matched with the target representation to obtain attention signal, and finally the target mixed text representation for stance classification is formed.
- *T-DAN*: The two-stage deep attention neural network (TDAN) proposed by Yang et al. [29] for target-specific stance detection. The model decompose the stance classification problem into two binary classification problems to mitigating the imbalanced distribution of labels. In the first stage, they find out the tweet is neutral or subjective about the specific target. In the second stage, they classify the stance of a given subjective tweet's stance.
- *BERT*: The Chinese BERT pre-training model *"BERT-Base, Chinese"* released by Google. The model uses 12 layers of transformer, outputs 768 dimension vectors, the head number of multi-head Attention is 12. The total number of trainable parameters of BERT and our model are the same (110M).

[1] http://tcci.ccf.org.cn/conference/2016/pages/page05_evadata.html.

**Evaluating Metrics.** We use Precision ($p$), Recall ($r$) and F-score as the evaluation metric stance detection. $F_{Favor}$ and $F_{Against}$ stand for the F-score of the "*Favor*" class and "*Against*" class respectively. $F_{Avg}$ is the macro-average of $F_{Favor}$ and $F_{Against}$. For more detail about the evaluation metrics, please refer to the previous work [28].

**Experimental Setup.** We perform stance detection experiments according to the following steps. We first train our model on training data, and save the best performance our proposed model. For our model and BERT, we use the same settings, in which the learning rate is set to $2e-5$ and the batch size is set to 8. We run the model for several iterations until it convergences. Please note that, we set a target-text pair can only have two entities that extend triple knowledge. For all other baseline methods, we directly obtain the results reported in their papers, since we conduct experiments under the same dataset and settings.

### 3.3   Experimental Results

**Comparison to Other Methods.** In this part, first of all, we compare the F-score our model with the four baselines obtained, using the same dataset and segmentation method, and the experimental results are shown in Table 3. We can come to the following conclusions: (1) In the task of stance detection, our K-BERT-MSD model obviously outperforms to the three state-of-the-art neural methods. (2) Compared with other baselines (including traditional machine learning methods and deep neural models), BERT is effective in learning semantics, and achieves the second best result. Our model K-BERT-MSD enhances BERT model by fusing knowledge, and obtains more competitive results with BERT.

**Table 3.** The performance of our model compared with the baseline.

| Model | $F_{Favor}$ | $F_{Against}$ | $F_{Avg}$ |
|---|---|---|---|
| RUC MMC | 0.697 | 0.724 | 0.711 |
| ATA | 0.762 | 0.671 | 0.717 |
| T-DAN | 0.762 | 0.702 | 0.741 |
| BERT | 0.821 | 0.804 | 0.813 |
| K-BERT-MSD | **0.834** | **0.815** | **0.825** |

We also compare the results of our model and baselines for every target separately, as shown in Table 4.

**Ablation Experiment.** In order to analyze the influence of knowledge, we also conduct two ablation experiments by controlling the specific parameters of our model. One is the number of entities introduced in the sentence tree(max_entities for short), and the other is the size of the external knowledge graph. We set the value of *max_entities* from

**Table 4.** The results of training five targets separately.

| Targets | RUC MMC | ATA | T-DAN | BERT | K-BERT-MSD |
|---------|---------|-----|-------|------|------------|
| IPhone SE | 0.615 | 0.600 | 0.732 | 0.737 | **0.769** |
| SF | 0.782 | 0.801 | 0.664 | 0.803 | **0.835** |
| RA | 0.720 | 0.563 | 0.543 | 0.733 | **0.784** |
| TP | 0.847 | 0.818 | 0.693 | 0.838 | **0.895** |
| PM | 0.776 | 0.807 | 0.761 | 0.777 | **0.810** |
| Overall | 0.711 | 0.717 | 0.741 | 0.813 | **0.825** |

0 to 3, to test the effect of number of introduced knowledge for each target-post instance on the final result.

Experimental results in Table 5 show that the performance of the F-score improves when *max_entities* increases from 0 to 2. When *max_entities* is equal to 2, K-BERT-MSD achieves the best results. The results decrease when *max_entities* is larger than 2. The results demonstrate that the increase of knowledge can improve the effect of stance detection, but too much knowledge will also bring noise. Next, we perform experiments 0%, 25%, 50% of knowledge graph and the full knowledge graph. The experimental results are shown in Table 6. Therefore, we can draw the following conclusions from the results in Table 6: (1) By adding external knowledge within a certain range, the performance of our model can be improved. (2) Because a large knowledge graph contains more information, it helps to detect the stance of target-post pairs.

**Table 5.** F-score with *max_entities* ranging from 0 to 3.

| max_entities | 0 | 1 | 2 | 3 |
|--------------|---|---|---|---|
| F-score | 0.813 | 0.824 | **0.831** | 0.820 |

**Table 6.** F-score with size of KG ranging from 0 to 1.

| Size of KG | 0 | 25% | 50% | 100% |
|------------|---|-----|-----|------|
| F-score | 0.813 | 0.820 | 0.826 | **0.831** |

# 4 Related Work

## 4.1 Stance Detection

So far, a considerable amount of literature has been published on microblog stance detection [7,29,31]. The proposed methods can be roughly divided into two categories: traditional machine learning methods and deep neural models.

Traditional machine learning methods focus on how to select appropriate features for stance detection. Besides simple textual features like bag-of-words(BoW) model, Xu et al. [27] used para2vec, LDA and LSA to represent the semantic information in tweets, and compared the effect of different machine learning algorithms such as random forest and support vector machine (SVM) in stance detection. Ebrahimi et al. [9] integrated sentiment polarity into target and stance, and modeled the interaction of target, stance label and sentiment words in a probabilistic graph model.

Deep learning based methods for stance detection make attempts to learn the representations of target and text, and then perform text classification based on the representations. Early stage of deep learning method, Augenstein et al. [3] proposed a neural network architecture based on conditional encoding. A LSTM network is used to encode the target, followed by a second LSTM that encodes the tweet using the encoding of the target as its initial state. Experimental results showed that the model performed better than coding tweets and targets separately, which is consistent with the work of Luo et al. [16] and Du et al. [8]. With the introduction of attention mechanism, Bai et al. [4] proposed a BiLSTM-CNN model based with attention mechanism to focus on the target and text respectively. When PLMs are on the stage of deep learning, Wang et al. [25] proposed a stance detection model BERT-condition-CNN. They use BERT pre-trained model to obtain the representation vector of the text, and the relationship matrix condition layer between the targets and the text vector is constructed. Finally, CNN was used to extract the features of the condition layer to perform the classification.

### 4.2  Pre-trained Language Model

Since BERT was proposed by Devlin et al. [6], a lot of models like RoBERTa [15], BERT WWM [5], TinyBERT [12] and so on had made great efforts on the optimization of the pre-trained process in different ways. Integrating structured knowledge, such as knowledge graph, into deep learning models can improve the effectiveness of information retrieval [20], but there is little research on this aspect. Zhang et al. [32] make the first attempt to integrate entity information to cover the weakness of BERT in data dependence, but they ignored the relationship between entities. Liu et al. [14] proposed K-BERT, which enriched the representation by injecting triples in knowledge graph into sentences. Liu et al. [15] encoded the knowledge graph with a graph attention model and encoded the text using RoBERTa as the language model. Inspired by [14], we modify the K-BERT model and apply it to the task of stance detection, and finally provide state-of-the-art (SOTA) results.

## 5  Conclusion

In this paper, we propose a microblog stance detection model with knowledge enhanced BERT. The K-BERT-MSD model incorporates knowledge graph into the BERT architecture through a tree-structured sentence encoding mechanism and takes over the advantages of the both. We design experiments to evaluate our model against several state-of-the-art models. By comparing BERT with baseline methods including RUC

MMC, ATA and T-DAN, we found that pre-trained language model like BERT significantly helped in this task. In addition, we found it beneficial to incorporate external knowledge by comparing our model with a basic BERT model. Finally, we also tested with different settings of K-BERT-MSD and found that both external knowledge is conducive to stance detection, but too much external knowledge will bring noise, and too little external knowledge will not improve the results.

**Acknowledgments.** We thank the anonymous reviewers for their constructive suggestions. This work was supported by the National Natural Science Foundation of China [61806049] and the Heilongjiang Postdoctoral Science Foundation [LBH-Z20104].

# References

1. Aldayel, A., Magdy, W.: Your stance is exposed! Analysing possible factors for stance detection on social media. Proc. ACM Hum. Comput. Interact. **3**(CSCW), 1–20 (2019)
2. AlDayel, A., Magdy, W.: Stance detection on social media: state of the art and trends. arXiv preprint arXiv:2006.03644 (2020)
3. Augenstein, I., Rocktäschel, T., Vlachos, A., Bontcheva, K.: Stance detection with bidirectional conditional encoding. arXiv preprint arXiv:1606.05464 (2016)
4. Bai, J., Li, F., Ji, D.: Attention based BiLSTM-CNN Chinese microblog stance detection model. Comput. Appl. Softw. **35**(3), 266–274 (2018)
5. Cui, Y., et al.: Pre-training with whole word masking for Chinese BERT. arXiv preprint arXiv:1906.08101 (2019)
6. Devlin, J., Chang, M.W., Lee, K., Toutanova, K.: BERT: pre-training of deep bidirectional transformers for language understanding. arXiv preprint arXiv:1810.04805 (2018)
7. Dian, Y., Jin, Q., Wu, H.: Stance detection in Chinese microblogs via fusing multiple text features. Comput. Eng. Appl. **53**(021), 77–84 (2017)
8. Du, J., Xu, R., Gui, L., Wang, X.: Leveraging target-oriented information for stance classification. In: Gelbukh, A. (ed.) CICLing 2017. LNCS, vol. 10762, pp. 35–45. Springer, Cham (2018). https://doi.org/10.1007/978-3-319-77116-8_3
9. Ebrahimi, J., Dou, D., Lowd, D.: A joint sentiment-target-stance model for stance classification in tweets. In: Proceedings of COLING 2016, The 26th International Conference on Computational Linguistics: Technical Papers, pp. 2656–2665 (2016)
10. Elfardy, H., Diab, M.: CU-GWU perspective at SemEval-2016 task 6: ideological stance detection in informal text. In: Proceedings of the 10th International Workshop on Semantic Evaluation, SemEval-2016 (2016)
11. Hasan, K.S., Ng, V.: Extra-linguistic constraints on stance recognition in ideological debates. In: Proceedings of the 51st Annual Meeting of the Association for Computational Linguistics (Volume 2: Short Papers) (2013)
12. Jiao, X., et al.: TinyBERT: distilling BERT for natural language understanding. arXiv preprint arXiv:1909.10351 (2019)
13. Liu, H., Singh, P.: ConceptNet–a practical commonsense reasoning tool-kit. BT Technol. J. **22**(4), 211–226 (2004). https://doi.org/10.1023/B:BTTJ.0000047600.45421.6d
14. Liu, W., Zhou, P., Zhao, Z., Wang, Z., Wang, P.: K-BERT: enabling language representation with knowledge graph. Proc. AAAI Conf. Artif. Intel. **34**(3), 2901–2908 (2020)
15. Liu, Y., et al.: RoBERTa: a robustly optimized BERT pretraining approach. arXiv preprint arXiv:1907.11692 (2019)

16. Luo, W., Liu, Y., Liang, B., Xu, R.: A recurrent interactive attention network for answer stance analysis. In: Proceedings of the 19th Chinese National Conference on Computational Linguistics, pp. 698–706 (2020)
17. Mohammad, S., Kiritchenko, S., Sobhani, P., Zhu, X., Cherry, C.: A dataset for detecting stance in tweets. In: Proceedings of the 10th International Conference on Language Resources and Evaluation, LREC 2016, Portorož, Slovenia, pp. 3945–3952. European Language Resources Association (ELRA) (May 2016). https://www.aclweb.org/anthology/L16-1623
18. Murakami, A., Raymond, R.: Support or oppose? Classifying positions in online debates from reply activities and opinion expressions. In: 23rd International Conference on Computational Linguistics, COLING 2010, Posters, Beijing, China, 23–27 August 2010 (2010)
19. Radford, A., Narasimhan, K., Salimans, T., Sutskever, I.: Improving language understanding by generative pre-training (2018)
20. Sheth, A., Kapanipathi, P.: Semantic filtering for social data. IEEE Internet Comput. **20**(4), 74–78 (2016)
21. Siddiqua, U.A., Chy, A.N., Aono, M.: Stance detection on microblog focusing on syntactic tree representation (2018)
22. Somasundaran, S., Wiebe, J.: Recognizing stances in online debates. In: Proceedings of the Joint Conference of the 47th Annual Meeting of the ACL and the 4th International Joint Conference on Natural Language Processing of the AFNLP (2009)
23. Sun, T., et al.: CoLAKE: contextualized language and knowledge embedding. arXiv preprint arXiv:2010.00309 (2020)
24. Vrandečić, D., Krötzsch, M.: Wikidata: a free collaborative knowledgebase. Commun. ACM **57**(10), 78–85 (2014)
25. Wang, A., Huang, k., Lu, l.: Chinese microblog stance detection based on BERT condition CNN. Comput. Syst. Appl. **28**(11), 45–53 (2019)
26. Xu, B., et al.: CN-DBpedia: a never-ending Chinese knowledge extraction system. In: Benferhat, S., Tabia, K., Ali, M. (eds.) IEA/AIE 2017. LNCS (LNAI), vol. 10351, pp. 428–438. Springer, Cham (2017). https://doi.org/10.1007/978-3-319-60045-1_44
27. Xu, J., Zheng, S., Shi, J., Yao, Y., Xu, B.: Ensemble of feature sets and classification methods for stance detection. In: Lin, C.-Y., Xue, N., Zhao, D., Huang, X., Feng, Y. (eds.) ICCPOL/NLPCC-2016. LNCS (LNAI), vol. 10102, pp. 679–688. Springer, Cham (2016). https://doi.org/10.1007/978-3-319-50496-4_61
28. Xu, R., Zhou, Y., Wu, D., Gui, L., Xue, Y.: Overview of NLPCC shared task 4: stance detection in Chinese microblogs. In: International Conference on Computer Processing of Oriental Languages National CCF Conference on Natural Language Processing and Chinese Computing (2016)
29. Yang, Y., Wu, B., Zhao, K., Guo, W.: Tweet stance detection: a two-stage DC-BILSTM model based on semantic attention. In: 2020 IEEE 5th International Conference on Data Science in Cyberspace (DSC), pp. 22–29. IEEE (2020)
30. Yang, Z., Dai, Z., Yang, Y., Carbonell, J.G., Salakhutdinov, R., Le, Q.V.: XLNet: generalized autoregressive pretraining for language understanding. CoRR abs/1906.08237 (2019). http://arxiv.org/abs/1906.08237
31. Yue, T., Zhang, S., Yang, L., Lin, H., Kai, Y.: Stance detection method based on two-stage attention mechanism. J. Guangxi Norm. Univ. (Nat. Sci. Edn.) **37**(1), 42–49 (2019)
32. Zhang, Z., Han, X., Liu, Z., Jiang, X., Sun, M., Liu, Q.: ERNIE: enhanced language representation with informative entities. arXiv preprint arXiv:1905.07129 (2019)

# Towards a Two-Stage Method for Answer Selection and Summarization in Buddhism Community Question Answering

Jiangnan Du[1], Jun Chen[1], Suhong Wang[2], Jianfeng Li[1], and Zhifeng Xiao[3(✉)]

[1] Ping An Technology (Shenzhen) Co., Ltd., Shenzhen, China
{DUJIANGNAN930,CHENJUN722,LIJIANFENG777}@pingan.com.cn
[2] Baixing AI Lab, Beijing, China
swsuh@sina.com
[3] Penn State Erie, The Behrend College, Erie, USA
zux2@psu.edu

**Abstract.** This paper proposes a two-stage learning pipeline for CQA in the Buddhism domain. In the first stage, we trained an answer selection model through Keywords-BERT that performs a deep semantic match for QA pairs. Given a question, our algorithm selects the answer with the highest relatedness score. Stage two also employs the trained Keywords-BERT model to eliminate redundant information and only keep the most relevant sentences of an answer for summary extraction. Our method only requires standard QA pairs for training, significantly reducing the annotation cost and the knowledge threshold for annotators. We tested our model on a self-created Buddhism CQA dataset. Results show that the proposed pipeline outperforms state-of-the-art methods like BERT-Sum in terms of summary quality and model robustness.

**Keywords:** Community question answering · BERT · Answer selection · Extractive summary · Buddhism

## 1 Introduction

Current answer selection methods exploit the semantic correlation between questions and answers via different deep neural architectures. Zhang et al. [10] designed an attentive and interactive neural network to learn interactions of each paired QA. Zhou et al. [11] combined a convolutional neural network a recurrent neural network to capture the semantic matching of a QA pair. Xie et al. [8] presented an Attentive User-engaged Adversarial Neural Network to incorporate users into the learning pipeline for answer selection. These prior studies, however, did not address the importance of domain knowledge in domain-specific CQA systems.

Text summarization techniques generally fall into two categories: extractive and abstractive summarization. The former aims to extract key sentences from

© Springer Nature Switzerland AG 2021
L. Fang et al. (Eds.): CICAI 2021, LNAI 13070, pp. 251–260, 2021.
https://doi.org/10.1007/978-3-030-93049-3_21

the source text, while the latter may use external vocabulary to construct sentences not belonging to the source text. In the context of CQA, both extractive [6,7] and abstractive [1,7,9] methods were developed. Deng et al. [6] propose a joint learning model that tackles answer selection and summarization via an attentive pointer-generator network. Zhang et al. [9] model a long sentence as a graph, where vertices represent domain concepts, and edges are identified via keywords/entities that co-occur between sentences. The constructed graph convolutional network is equipped with a question-focused dual attention module that allows the model to generate summaries relevant to questions. We emphasize that a competent summarization model in CQA should exploit the core information in answers and the semantic correlation between QA pairs to generate concise and accurate answer summaries.

To address these challenges, we propose a matching-based two-stage learning pipeline to tackle the CQA task in the Buddhism domain. In particular, the proposed learning pipeline includes answer selection and extractive summarization. The method utilized Keywords-BERT [5] for deep semantic matching between QA pairs. Attended by domain keywords, we trained a QA model to predict the relatedness of the answer with a given question. Both stages adopted the QA model. For answer selection, the most relevant answer was selected; while for answer summarization, we split the selected answer into sentences that were re-matched against the corresponding question to obtain a rank for each sentence, then only kept the top-k ranked sentences to form a summary. Unlike conventional summarization methods that require costly annotation, i.e., human-written reference summaries, our strategy only needs annotated QA pairs for training, significantly reducing the annotation cost. Below we provide an example Buddhism question with an answer of over 200 characters, while the answer summary only contains 40 characters and is concise enough to comprehend.

**Question:** 一念嗔心起百万障门开什么意思？ (What does it mean when one thought of anger raises a million barriers?)
**Answer summary:** 只要片刻之间生起对他人的仇恨和愤怒，那么千千万万个妨碍修行的罪恶就会滋生出来。(As soon as a moment of hatred and anger against others arises, millions of sins that hinder practice will grow.)

This paper makes the following contributions:

- We propose a two-stage method for the CQA problem in the Buddhism domain. The first stage, answer selection, trains a QA model based on Keywords-BERT to match the most relevant answer given a question. The second stage, extractive summarization, reuses the QA model from stage one to choose the most relevant sentences from the selected answer to form a summary. Our model training is featured by a deep semantic matching between QA pairs attended by domain keywords, exploiting the context information and domain knowledge for the task.
- We create a Buddhism CQA dataset used for both answer selection and summarization tasks. We also evaluated the impact of different training methods

on subsequent summarization. Experimental results show that the proposed method can identify the best answer and generate accurate answer summaries, outperforming other state-of-the-art (SOTA) methods like BERT-Sum in terms of summary quality and robustness.

## 2 Method

### 2.1 Problem Definition

Let $Q$ and $A$ denote the questions and answers sets, respectively. Given a question $q \in Q$, we are tasked to select the best answer $a^* \in A$ and then perform extractive summarization for $a^*$. In particular, let $S_{a^*} = \{s_1, s_2, \ldots, s_n\}$ be a list of sentences split from $a^*$, our model needs to extract a subset of sentences from $S_{a^*}$ to form an answer summary, denoted by $\beta^*$, which does not exceed a length of $l$ characters.

### 2.2 Keywords-BERT

Keywords-BERT [5] adds a keyword-attentive layer that highlights the domain keywords to enhance the semantic interaction of the sentence pair supplied during training. A well-trained Keywords-BERT model takes as input a sentence pair and outputs a score that quantifies the relatedness of the sentence pair. A set of keywords should be collected to facilitate training. Specifically, Keywords-BERT modifies BERT in two ways. First, keywords are masked to participate the training using masked language modeling [2]. For our case, the point-wise training takes a Q-A pair as input, where keywords in the answer sentence are masked so that tokens in the question can attend to the keywords in the answer. Second, to enhance/decrease the semantic relatedness for positive/negative samples, a keyword-attentive layer is added in parallel with the last Transformer layer to compute a keyword difference vector, which is a concatenation of the differences of average-pooled representations of the input sentence pair. With this change, the last Transformer layer of BERT is basically replaced by the keyword-attentive layer so that keywords information can be injected closer to the detection head, which turned out to be empirically effective.

### 2.3 Keywords Collection

We applied a NER method to extract Buddhism keywords used for the Keywords-BERT training. The pre-defined named entity types in Buddhism include PER (person), ORG (organization), NW (work name), LOC (place name), and NZ (other proper nouns). For each entity type, we list a few instance examples as follows.

- **PER:** 文殊菩萨 (Manjushri), 释迦摩尼 (Shakyamuni)
- **ORG:** 天台宗 (Tiantai Sect), 婆罗门教 (Brahmanism)
- **NW:** 金刚般若经 (Vajrapani Sutra), 无量寿经 (Immeasurable Life Sutra)

- **LOC:** 法华山 (Mount Fahua), 释迦寺 (Sakyamuni Temple)
- **NZ:** 业果 (Karma), 行禅 (Walking meditation)

We then scraped over 20k Buddhism words from encyclopedias, sutras, and Buddhist websites to create a small annotated corpus, which was utilized to fine-tune the Baidu LAC base NER model [3]. This way, the Buddhism domain knowledge can be incorporated into the resulting NER tagger. In addition, we selected 700 representative keywords from the 20k-word list to create a gazetteer. The final gazetteer-assisted NER tagger was employed to detect domain keywords in the QA pairs.

### 2.4   Training an Answer Selection Model

We employed point-wise and pair-wise approaches for training.

The input of the point-wise model is a QA pair that belongs to a training set $D = \{(x_i, y_i)\}_{i=1}^m$, where $x_i = (q, a)$, $q \in Q$, $a \in A$, $y_i \in \{1, 0\}$, and $m$ is the number of training examples. $y_i = 1$ means a match for $q$ and $a$, indicating a positive example, and vice versa. We applied the binary cross entropy loss function as follows:

$$L_{\text{point-wise}} = -\sum_{i \in I_{pos}} \log r_i - \sum_{i \in I_{neg}} \log(1 - r_i)$$

in which $I_{pos}$ and $I_{neg}$ are the sets of indices of the positive and negative examples, and $r_i = \text{Keywords-BERT}(x_i)$ is the model's output, which can be regarded as the probability of answer $a_i$ being a match of question $q$.

We also trained a pair-wise model on set $D' = \{(q, a_+, a_-) \mid q \in Q, a_+ \in A, a_- \in A\}$, in which $a_+$ and $a_-$ represent matched and mismatched answers for question $q$, respectively. A triplet loss function was adopted:

$$L_{\text{pair-wise}} = \max(0, \alpha - r_+ + r_-)$$

where $r_+ = \text{Keywords-BERT}((q, a_+))$, $r_- = \text{Keywords-BERT}((q, a_-))$, and $\alpha$ is a hyperparameter that represents the margin between positive and negative QA pairs.

### 2.5   Extractive Summarization

Given question $q$, we took the highest scoring answer $a^*$ and split it into sentences $S = \{s_1, s_2, \ldots, s_n\}$. For each sentence $s_i \in S$, we formed a q-$s_i$ pair and sent it through the trained selection model to obtain a score $r_i = \text{Keywords-BERT}((q, s_i))$. Based on the scores, we re-ranked the sentences in descending order and kept the top $k$ sentences such that the total length of the $k$ sentences does not exceed $l$. Finally, the $k$ sentences were reordered based on their original positions within the answer $a^*$ to form an extractive summary $\beta^*$. Figure 1 describes the process of our answer summary method.

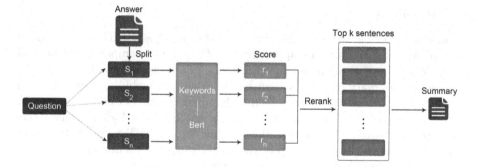

**Fig. 1.** Stage two: extractive summarization

# 3    Dataset and Experimental Setting

## 3.1    Dataset

The entire dataset contains the following components.

- $D_{\text{train}}$: to train the Keywords-BERT model, we gathered a collection of questions and answers by crawling a Buddhist community Q&A website and constructed a total of 50,000 QA pairs with a 50/50 split of positive and negative examples.
- $D_{\text{test-sel}}$: to test the answer selection model, we created a dataset with 3,553 QA pairs with 1,119 distinct questions. In this test set, correct answers were selected and labeled by human annotators.
- $D_{\text{test-sum}}$: to test the answer summarization, we created a test set with 2,166 QA pairs that were all positive samples. For each answer in the test set, we manually wrote a reference summary using the sentences from the original answer. Samples in the training and test sets were collected from the same source.
- $D_{\text{test-sum-OOS}}$: to further test the model's robustness, we gathered an additional out-of-sample (OOS) test set from a different source with 207 QA pairs, where each question was answered by Venerable Yin Guang.

## 3.2    Negative Sampling

High-quality negative samples are important for training a robust model. In this study, we followed the negative sampling method in [5] to auto-generate negative samples based on the positive samples available to us. In particular, given an answer $a'$, we first generated a keyword-augmented answer $a'||k_{a'}$ that was a concatenation of $a'$ and the domain keywords found in $a'$. We then performed a search for $a'||k_{a'}$ within the question set using Elasticsearch. The search returned a list of questions that were ranked by the similarity between the QA pair. For each candidate question $q'$ in the result list, we determined $(q', a')$ as a negative sample pair if 1) $a'$ was not marked as an answer to $q'$ in the positive examples, and 2) the similarity between $q'$ and $a'||k_{a'}$ was below a pre-defined threshold.

### 3.3   Baselines

For answer selection, we compared our method to the base BERT model trained using point-wise and pair-wise approaches on $D_{\text{test-sel}}$ and compared our method to the Joint Learning model on an open domain dataset called WikiHowQA [1]. For answer summary, we picked two baseline models, including Lead-100 and BERT-Sum [4]. Lead-100 takes the first hundred characters of the answer to form a summary; the reason to set a maximal length of one hundred characters is due to the display limit of the client mobile App. BERT-Sum, on the other hand, stacks several layers on top of the BERT output for summarization. The latter achieved SOTA results in several public QA datasets.

### 3.4   Experimental Setting

We utilized the Chinese BERT-base with a maximum document length of 256. In the experiment, a longer input led to performance degradation, potentially due to the fact that the key part of an answer generally appears early in the whole answer text, and the following redundant text that carries less critical information would have an unnecessary impact on learning. Also, we adopted a learning rate of 1e−5. For the pair-wise training, the margin was set to 0.6. The model was trained for three rounds on an NVIDIA Tesla V100 GPU. In the subsequent summary extraction, we removed some useless interrogative sentences and repeated sentences that were the same or highly similar to the question, and kept the summary length between 50 and 100 characters.

## 4   Results and Discussion

Experiments were conducted to evaluate the performance of the proposed model in both answer selection and summarization.

### 4.1   Answer Selection Evaluation

Two datasets, including our dataset $D_{\text{test-sel}}$ and an open domain WikiHowQA dataset [1], were used to evaluate answer selection. For $D_{\text{test-sel}}$, precision at one (P@1) and mean reciprocal rank (MRR) were used as performance metrics. The reason to use P@1 was that the human annotators only marked one correct answer (the gold standard) per question. Also, MRR was used to evaluate how well a model can rank the answers in relatedness. For the WikiHowQA dataset, mean average precision (MAP) was used instead of P@1, because each question in the dataset could have multiple correct answers.

Results in Table 1 show that the Keywords-BERT model outperformed the base BERT in answer quality on test set $D_{\text{test-sel}}$, verifying that adding a keyword-attentive module to BERT can enhance the semantic interaction between a QA pair, which allowed the model to find a more relevant answer.

The WikiHowQA dataset contains 203,596 questions and 1,188,189 QA pairs. We compared our model with ASAS [1], which was considered as the SOTA for

**Table 1.** Answer selection evaluation on $D_{\text{test-sel}}$

| Model | P@1 | MRR |
|---|---|---|
| BERT-Point-wise | 0.3083 | 0.5564 |
| BERT-Pair-wise | 0.3065 | 0.5545 |
| Keywords-BERT-Point-wise | 0.3834 | 0.6071 |
| Keywords-BERT-Pair-wise | **0.3941** | **0.6173** |

this task. Table shows that our method outperformed ASAS by 11.1% in MRR but under-performed ASAS by 4.7% in MAP, meaning that our method did a better job in ranking the answers and performed worse in classifying positive and negative samples. The relatively low MAP of our method could be resulted by the quality of keywords gathered on the WikiHowQA dataset, which is open domain and covers a variety of topics. It is challenging to collect high-quality keywords in an open domain dataset, and our experiment only utilized a basic NER tagger from NLTK for keywords collection, which may affect the performance.

**Table 2.** Answer selection evaluation on WikiHowQA

| Model | MAP | MRR |
|---|---|---|
| Joint learning (ASAS) [1] | **0.5522** | 0.5686 |
| Keywords-BERT-Pair-wise | 0.5051 | **0.6790** |

## 4.2 Extractive Summary Evaluation

We took the trained Keywords-BERT model and performed summary extraction on $D_{\text{test-sum}}$. Results were reported in ROUGE scores in Table 3. It is observed that our approach based on Keywords-BERT is not advantageous. There are three reasons. First, the majority of questions collected on the Buddhist CQA website were factual questions that asked for noun explanation. Key information to answer these questions mostly appeared in the first few sentences of an answer, which was in favor of Lead-100. Second, as mentioned earlier, the annotators did not have much knowledge of Buddhism, so that they may not fully understand a lengthy answer and tended to make a reference summary using the first couple of sentences of the answer, leading to a performance drop on the ROUGE scores for both BERT-Sum and our approach. Lastly, for certain Buddhism questions, there is more than one summary that can be extracted from a lengthy answer.

For these reasons, we conducted a human evaluation for the tested models on $D_{\text{test-sum}}$. The human evaluators had some knowledge of Buddhism, and their job was to tell whether or not the generated summary of a model could answer the question. We had four human evaluators conduct this test to reduce the impact of subjective opinions on the results, shown in Table 4. It is observed

258    J. Du et al.

**Table 3.** Extractive summary evaluation

| Model | R-1 | R-2 | R-L |
|---|---|---|---|
| Lead-100 | **54.50** | **46.25** | **51.10** |
| BERT-Sum | 51.01 | 39 | 45.31 |
| Keywords-BERT-Point-wise | 48.79 | 38.39 | 44.10 |
| Keywords-BERT-Pair-wise | 49.12 | 38.41 | 44.16 |

that the Keywords-BERT-Pair-wise model performed the best, with 90.1% of the generated summaries being able to answer the corresponding question because of the ranking ability introduced by the pair-wise model. Lead-100 did poorly in this test since critical pieces of the answer may spread across the entire answer, not just in the first few sentences. BERT-Sum was in line with our method.

**Table 4.** Human evaluation results on $D_{\text{test-sum}}$

| Model | Answerability |
|---|---|
| Lead-100 | 0.729 |
| BERT-Sum | 0.897 |
| Keywords-BERT-Point-wise | 0.889 |
| Keywords-BERT-Pair-wise | **0.901** |

Due to the annotation problem mentioned earlier, BERT-Sum is prone to overfit. As such, we conducted an OOS testing on $D_{\text{test-sum-OOS}}$ and report the results in Table 5. The test data contained 207 QA pairs from Venerable Yin Guang's QA records, which is different from the source from where $D_{\text{train}}$ was taken. Due to the lack of reference summary, we only did the human evaluation. Results show a noticeable performance drop for BERT-Sum, indicating an overfitting issue. This result further verifies that the interaction of questions and answers during BERT's training, along with a keyword attentive mechanism, is crucial to extract key information for a high-quality answer summary.

**Table 5.** Human evaluation results on $D_{\text{test-sum-OOS}}$

| Model | Answerability |
|---|---|
| Lead-100 | 0.677 |
| BERT-Sum | 0.778 |
| Keywords-BERT-Point-wise | 0.847 |
| Keywords-BERT-Pair-wise | **0.884** |

We display an example below to show that Keywords-BERT-pair-wise does better in locating the key information of the answer to generate a concise and accurate summary. The Mage's answer is lengthy, while the summary given by BERT-Sum fails to capture the key information.

> **Question:** 无量寿经现存五译，又有会集本，应以何本为准则？(There are five existing translations of the Infinite Life Sutra, and there are also collected texts, what should be the guideline?)
>
> **Mage's answer:** 无量寿经有五译。初译于后汉月支支娄迦谶，三卷，文繁，名佛说无量清净平等觉经。次译于吴月支支谦，有二卷，名佛说阿弥陀经。以日诵之经，亦名佛说阿弥陀经...(400个字) (There are five translations of the Sutra of Infinite Life. Originally translated from Lou Jiazhen, the Moon Branch of the Later Han Dynasty, three juan, literary complex, the name Buddha said boundlessly pure and equal enlightenment. The second translation is in Wuyue Zhizhiqian, there are two volumes, the name Buddha said Amitabha Sutra. The sutras recited in the sun, also known as the Buddha Amitabha Sutra... (400 characters))
>
> **BERT-Sum:** 无量寿经有五译。次译于吴月支支谦，有二卷，名佛说阿弥陀经。以日诵之经，亦名佛说阿弥陀经，故外面加一大字以别之。(There are five translations of the Sutra of Infinite Life. The second translation is in Wuyue Zhizhiqian, there are two volumes, the name Buddha said Amitabha Sutra. The sutras recited in the sun are also called Amitabha Sutras, so a large character is added to the outside.)
>
> **Keywords-BERT-pair-wise:** 就中无量寿如来会，文理俱好，而末后劝世之文未录，故皆以康僧铠之无量寿经为准则焉。(In the case of the infinite life of the Buddha, the text is good, but the text of the last exhortation is not recorded, so **all of them are based on Kang Shengjia's infinite life sutra as a guideline.**)

## 5 Conclusion

This paper presented a matching-based two-stage method for answer selection and extractive summarization in CQA. The method utilized Keywords-BERT for deep semantic matching between QA pairs. Attended by domain keywords, we trained a QA model to predict the relatedness of the answer with the given question. Both stages adopted the QA model. For answer selection, the most relevant answer was selected; while for answer summarization, we split the selected answer into sentences that were re-matched against the corresponding question to obtain a rank for each sentence, then only kept the top-k ranked sentences to form a summary. We compared the proposed method with SOTA approaches in the quality of answer selection and summarization. Results demonstrated that our method showed strong ranking ability in answer selection and generated a high-quality summary through extracting key sentences, especially in the OOS testing, showing the model's robustness. We also realize that the model's performance is limited by the keywords quality in an open domain dataset, which is a challenge to be addressed in our future work.

# References

1. Deng, Y., et al.: Joint learning of answer selection and answer summary generation in community question answering. In: Proceedings of the AAAI Conference on Artificial Intelligence, vol. 34, pp. 7651–7658 (2020)
2. Devlin, J., Chang, M.W., Lee, K., Toutanova, K.: BERT: pre-training of deep bidirectional transformers for language understanding. arXiv preprint arXiv:1810.04805 (2018)
3. Jiao, Z., Sun, S., Sun, K.: Chinese lexical analysis with deep Bi-GRU-CRF network. arXiv preprint arXiv:1807.01882 (2018)
4. Liu, Y.: Fine-tune BERT for extractive summarization. arXiv preprint arXiv:1903.10318 (2019)
5. Miao, C., Cao, Z., Tam, Y.C.: Keyword-attentive deep semantic matching. arXiv preprint arXiv:2003.11516 (2020)
6. Song, H., Ren, Z., Liang, S., Li, P., Ma, J., de Rijke, M.: Summarizing answers in non-factoid community question-answering. In: Proceedings of the 10th ACM International Conference on Web Search and Data Mining, pp. 405–414 (2017)
7. Su, D., Xu, Y., Yu, T., Siddique, F.B., Barezi, E., Fung, P.: CAiRE-COVID: a question answering and query-focused multi-document summarization system for COVID-19 scholarly information management. In: Proceedings of the 1st Workshop on NLP for COVID-19 (Part 2) at EMNLP 2020. Association for Computational Linguistics (December 2020). Online
8. Xie, Y., Shen, Y., Li, Y., Yang, M., Lei, K.: Attentive user-engaged adversarial neural network for community question answering. In: Proceedings of the AAAI Conference on Artificial Intelligence, vol. 34, pp. 9322–9329 (2020)
9. Zhang, N., Deng, S., Li, J., Chen, X., Zhang, W., Chen, H.: Summarizing Chinese medical answer with graph convolution networks and question-focused dual attention. In: Proceedings of the 2020 Conference on Empirical Methods in Natural Language Processing: Findings, pp. 15–24 (2020)
10. Zhang, X., Li, S., Sha, L., Wang, H.: Attentive interactive neural networks for answer selection in community question answering. In: Proceedings of the AAAI Conference on Artificial Intelligence, vol. 31 (2017)
11. Zhou, X., Hu, B., Chen, Q., Wang, X.: Recurrent convolutional neural network for answer selection in community question answering. Neurocomputing **274**, 8–18 (2018)

# Syllable Level Speech Emotion Recognition Based on Formant Attention

Abdul Rehman[1,2,3]($\boxtimes$) (iD), Zhen-Tao Liu[1,2,3] (iD), and Jin-Meng Xu[1,2,3]

[1] School of Automation, China University of Geosciences, Wuhan 430074, China
{abdulrehman,liuzhentao,2236245897}@cug.edu.cn

[2] Hubei Key Laboratory of Advanced Control and Intelligent Automation
for Complex Systems, Wuhan 430074, China

[3] Engineering Research Center of Intelligent Technology for Geo-Exploration,
Ministry of Education, Wuhan 430074, China

**Abstract.** The performance of speech emotion recognition (SER) systems can be significantly compromised by the sentence structure of words being spoken. Since the relation between affective content and the lexical content of speech is difficult to determine in a small training sample, the temporal sequence based pattern recognition methods fail to generalize over different sentences in the wild. In this paper, a method to recognize emotion for each syllable separately instead of using a pattern recognition for a whole utterance is proposed. The work emphasizes the preprocessing of the received audio samples where the skeleton structure of Mel-spectrum is extracted using formant attention method, then utterances are sliced into syllables based on the contextual changes in the formants. The proposed syllable onset detection and feature extraction method is validated on two databases for the accuracy of emotional class prediction. The suggested SER method achieves up to 67% and 55% unweighted accuracy on IEMOCAP and MSP-Improv datasets, respectively. The effectiveness of the method is proved by the experimentation results and compared to the state-of-the-art SER methods.

**Keywords:** Speech emotion recognition · Syllables · Feature extraction

## 1 Introduction

Speech Emotion Recognition (SER) is an experimental field that has future prospects of aiding language learning of computers. Currently, the automatic

This work was supported in part by the National Natural Science Foundation of China under Grant 61976197, 61403422 and 61273102, in part by the Hubei Provincial Natural Science Foundation of China under Grant 2018CFB447 and 2015CFA010, in part by the Wuhan Science and Technology Project under Grant 2020010601012175, in part by the 111 Project under Grant B17040, and in part by the Fundamental Research Funds for National University, China University of Geosciences, Wuhan, under Grant 1910491T01.

L. Fang et al. (Eds.): CICAI 2021, LNAI 13070, pp. 261–272, 2021.
https://doi.org/10.1007/978-3-030-93049-3_22

speech recognition requires labeling of spoken words and their meanings to understand those words. However, there are many aspects of language that can not be labelled as certain words, but they carry important symbolic meanings. SER is one of the aspects of understanding the spoken symbols in speech that can help the computers to recognize important messages with better accuracy.

There are three major categories of deep learning models to learn speech emotions. The LLD (Low Level Descriptors) based models usually use prosodic features for learning patterns in a non-sequential way [4,7], the CNN based models use the spectrograms find patterns in the spectra of speech signals (e.g., [11,21]), and thirdly the LSTM based models either use the LLDs or the spectral features of the frame-by-frame windows of speech signal to learn sequential patterns (e.g., [20,22]). Some related works have proposed attention based learning by removing redundant information to improve recognition speed without compromising precision [5,12,18]. Our focus of this paper is to find a way to use the inherently occurring separations in the speech signals such that a modular unit of speech (i.e., syllables) is used rather than a fixed sampling window size or few seconds long utterances.

To solve the above problems, we propose a model that can recognize syllable separations in the speech signal. We present a syllable level recognition model that splits the Mel-spectrum into constituent syllables by the cues of the changing speech sounds such that it helps to distinguish separate units of speech syllables in order to learn the syllable level features. This method helps to learn the relation between of the syllables of variable time duration and emotional categories without having to compromise between recognition accuracy or speed. The experimental results prove the effectiveness, leading to improved performance on two databases with four categorical emotional labels.

The rest of this paper is organized as follows. A syllable level features extraction method is explained in Sect. 2. In Sect. 3, the experimental setup and results are presented along with a brief discussion and comparison to other works. Then we conclude the paper in Sect. 4.

## 2    Syllable Level Emotion Recognition

The general framework for SER starts with the segmentation of audio signal into 20–100 millisecond frames, then spectrogram or audio features of each frame are passed as the input to a deep neural network or any other machine learning classification tool. The RNN based neural networks use LSTM or similar machine learning methods to learn the temporal order of frames (e.g., [8,14]). There are few other methods to learn the temporal cues without using recurrent neural networks (e.g., [10,23]). The number of audio features is usually within the range of 20 to 1000 per few milliseconds frame. Spectrogram based methods use varying sizes of image inputs per frame. Some related works have tried to decrease the input size of the network by using techniques such as discriminative dimension reduction or the attention based deep learning [5,6]. Decreasing the input size increases the risk of information loss, whereas increasing the input size increases the risk of over-fitting, therefore it is important that the automated feature selection process select features with a careful analysis and validation across variety of speakers and datasets. In this paper, however, we propose a static feature selection algorithm that focuses on the preprocessing stage and extracts only the statistical syllable features from a Mel-spectrum of an utterance.

As an alternative to the conventional approach, the proposed method has a highly granular time-step, which processes the speech syllable-by-syllable and decomposes the Mel-spectrum into easily digestible syllable features. In a recent paper, a phoneme type converge method was proposed that assigns phoneme labels to each 25 ms speech frame [17]. That method had few shortcomings such as that phoneme duration was fixed and no sequencing was taken into account. In this paper, using the same formant extraction method, we maintain the granularity from speech signal input to the label prediction at the syllable level. The core idea is to distinguish syllables as separate units of speech which are then used as input features for training a simpler and more generalizable neural network. An overview of the proposed model is given in Fig. 1. Parts of the method are explained in the following subsections.

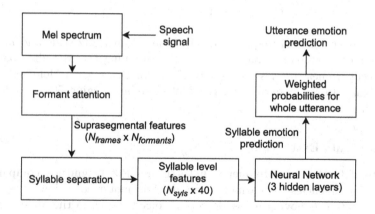

**Fig. 1.** An overview of the proposed method.

## 2.1  Mel-Spectrum Extraction

Similar to conventional methods, we start with creating a mel-spectrogram of the speech signal. A Mel-spectrogram is a spectrogram where the frequencies are converted to the Mel scale. On Mel scale, each unit step is such that pitch sounds equally distinct to the listener. In other words, on Hertz scale pitch units are objectively equidistant, but on Mel-scale pitch units are perceptively equidistant. This is because humans can perceive pitch differences at lower frequencies better than at higher frequencies.

The speech signal is preprocessed by dividing it into a few millisecond frames such that each frame has minimum variation within its timeframe. A contextual

frame window of time duration $T_w$ is iterated through the speech signal with a
stride of $T_s$. Then a Hamming window is applied to each window

$$x_t(n) = (0.54 - 0.46\cos(\frac{2\pi n}{W-1}))s_t(n) \tag{1}$$

where $s_t$ is the input signal of frame $t$, $x_t$ is the windowed frame, $0 \le n \le W-1$,
and $W$ is the size of window ($T_w$ times sampling rate). Then power spectrum
of each frame is calculated by taking Short-term Fourier Transform ($STFT$) of
$x_t$. Then a triangular Mel-filter is applied to the power spectrum that coverts
linear Hertz to a non-linear log scale, which is commonly used for many speech
recognition methods due to similarity with the human ear perception. The Mel
scale frequency can be converted to Hertz scale by

$$m = 2596\log_{10}(1 + \frac{f}{700}) \tag{2}$$

where $m$ is Mel frequency and $f$ is the Hertz scale frequency. Then the central
frequencies of Mel-filter banks can be calculated as

$$f(l) = 700(10^{(m_l - m_{l+1})/5190} - 1) \tag{3}$$

where $f(l)$ is the central frequency of filter bank $l$ on Hertz scale and $m$ is the
lower limit of filter bank $l$ on Mel scale. By adjoining the Mel-filter banks of
few adjacent frames (frames are usually 25 ms long), we get a Mel-spectrum of
speech signal. In our experiments, 128 Mel filter bin ranging 50 Hz to 4000 Hz
were used.

## 2.2   Formants Extraction

Formants of fundamental frequency have the highest magnitude compared to
the rest of frequencies for harmonic sounds. Formant recognition is useful for
estimating pitch, removing noise, detecting voiced speech in the Mel-spectrum.
We consider the top six Mel-filter banks with the highest magnitude as the
top six formants. Formants (i.e., high amplitude frequency bands) are usually
separated from each other by low energy frequency bands. Formants are detected
by comparing the local maxima and minima of amplitude of Mel-filter banks with
each other.

$$p_h = \max_{h=0|p(l) \le p_{h-1}} p(l) \tag{4}$$

where $p_h$ is the power amplitude of $h^{th}$ the highest amplitude formant and $p(l)$ is
the amplitude of filter bank $l$. Similarly, the Mel-scale frequencies of top formants
can be calculated as

$$f_h = \underset{h=0|p(l) \le p_{h-1}}{\arg\max} p(l) \tag{5}$$

where $f_h$ is the Mel-scale frequency of $h^{th}$ the highest amplitude formant.
Equation 5 gives us the central frequencies of formants, however, formants do

not have a precise narrow frequency band, instead, the width of the formant band is an important measure of sound quality. Therefore, we calculate the span formant frequency range (from minima to minima) as

$$s_h(t) = |\arg\min_{l<f_h} p(l) - \arg\min_{l>f_h} p(l)| \tag{6}$$

where $s_h$ is the frequency domain span of formant $h$. The width of formant represents the spread of pitch, which is perceived as the sharpness of voice. The wider vertical span of a formant at a higher frequency is an indicator of voice affects such as breath sounds or gasping expression. Similarly, the span of the lower frequency formants is helpful in discriminating male and female voices.

## 2.3  Formant Matching Index

In order to link the formants of supra-segments across a speech segment to distinguish syllables, a matching index $I_{a,b}$ between any two formants $(h_a, h_b)$ of any two frames $(t_a, t_b)$ is calculated as

$$I_{a,b} = \frac{10}{t_b - t_a} + (10 - (f_b - f_a)^2) + L_a \frac{\min_p(p_a, p_b)}{\max_p(p_a, p_b)} \tag{7}$$

where first term accounts the temporal distance and the frequency difference such that farther the distance or difference, lower the match index. The second term multiplies $L_a$, the number of already linked formants to the formant $h_a$, to the ratio of power of both formants given that the $t_a < t_b$, $tb - ta < 10$, and $fb - fa < 10$.

## 2.4  Syllable Onset and Offset Detection

Syllable separation is a difficult task because there is usually no clear boundary between the two parts of the words or even between two adjacent words. In our method, we propose a technique that uses the maxima and minima of the amplitude of the formants to separate phonemes. One of the clear signs of syllable separations is a silent pause in speech, but that only helps to find the end of the sentences or utterances. Besides that, some parts of speech such as nasal sounds have a clear minimum at the center while maxima at the edges, which makes it difficult to program a hard coded rule for the phoneme separation. Our proposed solution is to use an amplitude hysteresis that adjusts the thresholds for the syllable-ending parameters based on the ad-hoc amplitude threshold of the recent speech frames. Most of the time, there are multiple formants contributing to the energy of the frame, so the formant's center energy can't be taken as the perceived volume, but at the same time the overall energy of the frame is likely to include noise which is what we are trying to avoid. Therefore, we calculate a composite energy that sums up the energy of only the top ten major formants at each frame t as

$$e_c(t) = \sum_{h=0}^{6} e_h(t) + [1 + f_h(t) + H_E] \tag{8}$$

where $e_c$ is the composite energy at frame $t$, $f_{t,h}$ is the $h^{th}$ formant's frequency and $H_E < 0.1$ is an emphasis constant for raising the weights of the energies of the higher frequency formants because higher frequencies carry more energy than the lower frequencies if the amplitudes are kept the same. Using only the top 6 formants prevents low energy formants to be considered as the voiced sound energy since the low energy formants are likely to be noise or the echo of the actual speech. This frame energy is used to distinguish pauses in the speech. When the composite energy is lower than a certain threshold for a certain number of frames, then the syllables or the speech segment is truncated. A longer pause is used to separate utterances into multiple segments, while shorter pauses of at least 2 consistent frames (<50 ms) are used to separate syllables. The syllable separation algorithm initializes at the stage of looking for the rising edge of the amplitude. Once a rising edge is found, it moves to the second stage where is looks for at least 50% drop in energy $e_c$, then it moves to the third stage where it looks for the lower threshold of the amplitude that is set according to the contextual amplitude maxima. An example of separated syllables is shown in Fig. 2. The syllables' onsets are marked by their indices along the horizontal axis.

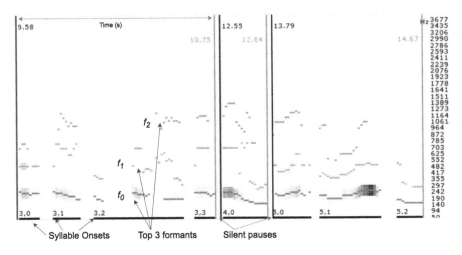

**Fig. 2.** A spectrum of formants showing the syllable onsets and word separations by longer pauses. Each syllable onset is market by its index (word index, syllable index).

## 2.5   Syllable Level Feature Extraction

Syllables come in various shapes and sizes. The spectrum representation of a syllable are more dependent on the lexical content rather than the affective content, therefore the proposed strategy here is to extract statistical features instead of sequential features. Based on the experimental evaluations, a set of 40 features is proposed for syllable level emotion recognition. Table 1 shows the list of features with their brief Introduction.

**Table 1.** Description of formant based syllable features. All features except for the syllable duration are calculated separately for top 3 formants of syllables. Mean, standard deviation, and sum are taken across multiple suprasegmental frames (25 ms) within a syllable.

| Count | Type | Brief description |
|---|---|---|
| 1 | Time axis | Duration of syllable in seconds |
| 3 | Frequency | Mean of non-zero formant frequencies in syllable context |
| 3 | Frequency | Standard deviation of non-zero formant frequencies |
| 3 | Time axis | Voiced duration of formant |
| 3 | Power | Formant energy/syllable duration |
| 3 | Power | Formant energy/formant voiced duration |
| 3 | Bandwidth | Mean of formant span |
| 3 | Numeric | Break counts due to unmatched formants in a syllable |
| 3 | Accent | $\sum$ formant frequency increments relative to previous frame |
| 3 | Accent | $\sum$ formant frequency decrements relative to previous frame |
| 3 | Stress | Count of power maxima within the syllable |
| 3 | Stress | Mean of power of formant power maxima |
| 3 | Stress | Standard deviation of formant power maxima |
| 3 | Stress | Ratio of mean of formant power maxima over average power |
| 40 | | |

In this method, each formant is treated separately to estimate the timbre, tone, accent, and stress of syllable rather than the overall pitch. There are essentially five types of features, i.e., formant frequencies, power, span, accent, and stress. Formant frequencies, power, and span at syllable level can be calculated by taking the mean and standard deviation of these features at suprasegmental level as given in Sect. 2.2. The accent features measure the overall declination or inclination of formant pitch which is also an indication of increase or decrease in pitch of a syllable from it's starting frame to the last frame. The stress features measure the stress within a syllable by comparing the energy maxima to the rest of the frames within a syllable.

## 3 Experimentation

We evaluated our SER model on two databases based on unweighted accuracy for the classification. All experiments were carried out in Python 3.7.10 (64-bit) environment on a computer of the 64-bit Windows 10 system with 16G memory and Intel Core i7-8550U processor. The syllable features were extracted using the method described in Sect. 2, then the Keras python library was used to incorporate the neural network in our method.

## 3.1   Databases

Two databases were used, i.e., IEMOCAP (Interactive emotional dyadic motion capture database) [2] and MSP-Improv [3]. Both the databases have 4 common categorical emotional labels i.e., Happiness, Sadness, Anger, and Neutral from multiple raters. We used these 4 common emotions for the IEMOCAP and MSP-Improv. In IEMOCAP, there are only 2942 utterances and in MSP-Improv there are 3476 utterances with more than 67% agreement among raters. Table 2 shows the sample counts and duration of utterances before the syllable extraction. Note that the total duration of is almost the half for syllables, its because leading and ending silences and short pauses (more than 250 ms) in between the words are removed while separating syllables.

**Table 2.** Sample counts and total duration (minutes) of raw utterances and the extracted syllables for each label in databases.

| Label | IEMOCAP | | | | MSP-Improv | | | |
|---|---|---|---|---|---|---|---|---|
| | Utterances | | Syllables | | Utterances | | Syllables | |
| | Count | Minutes | Count | Minutes | Count | Minutes | Count | Minutes |
| Angry | 289 | 21 | 1790 | 15.3 | 362 | 21 | 1956 | 12.4 |
| Happy | 946 | 61 | 5311 | 42.1 | 1205 | 65 | 6495 | 41.5 |
| Neutral | 1099 | 74 | 7013 | 46.3 | 1489 | 95 | 8591 | 50.4 |
| Sad | 608 | 51 | 3719 | 23.3 | 420 | 32 | 2749 | 15.82 |
| Total | 2942 | 207 | 17833 | 127 | 3476 | 213 | 19791 | 120.12 |
| Avg. duration | 4.2 s | | 0.43 s | | 3.8 s | | 0.36 s | |

## 3.2   Training and Validation

The parameters for the feature extraction process are quite few as compared to the parameters required to tune the neural network. We used 25 ms frame windows with 15 ms steps, 4 kHz maximum frequency, 256 FFT bins and 128 Mel bins for the feature extraction process. Syllable separation also have a parameter for pause length and minimum syllable length which was set to 100 ms (4 frames). A neural network with three hidden layers ($units = [100, 100, 32]$, with ReLU activation) that are followed by a softmax layer was used for all experiments. The training was performed with a learning rate of 0.01, decay rate of 0.001 and the batch size of 100. The model was trained for maximum 100 epochs for both databases, however the minimum validation loss was reached within the first 50 epochs. It is also worth mentioning that the IEMOCAP reached minimum validation loss at around 25 epochs while MSP-Improv took around 50 epochs.

For training the network to predict the emotional classes we used the categorical cross-entropy function given as

$$Loss = -\sum_{i=1}^{Ne} y_i \cdot \log \widehat{y}_i \qquad (9)$$

where $N_e$ is the total number of the emotion classes, $\widehat{y}_i$ is the $i^{th}$ scalar value in the model output, and $y_i$ is the corresponding target value of the model output.

For predicting labels at utterance level, the probabilities of emotional classes at syllable levels are summed up for all syllables in a given utterance as

$$C_{k,u} = \sum_{s=0}^{N_{syls}} \sqrt{T_s} P_{k,s} \tag{10}$$

where $C_{k,u}$ is the cumulative confidence in emotional class $k$ for an utterance $u$, $N_{syls}$ is the number of syllables in the utterance, $P_{k,s}$ is the predicted probability by the softmax for syllable $s$, and $T_s$ is the duration of the syllable $s$ in seconds. The utterance is assigned with a label with the highest cumulative confidence.

For validation, two types of schemes were used for a better comparison with related works. In k-fold scheme, all samples are first shuffled randomly then divided into various folds. In LOSO (Leave One Speaker Out) scheme, samples for one speaker are used for validation while all other samples are used for training, this process is repeated for each individual speaker as the validation fold, then the average of all folds is taken as the overall UAR. The validation results are given in Table 3.

The results in Table 3 show that the unweighted average recall (UAR) for the proposed model is between 45% to 67% for the baseline comparison databases of IEMOCAP and MSP-Improv. The unweighted accuracy takes into account the unbalanced samples for four emotions, and takes the average of recall precision for each emotion. Various validation schemes result into various results therefore unweighted accuracies of four schemes are reported. The results differed from minimum 52% to maximum 69% for each individual validation speaker of IEMOCAP during LOSO validation.

Other related works have reported the unweighted accuracy (or Unweighted average recall) within the same range for these datasets using the LOSO validation. Related works have reported similar accuracy for IEMOCAP database, such as 65.73% by [6], 60.89% by [9], 63.5% by [18] and 63.9% by [15]. A very similar accuracy of 62% for IEMOCAP and 56% for MSP-Improv has been reported by using a graph attentive GRU based method [19]. The best performance models use deep learning methods such as CNN and LSTM along with various feature finding techniques to predict the emotional categories.

The lack of difference in the accuracies of all these models shows that the accuracy as a metric of comparison is not enough, therefore some other pros and cons should also be compared to judge the effectiveness of the model. One of the advantages of our model is that it uses the shallow neural network model, which means it is not a deep learning model therefore it does not require long hours of training for few hours of speech. It also shows that the feature learning by deep learning predicts with as much accuracy as our proposed (rather simpler) method. The second advantage of the proposed method is the small sampling window for syllable level prediction. Almost all other related works have used methods that apply to a complete utterance of a few seconds. The smaller the sampling window, the poorer the accuracy. Although there is no available comparison for the syllable level predictions (<0.5 s), the syllable level predictions can be useful in real-time prediction with almost no latency.

An interesting trend to be noted in Table 3 is that the UAR of syllable level predictions increases with the increase in the validation sample ratio. Whereas, the general trend for utterance level prediction is the decrease in UAR with the increase in validation sample ratio. This trend hints towards the better ability of the syllable level prediction model to perform better with relatively smaller training samples.

**Table 3.** Comparison of the UAR% at utterance level using our method and results reported by the other comparative works. The prediction UAR at syllable level is given only for the proposed method.

| Ref. | Method | DB | Validation | UAR% | |
|------|--------|-----|-----------|------|---------|
| | | | | Utterance | Syllable |
| [1] | CNN | IEMOCAP | LOSO | 61.8 | NA |
| [13] | CNN-GRU | IEMOCAP | LOSO | 61 | NA |
| [9] | CNN-LSTM | IEMOCAP | LOSO | 60.9 | NA |
| [6] | Gaussian-NN | IEMOCAP | LOSO | 65.7 | NA |
| [18] | RNN | IEMOCAP | LOSO | 63.5 | NA |
| [15] | RNN | IEMOCAP | LOSO | 63.9 | NA |
| Ours | Proposed | IEMOCAP | LOSO | 63.1 | 52.1 |
| Ours | Proposed | IEMOCAP | 5-folds | 67.2 | 56.7 |
| [16] | Triplet | MSP-Improv | LOSO | 46.2 | NA |
| [1] | CNN | MSP-Improv | LOSO | 52.6 | NA |
| [19] | GRU | MSP-Improv | LOSO | 56 | NA |
| Ours | Proposed | MSP-Improv | LOSO | 52.5 | 48.3 |
| Ours | Proposed | MSP-Improv | 5-folds | 55.6 | 52.6 |

## 4   Conclusion

One of the challenges in SER is to learn to predict very specific labels from highly overlapping samples. There is a need for more discriminative yet generalizable methods. To solve this problem, we proposed an SER method that tries to predict categorical emotional label for spoken syllables. The idea behind the method was that the highly granular predictions will increase the generalizability. We used Mel-filter banks to extract speech formants which are used to recognize the syllable separations in order to extract syllable level features. These syllable features were then used to train a simple neural network to recognize categorical emotions. The effectiveness of the proposed model was evaluated on two standard benchmark databases (IEMOCAP and MSP-Improv). Our method achieves

UAR of 67% on the IEMOCAP database and 55% on MSP-Improv with 5-folds validation at utterance level, whereas it achieved a relatively lower accuracy for each individual syllable.

In the future, we plan to improve the proposed model to perform experiments on an even wider range of databases. There is a need for automatic recognition of differences among the databases. Even though most emotion labels are usually common, a recognition model that recognizes the differences other than emotion labels will greatly improve the cross-corpus SER. We plan to create such a model that requires minimum labeling to perform a multi-category classification task. We hope to improve the applicability of SER further so that it can be employed in more real-world applications such as medical assistance, social media, and online interfaces.

# References

1. Aldeneh, Z., Provost, E.M.: Using regional saliency for speech emotion recognition. In: 2017 IEEE International Conference on Acoustics, Speech and Signal Processing (ICASSP), pp. 2741–2745. IEEE (2017)
2. Busso, C., et al.: Iemocap: interactive emotional dyadic motion capture database. Lang. Resour. Eval. **42**(4), 335 (2008)
3. Busso, C., Parthasarathy, S., Burmania, A., AbdelWahab, M., Sadoughi, N., Provost, E.M.: MSP-IMPROV: an acted corpus of dyadic interactions to study emotion perception. IEEE Trans. Affect. Comput. **8**(1), 67–80 (2016)
4. Cao, H., Verma, R., Nenkova, A.: Speaker-sensitive emotion recognition via ranking: studies on acted and spontaneous speech. Comput. Speech Lang. **29**(1), 186–202 (2015)
5. Chen, M., He, X., Yang, J., Zhang, H.: 3-d convolutional recurrent neural networks with attention model for speech emotion recognition. IEEE Signal Process. Lett. **25**(10), 1440–1444 (2018)
6. Daneshfar, F., Kabudian, S.J., Neekabadi, A.: Speech emotion recognition using hybrid spectral-prosodic features of speech signal/glottal waveform, metaheuristic-based dimensionality reduction, and gaussian elliptical basis function network classifier. Appl. Acoust. **166**, 107360 (2020)
7. Dave, N.: Feature extraction methods LPC, PLP and MFCC in speech recognition. Int. J. Adv. Res. Eng. Technol. **1**(6), 1–4 (2013)
8. Etienne, C., Fidanza, G., Petrovskii, A., Devillers, L., Schmauch, B.: Cnn+ lstm architecture for speech emotion recognition with data augmentation. arXiv preprint arXiv:1802.05630 (2018)
9. Fayek, H.M., Lech, M., Cavedon, L.: Evaluating deep learning architectures for speech emotion recognition. Neural Netw. **92**, 60–68 (2017)
10. Hajarolasvadi, N., Demirel, H.: 3d CNN-based speech emotion recognition using k-means clustering and spectrograms. Entropy **21**(5), 479 (2019)
11. Issa, D., Demirci, M.F., Yazici, A.: Speech emotion recognition with deep convolutional neural networks. Biomed. Signal Process. Control **59**, 101894 (2020)
12. Koduru, A., Valiveti, H.B., Budati, A.K.: Feature extraction algorithms to improve the speech emotion recognition rate. Int. J. Speech Technol. **23**(1), 45–55 (2020)
13. Lakomkin, E., Weber, C., Magg, S., Wermter, S.: Reusing neural speech representations for auditory emotion recognition. arXiv preprint arXiv:1803.11508 (2018)

14. Le, D., Aldeneh, Z., Provost, E.M.: Discretized continuous speech emotion recognition with multi-task deep recurrent neural network. In: INTERSPEECH, pp. 1108–1112 (2017)
15. Lee, J., Tashev, I.: High-level feature representation using recurrent neural network for speech emotion recognition. In: Sixteenth Annual Conference of the International Speech Communication Association (2015)
16. Lee, S.w.: Domain generalization with triplet network for cross-corpus speech emotion recognition. In: 2021 IEEE Spoken Language Technology Workshop (SLT), pp. 389–396. IEEE (2021)
17. Liu, Z.T., Rehman, A., Wu, M., Cao, W.H., Hao, M.: Speech emotion recognition based on formant characteristics feature extraction and phoneme type convergence. Inf. Sci. **563**, 309–325 (2021)
18. Mirsamadi, S., Barsoum, E., Zhang, C.: Automatic speech emotion recognition using recurrent neural networks with local attention. In: 2017 IEEE International Conference on Acoustics, Speech and Signal Processing (ICASSP), pp. 2227–2231. IEEE (2017)
19. Su, B.H., Chang, C.M., Lin, Y.S., Lee, C.C.: Improving speech emotion recognition using graph attentive bi-directional gated recurrent unit network. Proc. Interspeech **2020**, 506–510 (2020)
20. Wang, J., Xue, M., Culhane, R., Diao, E., Ding, J., Tarokh, V.: Speech emotion recognition with dual-sequence LSTM architecture. In: ICASSP 2020–2020 IEEE International Conference on Acoustics, Speech and Signal Processing (ICASSP), pp. 6474–6478. IEEE (2020)
21. Yao, Z., Wang, Z., Liu, W., Liu, Y., Pan, J.: Speech emotion recognition using fusion of three multi-task learning-based classifiers: HSF-DNN, MS-CNN and LLD-RNN. Speech Commun. **120**, 11–19 (2020)
22. Zhang, S., Zhao, X., Tian, Q.: Spontaneous speech emotion recognition using multiscale deep convolutional LSTM. IEEE Transactions on Affective Computing, p. 1 (2019). https://doi.org/10.1109/TAFFC.2019.2947464
23. Zhang, S., Zhang, S., Huang, T., Gao, W.: Speech emotion recognition using deep convolutional neural network and discriminant temporal pyramid matching. IEEE Trans. Multimedia **20**(6), 1576–1590 (2017)

# Judging Medical Q&A Alignments in Multiple Aspects

Pengda Si[1], Qiang Deng[2], Yiru Wang[2], Bin Zhong[2], Jin Xu[2(✉)], and Yujiu Yang[1(✉)]

[1] Shenzhen International Graduate School, Tsinghua University, Beijing, China
spd18@mails.tsinghua.edu.cn, yang.yujiu@sz.tsinghua.edu.cn
[2] Tencent Inc., Shenzhen, China
{calvindeng,harryzhong,jinxxu}@tencent.com

**Abstract.** Question and answer (Q&A) matching is a widely used task, and there have been many works focusing on this. Previous works tend to give an overall label indicating whether the question matches the answer. However, this method mainly relies on detecting identical or similar keywords in Q&A, which is inappropriate for medical text data. Based on a drug, patients' questions may vary, such as usage, side effects, symptoms, and price. Thus, it is absurd to judge the answer containing the same drug as a matching answer. We argue a better solution is to judge alignments both in entity and intention aspects. To this end, we propose a novel model, which consists of two modules. Specifically, an extractor module gets matching features from text inputs, and then a discriminator module gives alignment labels in both aspects. An adversarial mechanism is designed to disentangle entity matching feature and intention matching feature, which reduces mutual interference. Experimental results show our method outperforms other baselines, including BERT. Further analysis indicates the effectiveness and interpretability of the proposed method.

**Keywords:** Q&A matching · Adversarial disentangle · Multi-aspects

## 1 Introduction

Question and answer (Q&A) matching is widely used in the search, community question answering, reading comprehension, and other scenarios. Some works aim to judge the alignment between Q&A, which usually gives a label indicating whether the question matches the answer. However, those whole matching methods are not suitable for medical question answering scenarios.

As we all know, an overall alignment label between a text pair depends on text similarity, where the similarity of the keywords occupies a large weight. There are many keywords in medical texts called medical entities, such as diseases,

This research was supported by the Guangdong Basic and Applied Basic Research Foundation (No. 2019A1515011387).

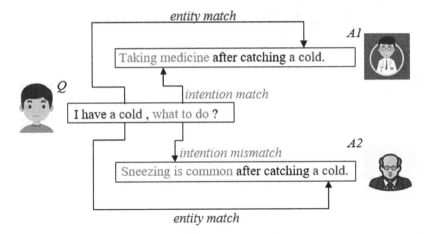

**Fig. 1.** An Example about Medical Q&A. For the question $Q$, both $A1$ and $A2$ contain the same entity " cold" as $Q$. But only $A1$ is a good answer while $A2$ is not. Thus, judging medical Q&A matching needs to consider both the entity and the intention. (Color figure online)

drugs, etc. However, questions and answers containing the same entity may not match because various questions exist based on the same entity. As far as a drug-related issue is concerned, patients may ask questions in different aspects, such as usage, applicable conditions, side effects, and price.

Therefore, we argue that medical Q&A text contains two critical attributes: entity and intention. A toy example is shown Fig. 1, and we mark entity in green and intention in orange. The entity in $Q$ is " cold", and the intention is "what to do". Both $A1$ and $A2$ have the entity " cold", but only $A1$ is a good answer because it gives the treatment while $A2$ describes the symptoms. Obviously, for a medical question, its good answers must match it in both entity aspect and intention aspect.

Unlike previous works, we argue a better solution is to give alignment labels in each aspect, respectively. To this end, we propose a novel fine-grained matching framework to address this issue, which consists of an encoder module and a discriminator module. Specifically, the extractor module utilizes a BERT-based network to get the matching features of question and answer. Then, the discriminator module makes judgments by a series of sub-discriminators. To reduce the mutual interference of matching features in two aspects, we apply an adversarial mechanism to disentangle them.

We conduct our experiment on a medical Q&A dataset, and the results show our method outperforms other baselines, including BERT [4]. Further analysis indicates the effectiveness and interpretability of our method.

We summarize the main contributions of this paper into three points:

– We propose to decompose the judging medical Q&A alignment task into entity aspect and intention aspect.

- We present a novel model, which could give alignment labels for each aspect.
- The experimental results show that our outperforms other baselines, including BERT, indicating its effectiveness and interpretability.

## 2   Related Work

Previous text matching models can be roughly divided into two types: siamese structures and attention structures. Siamese models first encode text pairs and then compute their similarity, such as InferSent [3], SSE [15], SiamCNN [14] and Multi-view [23]. This method ignores the interaction between text pairs. Attention models directly design various structures to model interaction features between text pairs, such as DecAtt [16], ESIM [1], PWIM [6], DAM [24] and HCAN [18]. Although attention models work better than Siamese models, they are surpassed by pre-trained models, such as BERT [4], XLNet [22], RoBERTa [12] and other models [19,21]. However, all these methods give an overall alignment label for a Q&A pair and fine-grained matching labels are lacked.

Disentanglement is a widely used method in the Computer Vision (CV) domain, which usually separates different attributes of images [2,13,17]. For NLP tasks, there have been some works that utilize disentangle framework for style transfer task [5,9,20]. Inspired by this, we implement an adversarial framework to disentangle matching features in multiple aspects for our fine-grained matching method. As far as we know, we are the first work that applies disentanglement to the text matching task.

## 3   Method

### 3.1   Task Definition and Pipeline

Q&A matching mask could be defined as follows: **Given a Q&A pair** $(q, a)$**, we aim to give a label** $z$ **indicating whether** $q$ **matches** $a$. As discussed above, medical text contains two attributes: entity and intention. Thus, we aim to give two alignment labels, $z_e$ for matching in the entity aspect and $z_i$ for matching in the intention aspect.

Our method's pipeline is shown in Fig. 2, which could be split into two steps. First, to get matching features $f_e$ and $f_i$ in two aspects respectively, we construct a BERT-based extractor module. Secondly, to disentangle matching features and get matching labels, we construct a discriminator module and apply the adversarial mechanism to it.

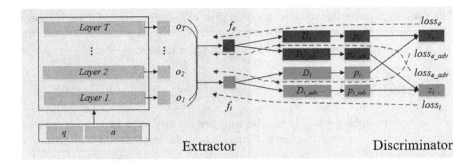

**Fig. 2.** Pipeline of our proposed method. On the left is the extractor module, and on the right is the discriminator module. We use the dashed lines to indicate the back propagation of the gradients of four losses. Red dashed lines represent positive gradients, and blue dashed lines represents negative gradients. (Color figure online)

### 3.2 Extractor

We apply inner attention [11] to vanilla BERT, and conduct our extractor module $E$. The process is shown as follows.

$$o_{ave} = \frac{\sum_{t=1}^{T} o_i}{T} \tag{1}$$

$$f_e = \sum_{t=1}^{T} a_{te} o_t, \quad f_i = \sum_{t=1}^{T} a_{ti} o_t \tag{2}$$

$$a_{te} = \frac{exp(W_e([o_{ave}, o_t]) + b_e)}{\sum_{k}^{T} exp(W_e([o_{ave}, o_k]) + b_e)} \tag{3}$$

$$a_{ti} = \frac{exp(W_i([o_{ave}, o_t]) + b_i)}{\sum_{k}^{T} exp(W_i([o_{ave}, o_k]) + b_i)} \tag{4}$$

where $T$ is the number of layers in BERT [4]. After feeding $(q, a)$ into BERT, we get output of each layer, $o_1, o_2, \cdots, o_T$. Then, we use inner attention twice to get $f_e$ and $f_i$, separately. $o_{ave}$ is the average of outputs of all layers. $W_e, b_e$ and $W_i, b_i$ are parameters of two linear networks, respectively. $a_{te}$ and $a_{ti}$ are attention weight of $t$-th layer in entity and intention aspect, respectively.

Compared to vanilla BERT, which regards the last layer's output as output, our method could extract text information from all layers. Under the assumption that features of different attributes are distributed in different layers, our method is more effective. We prove this assumption in Sect. 5.4.

### 3.3 Discriminator

Our discriminator module consists of four sub-discriminators, and each of them is implemented by a linear network. $D_e$ and $D_i$ utilize matching features to obtain corresponding matching probability $p_e$ and $p_i$. We use the cross-entropy between them and the ground truth label $z_e$ and $z_i$ as the loss function, as follows:

$$loss_e = cross\_entropy(p_e, z_e) \tag{5}$$

$$loss_i = cross\_entropy(p_i, z_i) \tag{6}$$

Gradients of $loss_e$ and $loss_i$ are used to update $D_e$, $D_i$, and the extractor module $E$. Inspired by adversarial learning in style transfer [9], we construct two more discriminators $D_{e\_adv}$ and $D_{i\_adv}$ to disentangle $f_e$ and $f_i$. The purpose is to reduce the mutual influence between the two features. Specifically, $D_{e\_adv}$ get intention alignment probability $p_{i\_adv}$ based on entity matching feature $f_e$, and then cross entropy between $p_{i\_adv}$ and the ground truth label $z_i$ is regarded as the loss function. Similarly, $D_{i\_adv}$ gets $p_{e\_adv}$ based on $f_i$. Thus, the adversarial losses are shown as follows:

$$loss_{e\_adv} = cross\_entropy(p_{e\_adv}, z_e) \tag{7}$$

$$loss_{i\_adv} = cross\_entropy(p_{i\_adv}, z_i) \tag{8}$$

Gradients of $loss_{e\_adv}$ and $loss_{i\_adv}$ are used to update $D_{e\_adv}$ and $D_{i\_adv}$, while the negative gradients are used to update the extractor module. To better illustrate the process, we give our training algorithm here.

---

**Algorithm 1.** Training algorithm of our method

---
    **for** Each minibatch **do**
        Calculate all losses.
        Get gradient $G$ of $loss_e + loss_i$, gradient $G_{adv}$ of $loss_{e\_adv} + loss_{i\_adv}$.
        Update $D_e$, $D_i$ with $G$.
        Update $D_{adv\_e}$, $D_{adv\_i}$ with $G_{adv}$.
        Update $E$ with $G$ - $G_{adv}$.
    **end for**

---

In this way, we conduct an adversarial relationship between two adversarial discriminators and the extractor. $D_{e\_adv}$ tries to get the true intention alignment label, while the extractor prevents the process. As a result, intention matching information in the entity matching feature $f_e$ is reduced. The adversarial discriminator $D_{i\_adv}$ functions in the same way. Therefore, $f_e$ and $f_i$ are disentangled.

## 4    Experiment

### 4.1    Dataset and Evaluation Metrics

We crawled about 1M medical question-answer pairs from some online Chinese medical forums. Then, we randomly sample 50K pairs and hire several medical experts for annotation. After removing pairs which text is not complete, there are 42793 pairs left. Annotators are required to give two types of matching labels: entity matching label and intention matching label. To ensure the agreement of different annotators, each pair is annotated by at least three human experts. In this way, we get a new medical Q&A dataset, and its information is presented in Table 1. As far as we know, our dataset is the first Q&A dataset in which matching labels are in multiple aspects. We hope it could help further research on multi-aspects text matching tasks.

**Table 1.** Statistical characteristics of the dataset constructed in this work

| Item | Information |
|------|-------------|
| Total Q&A pairs | 42793 |
| Match | 27607 |
| Entity mismatch | 2474 |
| Intention mismatch | 12712 |
| Avg sentences in question | 1.05 |
| Avg sentences in answer | 3.06 |
| Avg words in question | 8.12 |
| Avg words in answer | 85.80 |

During experiments, we randomly sample 1500 for the test, 1500 for the validation, and use the rest as train data. As mismatch pairs account for a smaller percentage, we use the F1 score of mismatched pairs as the evaluation criterion, which is consistent with the purpose of filtering mismatched Q&A pairs in the actual application scenario.

### 4.2   Baselines

We use BERT [4] as our baseline. We also conduct a baseline model named two-BERT based on vanilla BERT. Specifically, we finetune two BERT models, one for entity matching judging and the other for intention. In addition to pretrain models, we utilize some classic text matching models as baselines, as below:

- **Infersent:** It uses a LSTM architecture [7] get features of sentence pair [3].
- **SSE:** It uses stacked bidirectional LSTM with shortcut connections as encoder [15].
- **PWIM:** It proposes a novel similarity focus mechanism [6].
- **DecAtt:** It uses attention to decompose the problem into sub-problems [16].
- **ESIM:** It enhances chain LSTMs and introduces recursive architectures [1].

For InferSent, we utilize its original code, which is implemented on PyTorch. For the other four baselines, [10] realized them on PyTorch, and we directly utilize their codes.

### 4.3   Implementation Details

For our model, we build our extractor module based on 12-layer Chinese BERT codes. Every sub-discriminator network in the discriminator module is implemented with a fully connected layer. The overall framework is implemented on Tensorflow. We set hidden size, batch size, max length to 768, 32, 256, respectively. The Adamer optimizer with the learning rate of 2e−5 is hired to train our model. We use early stopping strategy during the training process, and the stop epoch is set to 3.

## 5   Evaluation

### 5.1   Evaluation Results

The evaluation results are shown in Table 2. We could see that two-BERT gets the highest F1 score, higher than BERT. This proves our assumption that the two matching features will interfere with each other. Thus, deposing the medical Q&A matching task in entity and intention aspect is necessary. Our model gets the second-highest F1 score, about 0.03 higher than BERT, which indicates the effectiveness of our framework.

**Table 2.** Evaluation results of all models

| Model | Precision | Recall | F1 |
| --- | --- | --- | --- |
| Infersent | 0.448 | 0.445 | 0.447 |
| SSE | 0.515 | 0.421 | 0.463 |
| PWIM | 0.684 | 0.427 | 0.526 |
| DecAtt | 0.558 | 0.424 | 0.482 |
| ESIM | 0.494 | 0.653 | 0.563 |
| BERT | 0.708 | 0.784 | 0.744 |
| Two-BERT | 0.749 | 0.811 | **0.779** |
| Ours | 0.748 | 0.800 | 0.773 |

Although our method doesn't get the highest score, the score is very close to the result of two-BERT, which demonstrates that our framework could reduce the mutual interference of matching features on two aspects. We also present the parameters and training time of different models in Table 3. We could see, two-BERT contains almost twice as many parameters as ours. Under the same computing resources, two-BERT takes approximately twice as long as our model to train an epoch. If the matching features of Q&A are distributed in more aspects, training a BERT model for each aspect will consume more computing resources. Therefore, our model is a better choice with limited computing resources.

## 5.2  Ablation Study

To analysis our method further, we remove the adversarial discriminators $D_{e\_adv}$ and $D_{i\_adv}$ in our framework and construct an ablation model named w.o adv. The results are presented in Table 4. Our w.o adv model gets a 0.754 F1 score, lower than our method, which shows that the adversarial disentangle mechanism is vital for the medical Q&A matching task. Meanwhile, w.o.adv model outperforms BERT, demonstrating the effectiveness our extractor module.

**Table 3.** Parameters and training time.

| Model | Parameters | Time |
|---|---|---|
| Ours | 103.5 M | 30 m |
| BERT | 102.3 M | 29 m |
| Two-BERT | 204.6M | 58 m |

**Table 4.** Results of the ablation model.

| Model | Precision | Recall | F1 |
|---|---|---|---|
| Ours | 0.748 | 0.800 | **0.773** |
| w.o adv | 0.709 | 0.805 | 0.754 |

To verify the effectiveness of our adversarial mechanism, we apply it to our two baseline models: Infersent and DecAtt, and construct two models: Infersent-dis and DecAtt-dis. The results are given in Table 5. Obviously, these two baselines get better scores with our mechanism, indicating that our adversarial disentangle framework is generic and effective.

**Table 5.** Evaluation results of two baseline models with our disentangle mechanism.

| Model | Precision | Recall | F1 |
|---|---|---|---|
| Infersent | 0.448 | 0.445 | 0.447 |
| Infersent-dis | 0.488 | 0.371 | 0.421 |
| DecAtt | 0.558 | 0.424 | 0.482 |
| DecAtt-dis | 0.589 | 0.536 | 0.562 |

## 5.3  Case Study

To further explore how our method judges Q&A pairs, we give two examples in Table 6, with the output labels of our model and BERT.

For the first Q&A pair, its answer is just a repeat of the question. However, BERT gives the wrong label that the answer matches the question. It shows that BERT makes judgments based on text similarity sometimes. Our model gives matched label in the entity aspect and mismatched label in the intention aspect, consistent with human judgment.

**Table 6.** Two cases of Q&A pairs. We present the matching labels that given by our method and BERT.

|  | case 1 | case 2 |
|---|---|---|
| **Question** | Does a low fever cause seizures? | What are the consequences of taking contraceptives? |
| **Answer** | Does a low fever cause seizures? | Hello, your situation could be irregular menstruation caused by contraceptives. |
| **result of BERT** | match ✓ | mismatch × |
| **result of ours** | entity match ✓ intention mismatch × | entity match ✓ intention match ✓ |

For the second Q&A pair, the question and answer contain the same entity "contraceptives", so the Q&A pair matches in entity aspect. Meanwhile, "irregular menstruation" is a type of "consequence" caused by contraceptives. Therefore, the Q&A pair matches in the intention aspect, too. We could see our method gives correct matching labels while BERT does not. We infer BERT may be affected by the mutual interference of matching features in two aspects.

## 5.4   Attention Visualization

For test data, we visualize attention weights of each layer in the extractor module, and the result is shown in Fig. 3. The main attention of $f_e$ is distributed at the 10th layer, while the main attention of $f_i$ is distributed at the 12th layer.

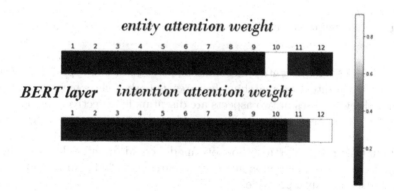

**Fig. 3.** Attention weights in different layers. The darker color means the greater attention weight.

For medical Q&A pairs in our experiment, this demonstrates that the entity matching information and the intention matching information in BERT are distributed on different layers. Therefore, our extractor structure is necessary for getting the proper matching features. Furthermore, this also indicates that our extractor module could obtain matching features in two aspects simultaneously.

What's more, entity matching features are extracted on lower layers while intention matching features on higher layers. We assume this is because judging whether two sentences have the same entity a simpler task than judging intention, as [8] proves before.

### 5.5   Matching Features Visualization

To verify whether our proposed framework successfully disentangles matching features in two aspects, we conduct a further analysis experiment. Specifically, we choose two types of Q&A pairs in our test data. The first type matches in both aspects, while the second type matches only in the entity aspect. We get matching features $f_e$ and $f_i$ of them and then use t-SNE to transform them into two-dimensional vectors. The visualization results are shown in Fig. 4, and we mark two types of Q&A pairs in red and blue separately.

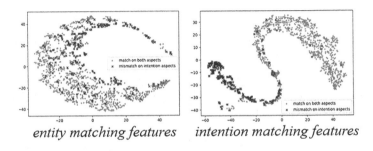

entity matching features      intention matching features

**Fig. 4.** Distribution of matching features in two aspects. (Color figure online)

It could be seen that entity matching features $f_e$ of two types of Q&A pairs are mixed while intention matching features $f_i$ have a clear margin. This indicates matching features in two aspects are disentangled effectively by our framework. The result also explains why our adversarial disentangle mechanism is valid.

All analysis results further show our model could disentangle matching features and give correct matching labels. Compared to BERT, our method is more effective and more interpretable.

## 6   Conclusion

We argue that giving an overall alignment label is not proper for the medical Q&A matching task because medical text contains two attributes: entity and intention. Thus, we propose a fine-grained matching method, which gives alignment labels in each aspect. Our model consists of an extractor module and a discriminator module. The extractor module is constructed on vanilla and could get matching features in two aspects. And the discriminator module contains a series of discriminators, which gets matching labels. We also apply an adversarial mechanism to disentangle matching features. Experiment results on a medical Q&A dataset show our method outperforms other baselines, including BERT. Further analysis indicates our method is more interpretable. Our feature work is to improve our framework and use it for Q&A matching tasks in other domains.

# References

1. Chen, Q., Zhu, X., Ling, Z., Wei, S., Jiang, H., Inkpen, D.: Enhanced LSTM for natural language inference. In: Proceedings of the 55th Annual Meeting of the Association for Computational Linguistics, ACL 2017, Volume 1: Long Papers, pp. 1657–1668. Association for Computational Linguistics (2017)
2. Chen, X., Duan, Y., Houthooft, R., Schulman, J., Sutskever, I., Abbeel, P.: Infogan: interpretable representation learning by information maximizing generative adversarial nets. In: Lee, D.D., Sugiyama, M., von Luxburg, U., Guyon, I., Garnett, R. (eds.) Advances in Neural Information Processing Systems 29: Annual Conference on Neural Information Processing Systems, NeurIPS2016, pp. 2172–2180 (2016)
3. Conneau, A., Kiela, D., Schwenk, H., Barrault, L., Bordes, A.: Supervised learning of universal sentence representations from natural language inference data. In: Proceedings of the 2017 Conference on Empirical Methods in Natural Language Processing, EMNLP 2017, pp. 670–680. Association for Computational Linguistics (2017)
4. Devlin, J., Chang, M.W., Lee, K., Toutanova, K.: Bert: Pre-training of deep bidirectional transformers for language understanding. In: NAACL-HLT, no. 1 (2019)
5. Fu, Z., Tan, X., Peng, N., Zhao, D., Yan, R.: Style transfer in text: Exploration and evaluation. In: Proceedings of the Thirty-Second AAAI Conference on Artificial Intelligence, (AAAI-18), pp. 663–670. AAAI Press (2018)
6. He, H., Lin, J.J.: Pairwise word interaction modeling with deep neural networks for semantic similarity measurement. In: NAACL HLT 2016, The 2016 Conference of the North American Chapter of the Association for Computational Linguistics: Human Language Technologies, pp. 937–948. The Association for Computational Linguistics (2016)
7. Hochreiter, S., Schmidhuber, J.: Long short-term memory. Neural Comput. $9(8)$, 1735–1780 (1997)
8. Jawahar, G., Sagot, B., Seddah, D.: What does BERT learn about the structure of language? In: Proceedings of the 57th Conference of the Association for Computational Linguistics, ACL 2019, Volume 1: Long Papers, pp. 3651–3657. Association for Computational Linguistics (2019)
9. John, V., Mou, L., Bahuleyan, H., Vechtomova, O.: Disentangled representation learning for non-parallel text style transfer. In: Proceedings of the 57th Conference of the Association for Computational Linguistics, Volume 1: Long Papers, pp. 424–434. Association for Computational Linguistics (2019)
10. Lan, W., Xu, W.: Neural network models for paraphrase identification, semantic textual similarity, natural language inference, and question answering. In: Proceedings of the 27th International Conference on Computational Linguistics, pp. 3890–3902 (2018)

11. Liu, Y., Sun, C., Lin, L., Wang, X.: Learning natural language inference using bidirectional lstm model and inner-attention. arXiv preprint arXiv:1605.09090 (2016)
12. Liu, Y., et al.: Roberta: A robustly optimized BERT pretraining approach. CoRR abs/1907.11692 (2019)
13. Luan, F., Paris, S., Shechtman, E., Bala, K.: Deep photo style transfer. In: 2017 IEEE Conference on Computer Vision and Pattern Recognition, CVPR 2017, pp. 6997–7005. IEEE Computer Society (2017)
14. Mueller, J., Thyagarajan, A.: Siamese recurrent architectures for learning sentence similarity. In: Proceedings of the Thirtieth AAAI Conference on Artificial Intelligence, pp. 2786–2792. AAAI Press (2016)
15. Nie, Y., Bansal, M.: Shortcut-stacked sentence encoders for multi-domain inference. In: Proceedings of the 2nd Workshop on Evaluating Vector Space Representations for NLP, RepEval@EMNLP 2017, pp. 41–45. Association for Computational Linguistics (2017)
16. Parikh, A., Täckström, O., Das, D., Uszkoreit, J.: A decomposable attention model for natural language inference. In: Proceedings of the 2016 Conference on Empirical Methods in Natural Language Processing, pp. 2249–2255 (2016)
17. Park, T., Efros, A.A., Zhang, R., Zhu, J.-Y.: Contrastive learning for unpaired image-to-image translation. In: Vedaldi, A., Bischof, H., Brox, T., Frahm, J.-M. (eds.) ECCV 2020. LNCS, vol. 12354, pp. 319–345. Springer, Cham (2020). https://doi.org/10.1007/978-3-030-58545-7_19
18. Rao, J., Liu, L., Tay, Y., Yang, H., Shi, P., Lin, J.: Bridging the gap between relevance matching and semantic matching for short text similarity modeling. In: Proceedings of the 2019 Conference on Empirical Methods in Natural Language Processing, EMNLP-IJCNLP 2019, pp. 5369–5380. Association for Computational Linguistics (2019)
19. Reimers, N., Gurevych, I.: Sentence-bert: Sentence embeddings using siamese bert-networks. In: Proceedings of the 2019 Conference on Empirical Methods in Natural Language Processing, EMNLP-IJCNLP 2019, pp. 3980–3990. Association for Computational Linguistics (2019)
20. Shen, T., Lei, T., Barzilay, R., Jaakkola, T.S.: Style transfer from non-parallel text by cross-alignment. In: Advances in Neural Information Processing Systems 30: Annual Conference on Neural Information Processing Systems, NeurIPS2017, pp. 6830–6841 (2017)
21. Shonibare, O.: ASBERT: siamese and triplet network embedding for open question answering. CoRR abs/2104.08558 (2021)
22. Yang, Z., Dai, Z., Yang, Y., Carbonell, J.G., Salakhutdinov, R., Le, Q.V.: Xlnet: generalized autoregressive pretraining for language understanding. In: Advances in Neural Information Processing Systems 32: Annual Conference on Neural Information Processing Systems, NeurIPS2019, pp. 5754–5764 (2019)
23. Zhou, X., et al.: Multi-view response selection for human-computer conversation. In: Proceedings of the 2016 Conference on Empirical Methods in Natural Language Processing, EMNLP 2016, pp. 372–381. The Association for Computational Linguistics (2016)
24. Zhou, X., et al.: Multi-turn response selection for chatbots with deep attention matching network. In: Proceedings of the 56th Annual Meeting of the Association for Computational Linguistics, ACL 2018, Volume 1: Long Papers, pp. 1118–1127. Association for Computational Linguistics (2018)

# DP-BERT: Dynamic Programming BERT for Text Summarization

Shiyun Cao and Yujiu Yang[✉]

Shenzhen International Graduate School, Tsinghua University, Beijing, China
csy18@mails.tsinghua.edu.cn, yang.yujiu@sz.tsinghua.edu.cn

**Abstract.** Extractive summarization aims to extract sentences containing critical information from the original text, one of the mainstream methods for summarization. Generally, extractive summarization is regarded as a sentence binary classification task in many works. Still, the positive samples selected by these methods are incomplete, and the negative samples are composed of random single sentences, which leads to unsatisfactory classification results and incomplete abstract sentences. To address this issue, we propose a Dynamic Programming BERT (DP-BERT), which can dynamically select the positive example with the closest meaning of the reference abstract and adjusts the corresponding negative samples. Specifically, we design a selector responsible for the dynamic selection of positive and negative samples and then utilize the BERT pre-training model to fine-tune the sentence classifier. Extensive experiments show that DP-BERT can better extract the original text's key sentences and achieve state-of-the-art performance on two widely-used benchmarks.

**Keywords:** Extractive summarization · Dynamic programming · Sentence selection

## 1 Introduction

Text summarization is a classic natural language processing task; due to the explosive growth of information and mobile applications' popularity, the demand for short and refined text to transfer knowledge increases. The task of text summarization has also become compelling.

Generally, text summary generation is divided into extractive summarization and abstractive summarization: the former is to select several vital sentences from the original text directly and sort and reorganize them to form a summary; the latter is similar to the process of human writing abstracts, compress and integrate vital sentences in the original text to form a new and more comprehensive summary.

According to Lebanoff Logan [10], most of the manual summary sentences are composed of multiple sentences in the article; if these sentences can be found and remove the redundant parts, they can be close to the hand-written abstract.

© Springer Nature Switzerland AG 2021
L. Fang et al. (Eds.): CICAI 2021, LNAI 13070, pp. 285–296, 2021.
https://doi.org/10.1007/978-3-030-93049-3_24

Therefore, to reduce the difficulty for machines to understand long texts and generate high-quality summaries, such as [10, 12, 22], they adopt a two-stage method: the first stage is an extractive summarization, which selects essential sentences from the article; the second stage is a generative summarization, which rewrites the sentences obtained in the previous stage into more concise sentences. In this way, generating abstracts is completed in stages. The model must accurately and comprehensively extract sentences containing critical information in the article in the first stage to better the final abstract.

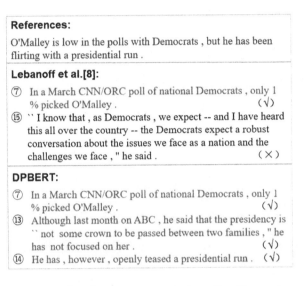

**Fig. 1.** The positive sample of sentence extraction. For the sentence in the reference(marked in brown), the sentences extracted by Lebanoff Logan [10] from the original text are 7, 15, the matching sentences extracted by DPBERT is 7, 13, 14. we marked the better-extracted sentences in yellow, and the rest were marked in black. (Color figure online)

But, the existing works limit the number of sentences when selecting positive sample sentences similar to the reference abstract, resulting in incomplete abstract sentence extraction and an inadequate abstract generation. As shown in Fig. 1, a sentence or two random sentences in the text are directly packaged as positive samples [10] and then classified to determine whether it is a summary sentence. Undoubtedly, this extraction method makes each summary sentence only come from a single sentence or two sentences in the text, which is relatively simple and violent. These methods, which limit the number of sentences, are easy to calculate. Still, they easily lead to incomplete semantics and further deviate from expression and classification because some summary sentences may come from a complete summary of more than two sentences.

Based on the above observations, we focus on making the extracted sentences more accurate and comprehensive and introduce the Dynamic Programming

BERT (DP-BERT) based on the dynamic programming algorithm, which can optimally select sentence combinations that are close to hand-written summaries and do not dropout key text information, and adjust the settings of the positive and negative examples of the classifier to avoid too short samples and increase the hard samples of classifier learning. The experimental results show that the abstract sentences extracted by the proposed model are more reasonable and closer to the content of manual abstracts, and the DP-BERT outperforms the state-of-the-art (SOTA) model randomly initialized counterpart by 1.70 ROUGE on the CNN/Dailymail dataset and by 0.16 ROUGE on the XSUM dataset.

## 2 Related Work

The research on extractive summarization generation has a long history. Generally speaking, the current mainstream methods can be divided into graph method, ranking method, matching method, classification method, etc.

To be specific, Some graph-based extractive summarizations [3,6,13,19,23] consider that the keywords and sentences are the primary nodes in the graph, so the sentences contained in the primary node are extracted as abstracts, this method of considering the structure of the article can make the abstract more logical to a certain extent, but it is computationally intensive and difficult to learn. There are also some work regard the abstract as a sentence ranking problem [2,9,11,17,18], all the sentences are sorted by importance, and the most of critical TOP-$N$ sentences are extracted as abstracts; this method is closer to the way humans write abstracts. However, it is difficult to define 'importance' and the $N$ value. Besides, in the deep neural networks framework, Ming Zhong [24] treat extractive summarization as a semantic matching problem and directly extract candidate summary (several sentences) instead of sentence-level (one by one) extraction, this semantic unit-based matching method has lower redundancy than sentence-level matching and higher semantic accuracy than word-level matching. Still, it is controversial whether the definition of Pearl-Summary in the reference summarization is accurate or not. Zhengyuan Liu [14] introduces information in a specific field to help extract key sentences, this method works well in specific fields, but the general-purpose type is poor.

At the same time, some summarization methods treat the abstract problem as a sentence two-category problem [1,2,10,15,21,22] by training the classifier model, the sentences in the article that are suitable for the summary are classified as positive samples, the redundant sentences are negative samples. But, how to find positive samples and set negative samples is the key.

Compared with previous two-category works, we combine the advantages of semantic mining and sentence classification problems; the proposed DP-BERT matches each reference abstract sentence semantically, improves abstract extraction accuracy, and helps generate high-quality abstracts.

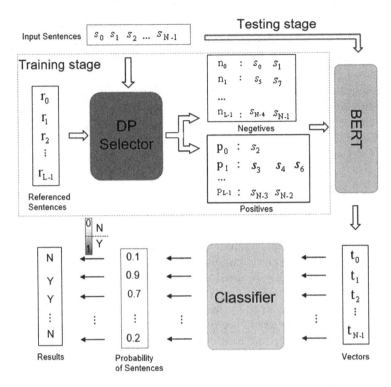

**Fig. 2.** The framework of the DP-BERT. In the training phase, the text sentences that match the reference abstract are marked as positive. Then the labeled positive examples and random negative examples are input to the classifier.

## 3   DP-BERT Model

Traditionally, extractive summarization can be regarded as a sentence binary classification task. The sentences containing critical text information are viewed as positive samples, and the rest are negative samples. A reference sentence corresponds to at most two sentences in the original text in the past work. Still, according to the findings of Lebanoff Logan [10], in most cases, two sentences cannot fully express the meaning of the reference abstract. Unlike the previous method, we introduce dynamic programming to select positive samples, named DP-BERT, which can match sentences with similar semantics to the reference abstract as positive samples to the greatest extent and ensure the abstract's semantic integrity. As shown in Fig. 2, DP-BERT is mainly divided into the DP-selector and a classifier.

### 3.1   Problem Formulation

Suppose there are $N$ sentences $(s_0, s_1, ..., s_{N-1})$ and $L$ reference abstract sentences $(r_0, r_1, ..., r_{L-1})$ in an article, our goal is to find the most similar sentence

---

**Algorithm 1.** The DP Selector Algorithm

---

**Input:** Given document sentence $(s_0, s_1, ..., s_{N-1}) \in S$ ,
   reference sentence $(r_0, r_1, ..., r_{L-1}) \in R$ and let $x = 0$, $max_{sim} = 0$
**Output:**

1: **for** $r_i \in R$, $i = 0$ to $L - 1$ **do**
2:    $j = x$
3:    **while** $j \leq N - 1$ **do**
4:       Calculate the similarity between $r_i$ and $s_j$
5:       $max_{sim} = max(similarity, max_{sim})$
6:       $j+ = 1$
7:    Select the sentence combination $(s_a, s_b, ..., s_c)$ with the highest similarity as the positive example of $r_i$, $k \leq len(r_i)$, $x = max(a, b, c)$.

---

combination to the reference abstract in the original text. In addition, people usually read the article from the beginning to the place in the order of sentences. Therefore, it is more in line with human reading habits to summarize the article according to the article's sentence order.

### 3.2 DP Selector

DP selector is designed to select multiple suitable sentences for merging as positive samples. Accurate positive and negative samples can improve the accuracy of the classifier.

In the task of generating extractive abstracts, for each reference sentence, it is necessary to find the best matching sentence in the original text, which means that the model needs to make a decision when matching each reference sentence. Therefore, the abstract generation can be divided into different reference sentence matching stages. Decisions are made at each stage (the most matching sentence or sentence combination is found in the article). As a result, the entire abstract can achieve the best matching effect. Furthermore, when the decision at each stage is determined, each reference sentence selects the most matching original sentence or sentence combination to form a decision sequence, thus confirming the selection of the overall abstract sentence. Therefore, in order to achieve the overall best decision, it is natural to think of using dynamic programming algorithms to solve the problem of multi-stage decision-making optimization.

The algorithm flow chart of DP selector is shown in Algorithm 1. For the $x_{th}$ reference summary sentence $r_x$ , the revenue of the $i_{th}$ sentence $s_i$ is defined as fellows,

$$\widehat{P_{r_x}}(s_i) = \max_{0 \leq j < i} [P_{r_x}(s_j) + A(s_j, s_i)]$$

$$P_{r_x}(s_i) = \max_{0 < i \leq N-1} (P_{r_x}(s_{i-1}), \widehat{P_{r_x}}(s_i)) \tag{1}$$

where $P_{r_x}(s_1) = A(s_0, s_1)$, and $A(s_j, s_i)$ means the maximum benefit of the sentence selected from the $j_{th}$ sentence to the $i_{th}$ sentence, which can be computed by the following equation:

$$A(s_j, s_i) = \max_k [A(s_j, s_k) + L((s_j, s_k), s_i) \atop - R(s_j, s_k, s_i)] \tag{2}$$

$$L((s_j, s_k), s_i) = cosine((s_j, s_k), s_i) \tag{3}$$

$$R((s_0, s_k), s_i) = cosine((s_0, s_k), s_i) \tag{4}$$

where, $L(s_j, s_k, s_i)$ represents the amount of information increased by $s_i$ compared to the sentence selected in the $(j, k)$ interval , $R(s_j, s_k, s_i)$ represents the redundancy of the $i_{th}$ sentence and the sentences in the $(j, k)$ interval and the sentence selected by $P(s_j)$ , computed as the cosine similarity of these sentences.

Regarding calculating the similarity between the reference abstract sentence and the article sentence, we first remove the stop words in the sentence. If the same words in the two sentences are greater than or equal to three, we use the cosine similarity to judge whether the two sentences are similar. On the contrary, if two sentences have less than three identical words after removing the stop words, we do not include the sentence to consider possible abstract sentences. Therefore, for a reference abstract sentence of length $W$, it corresponds to at most $W/3$ sentences in the original text. To reduce the computational complexity while selecting sentences similar to the reference summary sentence as complete as possible, we limit the number of sentences corresponding to each reference summary sentence in the original text to $W/3$. As shown in Fig. 3, for each reference sentence, traverse all sentences in the article and calculate sentence similarity. When the number of sentences is limited to 1, only the reference abstract sentence and the original sentence need to be compared for the sentence similarity one by one, and select the sentence with the highest similarity to the reference abstract sentence as the candidate sentence; when the number of determinate sentences is bigger than 1, select the sentence combined with the highest similarity under the number of sentences. For example, $(0, 1)0.28$ in the dashed frame means that when $r_0$ is traversed to sentence $s_2$ under the condition that the number of sentences is limited to 2, the selected sentence is $(0, 1)$, and the sentences similarity at this time is 0.28. Experimental results show that if the number of abstract sentences is too low, a certain amount of information will be lost.

## 3.3   Classifier

The classifier is obtained after fine-tuning the specificity of the summary task on the following sentence prediction task of BERT [4], which is roughly the same as Lebanoff Logan [10], and will not be repeated here. The difference is that because we used a DP selector before the classifier, the resulting favorable sample

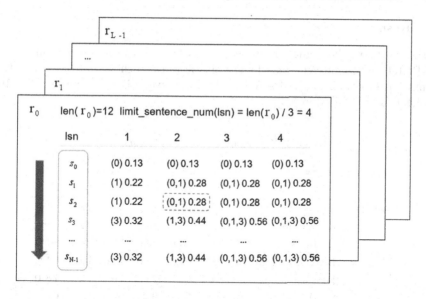

**Fig. 3.** An example of DP selector.

sentence combination may be more than two sentences. Still, the classifier inputs up to two sentences, so we cut the positive samples of more than two sentences into the form of sentence pairs. For example, in the second summary sentence in Fig. 1, the corresponding sentence selected by the DP selector is the 7-th, 13-th, and 14-th sentence, so the positive sample is {7-th,13-th,14-th}, we cut it into two positive examples {7-th,13-th} and {13-th,14-th}.

Actually, the framework designed by Lebanoff Logan [10] ignores the balance of positive and negative samples; that is, accurate positive samples are essential, but appropriate negative samples are also necessary. After analyzing the examples, we found that if the negative sample sentence's length is too short and the amount of information is too small. It is difficult for the classifier to learn the difference between the positive and negative samples. Therefore, to ensure that the classifier is not affected by the sentence's length, we force the negative sample to be a random combination of two sentences in the original text.

## 4 Experiments

### 4.1 Datasets

Our experiments are mainly based on both the CNN/Dailymail and the Xsum data sets. The CNN/Dailymail (CNN/DM) data set [8] is a commonly used long text summary data set containing $287k$ text-summary training examples and $11k$ test examples. XSUM [16] is a relatively new summary data set. Its text summary is relatively short and usually consists of one sentence containing $204k$ text-summary training examples and $11k$ test examples.

## 4.2 Results

In this part, we did some comparative experiments to analyze the impact of DP-BERT. We mainly compared some classic work, such as SumBasic [20], KL-Summ [7], LexRank [5], and some of the latest work BERT-SingPairMix [10], HIBERT [22], BERTsum [24].

**Table 1.** The result of extracted abstract.

| Dataset | Methods | Primary | | | Secondary | | |
|---|---|---|---|---|---|---|---|
| | | P | R | F | P | R | F |
| CNN/DM | SumBasic | 15.2 | 17.3 | 16.2 | 5.3 | 15.8 | 8.0 |
| | KL-Summ | 15.7 | 17.9 | 16.7 | 5.4 | 15.9 | 8.0 |
| | LexRank | 22.0 | 25.9 | 23.8 | 7.2 | 21.4 | 10.7 |
| | BERT-SingPairMix | 33.6 | **67.1** | 44.8 | 13.6 | 70.2 | 22.8 |
| | DP-BERT | **84.6** | 61.7 | **71.4** | **79.1** | **76.9** | **78.0** |
| XSUM | SumBasic | 8.7 | 9.7 | 9.2 | 5.0 | 8.9 | 6.4 |
| | KL-Summ | 9.2 | 10.2 | 9.7 | 5.0 | 8.9 | 6.4 |
| | LexRank | 9.7 | 10.8 | 10.2 | 5.5 | 9.8 | 7.0 |
| | BERT-SingPairMix | 33.2 | **56.0** | 41.7 | 24.1 | **65.5** | 35.2 |
| | DP-BERT | **47.7** | 53.9 | **50.6** | **34.9** | 63.8 | **45.1** |

We use the evaluation method of Lebanoff Logan [10] to evaluate the quality of the abstract generated by Precision (**P**), Recall (**R**), and F1-score (**F**), and compare the abstract generated by the model and the positive samples of DP selector selection. Table 1 shows the **P**, **R**, and **F** results of the sentences extracted by the DP-BERT on the CNN/DM and XSUM data sets. The results show that whether it is primary or secondary in both data sets, the **P** and **F** measures of the proposed model greatly exceed the previous baseline. Simultaneously, the **R** index is also slightly different from the baseline, which means that our model can more accurately extract sentences that match the meaning of the reference abstract.

To avoid the deviation caused by the data set, we test the model's effectiveness on two data sets. Table 2 shows the extractive summary (Ext) results of the rouge value on the CNN/DM and XSUM dataset. Obviously, whether it is the matching degree of 1-gram, 2-gram, or the longest matching substring, we have greatly exceeded the previous model on the CNN/DM dataset. We have achieved a new state-of-the-art extractive summary. On the XSUM dataset, except that the ROUGE-2 indicator is slightly lower than the baseline in the extractive summary, the remaining indicators are higher than before; this means that our model is also effective for generating short summaries.

In addition to using automatic indicators to evaluate the abstract's quality, we invited scholars with undergraduate education and above to manually

**Table 2.** The result on CNN/DM and XSUM dataset.

| Dataset | Methods | CNN/DM | | | XSUM | | |
|---------|---------|--------|--------|--------|--------|--------|--------|
| | | R-1 | R-2 | R-L | R-1 | R-2 | R-L |
| Ext | SumBasic | 34.11 | 11.13 | 31.14 | 18.56 | 2.91 | 14.88 |
| | KL-Summ | 29.92 | 10.50 | 27.37 | 16.73 | 2.83 | 13.53 |
| | LexRank | 35.34 | 13.31 | 31.93 | 17.95 | 3.00 | 14.30 |
| | BERT-SingPairMix-Ext | 41.13 | 18.68 | 37.75 | 23.53 | **4.54** | 17.23 |
| | HIBERT | 42.37 | 19.95 | 38.83 | / | / | / |
| | BERTsum | 44.41 | 20.86 | 40.55 | / | / | / |
| | DP-BERT | **46.11** | **24.05** | **42.77** | **23.69** | 4.48 | **17.75** |

**Table 3.** The result of human evaluation.

| | Systerm | Informativeness | Fluency | Succinctness |
|---------|---------|-----------------|---------|--------------|
| CNN/DM | BERT-SingPairMix | 35.8% | 36.3% | 24.2% |
| | DP-BERT | **50.0%** | **42.9%** | **41.3%** |
| Xsum | BERT-SingPairMix | 27.5% | 36.1% | **37.7%** |
| | DP-BERT | **44.6%** | **39.6%** | 34.8% |

evaluate our abstract and evaluate the abstract from three aspects: Informativeness, Fluency, and Succinctness. Table 3 shows the evaluation results. The results show that in the CNN/DM data set, we comprehensively surpassed the previous baseline in the above three aspects. Simultaneously, in the XSUM data set, we are also ahead of the baseline but slightly worse in succinctness. Overall, DP-BERT is superior to the existing approaches.

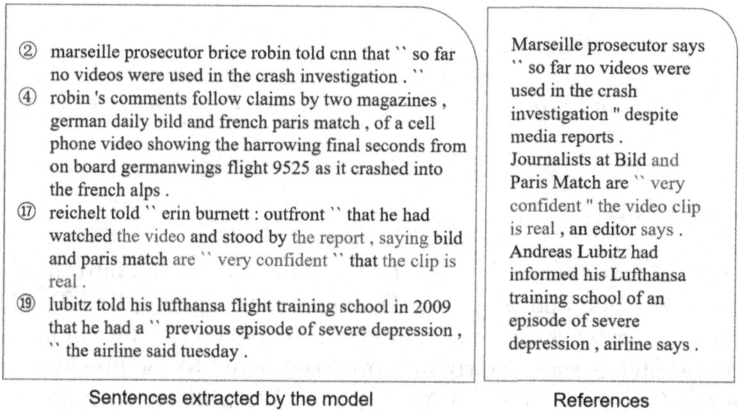

| |
|---|
| ② marseille prosecutor brice robin told cnn that `` so far no videos were used in the crash investigation . '' <br> ④ robin 's comments follow claims by two magazines , german daily bild and french paris match , of a cell phone video showing the harrowing final seconds from on board germanwings flight 9525 as it crashed into the french alps . <br> ⑰ reichelt told `` erin burnett : outfront '' that he had watched the video and stood by the report , saying bild and paris match are `` very confident '' that the clip is real . <br> ⑲ lubitz told his lufthansa flight training school in 2009 that he had a `` previous episode of severe depression , '' the airline said tuesday . |

Marseille prosecutor says `` so far no videos were used in the crash investigation " despite media reports . Journalists at Bild and Paris Match are `` very confident " the video clip is real , an editor says . Andreas Lubitz had informed his Lufthansa training school of an episode of severe depression , airline says .

Sentences extracted by the model                    References

**Fig. 4.** Case study

At the same time, as shown in Fig. 4, we did a case study. We compared the abstract sentences extracted by the model with the reference abstracts and marked the same words in the sentences with the same color. The case shows that the sentences extracted by the model can express the general meaning of the reference abstract, but there are also a few redundant sentences.

Moreover, we tested the impact of the number of extracted sentences on the quality of the abstract. As shown in Fig. 5, we tested the effects of a single sentence, double sentences, three sentences, and $W/3$ sentences on the CNNDM dataset. Among them, the single sentence test is to directly compare all the sentences in the original text with the reference abstract, and the two sentences adopt the method of Lebanoff Logan [10]; the three sentences and $W/3$ are all extracted by DPBERT, and the number of extracted sentences is limited to 3 and $W/3$ respectively. The results show that as the number of restricted extracted sentences increases, the three groups' Rouge values all increase to a certain extent. However, the limited number of extracted sentences is too large, and the Rouge value is not improved much. Therefore, to balance the summary effect and calculation difficulty to a certain extent, we will link the limited number of extracted sentences with the sentence length, taking $W/3$.

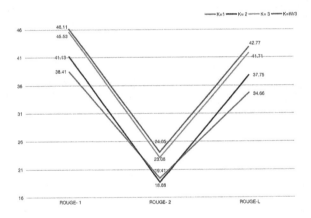

**Fig. 5.** The impact of different sentence limits on the summary results.

## 5   Conclusion

In this work, we introduce a dynamic programming algorithm into the summary sentence classification and propose a Dynamic Programming BERT, which can extract abstract sentences more accurately and completely. The summary generated by DP-BERT is significantly improved compared to the previous baselines based on the CNN/DM and XSUM data sets. We will further improve redundancy in the abstract in future work, hoping to produce a more concise summary.

**Acknowledgement.** This research was partially supported by the Key Program of the National Natural Science Foundation of China under Grant No. U1903213, the Guangdong Basic and Applied Basic Research Foundation (No. 2019A1515011387). In addition, thanks to Tencent Pattern Recognition Center and Yankai Lin for their support in the research process of this article.

# References

1. Chen, Y.C., Bansal, M.: Fast abstractive summarization with reinforce-selected sentence rewriting. In: Proceedings of the 56th Annual Meeting of the Association for Computational Linguistics (Volume 1: Long Papers), pp. 675–686. Association for Computational Linguistics (2018)
2. Cheng, J., Lapata, M.: Neural summarization by extracting sentences and words. In: Proceedings of the 54th Annual Meeting of the Association for Computational Linguistics (Volume 1: Long Papers), pp. 484–494. Association for Computational Linguistics (2016)
3. Cui, P., Hu, L., Liu, Y.: Enhancing extractive text summarization with topic-aware graph neural networks. In: Proceedings of the 28th International Conference on Computational Linguistics, COLING 2020, pp. 5360–5371. International Committee on Computational Linguistics (2020)
4. Devlin, J., Chang, M.W., Lee, K., Toutanova, K.: BERT: pre-training of deep bidirectional transformers for language understanding. In: Proceedings of the 2019 Conference of the North American Chapter of the Association for Computational Linguistics: Human Language Technologies, Volume 1 (Long and Short Papers), pp. 4171–4186. Association for Computational Linguistics, Minneapolis, Minnesota (June 2019)
5. Erkan, G., Radev, D.R.: Lexrank: graph-based lexical centrality as salience in text summarization. J. Artif. Intell. Res. **22**, 457–479 (2004)
6. Fernandes, P., Allamanis, M., Brockschmidt, M.: Structured neural summarization. In: 7th International Conference on Learning Representations, ICLR 2019. OpenReview.net (2019)
7. Haghighi, A., Vanderwende, L.: Exploring content models for multi-document summarization. In: Proceedings of Human Language Technologies: The 2009 Annual Conference of the North American Chapter of the Association for Computational Linguistics, pp. 362–370 (June 2009)
8. Hermann, K.M., et al.: Teaching machines to read and comprehend. In: Advances in Neural Information Processing Systems, vol. 28, pp. 1693–1701. Curran Associates, Inc. (2015)
9. Jones, K.S.: Automatic summarising: the state of the art. Inf. Process. Manag. **43**(6), 1449–1481 (2007)
10. Lebanoff, L., et al.: Scoring sentence singletons and pairs for abstractive summarization. In: Proceedings of the 57th Annual Meeting of the Association for Computational Linguistics, pp. 2175–2189. Association for Computational Linguistics (2019)
11. Liu, Y.: Fine-tune BERT for extractive summarization. CoRR abs/1903.10318 (2019)
12. Liu, Y., Lapata, M.: Text summarization with pretrained encoders. In: Proceedings of the 2019 Conference on Empirical Methods in Natural Language Processing and the 9th International Joint Conference on Natural Language Processing (EMNLP-IJCNLP), pp. 3730–3740. Association for Computational Linguistics (November 2019)

13. Liu, Y., Titov, I., Lapata, M.: Single document summarization as tree induction. In: Proceedings of the 2019 Conference of the North American Chapter of the Association for Computational Linguistics: Human Language Technologies, Volume 1 (Long and Short Papers), pp. 1745–1755 (2019)

14. Liu, Z., Shi, K., Chen, N.F.: Conditional neural generation using sub-aspect functions for extractive news summarization. In: Cohn, T., He, Y., Liu, Y. (eds.) Proceedings of the 2020 Conference on Empirical Methods in Natural Language Processing: Findings, EMNLP 2020, pp. 1453–1463. Association for Computational Linguistics (2020)

15. Nallapati, R., Zhai, F., Zhou, B.: Summarunner: a recurrent neural network based sequence model for extractive summarization of documents. In: Proceedings of the Thirty-First AAAI Conference on Artificial Intelligence, pp. 3075–3081. AAAI Press (2017)

16. Narayan, S., Cohen, S.B., Lapata, M.: Don't give me the details, just the summary! topic-aware convolutional neural networks for extreme summarization. In: Proceedings of the 2018 Conference on Empirical Methods in Natural Language Processing, pp. 1797–1807. Association for Computational Linguistics (2018)

17. Narayan, S., Cohen, S.B., Lapata, M.: Ranking sentences for extractive summarization with reinforcement learning. In: Proceedings of the 2018 Conference of the North American Chapter of the Association for Computational Linguistics: Human Language Technologies, Volume 1 (Long Papers), pp. 1747–1759. Association for Computational Linguistics (2018)

18. Shen, D., Sun, J., Li, H., Yang, Q., Chen, Z.: Document summarization using conditional random fields. In: Veloso, M.M. (ed.) IJCAI 2007, Proceedings of the 20th International Joint Conference on Artificial Intelligence, pp. 2862–2867 (2007)

19. Tan, J., Wan, X., Xiao, J.: Abstractive document summarization with a graph-based attentional neural model. In: Proceedings of the 55th Annual Meeting of the Association for Computational Linguistics (Volume 1: Long Papers), pp. 1171–1181. Association for Computational Linguistics (2017)

20. Vanderwende, L., Suzuki, H., Brockett, C., Nenkova, A.: Beyond sumbasic: task-focused summarization with sentence simplification and lexical expansion. Inf. Process. Manage. **43**(6), 1606–1618 (2007)

21. Yuan, R., Wang, Z., Li, W.: Fact-level extractive summarization with hierarchical graph mask on BERT. In: Proceedings of the 28th International Conference on Computational Linguistics, COLING 2020, pp. 5629–5639. International Committee on Computational Linguistics (2020)

22. Zhang, X., Wei, F., Zhou, M.: HIBERT: Document level pre-training of hierarchical bidirectional transformers for document summarization. In: Proceedings of the 57th Annual Meeting of the Association for Computational Linguistics, pp. 5059–5069. Association for Computational Linguistics, Florence, Italy (July 2019)

23. Zheng, H., Lapata, M.: Sentence centrality revisited for unsupervised summarization. In: Korhonen, A., Traum, D.R., Màrquez, L. (eds.) Proceedings of the 57th Conference of the Association for Computational Linguistics, ACL 2019, Volume 1: Long Papers, pp. 6236–6247. Association for Computational Linguistics (2019)

24. Zhong, M., Liu, P., Chen, Y., Wang, D., Qiu, X., Huang, X.: Extractive summarization as text matching. In: Proceedings of the 58th Annual Meeting of the Association for Computational Linguistics, ACL 2020, pp. 6197–6208 (2020)

**Robotics**

# Research on Obstacle Avoidance Path Planning of Manipulator Based on Improved RRT Algorithm

Tianying Hu[✉] [iD]

Beijing University of Technology, Beijing 100124, China
hutianying@emails.bjut.edu.cn

**Abstract.** In the field of obstacle avoidance path planning, the traditional Rapidly-Exploring Random Tree (RRT) algorithm has many problems, such as no direction and low efficiency. So it is often used to adjust the growth direction of random tree nodes by introducing a target bias strategy to decrease the search blindness. On this basis, the end movement distance and the variation range of each joint during the manipulator trajectory planning process have been focused on in this paper. Considering the requirements of the speed of the planning executable trajectory and the smoothness of the moving process, a cost function about the path length and the smooth change of the joints has been designed. Then, under the premise of the stability of the path planning results, an improved RRT algorithm on dynamical adjustment of the new nodes generation has been proposed to increase the planning efficiency obviously. Its feasibility and effectiveness have been verified fully by a series of simulation experiments based on MATLAB platform.

**Keywords:** Joint space · Obstacle avoidance path planning · RRT algorithm · Dynamic step size control · Greedy strategy

## 1 Introduction

Nowadays, the manipulator occupies a large market in the field of industrial robots because of its simple structure and flexible work. In the application, how to ensure that the manipulator in the workspace quickly planning a smooth and collision-free path will be the key issue. So far, there have been a variety of path planning algorithms, such as Rapidly-Exploring Random Tree (RRT) algorithm [1], artificial potential field method [2], ant colony algorithm [3] and A* algorithm [4]. But there are still some problems in certain aspects. For example, RRT algorithm has high randomness and slow convergence speed in planning, so its planned path is not optimal. Due to the lack of global information, the artificial potential field method is easy to fall into the local minimum [5]. Ant colony algorithm is mostly used in two-dimensional space or path planning of

Supported by Beijing University of Technology.

L. Fang et al. (Eds.): CICAI 2021, LNAI 13070, pp. 299–310, 2021.
https://doi.org/10.1007/978-3-030-93049-3_25

mobile robots. With the growth of the search dimension, the optimization efficiency of the algorithm decreases. As a heuristic algorithm, the spatial growth of A* algorithm is exponential. Much more time would be taken to get higher accuracy. In comparison, for path planning in multi-joint high-dimensional space like manipulators, RRT algorithm can generally plan a solution to the problem. However, the planned solution is only a feasible solution, not an ideal or optimal path, so it is unavoidable to have too many unnecessary path segments.

In order to optimize the quality of RRT planning path, the idea of cost function from the path distance and motion smoothness [6] as the guidance of later path search has been adapted. To improve the utilization of sampling nodes and reduce the number of iterations, some scholars have proposed an improved RRT algorithm based on variable sampling domain and Map Compression Algorithm for mobile robots [7], and the combination of variable sampling domain and greedy strategy can improve the sampling efficiency and save time. Literature [8] has proposed a greedy heuristic search algorithm in the search process effectively to make the newly generated nodes expand towards the target point continuously. Literature [9] has added the idea of dynamic step size characteristic, aiming at making the random tree expand as far as possible toward the target point when there are no obstacles, but there is no concrete implementation method in this article. In subsequent studies, an adaptive step size strategy [10] has been proposed. Only when there is no collision, the target gravity strategy is adopted to expand the random tree. To solve the problem of large differences in search path results of RRT algorithm and many invalid traversals, literature [11] has proposed the RRT* algorithm of adding prior knowledge, providing a low-cost solution with fewer iterations.

In this paper, a new step size control strategy is proposed based on the probability bias RRT algorithm in which the cost function has been added. It can reduce the cost and blindness of the path planning of the manipulator in joint space, and shorten the search time and the length of the obstacle avoidance path. In the case that random points are the target points, the greedy idea is used to further improve the search efficiency.

## 2    Modeling and Kinematics Analysis of Manipulator

The research object of this paper is a manipulator with 3-DOF series joint structure. The three revolute joints are the lumbar (lower back), shoulder (upper arm) and elbow (lower arm).

The manipulator is modeled by using the standard Denavit-Hartenberg (D-H) notation, as shown in Fig. 1. Table 1 is a list of DH parameters based on the mechanical structure and the link coordinate system. Where $\alpha_i$ is the torsion angle, which is a constant determined by the properties of the connecting rod; $a_i$ is the length of connecting rod; $d_i$ represents the bias distance of the connecting rod; $\theta_i$ is the joint angle.

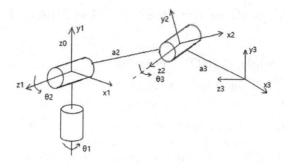

**Fig. 1.** DH modeling of the manipulator

**Table 1.** DH parameters of each link of the manipulator

| $i$ | $\alpha_i$ | $a_i$ | $d_i$ | $\theta_i$ |
|---|---|---|---|---|
| 1 | 90° | 0 | 0 | $\theta_1$ |
| 2 | 0 | 250 | 0 | $\theta_2$ |
| 3 | 0 | 250 | 0 | $\theta_3$ |

Since only forward kinematics [12] is used in the search process of RRT algorithm in joint space, only forward kinematics is analyzed. The homogeneous transformation matrix of two adjacent connecting rod coordinate systems is obtained through four standard motion transformations:

$$T_i^{i-1} = A_i = \text{Rot}\,(z_{i-1}, \theta_i) \times \text{Trans}\,(0, 0, d_i) \times \text{Trans}\,(a_i, 0, 0) \times \text{Rot}\,(x_i, \alpha_i)$$
$$= \begin{bmatrix} \cos\theta_i & -\sin\theta_i\cos\alpha_i & \sin\theta_i\sin\alpha_i & a_i\cos\theta_i \\ \sin\theta_i & \cos\theta_i\cos\alpha_i & -\cos\theta_i\sin\alpha_i & a_i\sin\theta_i \\ 0 & \sin\alpha_i & \cos\alpha_i & d_i \\ 0 & 0 & 0 & 1 \end{bmatrix} = \begin{bmatrix} n_x & o_x & a_x & p_x \\ n_y & o_y & a_y & p_y \\ n_z & o_z & a_z & p_z \\ 0 & 0 & 0 & 1 \end{bmatrix}$$

$$(1)$$

where $Rot(i, y)$ indicates that the previous joint rotates $y$ angles around the $i$ axis. $Trans(x, y, z)$ represents the translation along a certain axis of $x$, $y$ or $z$, and the non-zero parameter represents the translation distance. So $T_i^{i-1}$ represents the relative pose of the connecting rod $i$ in the member coordinate system $i - 1$.

For a serial manipulator with three joints, the homogeneous transformation matrices of each connecting rod are A1, A2 and A3 respectively, so the end pose of the manipulator arm can be calculated by Eq. (2).

$$T_3^0 = T_1^0 T_2^1 T_2^3 = A_1 A_2 A_3$$

$$(2)$$

In the formula, $T_3^0$ is the pose matrix of the end-effector coordinate system changing in the base coordinate system.

# 3 Search Algorithm for Obstacle Avoidance Path Planning

## 3.1 Basic Principle of RRT Algorithm

The idea of the RRT algorithm is to rapidly expand a group of tree-like paths to fill the search area. Figure 2 shows the expansion process.

- Put the starting point into the tree before searching. At this time, there is only one node $Qrand$ as the start of tree growth.
- In the configuration space of the manipulator, a random point $Qrand$ is generated. $Qrand$ is selected with a small probability as the target point.
- Traverse the existing nodes in the random tree. Select the node closest to $Qrand$ as $Qnearest$, which is the parent node of $Qnew$.
- Find the direction vector $(Qrand-Qnearest)/\|Qrand-Qnearest\|$ between $Qnearest$ and $Qrand$, and generate a new node $Qnew$ in this direction with a step $\Delta q$.
- Perform collision detection. If there is no collision with the obstacle, add $Qnew$ to the random tree.
- Repeat the above steps. When it reaches the target threshold, save the target point in the tree, and let the search stop.
- Backtrack the parent node from the target point to find a feasible path from the starting point to the target point.

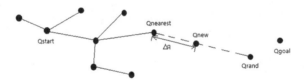

**Fig. 2.** Schematic diagram of random tree expansion

It can be seen from the above description that the growth direction of new nodes completely depends on the selection of random points. Therefore, there will be problems such as aimlessness and tortuous paths in the application.

## 3.2 Establishment of the Cost Function

In the joint space RRT algorithm, each step changes a set of joint angles of the manipulator. Therefore, the planning of the obstacle avoidance path should take two factors into account: the end-motion distance of the manipulator and the intensity of the angle change of each joint.

Assuming $F\_D$ is the parameter to measure the distance between the end of the manipulator, and the spatial pose of each joint angle is obtained through forward kinematics, then the distance between the end position of the manipulator and the random point can be expressed by Euclidean distance:

$$F\_D = \|\text{pxyzQrand} - \text{pxyzQtree}\| \tag{3}$$

where $pxyzQrand$ represents the spatial position of random points obtained by forward kinematics, that is, $p_x$, $p_y$, $p_z$ in homogeneous transformation matrix; $pxyzQtree$ represents the spatial position of each existing node in the tree. The smaller the value of $F\_D$, the shorter the distance between two points.

Assuming $F\_A$ is the standard to measure the joint change amplitude of the manipulator, in order to reach the joint angle represented by the random point, the joint angle variation of nodes in the tree can be expressed by Euclidean distance (Eq. (4)) or Manhattan distance (Eq. (5)).

$$F\_A = \|\theta_{Qtree} - \theta_{Qrand}\| \tag{4}$$

$$F\_A = \left| \sum_{i=1}^{3} \theta_{Qtree,i} - \theta_{Qrand,i} \right| \tag{5}$$

where $\theta_{Qtree,i}$ and $\theta_{Qrand,i}$ represent the joint angles of the $i(i = 1, 2, 3)$th joint of node and random point respectively. The smaller the $F\_A$ is, the more stable the change of planning joint is.

In the process of searching, the importance of the two evaluation criteria is determined by the actual demand. Different weights $W\_D$ and $W\_A$ are given to the distance change and angle change respectively, and the cost function is described in the form of Eq. (6).

$$F\_N = W\_D * F\_D + W\_A * F\_A \tag{6}$$

The smaller the value of the cost function is, the better the comprehensive evaluation result of the change of the end distance and the angle of each joint is. Thus, the node with the minimum function value is selected as $Qnearest$ node to be expanded in the tree.

### 3.3  Improved Dynamic Step Method to Generate $Qnew$

The generation of new nodes in traditional RRT algorithm depends on the direction vector from the nearest node to the random point and the step length growing along the direction of the random point. For the joint space, the calculation formula is shown in Eq. (7).

$$Qnew = Qnearest + \Delta\theta \times (Qrand - Qnearest)/\|Qrand - Qnearest\| \tag{7}$$

where $\Delta\theta$ is the step size increased each time. It can be seen that the generation of new nodes is completely determined by random points, which is too blind.

Therefore, the method of adding virtual attraction points can avoid the manipulator falling into the local minimum [13], and literature [14] has integrated it into RRT algorithm. The target gravity function is introduced to make the random tree grow towards the target point, it can effectively reduce unnecessary expansion during the search. The mathematical formula of adding the objective gravitational function is as follows:

$$\begin{aligned} Qnew =&\, Qnearest + \Delta\theta_1 \times (Qrand - Qnearest)/\|Qrand - Qnearest\| \\ &+ \Delta\theta_2 \times (Qgoal - Qnearest)/\|Qgoal - Qnearest\| \end{aligned} \tag{8}$$

where $\theta_1$ is the growth step in the direction of the random point, and $\theta_2$ is the growth step in the direction of the target point.

When $\theta_1$ is greater than $\theta_2$, the generation of new nodes tends to the direction of random points. When $\theta_2$ is greater than $\theta_1$, the random tree will expand to the target direction as much as possible. But for the environment with more obstacles, the target gravity method can not bypass the obstacles to reach the target point, and a collision occurs in the search process, and the obstacle avoidance path cannot be planned.

On how to control the new node as far as possible toward the direction of target generated without collision, this paper uses the basic idea of dynamic step: the generation of new node $Qnew$ is determined by two directions. Based on the idea of target gravity, a dynamic change quantity is set in each direction, makes the bias probability in the target direction increase iteratively, and the bias probability in the random point direction decrease correspondingly. Iteration stops when the first collision occurs. As shown in Eq. (9).

$$Qnew = Qnearest + (1 - t) \times \varDelta\theta_1 \times (Qrand - Qnearest)/\|Qrand - Qnearest\|$$
$$+ t \times \varDelta\theta_2 \times (Qgoal - Qnearest)/\|Qgoal - Qnearest\| \tag{9}$$

where $t$ is the proposed dynamic change. At first, $t = 0$, while $t \leq 1$, perform iterative judgment. During each iteration, collision detection is performed. When the pose of the connecting rod formed by the joint angle of the new node does not collide with obstacles, increase the value of $t$. The principle of increasing $t$ is: take the total number of judgments as $N$, $N$ is a positive integer, and $dt$ is the single change of $t$. Then, when $t \leq 1$ and collision detection is satisfied, there are:

$$\begin{cases} dt = 1/N \\ t = t + dt \end{cases} \tag{10}$$

In the process of iterative judgment, the generated node is a temporary node, which will not be stored in the random tree, and it will be overwritten by the nodes produced by the next round of judgment. Until $t = 1$ and there is no collision, or exit the cycle due to collision, the last generated new node is added to the tree. The principle of the improved dynamic step size method is shown in Fig. 3 (assuming $\varDelta\theta_1 = \varDelta\theta_2$).

### 3.4   Local Greedy Strategy

In order to further shorten the search time of the improved RRT algorithm, enhance the guidance of the random tree growth process, and reduce the number of path nodes generated, we consider adding greedy strategy in the early stage of the search. Literature [15] proposes a guided RRT path planner, which reduces the search time under dynamic constraints and avoids the expansion of most futile nodes. Considering that the greedy algorithm's node growth mode makes each branch longer, the planned path contains multiple linear path segments, and the result is unstable due to continuous rotation of joints in a specific direction,

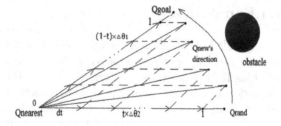

**Fig. 3.** Control principle of the improved dynamic step method

therefore, a greedy strategy is proposed to guide the generation of new nodes only when the sampling point is the target point in this paper. That is: in each iteration, when $Qrand = Qgoal$, jump into greedy search subroutine. In the greedy algorithm, the new node is calculated as follows:

$$Qnew1 = Qnearest + \Delta\theta_2 \times (Qgoal - Qnearest)/\|Qgoal - Qnearest\| \quad (11)$$

When $Qnew$ does not reach the threshold range of the target point, collision detection is carried out. If there is no collision, $Qnew$ is updated with Eq. (12). The new nodes generated in the intermediate process are not saved in the tree, they are represented as $Qnew1$.

$$Qnew1 = Qnew1 + \Delta\theta_2 \times (Qgoal - Qnew1)/\|Qgoal - Qnew1\| \quad (12)$$

When the collision occurs, take the last non-collision $Qnew$ and save it in the tree. Greedy subroutine exits. The calculation formula of $Qnew$ is as follows.

$$Qnew1 = Qnew1 - \Delta\theta_2 \times (Qgoal - Qnew1)/\|Qgoal - Qnew1\| \quad (13)$$

The principle of greedy strategy to add new nodes is shown in Fig. 4.

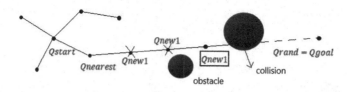

**Fig. 4.** Greedy strategy adds new nodes

When the random sampling point is not the target point, the improved dynamic step method is used to expand the random tree.

## 4    Simulation Experiment

The manipulator is modeled on MATLAB software platform. The two improved strategies proposed in this paper are compared with the existing algorithms, such as probability bias RRT algorithm and target gravity RRT algorithm, and the effect of optimization is analyzed from multiple perspectives.

In the three-dimensional space, obstacles are represented by the sphere envelope method. Each obstacle has a radius($R$) of 0.1 m; the collision safety distance is $R + 0.03$ meter; the starting point is $(0°, 0°, 0°)$; the target point is $(120°, 60°, -30°)$; the bias probability $p$ is 0.1; The step size is increased by 5° each time, and $N$ is set to 10. In the same obstacle environment and parameter settings, the four algorithms are run. Figure 5, 6, 7, 8, 9, 10, 11, 12, 13, 14, 15, Fig. 16 respectively show the simulation results of probability bias RRT, target gravity RRT algorithm, improved dynamic step RRT and local greedy-dynamic step RRT algorithms.

According to the situation results, the path length, the number of path nodes and the search time corresponding to each group of experiments are compared, and the data in Table 2 are obtained.

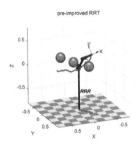 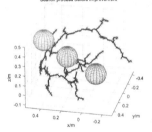 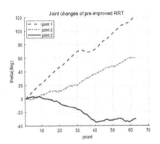

**Fig. 5.** Probability bias RRT

**Fig. 6.** Search process of probability bias RRT

**Fig. 7.** Joint changes of probability bias RRT

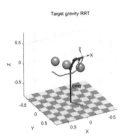 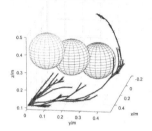 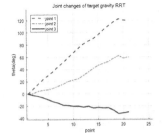

**Fig. 8.** Target gravity RRT

**Fig. 9.** Search process of target gravity RRT

**Fig. 10.** Joint changes of target gravity RRT

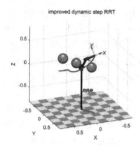

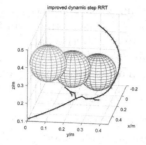

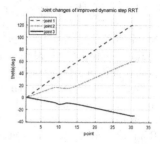

**Fig. 11.** Improved dynamic step RRT

**Fig. 12.** Search process of improved dynamic step RRT

**Fig. 13.** Joint changes of improved dynamic step RRT

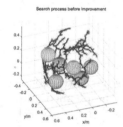

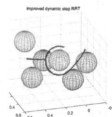

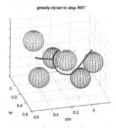

**Fig. 14.** Greedy-dynamic step RRT

**Fig. 15.** Search process of greedy-dynamic step RRT

**Fig. 16.** Joint changes of greedy-dynamic step RRT

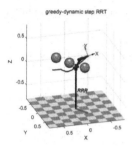

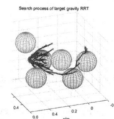

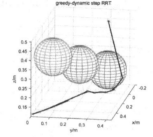

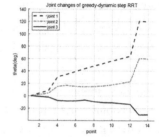

**Fig. 17.** Probability bias RRT

**Fig. 18.** Target gravity RRT

**Fig. 19.** Improved dynamic step RRT

**Fig. 20.** Greedy-dynamic step RRT

**Table 2.** Parameter comparison of four RRT algorithms

| Algorithm name | Path length/m | Number of nodes | Search time/s |
|---|---|---|---|
| Probability bias RRT | 1.2866 | 63 | 0.5939 |
| Target gravity RRT | 1.1774 | 22 | 0.3963 |
| Improved dynamic step RRT | 1.0670 | 31 | 0.2357 |
| Greedy-dynamic step RRT | 1.0397 | 14 | 0.1167 |

Each of the four algorithms runs 20 times, and the average parameters obtained by each algorithm are calculated to obtain the data in Table 3. It can be seen from the running results and the data in Table 3 that local greedy-dynamic step RRT algorithm has the shortest search time and the least path nodes. The average path length of the local greedy-dynamic step RRT algorithm is slightly smaller than that of the improved dynamic step RRT algorithm and much smaller than the other two algorithms. The smoothest change of joint angle is improved dynamic step RRT algorithm. Compared with the probability bias RRT algorithm and target gravity RRT algorithm, the performances of the two improved algorithms are all improved a lot.

When the obstacle environment is more complex, update the starting and target points. Improved RRT algorithm is more likely to jump out of the narrow channel, so as to avoid the search of a large number of redundant space, and the planned path is more smooth and simplified. The simulation results are shown in Fig. 17, 18, 19, Fig. 20.

**Table 3.** Average parameters of the four algorithms

| Algorithm name | Path length/m | Number of nodes | Search time/s |
|---|---|---|---|
| Probability bias RRT | 1.3123 | 64 | 0.4314 |
| Target gravity RRT | 1.1689 | 25 | 0.3753 |
| Improved dynamic step RRT | 1.0713 | 31 | 0.2326 |
| Greedy-dynamic step RRT | 1.0430 | 15 | 0.0967 |

## 5  Conclusion

The traditional RRT algorithm can explore all unknown regions with enough iterations. But its blindness and randomness are too large. The planned path is always not ideal. This paper designs a multi-objective cost function, which adjusts the weight of each factor independently according to the target demand. It can reduce the randomness of RRT algorithm partly. The path planned by the improved dynamic step method has the highest smoothness and greatly reduces the path length. Compared with target gravity RRT algorithm, the improved dynamic step RRT algorithm overcomes the disadvantage of weak obstacle avoidance ability, reduces the search of redundant space, and can jump out of the narrow exit quickly in the case of a complex obstacle environment, so as to ensure the success rate of path planning. At the same time, the greedy strategy has been adopted to ensure the smoothness of the path while sacrificing some randomness. The search speed has been accelerated because some intermediate nodes are avoided. Therefore, the improved methods proposed in this paper is effective and feasible. It has laid a theoretical foundation for the manipulator to realize the fast and stable trajectory execution without collision.

# References

1. LaValle, Rapidly-Exploring Random Trees: A New Tool for Path Planning. Technical Report Computer Science Department, Iowa State University, pp. 67–74 (1998)
2. Lee, M.C., Park, M.G.: Artificial potential field based path planning for mobile robots using a virtual obstacle concept. In: Proceedings 2003 IEEE/ASME International Conference on Advanced Intelligent Mechatronics (AIM 2003), vol. 2, pp. 735–740. Kobe, Japan (2003) . https://doi.org/10.1109/AIM.2003.1225434
3. Huadong, Z., Chaofan, L., Nan, J.: A path planning method of robot arm obstacle avoidance based on dynamic recursive ant colony algorithm. In: 2019 IEEE International Conference on Power, Intelligent Computing and Systems (ICPICS), pp. 549–552. Shenyang, China (2019) . https://doi.org/10.1109/ICPICS47731.2019.8942495
4. Ju, C., Luo, Q., Yan, X.: Path planning using an improved a-star algorithm. In: 2020 11th International Conference on Prognostics and System Health Management (PHM-2020 Jinan), pp. 23–26. Jinan, China (2020). https://doi.org/10.1109/PHM-Jinan48558.2020.00012
5. Hui, G., Lv, Z.: Research on inaccessible aim of artificial potential field method. Foreign Electron. Measur. Technol. **37**(01), 29–33 (2018)
6. Aravindan, A., Zaheer, S., Gulrez, T.: An integrated approach for path planning and control for autonomous mobile robots. In: 2016 International Conference on Next Generation Intelligent Systems (ICNGIS), pp. 1–6. Kottayam (2016). https://doi.org/10.1109/ICNGIS.2016.7854041
7. Wang, Z., Shan, L., Chang, L., Qiu, B., Qi, Z.: Variable sampling domain and map compression based on greedy RRT algorithm for robot path planning. In: 2020 39th Chinese Control Conference (CCC), pp. 3915–3919. Shenyang, China (2020) . https://doi.org/10.23919/CCC50068.2020.9188379
8. Chen, L., Yu, L., Libin, S., Jiwen, Z.: Greedy BIT* (GBIT*): greedy search policy for sampling-based optimal planning with a faster initial solution and convergence. In: 2021 International Conference on Computer, Control and Robotics (ICCCR), pp. 30–36. Shanghai, China (2021). https://doi.org/10.1109/ICCCR49711.2021.9349403
9. Lin, N., Zhang, Y.: An adaptive RRT based on dynamic step for UAVs route planning. In: 2014 IEEE 5th International Conference on Software Engineering and Service Science, pp. 1111–1114. Beijing, China (2014). https://doi.org/10.1109/ICSESS.2014.6933760
10. Xue, Y., Zhang, X., Jia, S., Sun, Y., Diao, C.: Hybrid bidirectional rapidly-exploring random trees algorithm with heuristic target graviton. In: 2017 Chinese Automation Congress (CAC), pp. 4357–4361. Jinan (2017). https://doi.org/10.1109/CAC.2017.8243546
11. Mashayekhi, R., Idris, M.Y.I., Anisi, M.H., Ahmedy, I., Ali, I.: Informed RRT*-connect: an asymptotically optimal single-query path planning method. In: IEEE Access, vol. 8, pp. 19842–19852 (2020). https://doi.org/10.1109/ACCESS.2020.2969316
12. Ayob, M.A., Zakaria, W.N.W., Jalani, J.: Forward kinematics analysis of a 5-axis RV-2AJ robot manipulator. In: 2014 Electrical Power, Electronics, Communicatons, Control and Informatics Seminar (EECCIS), pp. 87–92. Malang (2014). https://doi.org/10.1109/EECCIS.2014.7003725

13. Rafsanzani, A.R., Hidayat, R.C., Cahyadi, A.I., Herdjunanto, S.: Omnidirectional sensing for escaping local minimum on potential field mobile robot path planning in corridors environment. In: 2018 3rd International Seminar on Sensors, Instrumentation, Measurement and Metrology (ISSIMM), pp. 79–83. Depok, Indonesia (2018) . https://doi.org/10.1109/ISSIMM.2018.8727639
14. Otani, T., Koshino, M.: Applying a path planner based on RRT to cooperative multirobot box-pushing. Artif. Life Robot. **13**(2), 418–422 (2009). https://doi.org/10.1007/s10015-008-0592-7
15. Zhang, J., Wisse, M., Bharatheesha, M.: Guided RRT: a greedy search strategy for kinodynamic motion planning. In: 2014 13th International Conference on Control Automation Robotics & Vision (ICARCV), pp. 480–485. Singapore (2014). https://doi.org/10.1109/ICARCV.2014.7064352

# Visual Odometer Algorithm Based on Dynamic Region Culling

Hongwei Mo$^{(\boxtimes)}$ and Xifeng Zhang

College of Intelligent Systems Science and Engineering,
Harbin Engineering University, Harbin 150001, China

**Abstract.** There are many moving objects in the dynamic scenes. Due to excessive changes in feature points on dynamic objects, large positioning errors will be caused, which will have a great impact on mapping. Existing algorithms have well operating accuracy. However, when there are static and moving prior masks in the scene, the selection error of the dynamic region common occurs. To address the problem, we propose a method that fuses geometric feature and semantic segmentation, and the re-discriminant mechanism is used to dynamic scene recognition. In this work, the pyramid optical flow method is used to track the feature points between matching frames. The dynamic points will be filtered through geometric constraints. When the pre-selected dynamic points are in the prior mask, the pre-selected dynamic area is delineated. Then, when the number of dynamic points in the mask exceeds the threshold, the dynamic area is finally located. This eliminates the impact of dynamic objects on the accuracy of the system. Experiments on the TUM datasets show that the proposed algorithm can effectively improve the robustness and tracking accuracy than the ORB_SLAM2 system, meanwhile it can effectively solve the miss election of the dynamic region.

**Keywords:** Dynamic scene · Semantic segmentation · Epipolar geometric constraints · Dynamic region elimination

## 1 Introduction

For a long time, the constructed maps by SLAM technology are mostly static and pure environmental objects. In other words, a basic assumption of SLAM technology is that the environment is static. It can only handle a small amount of dynamic content, the solution is to mark them as outliers of static models [1–6]. However, moving objects like humans exist in many real scenes.

Existing algorithms are basically looking for data associations between the previous frame information and the environmental map that is assumed to be static. When a mobile robot is working in a complex dynamic scene, highly dynamic obstacles will cause errors in the corresponding relationship. These errors will cause huge inconsistencies in the adjacent frames and then greatly

This research is supported by the Fundamental Research Funds for the Central Universities, No.3072020CFT0402.

L. Fang et al. (Eds.): CICAI 2021, LNAI 13070, pp. 311–322, 2021.
https://doi.org/10.1007/978-3-030-93049-3_26

affect the process of system tracking and pose estimation. So the method originally designed to execute SLAM in a static environment can no longer handle serious dynamic scenes and provide enough effective information for subsequent work. Therefore, how to eliminate the impact of dynamic targets in the scene on the SLAM system, and improve the robustness of the SLAM system has become the major challenge in the practical application of SLAM systems. Among them, the identification of dynamic areas is a breakthrough in dealing with dynamic environments. The proposed methods in recent years mainly include: multi-view geometric constraints; scene flow; foreground or background detection methods; fusion deep learning methods, etc.

Tan et al. [7] used the prior adaptive random sampling consensus algorithm [8] to remove abnormal points in the scene, then estimated the more accurate camera poses. But this method is not suitable for high dynamic scenes. Li et al. [9] detected dynamic targets in actual scenes based on optical flow and image super-pixel segmentation. But the dense optical flow method requires a large amount of calculation, this is difficult to calculate in real time. And this method is greatly affected by changes in ambient lighting. Wang et al. [10] improved the traditional optical flow method to detect and process moving objects in the scene, thereby reducing the impact of dynamic objects. But the improvement of positioning and mapping accuracy is limited. And it uses the dense optical flow method, which makes the system unable to run in real time. Han et al. [11] proposed an improved PSPNet-SLAM based on ORB-SLAM2, which combined the PSPNet network and optical flow method for the detection of dynamic features. This method first filtered out the features with larger optical flow values, and then filtered out the features that are judged to be dynamic objects. This method achieved higher accuracy on the TUM data set, effectively reduced tracking drift and improved the robustness of visual SLAM in dynamic environments. In addition to the above methods, many algorithms applied to dynamic scenes have been proposed [12–18].

There are two more classic methods currently recognized for processing dynamic scenes: Chao Yu et al. [19] improved the traditional ORB-SLAM2 algorithm and proposed a complete semantic SLAM system (DS-SLAM) in dynamic scenarios. Berta et al. [20] proposed the DynaSLAM algorithm. Both of these algorithms have the characteristics of high precision and high robustness. However, they have some disadvantages. For example, DS-SLAM algorithm usually uses human-being as prior moving objects. And the feature points within the mask range of both will be eliminated when there are static and moving prior masks in the scence. DynaSLAM applies the Mask RCNN algorithm in semantic segmentation thread. But the running speed of this algorithm is slow and it cannot meet the real-time requirements. This method chooses to remove all objects with potential movement, for the prior mask that does not actually move, it is often removed for no reason. This may result in too few remaining stationary feature points and affect the camera pose estimation.

This research is based on a dynamic scene, and at the same time, this studies the situation that when there are both a moving and a non-moving prior mask in the scene, the recognition error causes the feature points in the static mask to be eliminated. It is to ensure that the robot is working in the camera's vision range and there are dynamic objects, it can avoid the interference of dynamic features and construct a static environment map that can be used by the robot to complete the subsequent required work.

## 2   Visual Odometer in Dynamic Scene

### 2.1   Algorithm Framework

Due to the flexible deformation and complex motion of moving objects such as the human body, it is difficult to extract a complete dynamic area contour with a simple motion consistency inspection method. And it takes a long time to complete the extraction of the entire contour. Using a semantic segmentation network, the complete outline of the target is easy to obtain. Therefore, the combination of motion consistency detection and deep learning is the currently ideal choice. Our algorithm is based on the depth camera, with ORB_SLAM2 as the main line for improvement. The semantic segmentation method and geometric constraint are combined to work together to detect dynamic objects, and then identify dynamic regions and add dynamic masks to eliminate dynamic points. The specific implementation process is shown in Fig. 1.

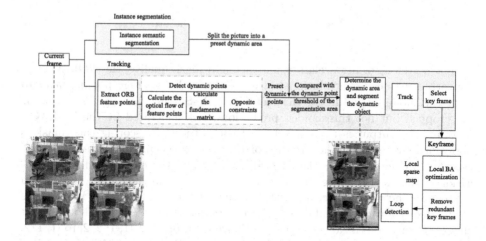

**Fig. 1.** System structure framework

Our algorithm mainly improves the fusion method and adds a re-discrimination mechanism. In the tracking thread, the F matrix is calculated between two frames. The RANSAC algorithm is used for classification first, and feature points are divided into candidate dynamic points and static points. Afterwards, it is detected whether the preselected dynamic point is within the range of the delivered prior mask. If dynamic feature points are within this range and total exceeds the set threshold, the position of the prior mask is regarded as the dynamic area and removed.

## 2.2   Semantic Segmentation

We chooses the YOLACT++ network [21] for instance segmentation, and its real-time performance and accuracy are relatively high. The YOLACT++ network is improved on the basis of the YOLACT network [22]. The YOLACT++ network has made great progress in capability. It can achieve real-time (>30fps) speed, so it can meets the real-time requirements of SLAM. The network structure of YOLACT is shown in Fig. 2.

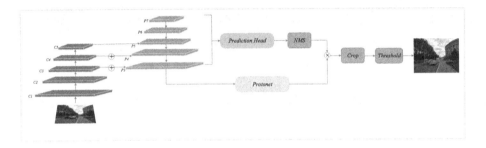

**Fig. 2.** YOLACT network structure

For better association class confidence and mask quality, the YOLACT++ network combines 6 convolutional layers and a global average pooling layer to form the model structure of the fast Mask Re-Scoring branch. The input is the cropped but not thresholded prediction mask generated by the YOLACT network, the output is the mask IoU of each object. Re-scoring each mask is performed by selecting the product of the IoU of the prediction mask of the category predicted by the classification header and the corresponding classification confidence. Replace the standard convolutions in the $C_3 \sim C_5$ layers in the ResNet network structure with 3*3 deformable convolutions. Do not use modulated deformable module to ensure detection speed. At the same time, it chooses to increase the anchor aspect ratio from the original [1, 1/2, 2] to [1, 1/2, 2/3, 3] while keeping the original scale unchanged. Compared with the original YOLACT, the number of anchors has increased by 5/3 times.

Our select the MS COCO [23] which is an open dataset created by the Microsoft team, as the training sample to train the network, and applies the network to the work of removing outliers. The network trained with this data set can recognize and detect objects in the dynamic environment set in this article.

# 3    Methods

## 3.1    Traditional Method

Traditional methods use the walking series data set in TUM for experimental testing. Figure 3(a) is the original picture. The mask $M$ is applied to the original image as shown in Fig. 3(b); then, as shown in Fig. 3(c), the dynamic area is eliminated. The system will only use these static points for subsequent pose estimation and other tasks. Such as the classic algorithm DynaSLAM, etc.

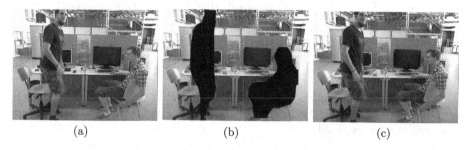

(a)                    (b)                    (c)

**Fig. 3.** Removal of dynamic points (a) Original image; (b) Dynamic area segmentation; (c) Removal of dynamic points (Color figure online)

But just like the shortcomings of the DynaSLAM algorithm, it chooses to remove all objects with potential movement, and almost eliminates the area where the person is in the image during the processing. For the prior mask area that does not actually move, it is often removed for no reason. When there are both a moving and a static prior object in the current image, the processing is not accurate. It can be seen from the figure above that the person on the right is actually sitting still, with only slight movements. For the whole environment, the person on the right is relatively static now. But the feature points on it are also eliminated, which is to some extent inaccurate.

## 3.2    Improved Method

**Dynamic Points Detection.** First, the image is converted to a grayscale image and extract key feature points to get a feature point set $p = \{p_1^t, p_2^t, ..., p_i^t\}$. $p_i^t$ represents the image coordinates of the i-th feature point on the t-th frame. From the tracking results of the image pyramid LK optical flow method, several pairs of feature points matching the current frame can be obtained. Here, a preprocessing is performed on the matching point pairs first, and the matching pairs that are too close to the edge of the image or the 3*3 image block pixels in the center of the matching pair have obvious differences are regarded as abnormal matches and preliminary eliminated. Then, the matched points after screening are constrained by epipolar geometry to determine whether they are dynamic feature points.

Then we calculate the distance from the tracked feature point to its corresponding epipolar line, and judge whether the tracked feature point is a dynamic feature point according to the principle of geometric constraint of the epipolar line. Different projection points produced by the same image point through different projection matrices must meet the following constraints:

$$x_2^T F x_1 = 0 \tag{1}$$

We use the eight-point method to calculate the fundamental matrix $F$. Now we supposed that given any pairs of matching points, the normalized coordinates are $x = (x, y, 1)^T, y = (x', y', 1)^T$, then substitute it into the formula (1)

$$\begin{bmatrix} x & y & 1 \end{bmatrix} \begin{bmatrix} f_{11} & f_{12} & f_{13} \\ f_{21} & f_{22} & f_{23} \\ f_{31} & f_{32} & f_{33} \end{bmatrix} \begin{bmatrix} x' \\ y' \\ 1 \end{bmatrix} = 0 \tag{2}$$

After the above formula is expanded, rendered $F$ in vector form and given a set of n groups of points, the motion relationship between two adjacent frames of images can be obtained:

$$l_2 = \begin{bmatrix} A \\ B \\ C \end{bmatrix} = F p_1 = F \begin{bmatrix} x_1 \\ y_1 \\ 1 \end{bmatrix} \tag{3}$$

In the formula, A, B, C represent the polar equation vector. In the actual environment, there are two reasons for the excessive distance: one is the mismatch of tracking feature point; the other is the existence of dynamic objects in the environment. Then we calculate the distance $d$ from the feature point $p_2$ to the corresponding epipolar line.

$$d = \frac{|p_2^T F p_1|}{\sqrt{||A||^2 + ||B||^2}} \tag{4}$$

We set the filtering threshold to $\varepsilon$, and the calculation method is as follows:

$$\varepsilon = \frac{\sum_{i=0}^{N} e^{-d_i}}{N} \tag{5}$$

In the formula, $N$ is the total number of extracted ORB feature points, and $d_i$ represents the distance from the i-th matching point to its corresponding epipolar line. According to the constraint of the distance $d$, both mismatching can be eliminated, and dynamic points caused by moving objects can also be eliminated. If $d_i > \varepsilon$, it is preset as a dynamic point.

As shown in Fig. 4(b), the red points are the detection results of dynamic feature points in the scene.

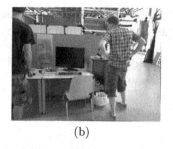

(a)                                                    (b)

**Fig. 4.** Dynamic point elimination (a) Original picture; (b) Dynamic point detection (Color figure online)

**Improved Fusion Method.** In our research, the fusion method is improved to improve the accuracy of system processing. First, we extract global ORB feature points. Through epipolar geometric constraints to choose feature points with a distance greater than *varepsilon* are temporarily set as dynamic points $O$. Meanwhile, the image is semantically segmented and the prediction mask $m$ is generated. Combined with the preset dynamic point information, when the points $o$ and mask $m$ in $O$ are at the same coordinate position of the image, the mask becomes the preselected dynamic area.

Then we increase the re-discrimination mechanism, when the number of dynamic feature points in the area contour is greater than the set threshold $\delta$, the final dynamic area is determined. Then dynamic area will be split. Considering the noise points that may appear in dynamic feature point detection, and the actual density of feature point extraction and other factors, the threshold $\delta$ is finally set to four.

Seen from the test results, our method can accurately and completely eliminate the entire dynamic area, and the robustness of the system is effectively enhanced.

(a)                                                    (b)

**Fig. 5.** Experimental results in a semi-dynamic scenarios (Color figure online)

It can be seen from Fig. 5(a) and 5(b) that the person on the left side of the image is moving, while the person on the right side is sitting quietly. Our improved algorithm can well extract feature points when the prior masks are both dynamic and static. Meanwhile, it also has a good performance when the prior masks are moved. As shown in Fig. 6, all dynamic areas are accurately eliminated.

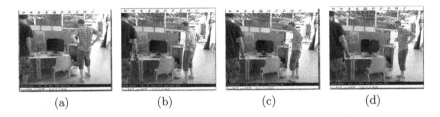

    (a)                (b)                (c)                (d)

**Fig. 6.** Experimental results in dynamic scenarios

From the experimrntal results, our algorithm has neither eliminated the feature points in the non-moving prior mask. At the same time, the dynamic area can be accurately eliminated in a semi-dynamic environment (there are both a dynamic and a static prior masks) or a high-dynamic environment. This proves the accuracy of our algorithm.

Compared with the DynaSLAM algorithm, it solves the shortcomings of choosing to remove all potentially moving objects, and can accurately segment moving objects when they are prior dynamic objects in a semi-dynamic environment.

## 4  Experimental Results and Analysis

### 4.1  Qualitative Analysis of Motion Trajectory

The experiment is tested under Ubuntu 16.04 system. The host environment is equipped with Intel i5-8500 processor and NVIDIA GeForce GTX 980 graphics card. To verify the validity of the algorithm, we select the dynamic frame sequence to test (respectively freiburg3_walking_xyz, freiburg3_walking_halfspere), and compares it with the original ORB-SLAM2 algorithm. The trajectory comparison chart is shown in Fig. 7.

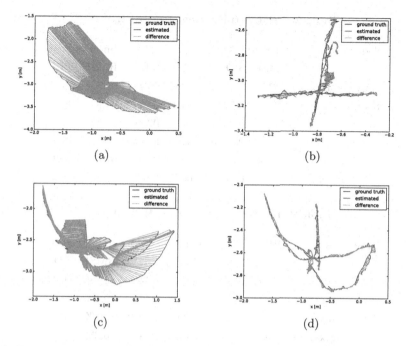

**Fig. 7.** Trajectory error comparison chart (a) ORB-SLAM2_xyz (b) Ours_xyz (c) ORB-SLAM2_half (d) Ours_half

Compared with the ORB-SLAM2 algorithm, the performance improvement of our algorithm can be calculated by the following formula:

$$\eta = (1 - \frac{\alpha}{\beta}) \times 100\% \tag{6}$$

Where, $\eta$ represents the degree of performance improvement; $\alpha$ indicates the calculated error of the algorithm in this paper; $\beta$ represents the calculated ORB-SLAM2 algorithm error.

**Table 1.** Absolute track error (ATE, m)

| Sequences | ORB-SLAM2 | | Ours | | Improvement | |
|---|---|---|---|---|---|---|
| | RMSE | S.D | RMSE | S.D | RMSE | S.D |
| fr3_walking_xyz | 0.8612 | 0.4375 | 0.0312 | 0.0182 | 96.3% | 95.8% |
| fr3_walking_half | 0.4824 | 0.2693 | 0.0501 | 0.0256 | 89.6% | 90.5% |
| fr3_walking_rpy | 0.8907 | 0.4167 | 0.4178 | 0.3275 | 53.1% | 21.4% |
| fr3_siting_xyz | 0.0223 | 0.0128 | 0.0171 | 0.0092 | 23.3% | 28.1% |
| fr3_siting_half | 0.0214 | 0.0109 | 0.0161 | 0.0095 | 24.7% | 12.8% |
| fr3_siting_static | 0.0087 | 0.00412 | 0.0073 | 0.0034 | 16.1% | 17.5% |

**Table 2.** Relative translation error (RPE, m/s)

| Sequences | ORB-SLAM2 | | Ours | | Improvement | |
|---|---|---|---|---|---|---|
| | RMSE | S.D | RMSE | S.D | RMSE | S.D |
| fr3_walking_xyz | 0.8117 | 0.4992 | 0.0362 | 0.0186 | 95.5% | 96.2% |
| fr3_walking_half | 0.8234 | 0.5024 | 0.0479 | 0.0174 | 94.2% | 96.5% |
| fr3_walking_rpy | 0.9241 | 0.4125 | 0.5712 | 0.2081 | 38.2% | 49.6% |
| fr3_siting_xyz | 0.0352 | 0.0176 | 0.0266 | 0.0152 | 24.4% | 13.6% |
| fr3_siting_half | 0.0368 | 0.0267 | 0.0252 | 0.0154 | 31.5% | 42.3% |
| fr3_siting_static | 0.0135 | 0.0074 | 0.0097 | 0.0056 | 28.1% | 24.3% |

**Table 3.** Relative rotation error (RPE, deg/s)

| Sequences | ORB-SLAM2 | | Ours | | Improvement | |
|---|---|---|---|---|---|---|
| | RMSE | S.D | RMSE | S.D | RMSE | S.D |
| fr3_walking_xyz | 8.9542 | 5.1502 | 0.9214 | 0.5283 | 89.7% | 89.7% |
| fr3_walking_half | 7.4834 | 5.3099 | 0.8133 | 0.3954 | 89.1% | 92.6% |
| fr3_walking_rpy | 9.3457 | 6.5917 | 5.6206 | 3.5548 | 39.8% | 46.1% |
| fr3_siting_xyz | 0.4669 | 0.2375 | 0.4216 | 0.2612 | 10.3% | – |
| fr3_siting_half | 0.6012 | 0.3215 | 0.5329 | 0.2848 | 11.3% | 11.4% |
| fr3_siting_static | 0.2953 | 0.1324 | 0.2813 | 0.1187 | 4.7% | 10.3% |

Compare the trajectory error obtained by running the algorithm with the ORB-SLAM2 algorithm, and the results are shown in Table 1, Table 2, and Table 3. In a high dynamic scenario, the RMSE of the absolute error has dropped by 80.0% on average, the RMSE of the relative translation error has dropped by 75.96%, and the RMSE of the relative rotation error dropped by 72.86% on average. It can be seen that compared with the ORB-SLAM2 algorithm, ours performs better on highly dynamic sequences. The accuracy and robustness have obviously improve. Because the ORB-SLAM2 algorithm has been able to solve the low dynamic and static scenes well, the room for improvement in accuracy is limited.

## 4.2  Time Efficiency

The following is a statistical analysis of the time consumed by each module of the improved visual front end, as shown in Table 4. Take the fr3_walking_xyz sequence as an example. In this article, ORB feature point extraction, detection of dynamic points, and instance segmentation are executed simultaneously on two different threads. As can be seen from the table below, the total time used by the instance segmentation thread and the tracking thread is similar, which is conducive to the real-time operation of the system.

**Table 4.** Time used by each module(ms)

| Thread | Module | Time | Total time |
|---|---|---|---|
| Tracking | ORB feature point extraction | 14.27 | 37.91 |
| | Detect dynamic points | 23.62 | |
| Instance segmentation | Instance semantic segmentation | 35.32 | 35.32 |

The Semantic Segmentation Network Mask R-CNN used in the DynaSLAM algorithm. He et al. [24] report that Mask R-CNN runs at 195ms per image on Nvidia Tesla M40 GPU, plus 15 ms CPU time for resizing outputs to original resolution. Note that Dyna-SLAM has not been optimized for real-time operation [20]. Compared with Dyna-SLAM, the algorithm strength of this paper requires only 35.32ms per frame for the segmentation thread, which greatly reduces the running time and can meet the real-time requirements.

## 5  Conclusion

Our algorithm adds a re-discrimination mechanism, which greatly reduces the miss election of dynamic regions and improves the accuracy of dynamic region elimination. The feature point can be extracted well when the object's prior masks are both dynamic and static. At the same time, it also has a good performance when there is movement in the prior mask, and can accurately remove the dynamic area. At the same time, the selected instance segmentation network has good real-time performance, which ensures that the SLAM system can run in real time. Experiments were conducted on public datasets. The experiments show that the proposed method has improved trajectory accuracy and is superior to other algorithms. But the algorithm still has some shortcomings. For example, the stability of the algorithm is not enough, and sometimes the feature points of some frames have tracking loss. Therefore, the algorithm still needs subsequent improvement.

## References

1. Davison, A.J., Reid, I.D., Molton, N.D., et al.: MonoSLAM: real-time single camera SLAM. IEEE Trans. Pattern Anal. Mach. Intell. **29**(6), 1052–1067 (2007)
2. Klein G, Murray D.: Parallel tracking and mapping for small AR workspaces. In:6th IEEE & Acm International Symposium on Mixed & Augmented Reality, ACM Publisher, New York (2008)
3. Engel, J., Schöps, T., Cremers, D.: LSD-SLAM: large-scale direct monocular SLAM. In: Fleet, D., Pajdla, T., Schiele, B., Tuytelaars, T. (eds.) ECCV 2014. LNCS, vol. 8690, pp. 834–849. Springer, Cham (2014). https://doi.org/10.1007/978-3-319-10605-2_54
4. Mur-Artal, R.T., Juan, D.: ORB-SLAM2: an open-source SLAM system for monocular, stereo and RGB-D cameras. IEEE Trans. Robot. **3**(2), 1–8 (2017)

5. Forster, C, Pizzoli M, Scaramuzza, D.: SVO: fast semi-direct monocular visual odometry. In: 2014 IEEE International Conference on Robotics and Automation (ICRA), pp. 15–22. IEEE, Hong Kong (2014)

6. Engel, J., Koltun, V., Cremers, D.: Direct sparse odometry. IEEE Trans. Pattern Anal. Mach. Intell. **40**(3), 611–625 (2017)

7. Zou, D., Tan, P.: CoSLAM: collaborative visual SLAM in dynamic environments. IEEE Trans. Pattern Anal. Mach. Intell. **35**(2), 354–366 (2013)

8. Chum, O., Ji, M., Kittler, J.: Locally Optimized RANSAC. In: Michaelis, B., Krell, G. (eds.) Pattern Recognition 2003, DAGM, vol. 2781, pp. 236–243. Springer, Heidelberg (2003) . https://doi.org/10.1007/978-3-540-45243-0-31

9. Li, X.Z., Xu, C.L.: Moving object detection in dynamic scenes based on optical flow and super pixels. In: IEEE International Conference on Robotics and Biomimetics, pp. 84–89. IEEE, Zhuhai (2015)

10. Wang, Y., Huang, S.: Towards dense moving object segmentation based robust dense RGB-D SLAM in dynamic scenarios. In: 13th International Conference on, pp. 1841–1846. IEEE, Singapore (2014)

11. Han, S., Xi, Z.: Dynamic scene semantics SLAM based on semantic segmentation. IEEE Access **8**, 43563–43570 (2020)

12. Zhou, D.F., Fremont, V., Quost, B., et al.: On modeling ego-motion uncertainty for moving object detection from a mobile platform. In: 2014 IEEE Intelligent Vehicles Symposium, pp. 1332–1338. IEEE, Dearborn (2014)

13. Yba, A., Tr, A., Xqy, A., et al.: Visual SLAM in dynamic environments based on object detection. Defence Technology (2020)

14. Azartash, H., Lee, K.R., Nguyen, T.Q.: Visual odometry for RGB-D cameras for dynamic scenes. In: 2014 IEEE International Conference, pp. 1280–1284. IEEE, Florence (2014)

15. Kim, D.H., Kim, J.-H.: Effective background model-based RGB-D dense visual odometry in a dynamic environment. IEEE Trans. Robot. **32**(6), 1565–1573 (2016)

16. Kim, D.-H., Han, S.-B., Kim, J.-H.: Visual odometry algorithm using an RGB-D sensor and IMU in a highly dynamic environment. In: Kim, J.-H., Yang, W., Jo, J., Sincak, P., Myung, H. (eds.) Robot Intelligence Technology and Applications 3. AISC, vol. 345, pp. 11–26. Springer, Cham (2015). https://doi.org/10.1007/978-3-319-16841-8_2

17. Sen, W., Ronald, C., Hongkai, W., Niki, T.: DeepVO: towards end-to-end visual odometry with deep recurrent convolutional neural networks. In: 2017 IEEE International Conference, pp. 2043–2050. IEEE, Singapore (2017)

18. Clark, R., Wang, S., Wen, H., et al.: VINet: visual-intertial odometry as a sequence-to-sequence learning problem. In: 31th AAAI Conference on Artifical Intelligence (2017)

19. Yu, C., Liu, Z., Liu, X., et al.: DS-SLAM: a semantic visual SLAM towards dynamic environments. In: 2018 IEEE/RSJ International Conference on Intelligent Robots and Systems (IROS), pp. 1168–117. IEEE, Madrid (2018)

20. Berta, B., Facil, J.M., Javier, C., et al.: DynaSLAM: tracking, mapping and inpainting in dynamic scenes. IEEE Robot. Autom. Lett. **3**(4), 4076–4083 (2018)

21. Bolya, D., Zhou, C., Xiao, F., et al.: YOLACT++: better real-time instance segmentation. IEEE Trans. Pattern Anal. Mach. Intell. **1**(1), 99 (2020)

22. Bolya, D., Zhou, C., Xiao, F., et al.: YOLACT: real-time instance segmentation. **1**(1), 99 (2019)

23. Fleet, D., Pajdla, T., Schiele, B., Tuytelaars, T. (eds.): ECCV 2014. LNCS, vol. 8693. Springer, Cham (2014). https://doi.org/10.1007/978-3-319-10602-1

24. He, K., Gkioxari, G., Piotr, D., et al.: Mask R-CNN. IEEE, Venice (2017)

# Viewing Angle Generative Model for 7-DoF Robotic Grasping

Xiang Gao[1]([📧])(iD), Wei Li[1,2], and Zhiqing Wen[1]

[1] Jihua Laboratory, Foshan, Guangdong, China
{gaoxiang,wenzq}@jihualab.com
[2] Academy for Engineering and Technology, Fudan University, Shanghai, China

**Abstract.** Grasping is the first step in most robotic manipulation tasks, and it is essential for applications of robots in real-life scenarios. For humans, grasping novel objects is a naturally gained ability, however, for robots, it is a challenging task due to complex object shapes and incomplete visual information. Many current grasp pose estimation methods need to first construct 3D models of the scene and generates a large pool of grasp candidates, and then perform a search for the best grasp. These methods rely on high quality 3D models, and their long pipeline makes them unfeasible for real-time processing. End-to-end grasp pose estimation methods mitigate these issues, but they can only deals with few DoF planar grasps that fail to cover many successful grasps. In this paper, we propose a viewing angle generative network (VAGN), an approach that bridges the aforementioned two main classes of methods. VAGN decouples 7-DoF grasp detection into two stages. In the first stage, it predicts the camera viewing angle, which is also the orientation of the gripper around the object from an RGBD frame. In the second stage, it generates a planar grasp pose by taking another RGBD image at the predicted viewing angle in stage 1. We trained VAGN on the Cornell dataset. Real robot experiments on a UR-10e robot with camera-in-hand show real-time processing speed and higher success rates compared to the state-of-the-art GR-ConvNet, in both single object scenes and cluttered scenes.

**Keywords:** Robotic grasping · 7-DoF · Viewing angle · Real-time

## 1 Introduction

Robotic grasping of unseen objects is fundamentally important to robot applications in unstructured environments such as flexible manufacturing, warehouse and household servicing. It is one of the key abilities that an intelligent robot should have. However, grasping novel objects remains a highly challenging task due to occlusions in a cluttered scene, complex object shapes and sensory noise and deficiency in visual perception. Traditional physical grasp analysis techniques rely on contact force analysis [1], object contour features [4] or template matching [7] to search for the optimal grasp. These approaches, however, are not

© Springer Nature Switzerland AG 2021
L. Fang et al. (Eds.): CICAI 2021, LNAI 13070, pp. 323–333, 2021.
https://doi.org/10.1007/978-3-030-93049-3_27

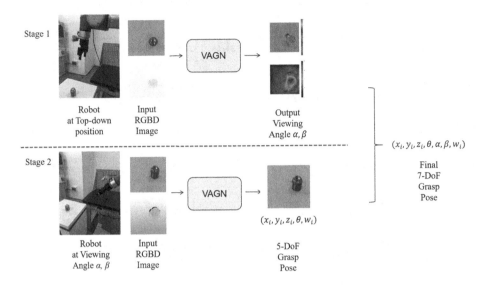

**Fig. 1.** The two-stage grasp prediction process of VAGN.

applicable to previously unseen objects. Besides, they usually require accurate object models which are not always available due to perception limitation of object appearance.

The development of computer vision and AI enables researchers to use learning-based or data-driven approaches to find 6-DoF grasp poses. 6-DoF grasping has attracted much attention, as it allows a 6-axis robot to approach the object from many different angles and thus fully utilize the robot's capability. In [16] and [14], researcher study 6D grasp estimation with pre-scanned object CAD models, which are methods that are unable to quickly generalize to unseen objects. Without knowing the object models, existing methods such as GPD [13], GraspNet [11] and PointNet++ [12] typically follow a object pose recognition, grasp candidates generation and sampling pipeline. However, these methods need to evaluate a huge number of grasp candidates to find suitable grasp poses which is time-consuming. In addition, point cloud data is not as stable as RGB data due to the limitations in consumer grade 3D camera.

In recent years, several benchmark grasp datasets such as Cornell [15] and Jacquard [3] greatly facilitated the researches and evaluations of end-to-end grasp learning algorithms [2,9,10]. These methods achieve high success rate and real-time performance on the benchmark datasets. However, the 2D planar grasps generated by these methods are greatly restricted compared to 6-DoF grasps: the gripper can only approach the object from a fixed direction. For example, the gripper can only grab the water cup from a top-down direction, which not only makes grabbing certain objects difficult, but also limits the next move of the robot such as pouring the water out of the cup.

In this paper, we propose an end-to-end 7-DoF viewing angle generative network (VAGN). The aim of VAGN is to bridge the gap between the 6-DoF grasp sampling methods and the end-to-end planar grasping methods, and thus taking advantages of both. In VAGN, a camera mounted in-hand first captures an RGBD image from the top-down direction. The robot then rapidly moves to a suitable viewing angle generated by VAGN and takes another RGBD shot. Finally, the robot executes the planar grasp generated by VAGN. VAGN is a lightweight end-to-end generative model that trained with publicly available grasp datasets. It does not require object CAD models or grasp simulations.

The contributions of this paper are as follows.

1. An end-to-end grasp detection model is proposed to detect grasps efficiently. The model performs viewing angle prediction and planar grasp generation in a single network.The model uses RGBD images as input and infers 7-DoF grasps for two-finger parallel grippers.
2. The results show the real-time performance and higher success rate over a state-of-the-art baseline method on a set of household items.

## 2    Problem Formulation

In this work, we define the problem of robotic grasping as predicting an optimal antipodal grasp pose for unknown objects from 4-channel RGBD images of the scene.

### 2.1    Grasp Representation

Inspired by the 6-DoF grasp representation [11], we improve the planar grasp representation Morrison et al. in [10] to a 7-DoF grasp representation in robot frame as

$$G_r = (x, y, z, r_x, r_y, r_z, w), \tag{1}$$

where $x, y$ and $z$ denote the position of the gripper, $r_x, r_y$ and $r_z$ denote the rotation and $w$ denotes the gripper width.

A grasp is detected from an 4-channel RGBD image $I = \mathbb{R}^{4 \times m \times n}$ with image height $m$ and width $n$, which can be defined as

$$G_i = (x_i, y_i, z_i, \theta, \alpha, \beta, w_i), \tag{2}$$

where $(x_i, y_i)$ corresponds to the center of grasp in image coordinates, $z_i$ is the depth value at the grasp center $(x_i, y_i)$, $\theta$ is the grasp rotation in image coordinates, $\alpha$ and $\beta$ are viewing angles in the horizontal and vertical directions as shown in Fig. 2, $w_i$ is the required width in image coordinates. Specifically, $\theta$ indicates the angular rotation required at each point of the image plane to grasp the object of interest and is a value in the range $[-\pi/2, \pi/2]$. Similarly, $\alpha$ indicates the horizontal viewing angle towards the object in the range of $[-\pi, \pi]$, and $\beta$ indicates the vertical viewing angle is a value in the range $[0, \pi/2]$. $w_i$ is the required width ranging from 0 to $W_{max}$ pixels, where $W_{max}$ is the maximum width of the two-finger gripper.

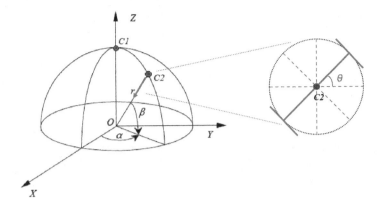

**Fig. 2.** Horizontal viewing angle $\alpha$, vertical viewing angle $\beta$ and in-plane rotation angle $\theta$ for object $O$. (Color figure online)

## 2.2   Grasp Quality Score

The grasp quality score $q$ is the quality of the grasp at every point in the image and is indicated as a score value between 0 and 1 where a value that is in proximity to 1 indicates a greater chance of grasp success. For every measurable points $(x_i, y_i, z_i)$ in the 3D space, our goal is to design a network that estimates a tuple consisting of five scalar values $(\theta, \alpha, \beta, w_i, q)$. Then the inference module finds the best grasp pose by locating the largest corresponding grasp quality score $q$.

To make the robot actually perform a grasp in its coordinates, we need to convert the grasp pose in image space $G_i$ into a grasp pose in robot space. Such translation can be performed given the robot hand-eye calibration and camera calibration results.

## 3   Method

Our proposed approach VAGN treats the grasp pose prediction problem as a two-stage regression problem (see Fig. 1). In stage 1, VAGN predicts the best viewing angle pair $(\alpha, \beta)$ by overlooking the object-to-be-grasp from an top-down position; and the robot adjusts to the best viewing angle accordingly. In stage 2, VAGN predicts the final suitable grasp poses for the objects in the camera's field of view. The robot's controller uses these grasp poses to plan and execute robot trajectories to perform antipodal grasps.

### 3.1   Stage 1: Viewing Angle Prediction

At this stage, VAGN predicts the pixel-wise camera viewing angles of the camera from an RGBD frame taken from a top-down position. Because we assume an eye-in-hand setup, where the camera is installed near the robot's end-effector, i.e. the gripper. The viewing angle also represents the approaching direction of the gripper. The blue arrow in Fig. 2 shows an example of the approaching direction. As such, the rotation of the gripper is decoupled into the viewing angle $(\alpha, \beta)$ and in-plane rotation $\theta$.

The input RGBD image is first pre-processed where it is cropped, resized, and normalized to a four channel image of the size $224 \times 224$. The depth channel is inpainted and aligned with the RGB channels to obtain corresponding depth values at each pixel. The $224 \times 224$ 4-channel image is then fed into the VAGN to obtain three images of the size $224 \times 224$, grasp quality score $q$, horizontal viewing angle $\alpha$ and vertical viewing angle $\beta$ as the output. Finally, an inference module infers an optimal viewing angles from the three output images.

## 3.2 Stage 2: Planar Grasp Pose Generation

After stage 1, the robot will move to the viewing pose defined by $(\alpha, \beta)$ to perform the planar grasp pose generation stage. The second stage is similar to the first stage, except that 1) the input RGBD image is taken from a viewing angle $(\alpha, \beta)$ rather than a top-down pose; 2) The output of VAGN are three images including grasp quality score $q$, in-plane rotation angle $\theta$ and gripper width $w_i$. All three images share the same size of $224 \times 224$. The role of VAGN in stage 2 is similar to those in GG-CNN [10] and GR-ConvNet [9]. The inference module returns the grasp pose with the highest quality score. The grasp pose is then converted from camera coordinates into robot coordinates using the transform calculated from hand-eye calibration. Further, the grasp pose in robot frame is used to plan a trajectory to perform the object picking action using inverse kinematics through the robot controller. The robot then executes the planned trajectory and grasp the object.

## 3.3 Model Architecture

Figure 3 shows the network architecture of the proposed VAGN model, which is inspired by GG-CNN [10] and GR-ConvNet [9]. VAGN is a generative architecture that generates pixel-wise grasps in the form of five images. The 4-channel processed RGBD image is first fed into the three convolutional layers, where features are extracted by learning a large number of filters automatically.

The convolutional layers are followed by five residual blocks. Each residual block contains two convolutional layers with skip connections, which is identical to the residual blocks in ResNet-34 [6] where no down-sampling happens. Using residual blocks enable us to gain higher accuracy from increased depth of the network, while avoiding difficulties such as vanishing gradients and performance degradation.

Followed by the residual blocks, we use convolution transpose layers to up-sample the images from $56 \times 56$ to $224 \times 224$ so as to retain spatial features. In other words, we obtain the same size of the output image as the size of the input. There are two separate ends in the convolution transpose layers. One generates

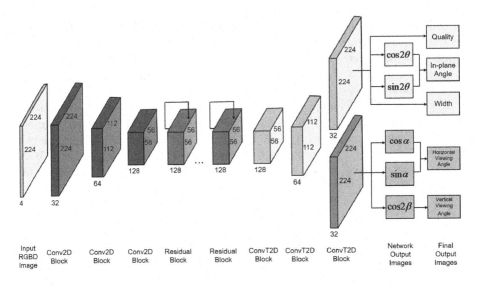

**Fig. 3.** Network architecture.

output images for viewing angle prediction (stage 1), including horizontal angle $\alpha$ and vertical angle $\beta$. Note that $\alpha$ is decoupled in the form of $\cos\alpha$ and $\sin\alpha$ for unique values in the range of $[-1, 1]$ that can be combined to form the required angle $\alpha$ in $[-\pi, \pi]$. The other end generates the grasp quality score, in-plane rotation angle as well as the width of the end-effector. The simple architecture of VAGN makes it efficient and fast in computation compared to other similar models [11,17].

### 3.4   Model Training

For a dataset having input scene images $I = \{I_1, I_2, ..., I_n\}$ and successful grasps in the $n$ images $G = \{G_1, G_2, ..., G_n\}$, where $G_i = \{g_1^i, g_2^i, ..., g_{m_i}^i\}, i = 1, 2, ... n$ means that for image $i$ there are $m_i$ successful grasps: $g_1^i, g_2^i, ..., g_{m_i}^i$. Then the mapping function $f(I) = G$ can be learned by minimizing the negative log-likelihood of $G$ given the input image scenes $I$,

$$-\frac{1}{n}\sum_{i=1}^{n}\frac{1}{m_i}\sum_{j=1}^{m_i}\log f(g_j^i|I^i). \tag{3}$$

The model was trained using the standard back-propagation with Adam optimizer [8]. We set the learning rate to $10^{-3}$ and the mini-batch size to 8.

### 3.5   Loss Function

We define two losses $L_1$ and $L_2$ for the two-head structure accordingly. For the head that generates a planar grasp pose, we define $L_1$ as

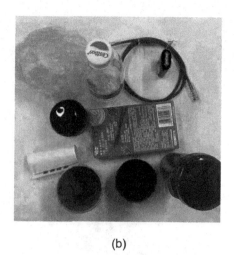

(a)                                                    (b)

**Fig. 4.** Experiment setup. (a) Robot experiment on a cluttered scene. A: UR-10e Robot; B: Robotiq two-finger parallel gripper; C: Intel RealSense RGBD camera; D: Graspable objects. (b) Household objects for the experiment.

$$L_1(G_i, \hat{G}_i) = \frac{1}{n} \sum^{k} l_k. \tag{4}$$

where $k$ is the mini-batch index and $l_k$ is given by

$$l_k = \begin{cases} 0.5(G_i - \hat{G}_i)^2, & \text{if } \left|G_i - \hat{G}_i\right| < 1 \\ \left|G_i - \hat{G}_i\right| - 0.5, & \text{otherwise} \end{cases} \tag{5}$$

$l_k$ is the smooth L1 loss (also known as Huber loss). $\left|G_i - \hat{G}_i\right|$ represents the summation of element-wise difference between the predicted planar grasp pose $\hat{G}_i$ and the ground truth $G_i$, in terms of $(x_i, y_i, \cos 2\theta, \sin 2\theta, w_i)$. The definition of $L_2$ is similar to $L_1$, except that the terms now become $(\cos \alpha, \cos \alpha, \cos 2\beta)$. Note that the ground truth grasps $G_i$ are manual annotations provided by a grasp dataset such as the Cornell dataset [15]. There is no guarantee that they will work perfectly with all types of two-finger grippers (Fig. 5).

## 4   Experiments

### 4.1   Training Dataset

We trained our VAGN model using the Cornell grasp dataset [15], which is one of the most popular robotic grasping datasets. The Cornell dataset contains 1035 RGBD images with a resolution of 640 × 480 pixels of 240 different real objects with 5110 positive and 2909 negative grasps. In each RGBD image, grasp

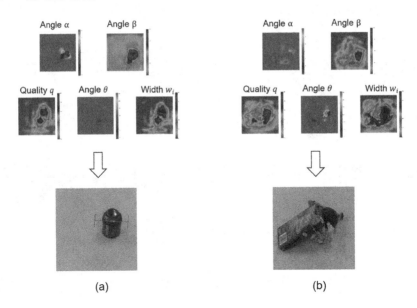

**Fig. 5.** 7-DoF grasp poses (bottom) are inferred based on the output images in (a) a single object scene; (b) a cluttered scene.

rectangles are used to annotate both positive and negative grasps per object. The annotations of grasp rectangles can only represent planar grasp poses, which are insufficient to train our VAGN model. Therefore, we manually label viewing angles $\alpha$ and $\beta$ along with each grasp rectangles. Uniformly distributed noise between $\pm\pi/30$ is added to the ground truths of $\alpha$ and $\beta$. To prepare sufficient training samples, we augmented the dataset using random crops, zooms, and rotations which effectively has 51k grasp examples. Only positively labeled grasps from the dataset were considered during training.

## 4.2 Real Robot Experiments

In this section, we show the experiments carried on a UR-10e robot. We compare our results with a baseline method GR-ConvNet [9] which reports state-of-the-art performance on Cornell dataset and analyze the differences.

The setup of our experiment is shown in Fig. 4. The experiment is carried on a Universal Robot UR-10e robot mounted with an Intel RealSense D435i camera and a Robotiq 2f-85 two-finger gripper. A computer with an Nvidia GTX 2080Ti GPU, an Intel i9-10900X CPU, 64G RAM and an Ubuntu 18.04 operating system is used to run our pipeline and control the robot. The trajectory is planed using UR's built-in controller. To ensure the robot is planning in a safe workspace, and avoid possible collision with other objects in the scene, we limit the vertical viewing angle $\alpha$ to $[0, \pi/3]$ rad.

*Experiment on Single Object Scenes.* We randomly select 10 household objects and place them on the table at random place with random orientation. Objects with different shapes and sizes are selected to analyze the performance of generalization. We conduct 10 experiments for each object and calculate the success rate. The results are shown in Table 1.

*Experiment on Cluttered Scenes.* We also conduct experiments on cluttered scenes. In each scene, we randomly select 3–5 objects from the 10 objects in Fig. 4(b) and randomly place them on the table. We repeatedly executed the optimal grasp using VAGN until all the objects are clear. The results are shown in Table 2.

## 5   Results

### 5.1   Single Objects

In the single objects experiment (Table 1), the robot performed 96 successful grasps of the total 100 grasp attempts on household objects, which outperforms the 81% success rate from GR-ConvNet. The advantages of VAGN over GR-ConvNet is that VAGN performs 7-DoF grasps. The additional DoF allows the robot to grab an object from the side, which we found was more stable for round objects or tall objects. The results obtained in Table 1 indicates that VAGN is able to generalize well to novel objects that it has never seen before.

**Table 1.** Grasp success rates on single object scenes.

| ID | Object | Type | GR-ConvNet [9] | Our method |
|---|---|---|---|---|
| 1 | Paper Cup | Unseen | 80% | 100% |
| 2 | Chewing Gum Box | Unseen | 70% | 90% |
| 3 | Cylinder Tea Can | Unseen | 70% | 90% |
| 4 | Toilet Paper Roll | Unseen | 80% | 100% |
| 5 | Water Bottle | Unseen | 60% | 90% |
| 6 | Plastic Bottle | Unseen | 60% | 90% |
| 7 | Transparent Plastic Bag | Unseen | 100% | 100% |
| 8 | Network Cable | Similar | 90% | 100% |
| 9 | Box | Similar | 100% | 100% |
| 10 | Orange | Similar | 100% | 100% |
| – | Overall | – | 81% | 96% |

### 5.2   Objects in Clutter

Although our model is trained on single isolated objects from Cornell dataset, it is also able to generate grasps for multiple objects in clutter. In this experiment, we let the robot perform 10 grasp attempts the cluttered scenes. Objects were grabbed one by one without replacement. If all objects in a scene were taken away, we reset the scene with the objects with different placement and orientations. We achieved an average success rate of 84% in all five scenes (Table 2). This demonstrates that VAGN is able to predict robust grasps for objects in clutter.

Table 2. Grasp success rates on cluttered scenes.

| ID | Objects | Attempts | GR-ConvNet [9] | Our method |
|---|---|---|---|---|
| 1 | 2,7,9,10 | 10 | 60% | 80% |
| 2 | 1,4,8 | 10 | 80% | 80% |
| 3 | 3,5,7,9 | 10 | 80% | 90% |
| 4 | 4,6,8,9,10 | 10 | 70% | 90% |
| 5 | 1,2,3,5 | 10 | 70% | 80% |
| – | Overall | – | 72% | 84% |

During the experiments, we observed that the cascade pipeline structure of VAGN did not necessarily result in accumulating errors, because the second stage of VAGN generates grasps for any viewing angle, and usually objects could be picked up from a wide range of viewing (approaching) angles.

Our VAGN model runs at a real-time speed of 16 ms per frame on our current platform. The simple pipeline and real-time performance make our model suitable for migration to embedded platforms. Moreover, our solution is cost-efficient given that no expensive 3D camera is needed.

## 6 Conclusion

We presented a two-stage solution for 7-DoF grasping novel objects using our Viewing Angle Generative Network (VAGN). It decouples the 7-DoF grasp pose into a viewing angle prediction step and a planar grasp pose generation step, where both are carried out using one model. We trained VAGN on the Cornell grasp dataset, and validated VAGN in cluttered scenes using a robotic arm. The results demonstrate that our approach can generate more accurate grasps for previously unseen objects than a baseline planar grasp algorithm. Also, the simple pipeline of our model makes it achieve a real-time performance at 60 fps, which is suitable for real-time dynamic grasping.

In future work, we would like to extend our solution to real-time close-loop grasping. We would also like to train and improve our model on GraspNet-1Billion [5], which is a recently published large-scale grasp dataset.

**Acknowledgement.** This study was supported by Jihua Laboratory through the Self-Programming Intelligent Robot Project (No. X190101TB190) and Funds for Young Scholar (No. X201181XB200), also by Guangdong Basic and Applied Basic Research Foundation (No. 2020A1515110267).

# References

1. Bicchi, A., Kumar, V.: Robotic grasping and contact: a review. In: Proceedings 2000 ICRA. Millennium Conference. IEEE International Conference on Robotics and Automation, vol. 1, pp. 348–353 (2000)
2. Chu, F., Xu, R., Vela, P.A.: Real-world multiobject, multigrasp detection. IEEE Robot. Autom. Lett. **3**(4), 3355–3362 (2018)
3. Depierre, A., Dellandrea, E., Chen, L.: Jacquard: a large scale dataset for robotic grasp detection. In: 2018 IEEE/RSJ International Conference on Intelligent Robots and Systems (IROS), pp. 3511–3516 (2018)
4. Detry, R., et al.: Learning object-specific grasp affordance densities. In: 2009 IEEE 8th International Conference on Development and Learning, pp. 1–7 (2009)
5. Fang, H.S., Wang, C., Gou, M., Lu, C.: Graspnet-1billion: A large-scale benchmark for general object grasping. In: Proceedings of the IEEE/CVF Conference on Computer Vision and Pattern Recognition (CVPR), pp. 11444–11453 (2020)
6. He, K., Zhang, X., Ren, S., Sun, J.: Deep residual learning for image recognition. In: 2016 IEEE Conference on Computer Vision and Pattern Recognition (CVPR), pp. 770–778 (2016)
7. Herzog, A., Pastor, P., Kalakrishnan, M., Righetti, L., Bohg, J., Asfour, T., Schaal, S.: Learning of grasp selection based on shape-templates. Autonomous Robots 36, January 2014
8. Kingma, D., Ba, J.: Adam: A method for stochastic optimization. In: International Conference on Learning Representations (2015)
9. Kumra, S., Joshi, S., Sahin, F.: Antipodal robotic grasping using generative residual convolutional neural network. In: IEEE/RSJ International Conference on Intelligent Robots and Systems (IROS), September 2020
10. Morrison, D., Corke, P., Leitner, J.: Learning robust, real-time, reactive robotic grasping. Int. J. Robot. Res. **39**(2–3), 183–201 (2020)
11. Mousavian, A., Eppner, C., Fox, D.: 6-dof graspnet: variational grasp generation for object manipulation. In: 2019 IEEE/CVF International Conference on Computer Vision (ICCV), pp. 2901–2910, May 2019
12. Ni, P., Zhang, W., Zhu, X., Cao, Q.: Pointnet++ grasping: learning an end-to-end spatial grasp generation algorithm from sparse point clouds. In: 2020 IEEE International Conference on Robotics and Automation (ICRA), pp. 3619–3625, March 2020
13. ten Pas, A., Gualtieri, M., Saenko, K., Platt, R.: Grasp pose detection in point clouds. Int. J. Robot. Res. **36**(13–14), 1455–1473 (2017)
14. Peng, S., Zhou, X., Liu, Y., Lin, H., Huang, Q., Bao, H.: Pvnet: pixel-wise voting network for 6dof object pose estimation. IEEE Trans. Pattern Anal. Mach. Intell., 1 (2020)
15. Yun Jiang, Moseson, S., Saxena, A.: Efficient grasping from rgbd images: Learning using a new rectangle representation. In: 2011 IEEE International Conference on Robotics and Automation (ICRA), pp. 3304–3311 (2011)
16. Zeng, A., et al.: Multi-view self-supervised deep learning for 6d pose estimation in the amazon picking challenge. In: 2017 IEEE International Conference on Robotics and Automation (ICRA), pp. 1386–1383 (2017)
17. Zhou, X., Lan, X., Zhang, H., Tian, Z., Zhang, Y., Zheng, N.: Fully convolutional grasp detection network with oriented anchor box. In: 2018 IEEE/RSJ International Conference on Intelligent Robots and Systems (IROS), pp. 7223–7230 (2018)

# RGB-D Visual Odometry Based on Semantic Feature Points in Dynamic Environments

Hao Wang[1,2], Yincan Wang[1,2], and Baofu Fang[1,2(✉)]

[1] Key Laboratory of Knowledge Engineering with Big Data, Ministry of Education,
Hefei University of Technology, Anhui 230009, China
`fangbf@hfut.edu.cn`
[2] School of Computer Science and Information Engineering,
HeFei University of Technology, Hefei 230009, Anhui, China

**Abstract.** Various algorithms of traditional visual Simultaneous Localization and Mapping (SLAM) can well match with static scenes, but mismatches will occur in dynamic scenes, which makes the positioning and mapping of the SLAM system produce large errors. Therefore, this paper proposed a visual odometry algorithm based on semantic feature points, which can improve the positioning accuracy in dynamic scenes. The algorithm combined semantic information to detect dynamic objects, and then detects and eliminates dynamic feature points. This paper conducted an extensive evaluation of the system and compared it with ORB-SLAM3 and other dynamic scene SLAM systems. The experimental results show that this method greatly improves the positioning accuracy of the camera and the robustness of the system in a highly complex dynamic environment, which verifies the advancement and effectiveness of the algorithm in this paper.

**Keywords:** Visual SLAM · Dynamic scenes · Semantic SLAM · Semantic feature points

## 1 Introduction

The SLAM of robot solves the problem of its positioning and mapping. In an unknown environment, the robot uses its own various sensors to collect information about its body, analyzes its own position through algorithms, and builds a map incrementally. Visual SLAM mainly uses cameras to obtain data sources

Supported by Fundamental Research Funds for Central Universities (No. ACAIM190102), the Project of Collaborative Innovation in Anhui Colleges and Universities (Grant No. GXXT-2019-003), the Open Fund of Key Laboratory of Flight Techniques and Flight Safety, (Grant No.2018KF06), Scientific Research Project of Civil Aviation Flight University of China (Grant No. J2020-125) and Open Fund of Key Laboratory of Flight Techniques and Flight Safety, CAAC (Grant No. FZ2020KF02).

© Springer Nature Switzerland AG 2021
L. Fang et al. (Eds.): CICAI 2021, LNAI 13070, pp. 334–344, 2021.
https://doi.org/10.1007/978-3-030-93049-3_28

and has been a popular research direction in robotics, unmanned driving and other fields in recent years [6].

During the past decades, researchers in the field of SLAM have developed many classic SLAM frameworks amid its development. Davison et al. [3] proposed MonoSLAM in 2007, which is a real-time monocular vision SLAM with extended Kalman filtering as the backend. In the same year, Klein et al. [8] proposed PTAM, which established a parallel thread for tracking and mapping, and introduced a key frame mechanism and a back-end nonlinear optimization method, which is of great significance. Forster et al. [5] proposed the SVO algorithm in 2014, which is a visual odometry based on the semi-direct method. The ORB-SLAM proposed by Artal and Tardos et al. [10] in 2015 is one of the successors of PTAM, which performs visual odometry and loop detection around ORB features. In 2017, they proposed an improved version of ORB-SLAM2 [11] and they launched the latest version of ORB-SLAM3 [2] to support for IMU sensors in 2020.

These SLAM systems are all based on static environments. However, there are inevitably dynamic objects in the actual environment, which will affect the accuracy and robustness of the SLAM system. Li and Lee et al. [9] calculated the points on the edge according to the depth map, and judged the possibility of them belonging to dynamic objects according to the weight of these points. Sun et al. [12] calculate the intensity difference between consecutive RGB images and complete the pixel classification by quantizing the segmentation of the depth map.

Some dynamic object detection algorithms are introduced above. Although their positioning in dynamic scenes has achieved good results, they can't identify objects in the scene and use semantic information to identify dynamic feature points in the scene. Therefore, researchers have proposed the use of deep learning to identify objects in the scene, in recent years.

Yu et al. [13] used SegNet to detect dynamic objects and combined with motion consistency to detect and filter dynamic feature points, but their algorithm only judged dynamic feature points by the distance between the feature point and the epipolar line. Fang et al. [4] used semantic descriptors combined with knowledge graph to detect and eliminate dynamic objects, thus improving the accuracy of pose estimation in medical places. However, its ability to recognize dynamic objects is limited by the scale of knowledge graph.

In this paper, we propose a novel dynamic objects detection and dynamic feature points detection method based on semantic feature points and sparse optical flow. The method fully considers the object instance, the possibility of object movement, the movement direction and distance of feature points. Furthermore, a robust visual odometry was proposed based on our dynamic feature points detection, which can work well in highly complex dynamic environments.

## 2   Our Works

This section will introduce algorithm in detail. The system structure is shown in Fig. 1. A segmentation thread is added based on the ORB-SLAM2 system structure. In the tracking thread, the structure of feature points is reformed, semantic information to form semantic feature points is added, and a dynamic object detection algorithm is designed based on semantic feature points.

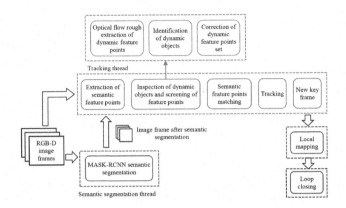

**Fig. 1.** The system flowchart of the algorithm.

## 2.1   Semantic Segmentation

As the algorithm in this paper aims to extract and match semantic feature points, the accuracy of semantic segmentation is particularly important. In the semantic segmentation thread, MASK R-CNN [7] is used for semantic segmentation.

The input RGB image in the semantic segmentation thread is semantically segmented, a two-dimensional matrix $P$ with the same scale as the original image frame $F$ is outputted, and the semantics of each pixel is recorded. For each pixel, there are two semantic attributes, that is, the number of the object it belongs to and the number of different individuals of the same object. We number the objects that MASK R-CNN can recognize from 0 to 80, where 0 represents the background. The number of vertical and horizontal pixels of the original image frame are $w, h$. $P$ is shown in formula (1).

$$P_{w,h} = \begin{bmatrix} hP_{11} & \cdots & P_{1h} \\ \vdots & \ddots & \vdots \\ P_{w1} & \cdots & P_{wh} \end{bmatrix}, P_{ij} = (m, n) \tag{1}$$

Among them, $m$ represents the type number of the object, and $n$ represents the number of different individuals of the same type of object. For example, there are two cups in the image, whose pixels are distinguished by the number $n$.

## 2.2   Detection of Dynamic Feature Points Based on Semantic Feature Points

The detection algorithm of dynamic feature points in this paper is divided into the following steps. The first is to construct semantic feature points. The second one is to track motion vectors of feature points using sparse optical flow. The third one is to roughly detect dynamic feature points set. The fourth one is to identify dynamic objects and correct the dynamic feature points set, combined with the possibility of object motion.

**Construct Semantic Feature Points.** The semantic feature points proposed in this article are expanded on the basis of ORB feature points. Semantic features are added on the basis of key points and descriptors of ORB feature points. According to the pixel-level semantic information matrix $P$ provided by the semantic segmentation thread, after the ORB feature points are extracted, the semantic information of the feature points is obtained from the matrix $P$. The semantic feature points $B_i$ is expressed as $B_i = \{keypoint, descriptor, semantic\}$. After the semantic feature points are extracted, the semantic feature points set $C$ is obtained, then a collection $M$ is created and classified by the feature points according to object instances, as in formula (2).

$$\begin{cases} M = \{Q_1, Q_2, \cdots, Q_n\} \\ Q_i = \{B_1, B_2, \cdots\}, 0 \leq i \leq n \end{cases} \tag{2}$$

Where $n$ represents the number of object instances, $Q_i$ represents the set of feature points on the $i^{th}$ object. $Q_i \subseteq C$.

**Track Motion Vectors of Feature Points Using Sparse Optical Flow.** The motion vectors of the feature points in the feature points set are detected by the pyramid Lucas optical flow method [1]. The set of motion vectors of all feature points is detected as $V = \{v_1, v_2, \cdots, v_n\}$.

**Roughly Detect Dynamic Feature Points Set.** According to the results of semantic segmentation, objects can be classified into 3 categories, that is, immovable objects, like tables, walls, floors, lights, etc., movable objects like, people, animals, cups, books, chairs, etc. and background objects, like objects not recognized by semantic segmentation.

After using optical flow method to track the motion vectors of the feature points, the motion vectors of the feature points marked as immovable objects and background objects are extracted according to the semantic information of the feature points. The average motion vector $d_b$ is calculated and shown in formula (3), among which, $s$ represents the average modulus length of all feature points, and $\mu$ represents the average movement direction of all feature points.

$$d_b = s \cdot \mu = \frac{1}{n} \sum_{i=1}^{n} \sqrt{v_{ix}^2 + v_{iy}^2} \cdot \sum_{i=1}^{n} \frac{v_i}{\sqrt{v_{ix}^2 + v_{iy}^2}} \tag{3}$$

In addition, the differences $d_m$ of modulus length of the feature points between the maximum and the minimum motion vector are calculated in the current frame. In order to reduce the error, the average value of the smallest five module lengths and the average of the largest five module lengths are taken as the minimum and maximum values, respectively, as shown in formula (4). Among them, $\{v_1, v_2, \cdots, v_n\}$ is sorted in ascending order of vector modulus length.

$$d_m = \frac{1}{5} \left( \sum_{i=v.size-4}^{v.size} \sqrt{v_{ix}^2 + v_{iy}^2} - \sum_{i=1}^{5} \sqrt{v_{ix}^2 + v_{iy}^2} \right) \qquad (4)$$

When the movement speed of the feature point is greater than the average background speed, or the angle between the movement direction of the feature point and the background movement direction is large, there will be obvious movement inconsistencies. Based on this principle, an algorithm is designed to identify dynamic feature points. The algorithm flow is shown in Fig. 2.

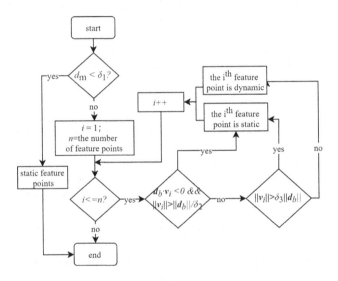

**Fig. 2.** Process of judging dynamic feature points.

It is more appropriated to assign $\delta_1, \delta_2$ and $\delta_3$ in the process as 15.0, 2.0, and 2.0 after experiments.

**Identify Dynamic Objects and Correct the Dynamic Feature Points Set.** After rough extraction of feature points, it combines semantic information to identify dynamic objects in the scene and correct incorrectly marked dynamic feature points, with the specific process as follows. According to the above dynamic feature points detection algorithm, the dynamic feature points set $C'$ is initially calculated. According to $C'$, the number of dynamic feature points extracted from each object instance in the calculation set $M$, which is saved in the set $U = \{u_{Q_1}, u_{Q_2}, \cdots, u_{Q_n}\}$, among which, $u_{Q_i}$ represents the number of dynamic feature points in the set $Q_i$. The calculation method of is as formula (5).

$$u_{Q_i} = \sum_{x=1}^{Q_i.size} f(B_x), B_x \in Q_i$$

$$f(X) = \begin{cases} 0, & X \in C' \\ 1, & X \notin C' \end{cases} \tag{5}$$

The three categories of immovable objects, movable objects, and background objects are numbered as 0, 1, and 2. A vector $L = [l_1, l_2, \cdots, l_n], n = M.size, l_i \in \{0, 1, 2\}$ is created, so as to identify the category of all objects in the collection $M$. Then the object category and the number of feature points on the object are combined to determine the dynamic object, and the static feature points for feature matching are screened out. The algorithm is as follows:

Where $Q_i, 1 \leq i \leq n$ in scanning collection $M$,

If $Q_i$ is an immovable object, the object is considered and marked as static. If there is a dynamic point in $Q_i$, then it can be considered as a result of semantic segmentation error or optical flow calculation error. The dynamic point will be corrected as a static point.

If $Q_i$ is a movable object, then it is judged whether it is a movable object or not according to $u_{Q_i}$. If $u_{Q_i} < \alpha \cdot size(Q_i)$, where $\alpha$ is the judgment threshold and it is more appropriate to assign 0.2 after experiment and $size(Q_i)$ represents the number of feature points in $Q_i$, the object is considered to be static, otherwise it is considered to be a dynamic object. The object is marked as a dynamic object and deleted from the collection.

If $Q_i$ is the background objects, all dynamic feature points are deleted directly in $Q_i$.

After the complement of the above steps, all semantic feature points in the set are static feature points. As shown in formula (6), the feature points in each object instance $Q_i$ are taken out in the set $M$ and put them into the set $W$, which can be used as a set of candidate feature points for feature matching.

$$W = Q_1 \cup Q_2 \cup \cdots \cup Q_n, \quad Q_i \in M \tag{6}$$

## 3    Experiments and Results

This article uses the data set in the dynamic scenes provided by the TUM data set for testing. The hardware parameters of the experimental computer are AMD 2990wx 32-core 3GHz CPU, 96GB memory, 1TB hard disk, NVIDIA 2080Ti discrete graphics card × 2 and Ubuntu 16.04 operating system. The test set includes two types: low dynamic environment (sitting) and high dynamic environment (walking). Test indicators include Relative Pose Error (RPE) and Absolute Trajectory Error (ATE). The RPE indicator evaluates the drift of the visual odometry by calculating the difference between cameras' pose changes of two same time stamp, including translation drift and rotation drift. The ATE indicator directly calculates the difference between the true value and the estimated value of the camera pose.

### 3.1   Feature Points Extraction Experiment

(a) and (d) in Fig. 3 show the recognition effect of the roughly detect dynamic feature points algorithm on fr3_walking_half and fr3_walking_xyz of the TUM dataset. The red lines represent the motion trajectory of the dynamic points, while the green points represent the motion trajectory of the static feature points. It can be seen from these figures that some feature points are incorrectly recognized. For example, the feature points extracted from the legs of a person are marked as static feature points, due to little motion, while some of the feature points extracted on the table are marked as dynamic feature points due to calculation errors.

(b) and (e) in Fig. 3 show the result of semantic segmentation. As shown in (c) and (f) in Fig. 3, the results of the correction dynamic feature points algorithm are shown after the rough extraction of dynamic feature points. It can be found that the incorrectly identified dynamic feature points on the table are corrected to static, while the static feature points of the walking person are corrected to be dynamic.

### 3.2   Experiment Results of VO in TUM Dataset

This article compares RPE, ATE indicators with ORB-SLAM3 and some other SLAM algorithms in dynamic scenes. The best and second performance data are represented in the bolded data and underlined data.

The comparison results of translation drift of RPE between ORB-SLAM3 and the algorithm in this paper are shown in Table 1, respectively, including root mean square error (RMSE) and standard deviation (S.D.). Table 2 shows the comparison results of ATE between ORB-SLAM3 and the algorithm in this paper. According to the comparison results, compared with ORB-SLAM3, the proposed algorithm greatly reduces the translation drift and trajectory error of the visual odometry in the dynamic environment, especially in a high dynamic environment, it eliminated the interference of dynamic objects on extraction and matching of feature point, with significant improvement effects.

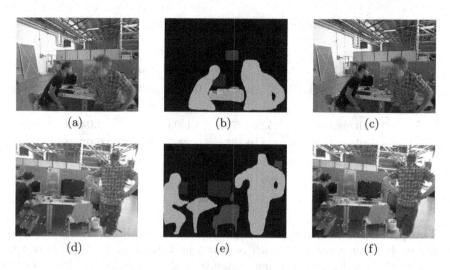

**Fig. 3.** Feature points extraction experiment.

**Table 1.** Comparisons of Translation drift of RPE between ORB-SLAM3 and Ours (unit m/s)

| Sequences | ORB-SLAM3 | | Ours | |
|---|---|---|---|---|
| | RMSE | S.D. | RMSE | S.D. |
| fr3_walking_static | 0.0172 | 0.0106 | **0.0121** | **0.0063** |
| fr3_walking_xyz | 0.1189 | 0.0597 | **0.0202** | **0.0099** |
| fr3_walking_rpy | 0.1754 | 0.1286 | **0.0488** | **0.0278** |
| fr3_walking_half | 0.2862 | 0.1910 | **0.0292** | **0.0139** |
| fr3_sitting_static | 0.0138 | 0.0078 | **0.0082** | **0.0042** |
| fr3_sitting_half | 0.0809 | 0.0579 | **0.0199** | **0.0093** |

**Table 2.** Comparisons of ATE between ORB-SLAM3 and Ours (unit m)

| Sequences | ORB-SLAM3 | | Ours | |
|---|---|---|---|---|
| | RMSE | S.D. | RMSE | S.D. |
| fr3_walking_static | 0.0271 | 0.0136 | **0.0101** | **0.0044** |
| fr3_walking_xyz | 0.1189 | 0.1417 | **0.0157** | **0.0077** |
| fr3_walking_rpy | 0.2836 | 0.1436 | **0.0456** | **0.0275** |
| fr3_walking_half | 0.4967 | 0.1972 | **0.0310** | **0.0154** |
| fr3_sitting_static | 0.0355 | 0.0165 | **0.0064** | **0.0034** |
| fr3_sitting_half | 0.1519 | 0.0701 | **0.0177** | **0.0081** |

**Table 3.** Comparisons of RMSE of RPE between Ours and Other Algorithms.

| Sequences | Translation drift, unit m/s | | | | |
|---|---|---|---|---|---|
| | Li [9] | Sun [12] | Yu [13] | Fang [4] | Ours |
| fr3_walking_static | 0.0327 | 0.0842 | **0.0102** | 0.0150 | 0.0121 |
| fr3_walking_xyz | 0.0651 | 0.1214 | 0.0333 | 0.0241 | **0.0202** |
| fr3_walking_rpy | 0.2252 | 0.1751 | 0.1503 | - | **0.0488** |
| fr3_walking_half | 0.0527 | 0.1672 | 0.0297 | 0.1369 | **0.0292** |
| fr3_sitting_static | 0.0231 | - | **0.0078** | 0.0115 | 0.0082 |
| fr3_sitting_half | 0.0389 | 0.0458 | - | **0.0189** | 0.0199 |

This paper also compares RPE and ATE indicators with some SLAM algorithms in dynamic scenarios, which is shown in Table 3 and Table 4. It can be seen from the results that the performance of the algorithm in this paper is significantly better than that of Li [9], Sun [12] and Fang [4] in both high and low dynamic environment. Compared with the algorithm of Yu [13], the performance is better in high-dynamic scenes, but the difference in RPE indicators is very small in some low-dynamic scenes, which can be considered as equal performance, because in this case dynamic objects have less influence on the estimation of camera pose estimation.

**Table 4.** Comparisons of RMSE of ATE between Ours and Other Algorithms.

| Sequences | Absolute Trajectory Error, unit m | | | | |
|---|---|---|---|---|---|
| | Li [9] | Sun [12] | Yu [13] | Fang [4] | Ours |
| fr3_walking_static | 0.0261 | 0.0656 | **0.0081** | 0.0104 | 0.0101 |
| fr3_walking_xyz | 0.0601 | 0.0932 | 0.0247 | 0.0164 | **0.0157** |
| fr3_walking_rpy | 0.1791 | 0.1333 | 0.4442 | - | **0.0456** |
| fr3_walking_half | 0.0489 | 0.1252 | **0.0303** | 0.0923 | 0.0310 |
| fr3_sitting_static | - | - | 0.0065 | 0.0065 | **0.0064** |
| fr3_sitting_half | 0.0432 | 0.047 | - | **0.0145** | 0.0177 |

The comparison of the camera trajectories between ORB-SLAM3 and the algorithm in this paper on the two dynamic environment sequences of the TUM dataset are shown in Fig. 4. It also shows the camera motion trajectory obtained by the ORB-SLAM3 algorithm without dynamic feature points detection and removal, and the camera motion trajectory after adding the algorithm in this paper. It can be seen from the figure that after adding the algorithm in this paper, the accuracy of motion trajectory estimation in a dynamic environment is significantly improved.

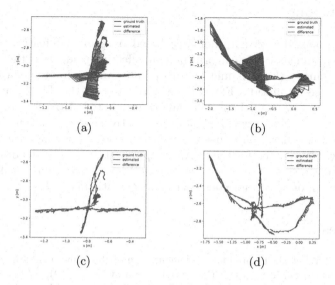

(a)                    (b)

(c)                    (d)

**Fig. 4.** Trajectories estimated by the proposed system.

## 3.3 Experiment in a Real Environment

To demonstrate the robustness and real-time performance of the algorithm in this paper, we conduct extensive experiments in a real environment. Images are captured by Kinect V2 camera with $480 \times 640$ resolution.

Meanwhile, the experiment with the algorithm in a real scene is also carried out in this paper. The experimental scene is a living room environment, which contains people walking around. Fig. 5 shows the removal effect of dynamic feature points. The green dots in the figure represent the feature points. Figure 5(a) shows the feature points extracted using the ORB feature points extraction algorithm, while Fig. 5(b) shows the segmentation results of the semantic segmentation thread. The different color masks in the figure represents different objects. Figure 5(c) shows the static feature points after the algorithm culls the dynamic feature points. It can be seen that the algorithm in this paper can remove the dynamic feature points well in the actual scene.

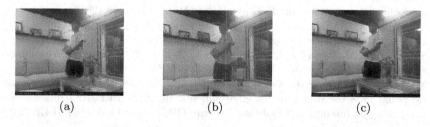

(a)                    (b)                    (c)

**Fig. 5.** The process of deleting dynamic feature points. (Color figure online)

## 4   Conclusions

This paper proposed a visual odometry based on semantic feature points in dynamic environments. The algorithm can effectively identify the dynamic objects and remove the dynamic feature points, and ultimately improve the accuracy of pose estimation. Extensive experiments on the TUM dataset and in the actual scene with ORB-SLAM3 and other related algorithms in dynamic scenarios were conducted, which shows the effectiveness and rationality of the algorithm in this paper. Following studies will be carried out and semantic feature points will be applied for the SLAM mapping stage. Since semantic feature points contain object motion state and instance information, which will provide effective input information for constructing a semantic map.

## References

1. Bouguet, J.Y., et al.: Pyramidal implementation of the affine lucas kanade feature tracker description of the algorithm. Intel Corporation **5**(1–10), 4 (2001)
2. Campos, C., Elvira, R., Rodríguez, J.J.G., Montiel, J.M., Tardós, J.D.: Orb-slam3: an accurate open-source library for visual, visual-inertial and multi-map slam. arXiv preprint arXiv:2007.11898 (2020)
3. Davison, A.J., Reid, I.D., Molton, N.D., Stasse, O.: Monoslam: real-time single camera slam. IEEE Trans. Pattern Anal. Mach. Intell. **29**(6), 1052–1067 (2007)
4. Fang, B., Mei, G., Yuan, X., Wang, L., Wang, Z., Wang, J.: Visual slam for robot navigation in healthcare facility. Pattern Recogn. **113**, 107822 (2021)
5. Forster, C., Pizzoli, M., Scaramuzza, D.: Svo: Fast semi-direct monocular visual odometry. In: 2014 IEEE International Conference on Robotics and Automation (ICRA), pp. 15–22. IEEE (2014)
6. Fuentes-Pacheco, J., Ruiz-Ascencio, J., Rendón-Mancha, J.M.: Visual simultaneous localization and mapping: a survey. Artif. Intell. Rev. **43**(1), 55–81 (2012). https://doi.org/10.1007/s10462-012-9365-8
7. He, K., Gkioxari, G., Dollár, P., Girshick, R.: Mask r-cnn. In: Proceedings of the IEEE International Conference on Computer Vision, pp. 2961–2969 (2017)
8. Klein, G., Murray, D.: Parallel tracking and mapping for small ar workspaces. In: 2007 6th IEEE and ACM International Symposium on Mixed and Augmented Reality, pp. 225–234. IEEE (2007)
9. Li, S., Lee, D.: Rgb-d slam in dynamic environments using static point weighting. IEEE Robot. Autom. Lett. **2**(4), 2263–2270 (2017)
10. Mur-Artal, R., Montiel, J.M.M., Tardos, J.D.: Orb-slam: a versatile and accurate monocular slam system. IEEE Trans. Rob. **31**(5), 1147–1163 (2015)
11. Mur-Artal, R., Tardós, J.D.: Orb-slam2: an open-source slam system for monocular, stereo, and rgb-d cameras. IEEE Trans. Rob. **33**(5), 1255–1262 (2017)
12. Sun, Y., Liu, M., Meng, M.Q.H.: Improving rgb-d slam in dynamic environments: a motion removal approach. Robot. Auton. Syst. **89**, 110–122 (2017)
13. Yu, C., Liu, Z., Liu, X.J., Xie, F., Yang, Y., Wei, Q., Fei, Q.: Ds-slam: a semantic visual slam towards dynamic environments. In: 2018 IEEE/RSJ International Conference on Intelligent Robots and Systems (IROS), pp. 1168–1174. IEEE (2018)

# Other AI Related Topics

# Robust Anomaly Detection from Partially Observed Anomalies with Augmented Classes

Rundong He[1], Zhongyi Han[1(✉)], Yu Zhang[1], Xueying He[2], Xiushan Nie[3], and Yilong Yin[1(✉)]

[1] Shandong University, Jinan, Shandong, China
ylyin@sdu.edu.cn
[2] Shandong University of Traditional Chinese Medicine, Jinan, Shandong, China
[3] Shandong Jianzhu University, Jinan, Shandong, China

**Abstract.** Anomaly detection is becoming increasingly ubiquitous in the society of data mining. Prominent anomaly detection works have achieved great success in theory and practice. However, they cannot handle the generalized semi-supervised scenario where there are only a handful of labeled anomalies, and plentiful unlabeled data that may bring in some instances of augmented anomaly classes but which are hard to be sampled. To solve this new problem, we propose a method called ACAD (Augmented Classes Anomaly Detection), which consists of three components. ACAD firstly suggests an augmented anomaly class discovery module that connects the isolation score and the similarity score to excavate the instances of hidden anomaly classes from unlabeled data accurately. ACAD then uses a specific cluster approach to compute useful similarity scores to separate reliable anomalous and normal instances among unlabeled data, respectively. ACAD finally builds a robust anomaly detector based on mined examples, successfully performing anomaly detection from partially observed anomalies with augmented classes. A series of empirical studies show that our algorithm remarkably outperforms state of the art on almost twenty datasets.

**Keywords:** Anomaly detection · Augmented classes · Positive unlabeled learning

## 1 Introduction

Anomaly detection is becoming increasingly ubiquitous, making the information-based society more safe, civilized, harmonious, beautiful. The application of anomaly detection plays a key role in our daily life, such as fraud detection [1], intrusion detection [27], anomaly vision detection [5], cardiac signal abnormality detection [17], rare disease detection [9] and video anomaly detection [34]. Anomaly detection techniques can help society find dangerous anomalies faster to reduce latent risks. For instance, anomaly detection techniques can be used to detect new emerging COVID-19 diseases effectively [31]. In addition, anomaly

© Springer Nature Switzerland AG 2021
L. Fang et al. (Eds.): CICAI 2021, LNAI 13070, pp. 347–358, 2021.
https://doi.org/10.1007/978-3-030-93049-3_29

detection is commonly used in security departments. Another example is surveillance video anomaly detection. By the technology of anomaly detection, the staff of traffic management department can more quickly detect traffic accidents or illegal driving on the road.

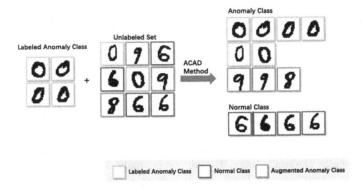

**Fig. 1.** Overview of anomaly detection from partially observed anomalies with augmented classes. '0' denotes labeled anomaly class. '8' and '9' denote augmented anomaly classes. '6' denotes normal class.

Prominent advanced algorithms have been achieved in anomaly detection. Dating back to 1887, Edgeworth [11] carried out a pioneering work to detect anomalies from statistical data. To date, anomaly detection has grown into a big tree with mainly three branches of approaches: unsupervised anomaly detection [19], supervised anomaly detection [13], and semi-supervised anomaly detection [28]. Unsupervised anomaly detection methods attempt to utilize the intrinsic statistical properties of data to find anomalies, including distance-based methods [23], density-based methods [3] and depth-based methods [26]. The branch of supervised anomaly detection obtains less attention since it can be easily resolved by traditional supervised classifiers, such as support vector machine [29], decision tree [30], and so on. Recently, semi-supervised anomaly detection, with partially observed anomalies, gets much attention because it meets actual needs in industry [15]. Zhang *et al.* [32] viewed semi-supervised anomaly detection as a positive unlabeled (PU) learning problem, which assumes that there are a handful of labeled positive (anomalous) examples and plentiful unlabeled data that include normal and anomalous instances. They then adopted a two-stage PU learning approach to perform anomaly detection. Zhang *et al.* [33] suggested that anomalies always involves multiple distinct classes, which are ignored by previous works. They thus built a multi-class anomaly detection model by dividing anomalies into multiple sub-categories.

While previous semi-supervised anomaly detection works have achieved great success, they cannot handle the generalized semi-supervised anomaly detection in which new (augmented) classes may appear in the unobserved anomalies.

That is, there are a few labeled positive (anomalous) examples and many unlabeled data that contain normal and anomalous instances that easily involve some instances of augmented anomaly classes. Let's take Fig. 1 as an example to illustrate the generalized semi-supervised anomaly detection scenario. We assume that '0' represents labeled anomaly class, '6' represents normal class, '8' and '9' represent the augmented anomaly classes. During the training stage, only labeled abnormal instances and unlabeled instances are provided, and unlabeled instances include known abnormal instances, normal instances and augmented anomaly instances. We define this generalized semi-supervised anomaly detection scenario as ADPOAC (Anomaly Detection from Partially Observed Anomalies with Augmented Classes).

In a nutshell, this new scenario is more practical, generalized, and challenging than existing settings. It is more practical because, for example, the anomalous data of COVID-19 diseases first appear in the unlabeled set and are not easily detectable. It is more generalized because anomalous data involve multiple sub-categories, and the sampling process easily misses a part of sub-categories due to the sampling bias. Meanwhile, the setting is appropriate for the common situation where the distribution of anomalies is relatively wide. Due to the sampling bias, the labeled anomalous instances only occupy a part of the whole distribution in a broad spectrum, such that lots of anomalous instances with different properties are missing into the labeled set. Moreover, it is exceptionally challenging because of the new emerging augmented classes. For example, in the detection of malicious URLs, new emerging malicious URLs are often difficult to detect.

In this paper, we propose a method called ACAD (Augmented Classes Anomaly Detection) to achieve robust anomaly detection from partially observed anomalies with augmented classes. As illustrated in Fig. 2, the framework of ACAD consists of three components. Given a set of training examples, ACAD firstly suggests an augmented anomaly class discovery module that connects the isolation score and the similarity score to discover the instances of augmented anomaly classes from unlabeled data. Secondly, ACAD attempts to mine useful information of the unlabeled instances by clustering them to select potential anomalies and reliable normal instances. Finally, ACAD builds a robust anomaly detector based on mined examples, successfully executing anomaly detection from partially observed anomalies with augmented class.

The main contributions of this paper include:

- To the best of our knowledge, it is the first time to study on anomaly detection from partially observed anomalies with augmented classes.
- To overcome this new open problem, we propose a first specific solution called ACAD (Augmented Classes Anomaly Detection). Specifically, we propose a Augmented classes discovering module which connects isolation score and similarity score to excavate the instances of augmented anomaly classes from unlabeled data.
- A series of empirical studies and experiment results have demonstrated the robustness and effectiveness of our method.

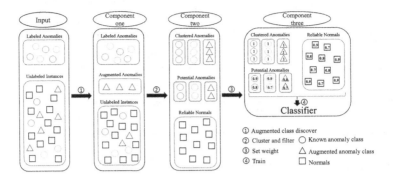

**Fig. 2.** Overview of ACAD for anomaly detection from partially observed anomalies with augmented classes.

## 2    Related Work

The setting we proposed in this paper lies on the intersection of anomaly detection, PU (Positive and Unlabeled) learning and new class discovery.

### 2.1    Anomaly Detection

Anomaly detection has attracted attention from various research areas and application domains. Anomaly detection is the task of finding patterns in data that do not conform to the expected behavior [34]. These inconsistent patterns are often referred to as outliers, inconsistent observations, anomalies or contaminants in different application domains.

Based on the different way of training, anomaly detection methods can be divided into three categories [4]: 1) Unsupervised methods, which detect anomalies based on intrinsic properties of the data samples without labels, assuming that normal instances have a higher occurrence frequency than anomalies. For example, a popular method is Isolation Forest [22]; 2) supervised methods, which infers functions from labeled training data sets. In this kind of methods, the training set generally contains both normal and anomalous samples. However, normal samples are more common than abnormal instances in practice, resulting suboptimal performance of supervised classifier. 3) semi-supervised methods, which provides a small number of labeled anomaly instances and a large number of unlabeled instances. Different from the existing work [33], in this paper, new anomaly classes are involved in the unlabeled data.

### 2.2    PU Learning

PU learning [18] is a special case of semi-supervised learning [6,35]. It is suitable when only positive and unlabeled data are available, and no negative sample is labeled. Existing methods can be divided into three types [33]: (1) Two-step method that firstly identifies reliable negative or positive examples and then adopts supervised learning techniques [21]. (2) Biased PU learning method [20]. unlabeled set is regarded as a negative sample set, and negative samples are given lower regularization weights, thereby allowing a certain amount of negative instances to be misclassified. (3) Class prior estimation method [18]. They estimates the class prior to positive classes.

However, one of the default assumptions of PU Learning is that labeled anomaly set only includes one type of class, which obviously does not meet actual needs. In [33], the application of PU learning is extended to the case where the marker set contains multiple classes. However, the existing research has not solved the problem: some abnormal categories that do not appear in the unmarked anomaly will appear in the unmarked data, which is a common but more complicated situation. Besides, recent works [7,24] show that including novel classes in the unlabeled set can hurt the performance compared to not using any unlabeled data. Based on this, in this paper, we aim to solve a more realistic and challenging task, requiring the model to discover augmented anomaly classes and normal class in the unlabeled data under the condition where the labeled data provided only contain known anomaly classes.

### 2.3  New Class Discovery

New class discovery refers to the process of learning to find categories that have not appeared before without labeled instances provided, which is a fundamental task for robust learning in open and dynamic environment. In [10], Da *et al.* proposed to exploit unlabeled data for the learning problem and design a novel algorithm by tuning the decision boundary to pass through low-density regions. In [16], Hsu *et al.* proposed to transfer predictive pairwise similarities from labeled to unlabeled data by posing the categorization problem as a surrogate same task problem. Deep Transfer Clustering extends the deep clustering framework by incorporating information about the known classes [14]. [12] trains the model by generating pseudo-labels of the unlabeled data using rank statistic. However, in our setting, the unlabeled data contain vast unknown normal class samples and a few new abnormal class samples, in addition to a few samples of known abnormal classes, which previous methods are not suitable to solve.

## 3  Proposed Method

### 3.1  Learning Set-Up

During the training phase, we are only provided a few labeled anomaly instances and plentiful unlabeled instances. We use $P^+ = \{(x_1, y_1), (x_2, y_2), \ldots, (x_m, y_m)\}$ to denote the labeled anomaly data set, and use $U = \{x_{m+1}, x_{m+2}, \ldots, x_{m+n}\}$ to denote the unlabeled set which contains vast normal instances and a few anomaly instances, where $m$ denotes the number of labeled instances and $n$ denotes the number of unlabeled instances respectively, and $n >> m$. We denote the set of ground-truth anomaly classes in $P^+$ and $U$ as $C^+$ and $C$, respectively. And we

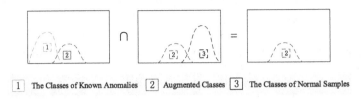

**Fig. 3.** Brief process of discovering augmented classes.

assume $C^+ \subset C$. We consider $C_N = C - C^+ \neq \emptyset$ as augmented anomaly classes. That is, augmented anomaly classes are classes that emerge in the unlabeled set $U$ and the test set but not emerge in the labeled data set $P^+$. The key of this problem is to excavate the augmented anomaly classes. To achieve that, we propose a first specific solution called ACAD (Augmented Classes Anomaly Detection), which contains three modules.

### 3.2  Proposed Method

**Augmented Classes Discovering.** The augmented class discovery method can effectively excavate the instances of augmented anomaly classes by connecting the isolation score [22] and the similarity score [33]. As illustrated in Fig. 3, we firstly acquire two score distributions and then take the intersection of them to get the instances of augmented classes. The two scores are introduced as follows.

*Isolation Score.* The concept of isolation score was first introduced in [22]. By using an extremely random forest, we can obtain the isolation score for each instance. Because the anomalous examples are few and different, the isolation score $IS$ is relatively high for the anomalous examples. The isolation score $IS(\boldsymbol{x})$ can be used to describe the probability of $\boldsymbol{x}$ being anomalous. We assume that there are $n$ instances, and

$$T(n) = H(n-1) - (2(n-1)/n) \tag{1}$$

denotes the average path length of unsuccessful search in the random trees. $H(n-1)$ is the harmonic number. The isolation score $IS(\boldsymbol{x})$ for instance $\boldsymbol{x}$ can be acquired by

$$IS(\boldsymbol{x}) = 2^{-\frac{E(h(\boldsymbol{x}))}{T(n)}}, \tag{2}$$

where $E(h(\boldsymbol{x}))$ is the average value of $h(\boldsymbol{x})$, which represents the path length of a point x in the isolation tree.

*Similarity Score.* The similarity score $SS(\boldsymbol{x})$ indicates the similarity of instance $\boldsymbol{x}$ to its nearest anomaly clustering center [33], and it is defined by

$$SS(\boldsymbol{x}) = \max_{k=1}^{N} e^{-(\boldsymbol{x} - \mu_k)^2}, \tag{3}$$

where $\mu_k$ represents the $k$-th clustering center and $N$ represents the number of clustering centers. The higher the $SS(x)$, the higher the probability of being anomalous.

Based on these two scores, on the one hand, we first obtain an isolation score distribution on unlabeled instances. Based on a predefined threshold $\beta_{is}$, we divide the distribution into two sets of $\{\tilde{P}^+, n\tilde{e}w\}$ and $\{\tilde{N}\}$ by

$$IS(x) > \beta_{is}, \tag{4}$$

such that we can get $\{\tilde{P}^+, n\tilde{e}w\}$. $\tilde{P}^+$, $n\tilde{e}w$, and $\tilde{N}$ represent possible anomalies of known classes, possible anomalies of augmented classes, and normal examples, respectively. On the other hand, we cluster the labeled anomalies to obtain the clustering centers. Then we compute the similarity score for each unlabeled instance to the nearest clustering center. Accordingly, we can acquire the similarity scores of all unlabeled instances. After setting a threshold $\beta_{ss}$, we can divide the similarity score distribution into $\{\tilde{P}^+\}$ and $\{n\tilde{e}w, \tilde{N}\}$ by

$$SS(x) < \beta_{ss}, \tag{5}$$

such that we can get $\{n\tilde{e}w, \tilde{N}\}$. Finally, we take the intersection between $\{\tilde{P}^+, n\tilde{e}w\}$ and $\{n\tilde{e}w, \tilde{N}\}$ to acquire the possible anomalies of augmented classes $\{n\tilde{e}w\}$.

**Unlabeled Anomalies Discovering.** The objective of this component is to separate reliable anomalous and normal instances among unlabeled data. Firstly, we obtain the new clustering centers by considering the newly discovered anomalies of augmented classes. The similarity score can be updated according to the new clustering centers. Then we combine it with the isolation score to get the weighted score $WS$ of each unlabeled instance by

$$WS(x) = \theta \times IS(x) + (1 - \theta) \times SS(x), \tag{6}$$

where $\theta \in [0, 1]$. Finally, we design two thresholds to select out reliable anomalous and normal instances, respectively, by

$$WS(x) > \alpha_w, \ WS(x) < \beta_w, \tag{7}$$

where $\alpha_w$ is the selection threshold of reliable anomalies, which is defined by

$$\alpha_w = \frac{1}{L + N_{new}} \sum_{i=1}^{L+N_{new}} WS(x_i), \tag{8}$$

where $L$ denotes the number of known anomalies and $N_{new}$ denotes the number of obtained reliable instances of augmented classes. Besides, $\beta_w$ is a predefined parameter. Note that these predefined thresholds $(\beta_{is}, \beta_{ss}, \beta_w)$ are set according to the class prior or the mean value of $WS$ respectively, following the existing work [33].

**Robust Weighted Classifier.** The objective of this component is to build a robust anomaly classifier. Firstly, we perform a weighting operation to obtain the weights of reliable anomalies, potential anomalies, and reliable normal instances, respectively. Specifically, we set the weights of anomalies in $P^+$ and anomalies in augmented classes obtained in the first component to one. For potential anomalies, we assign their weights according to the criterion:

$$w(\boldsymbol{x}) = \frac{WS(\boldsymbol{x}) - \alpha_w}{\max_x WS(\boldsymbol{x}) - \alpha_w}. \tag{9}$$

For reliable normal instances, we assign their weights according to the following distinct criterion:

$$w(\boldsymbol{x}) = \frac{\beta_w - WS(\boldsymbol{x})}{\beta_w - min_x WS(\boldsymbol{x})}, \tag{10}$$

where $0 < w(\boldsymbol{x}) \leq 1$.

Furthermore, we train a weighed classifier using all the above-mentioned instances and their weights. The objective of this classifier is given by

$$\ell(\boldsymbol{X}, \boldsymbol{Y}) = \sum_{i=1}^{n} w(\boldsymbol{x}_i)\ell(y_i, f(\boldsymbol{x}_i)), \tag{11}$$

where $w(\boldsymbol{x}_i)$ denotes the weight of instance $\boldsymbol{x}_i$, $n$ is the training instance number, and $\ell$ represents a general loss function.

## 4    Experiments

We evaluate our algorithm on twenty datasets against the state-of-the-art methods. The code about the proposed method will be available. The compared methods include unsupervised method (Isolation Forest [22]), semi-supervised methods (ADOA [33], PU learning [2]), and supervised method (PN learning [8]) that views $U$ set as negative set. We use the hinge loss of support vector machine as the basic loss function in Eq. (11) as same as [8]. All the experiments are repeated 20 times to obtain a fairer AUC score.

### 4.1    Experiments on Synthetic Data

We first verify the performance of our proposed method on a synthetic 32-dimensional dataset from Gaussian distributions. This dataset consists of four different Gaussian distributions $P_i = N(\mu_i, \Sigma_i), i \in \{0, 1, 2, 3\}$ that represent four different classes, respectively. $\mu_i \in R^{32}$ and $\Sigma_i \in R^{32 \times 32}$. Each element of $\mu_i$ belongs to the interval $[2, 4], [0, 1], [0, 2], [1, 3]$, respectively. And each diagonal element of $\Sigma_i$ belongs to the interval $[1, 8], [1, 2], [1, 2], [1, 2]$, respectively. We randomly select 100 samples respectively from $P_1$, $P_2$ to put them into $P^+$ as the set of labeled anomalies. We then put the remaining instances from $P_1$, $P_2$, the

**Table 1.** AUC on different datasets. Bold values indicate the best performance. Besides, • (○) means that ACAD is significantly better (worse) than the compared method. And Win/Tie/Loss are summarized in the last row.

| | ADOA | Unsupervised method | PU method | PN method | ACAD |
|---|---|---|---|---|---|
| Synthetics | 0.931 • | 0.784 • | 0.943 • | 0.797 • | **0.983** |
| Mnist | 0.858 • | 0.853 • | 0.808 • | 0.803 • | **0.996** |
| Arrhythmia | 0.669 • | 0.754 • | 0.686 • | 0.684 • | **0.785** |
| KDDCUP99 | 0.831 • | 0.865 • | 0.833 • | 0.776 • | **0.886** |
| Annthyroid | 0.914 • | 0.817 • | 0.807 • | 0.822 • | **0.934** |
| Synthetics | 0.989 • | 0.987 • | 0.985 • | 0.968 • | **0.990** |
| Http | **0.999** ○ | **0.999** ○ | 0.997 ○ | 0.996 ○ | 0.991 |
| Ionosphere | 0.934 • | 0.822 • | 0.887 • | 0.708 • | **0.948** |
| Letter | 0.671 • | 0.639 • | 0.629 • | 0.613 • | **0.810** |
| Mammography | 0.903 ○ | 0.890 • | **0.913** ○ | 0.695 • | 0.901 |
| Musk | **1.000** | 0.995 • | **1.000** | 0.979 • | **1.000** |
| Optdigits | 0.964 • | 0.719 • | 0.959 • | 0.959 • | **0.966** |
| Pendigits | 0.985 ○ | 0.956 • | **0.986** ○ | 0.869 • | 0.982 |
| Pima | 0.782 • | 0.712 • | 0.775 • | 0.716 • | **0.793** |
| Satellite | 0.803 • | 0.750 • | **0.837** ○ | 0.735 • | 0.820 |
| Satimage | 0.992 • | 0.990 • | 0.988 • | 0.962 • | **0.994** |
| Shuttle | 0.997 • | 0.994 • | 0.985 • | 0.983 • | **0.998** |
| Smtp | 0.902 • | 0.907 • | 0.876 • | 0.788 • | **0.956** |
| Speech | 0.644 • | 0.509 • | 0.550 • | 0.615 • | **0.661** |
| Thyroid | 0.994 • | 0.964 • | 0.996 • | 0.873 • | **0.997** |
| ACAD:W/T/L | 16/1/3 | 19/0/1 | 15/1/4 | 19/0/1 | Rank first 16/ 20 |

instances from $P_3$ (augmented anomalies), and the instances from $P_0$ (normal class) into $U$ set and the test set in a 4:1 ratio. The results on the synthetic dataset are located in the first raw of Table 1. ACAD remarkably outperforms compared methods in a large margin, which verifies its ability on generalized semi-supervised anomaly detection with augmented class discovery.

## 4.2 Experiments on Real-World Data

To verify the robustness of our method, we perform experiments on real-world datasets [25]. We strictly construct three generalized semi-supervised anomaly detection datasets based on three multi-class datasets of MNIST, arrhythmia, and KDDCUP99. We randomly define some classes as anomalies. Among them, partially anomaly classes are sampled as labeled anomaly classes, and the remaining anomaly classes are considered as augmented anomaly classes. Rows two to four of Table 1 report the results. Our algorithm significantly outperforms existing anomaly detection methods.

To verify the universality of our method, we perform experiments on the other datasets that only contain one anomaly class and one normal class. However, this anomaly class may implicitly contain some sub-categories of anomalies. We apply our algorithm to these datasets directly. Rows five to twenty of Table 1 report the results, which clearly demonstrate that our algorithm achieves the best performance than existing methods. These surprising results verify that our algorithm can apply to general anomaly detection scenarios.

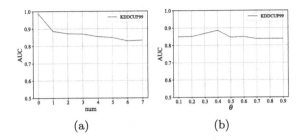

**Fig. 4.** The AUC of the (a) different numbers of augmented classes, and (b) different values of $\theta$ on KDDCUP99 dataset.

### 4.3 Ablation Study

*The Effect of the Number of Augmented Classes.* We dissect the strengths of augmented class discovery. Figure 4(a) reports the results of different numbers of augmented classes on KDDCUP99. The result of our method stably decreases with the increase of class numbers. This result fits our intuition that the richer the augmented classes, the harder the task.

*The Effect of the Weight.* $\theta$ Fig. 4(b) reports the results of different weights $\theta$ that play a key role in balancing the isolation score and similarity score. On the KDDCUP99 dataset, the weight of 0.4 achieves the best performance, while the other values also obtain stable results. This result verifies the effectiveness of our algorithm on discovering reliable anomalous and normal instances among unlabeled data.

## 5  Conclusions

We presented a new analysis of generalized semi-supervised anomaly detection, an under-explored but more realistic scenario. We also proposed the Augmented Class Anomaly Detection method tailored to the new problem. Specifically, we propose a Augmented classes discovering module which connects isolation score and similarity score to excavate the instances of augmented anomaly classes from unlabeled data. Extensive experiments show that the proposed method achieves the best performance compared to the state-of-the-art methods. However, the method proposed in this paper is based on traditional machine learning method, resulting in limited application. In the future, we will combine the idea of the method proposed in this paper with deep learning to deal with visual anomalies.

**Acknowledgement.** This work is supported by the National Natural Science Foundation of China (61876098), the National Key R&D Program of China (2018YFC0830100, 2018-YFC0830102).

# References

1. Abdallah, A., Maarof, M.A., Zainal, A.: Fraud detection system: a survey. J. Netw. Comput. Appl. **68**, 90–113 (2016)
2. Bekker, J., Davis, J.: Learning from positive and unlabeled data: a survey. Mach. Learn. **109**(4), 719–760 (2020). https://doi.org/10.1007/s10994-020-05877-5
3. Breunig, M.M., Kriegel, H.P., Ng, R.T., Sander, J.: LOF: identifying density-based local outliers. In: Proceedings of the 2000 ACM SIGMOD International Conference on Management of Data, pp. 93–104 (2000)
4. Chandola, V., Banerjee, A., Kumar, V.: Anomaly detection: a survey. ACM Comput. Surv. (CSUR) **41**(3), 1–58 (2009)
5. Chang, S., Du, B., Zhang, L.: BASO: a background-anomaly component projection and separation optimized filter for anomaly detection in hyperspectral images. IEEE Trans. Geosci. Remote Sens. **56**(7), 3747–3761 (2018)
6. Chapelle, O., Scholkopf, B., Zien, A.: Semi-supervised Learning (2006). IEEE Transactions on Neural Networks 20(3), 542–542 (2009)
7. Chen, Y., Zhu, X., Li, W., Gong, S.: Semi-supervised learning under class distribution mismatch. In: AAAI, pp. 3569–3576 (2020)
8. Cortes, C., Vapnik, V.: Support-vector networks. Mach. Learn. **20**(3), 273–297 (1995)
9. Cui, L., Biswal, S., Glass, L.M., Lever, G., Sun, J., Xiao, C.: CONAN: complementary pattern augmentation for rare disease detection. In: Proceedings of the AAAI Conference on Artificial Intelligence, vol. 34, pp. 614–621 (2020)
10. Da, Q., Yu, Y., Zhou, Z.H.: Learning with augmented class by exploiting unlabeled data. In: Proceedings of the AAAI Conference on Artificial Intelligence, vol. 28 (2014)
11. Edgeworth, F.Y.: XLI. On discordant observations. London Edinburgh Dublin Philos. Mag. J. Sci. **23**(143), 364–375 (1887)
12. Ehrhardt, S., Zisserman, A., Rebuffi, S., Han, K., Vedaldi, A.: Automatically discovering and learning new visual categories with ranking statistics. In: Proceedings of the 8th International Conference on Learning Representations, ICLR 2020. Schloss Dagstuhl-Leibniz-Zentrum für Informatik (2020)
13. Görnitz, N., Kloft, M., Rieck, K., Brefeld, U.: Toward supervised anomaly detection. J. Artif. Intell. Res. **46**, 235–262 (2013)
14. Han, K., Vedaldi, A., Zisserman, A.: Learning to discover novel visual categories via deep transfer clustering. In: Proceedings of the IEEE International Conference on Computer Vision, pp. 8401–8409 (2019)
15. Hasan, M., Choi, J., Neumann, J., Roy-Chowdhury, A.K., Davis, L.S.: Learning temporal regularity in video sequences. In: Proceedings of the IEEE Conference on Computer Vision and Pattern Recognition, pp. 733–742 (2016)
16. Hsu, Y.C., Lv, Z., Kira, Z.: Learning to cluster in order to transfer across domains and tasks. arXiv preprint arXiv:1711.10125 (2017)
17. Keller, F., Muller, E., Bohm, K.: HiCS: high contrast subspaces for density-based outlier ranking. In: 2012 IEEE 28th International Conference on Data Engineering, pp. 1037–1048. IEEE (2012)
18. Lee, W.S., Liu, B.: Learning with positive and unlabeled examples using weighted logistic regression. In: ICML, vol. 3, pp. 448–455 (2003)
19. Leon, E., Nasraoui, O., Gomez, J.: Anomaly detection based on unsupervised niche clustering with application to network intrusion detection. In: Proceedings of the 2004 Congress on Evolutionary Computation (IEEE Cat. No. 04TH8753), vol. 1, pp. 502–508. IEEE (2004)

20. Liu, B., Dai, Y., Li, X., Lee, W.S., Yu, P.S.: Building text classifiers using positive and unlabeled examples. In: Third IEEE International Conference on Data Mining, pp. 179–186. IEEE (2003)
21. Liu, B., Lee, W.S., Yu, P.S., Li, X.: Partially supervised classification of text documents. In: ICML, vol. 2, pp. 387–394 (2002)
22. Liu, F.T., Ting, K.M., Zhou, Z.H.: Isolation forest. In: 2008 Eighth IEEE International Conference on Data Mining, pp. 413–422. IEEE (2008)
23. Münz, G., Li, S., Carle, G.: Traffic anomaly detection using k-means clustering. In: GI/ITG Workshop MMBnet, pp. 13–14 (2007)
24. Oliver, A., Odena, A., Raffel, C.A., Cubuk, E.D., Goodfellow, I.: Realistic evaluation of deep semi-supervised learning algorithms. Adv. Neural. Inf. Process. Syst. **31**, 3235–3246 (2018)
25. Rayana, S.: ODDS library (2016)
26. Ruts, I., Rousseeuw, P.J.: Computing depth contours of bivariate point clouds. Comput. Stat. Data Anal. **23**(1), 153–168 (1996)
27. Sabahi, F., Movaghar, A.: Intrusion detection: a survey. In: 2008 Third International Conference on Systems and Networks Communications, pp. 23–26. IEEE (2008)
28. Sillito, R.R., Fisher, R.B.: Semi-supervised learning for anomalous trajectory detection. In: BMVC, vol. 1, pp. 035–1 (2008)
29. Tang, H., Cao, Z.: Machine learning-based intrusion detection algorithms. J. Comput. Inf. Syst. **5**(6), 1825–1831 (2009)
30. Wu, S.Y., Yen, E.: Data mining-based intrusion detectors. Expert Syst. Appl. **36**(3), 5605–5612 (2009)
31. Zhang, J., et al.: Viral pneumonia screening on chest X-ray images using confidence-aware anomaly detection (2020)
32. Zhang, Y.L., et al.: POSTER: a PU learning based system for potential malicious URL detection. In: Proceedings of the 2017 ACM SIGSAC Conference on Computer and Communications Security, pp. 2599–2601 (2017)
33. Zhang, Y.L., Li, L., Zhou, J., Li, X., Zhou, Z.H.: Anomaly detection with partially observed anomalies. In: Companion Proceedings of the Web Conference 2018, pp. 639–646 (2018)
34. Zhang, Y., Nie, X., He, R., Chen, M., Yin, Y.: Normality learning in multispace for video anomaly detection. IEEE Trans. Circuits Syst. Video Technol. **31**, 3694–3706 (2020)
35. Zhu, X.J.: Semi-supervised learning literature survey. Technical report, University of Wisconsin-Madison Department of Computer Sciences (2005)

# A Triple-Pooling Graph Neural Network for Multi-scale Topological Learning of Brain Functional Connectivity: Application to ASD Diagnosis

Zhiyuan Zhu[1,2,3], Boyu Wang[3], and Shuo Li[1(✉)]

[1] Department of Medical Imaging, Western University,
London, ON N6A5C1, Canada
[2] School of Artificial Intelligence, Beijing Normal University, Beijing 100875, China
[3] Department of Computer Science, Western University,
London, ON N6A5C1, Canada

**Abstract.** Brain functional connectivity (BFC) built from resting-state functional magnetic resonance imaging (rs-fMRI) has shown promising results in revealing the pathological basis of neurological disorders. However, a major problem is that existing approaches tend to limit analysis to a single scale, which unmatches the truth that modern neuroscience highlights BFC as a multi-scale topological architecture. Such a narrow view does lose representation of the inherent BFC topology and would weaken its performance. To solve this issue, we propose a novel triple-pooling graph neural network (TPGNN) to learn different scales of BFC topological knowledge in a task-adaptive way. Specifically, a pooling architecture with triple branches is designed to automate BFC analysis on the global scale, community scale, and region of interest (ROI) scale, respectively. We validate the diagnostic performance of TPGNN on an open autism spectrum disorder (ASD) dataset. Experimental results demonstrate that TPGNN outperforms the alternative state-of-the-art BFC analysis methods and provides potential biomarkers of different scales to benefit neuroscience.

**Keywords:** Multi-scale topological learning · Brain functional connectivity · Graph neural network

## 1 Introduction

Over the past decade, brain functional connectivity (BFC), which is constructed by the correlation between different regions of resting-state functional magnetic resonance imaging (rs-fMRI) series over time, has become a promising tool to investigate functional changes in patient populations [4,7]. Learning and understanding its intrinsic functional organization is essential for revealing the pathological basis of neurological disorders [9,17]. With the evolvement of practices in neuroscience, recently neuroscientists like Bassett et al. have shown evidence

© Springer Nature Switzerland AG 2021
L. Fang et al. (Eds.): CICAI 2021, LNAI 13070, pp. 359–370, 2021.
https://doi.org/10.1007/978-3-030-93049-3_30

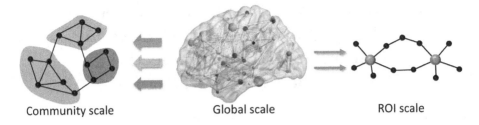

Community scale                Global scale                ROI scale

**Fig. 1.** Topologies of BFC at three different scales

that BFC is fundamentally multi-scale topological architecture [3]. However, the multi-scale reality has never been fully considered as a priority, which would weaken the effectiveness of brain disease diagnosis. The multi-scale topology of BFC focuses on the relations highlighted by different measurement methods [3]. The scales can range from individual regions to the brain as a whole. In neuroscience research, three scales are often concerned, namely the global scale, community scale, and region of interest (ROI) scale [2]. The specific schematic is shown in Fig. 1. Specifically, the global scale takes notice of the synergetic patterns all over the whole brain; community scale emphasizes the relationships between communities formed by clustering brain regions with similar properties together, such as the default mode community that is always active during people's passive rest [24]; ROI scale focuses on the relationship between brain regions associated with the task. Because of the close relationship between topological alterations and neurological diseases, each particular-scale topology could give its own unique insight [12,34], thus, developing an approach to bridge multiple scales of BFC will be an essential force in understanding the brain [16]. Multi-scale topological learning of BFC will help to better characterize its complex neural mechanisms and improve the diagnosis performance.

With the development of deep learning technologies [25,32], graph neural networks (GNNs) have shown superior performance in mining useful topological patterns of BFC for disease classification [1]. The main reason is that BFC can be seen as a graph consisting of a series of nodes and edges, GNN can explicitly capture the topological information by embedding and passing the nodes' information through edges in the graph. Although existing GNN-based approaches have proven effective in processing neuroimaging applications, they just focused on a single scale to identify the difference between patients and healthy controls (HCs) [6]. For example, Ktena et al. used a siamesed GNN to learn a global-based similarity metric with a supervised setting for disease diagnosis [18]; Yang et al. used an attention-based GNN to understand the causes of the disorders from its global topology [29]; Zhang et al. incorporates a spatial GNN to learn a global-scale embedding for classifying human brain activity under cognitive tasks which outperforms a multi-class support vector machine classifier [33]. None of the above GNN-based methods has sufficiently taken into account the multi-scale topological property learning. Such a narrow view does lose representation of the inherent BFC topology and would weaken its performance.

To overcome the issues, we proposed a novel triple-pooling graph neural network (TPGNN) to conduct multi-scale topological learning of BFC. As autism spectrum disorder (ASD) is a neurodevelopmental disorder closely related to potential dysfunction of the brain [19], our experiments are performed on the large challenging ABIDE database to distinguish ASDs from HCs. Our proposed TPGNN consists of three branches with different pooling mechanisms, each branch in TPGNN is designed to learn the above-mentioned three types of topological scales respectively. Our design achieves different-scale topological learning of BFC in a one-to-one corresponding architecture. Finally, we integrate information from different topological scales by a concatenate mechanism to realize a multi-scale representation of BFC.

Our contributions are summarized as follows: (1) Considering the multi-scale topological reality of BFC, we proposed TPGNN with three novel pooling branches to learn different scales of information respectively with a workable setting. (2) We applied the proposed method on a multi-center public ASD dataset for a classification task, achieving state-of-the-art performance (72.5% accuracy on the ABIDE dataset). (3) Our model is highly interpretable, different scales will provide their own insights into potential biomarkers.

## 2    Materials and Methods

### 2.1    Overview

The architecture of our proposed TPGNN computing framework is shown in Fig. 2. The three branches correspond to the topological learning for global scale, community scale, and ROI scale respectively. In Sect. 2.2, data processing was performed on each subject. With the BFC graphs constructed by the preprocessed fMRI data, the TPGNN framework was designed for the multi-scale topological learning of BFC (Sect. 2.3).

### 2.2    Image Acquisition and Preprocessing

**ABIDE Dataset.** The Autism Brain Imaging Data Exchange (ABIDE) aggregates data from 20 different sites and openly shares 1112 existing rs-fMRI datasets. The Preprocessed Connectomes Project (PCP) released preprocessed ABIDE data using several pipelines and calculated derivatives. We downloaded the time series data of brain regions processed by DPARSF (Data Processing Assistant for Resting-State fMRI) and AAL atlas, which comprises $R = 116$ cortical and subcortical regions. It is worth noting that the preprocessed data with missing values were excluded from our experiment. One can check the ABIDE Preprocessed website[1] for more details.

---

[1] http://preprocessed-connectomes-project.org/abide/download.html.

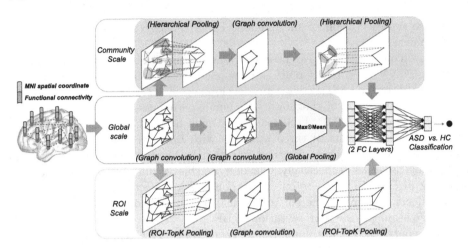

**Fig. 2.** A schematic diagram of the proposed TPGNN framework for ASD identification. Each of the three branches corresponds to the topological learning for global scale, community scale, and ROI scale respectively. Different graph pooling mechanisms are performed in each branch to construct and learn the corresponding-scale topology.

### 2.3   Triple-Pooling Graph Neural Network (TPGNN)

In this work, we focus on improving a multi-scale topological learning method for BFC, aiming at integrating topological information from different scales for better diagnosis of brain diseases. Specifically, a triple-pooling mechanism is proposed, with each pooling mechanism corresponding to each scale branch. For the global-scale branch, local-aggregation graph convolutions are first conducted to learn node embeddings, and then a global pooling technique is implemented to learn the topological representations of the whole brain. For the community-scale branch, a soft clustering graph pooling technology is exploited to carry out low-dimensional continuous community mining on the original BFC so as to realize the learning of the topological representations between communities. For the ROI scale, a projection-based pooling technology is adopted to extract the most indicative ROIs in a task-adaptive manner, thus establishing the topological representations between the task-related ROIs. Finally, We integrated information at different scales to realize multi-scale topological learning of BFC, and the concatenated features are then fed to two fully connected layers for classification.

**Graph Settings.** A graph $G$ with $n$ nodes is represented by $(V, X, A)$, where $V$ denotes the nodes, $X \in \mathbb{R}^{n \times d}$ denotes the nodes feature matrix that each node with a $d$-dimensional vector. $A \in \mathbb{R}^{n \times n}$ is the adjacency matrix. $A_{ij} = 1$ if there is an edge between $v_i$ and $v_j$, and 0 otherwise. Given a labeled graph dataset $\mathcal{D} = \{(G_1, y_1), (G_2, y_2), ...\}$, a graph classifier can be trained to learn a mapping relationship between graphs $G$ to the set of labels $y$.

From fMRI data to graph signals, the nodes are defined by the brain regions from a given atlas. For the adjacency matrix, Pearson's correlation coefficients (PCCs) of BOLD time-series between each pair of regions were used as the edges. Note that a sparse strategy of setting negative edges to zero was adopted because research shows that spurious negative correlation values would be introduced by the global signal regression preprocess [21]. Thus only the positive edges were retained as the adjacency matrices. For each node, the PCCs between the node and all of the others and the $(x, y, z)$ Montreal neurological institute (MNI) spatial coordinates were utilized as nodes features, where PCC reflects the functional connectivity strength, coordinates reflect the spatial position relationship. Since AAL atlas parcellates the whole brain into 116 regions, plus its spatial three-dimensional coordinates, nodes feature $X \in \mathbb{R}^{116 \times 119}$. It is worth noting that the features of each node are ordered and aligned.

**Graph Convolution.** Graph-based convolution tends to follow a message passing mechanism directly in the nodal domain; namely, each node sends its feature message to the nodes in its neighborhood; and then updates its feature representation by aggregating all messages received from the neighborhood. In our TPGNN framework, GraphSAGE is exploited as the basis for graph convolutions [14]. Specifically, GraphSAGE performs inductive learning by aggregating feature information from a node's local neighborhood with a trainable aggregation function. Each node $\mathrm{v} \in V$ has a feature vector $X_\mathrm{v} \in \mathbb{R}^d$, GraphSAGE layer infers a new vector representation $h_v^{(k+1)} \in \mathbb{R}^{d_2}$ for node v from its neighbors by:

$$h_v^{(k+1)} = g(W^{(k)}[h_v^{(k)} \odot h_{\mathcal{N}_v}^{(k+1)}]), \forall v \in V \tag{1}$$

where $[\odot]$ denotes a vector concatenation operation, $W^{(k)} \in \mathbb{R}^{d_2 \times d_1}$ is a weight matrix, $g$ is a non-linear activation function, $k$ is the layer index, and $h_v^{(0)} = X_\mathrm{v}$. $h_{\mathcal{N}_v}^{(k+1)}$ can be computed in several ways such as applying an element-wise max-pooling operation as follows:

$$h_{\mathcal{N}_v}^{(k+1)} = \max(\{g(W_{pool}h_u^{(k)} + b^{(k)}), \forall u \in \mathcal{N}_v\}), \forall v \in V \tag{2}$$

where $\mathcal{N}_v$ is defined as a fixed-size, uniformly drawn from the set of all neighbor nodes of $v$, and uniformly sampled differently through each layer.

**Global Pooling for Global-Scale Branch.** Inspired by [23] that leverages a global pooling technology to directly extract features from the unordered nodes as a task-related global representation, we performed the global max-pooling and the global mean-pooling on the embedded features of all nodes to select the most representative features in each feature channel as a global-scale representation. In detail, global max-pooling can effectively extract features that have a significant contribution to the task, and global mean-pooling can better preserve the localization features. Finally, they concatenate together to effectively learn the global topological representation of the brain.

**Hierarchical Pooling for Community-Scale Branch.** The hierarchical pooling layer is a differentiable graph pooling layer to build the community-scale topology. Specifically, we adopt the approach in [30] to perform the computation of the soft clustering assignments and learn hierarchical representations in an end-to-end manner with relatively few parameters. Specifically, from $k$-th layer to $(k+1)$-th layer, the hierarchical pooling uses $A^{(k+1)} = S^{(k)^T} A^{(k)} S^{(k)}$ and $X^{(k+1)} = S^{(k)^T} Z^{(k)}$ to coarsen the input graph by generating a new coarsened adjacency matrix $A^{(k+1)}$ and a new matrix of embeddings $X^{(k+1)}$ for each of the nodes/clusters in the coarsened graph. The assignment matrix $S^{(k)}$ and embedding matrix $Z^{(k)}$ can be respectively learned by two separate graph convolutions, this process can be represented as:

$$Z^{(k)} = GraphSAGE_{k,embed}(A^{(k)}, X^{(k)}) \tag{3}$$

$$S^{(k)} = softmax(GraphSAGE_{k,pool}(A^{(k)}, X^{(k)})) \tag{4}$$

The computed $Z^{(k)}$ and $S^{(k)}$ in the $k$-th layer can be used for the community construction in the $(k+1)$-th layer.

**ROI-topK Pooling for ROI-Scale Branch.** The ROI-topK Pooling Layer is designed for sparsing the original graph by selecting the top-$k$ nodes associated with the task [5,13]. Specifically, it implements downsampling on the graph by adaptively selecting nodes to form a smaller graph based on their scalar projection values on a trainable projection vector. Given a node $v$ with its feature vector $x_v$, the scale projection of $X_v$ on $p$ is $x_v p / \|p\|$, this measures the information reservation of node $v$ when projected onto the direction of $p$. The nodes with larger-scale projection values on direction $p$ can preserve the topology of the original graph to the greatest extent. Finally, the top-$k$ nodes are selected to form a new small graph for performing topological learning on the ROI scale.

**Pooling Layer Designing.** A triple-pooling framework is designed to build the learning of BFC at different scales. Especially for both the Community-scale and ROI-scale branched, they all have a continuous and gradual pooling behavior. Therefore, to make sure that these graph down-sampling layers behave idiomatically with respect to a wide class of graph sizes and structures, we adopt the approach of reducing the graph with a pooling ratio, $r \in (0, 1]$. Due to deeper graph neural network will result in over-smoothing effect which weakens the performance [20], a two-layer pool was designed on Community-scale and ROI-scale branches. Considering the typical numbers of communities discovered in previous literature that the brain exists in seven or eight ubiquitous communities [11,26], and brain disease usually occurs in a part of the brain regions [31], we set the pooling ratio $r = 0.25$ for the two continuous pooling branches. Thus after two pooling blocks, we will get eight ROIs and eight communities, respectively.

The architecture of TPGNN is depicted in Fig. 2. The number of feature maps for all the graph convolutional layers was set to 64, each convolution followed by ELU activation to increase non-linearity. A fusion block concatenated the features from all scales and then fed them into two fully connected (FC) layers. The FC layers were implemented with 32 and 2 channels, respectively, followed by a softmax function to drive a 2-class disease probability vector for each subject. Cross-entropy loss was employed on the labelled graphs for training the overall model. During each training process, all the subjects were randomly divided into ten subsets. Specifically, each time subjects in one subset were selected as the testing data, another one as the validating data, samples in the remaining eight subsets were used as the training data. The training was carried out for 300 epochs with a batch size of 16 and a learning rate at 0.001 for 100 epochs and then adjusted to 0.0001, using Adam optimizer. The weight decay parameter was 0.005. In this paper, we implemented the TPGNN model with PyTorch geometric [10], and train it on a GPU of Nvidia Tesla V100.

**Summary of Advantages.** (1) Our proposed TPGNN consists of three branches with different pooling layers, achieving different-scale topological learning of BFC in a corresponding one-to-one architecture. (2) Thanks to the advanced pooling architecture, the constructions of different scale topologies are done adaptively in a task-driven process, without any manual processing. (3) Our designed biologically significant pooling architecture consistent with real-world neuroscience guidance makes model results easier to interpret.

## 3   Results

We validate TPGNN on the large open ABIDE dataset. Comparative experiments and ablation experiments respectively proved the excellent performance of our method. Overall, our method achieved better performance in comparison to other incomplete-scale approaches in the ASDs vs. HCs classification task in terms of accuracy, f1-score, precision, and recall.

### 3.1   Comparison with State-of-the-Art Methods

TPGNN is compared with six state-of-the-art methods. These methods can be grouped into three categories: (1) Representative traditional machine learning (ML) methods including Random Forest (1000 trees), SVM (RBF kernel); (2) State-of-the-art CNN-based methods including DNN [28], ASD-DiagNet [8]; (3) State-of-the-art GNN-based methods including population-based GCN (P-GCN) [22], EGAT [29].

Table 1 shows the comparative results on the challenge ABIDE dataset. The traditional feature-based methods are to train and learn each edge of the BFC as a discrete feature, which loses the representation of its inherent topological structure. The methods based on deep learning can capture the topological information through a convolution mechanism, thus achieving better performance than traditional methods. Comparatively, graph-based methods (Parisot, EGAT, and

**Table 1.** Comparison with state-of-the-art methods for ASD identification using brain functional connectivity data.

| Methods | Accuracy (%) | Precision (%) | Recall (%) | F1-score (%) | Param. (k) |
|---|---|---|---|---|---|
| Random Forest | 61.41 ± 3.75 | 59.86 ± 6.00 | 76.46 ± 5.09 | 67.03 ± 5.02 | 4 |
| SVM | 65.76 ± 3.69 | 65.02 ± 3.84 | 75.12 ± 6.00 | 70.82 ± 3.99 | 4 |
| DNN [28] | 66.31 ± 3.01 | **73.14 ± 4.09** | 56.99 ± 4.78 | 68.92 ± 5.42 | 580 |
| ASD-DiagNet [8] | 67.28 ± 3.43 | 72.25 ± 3.69 | 64.25 ± 4.11 | 67.54 ± 5.74 | 500 |
| P-GCN [22] | 69.65 ± 4.28 | 71.90 ± 5.92 | 65.42 ± 7.45 | 68.25 ± 6.42 | 95 |
| EGAT [29] | 70.71 ± 3.75 | 71.57 ± 5.59 | 68.11 ± 4.47 | 71.88 ± 4.70 | 82 |
| Ours | **72.49 ± 3.65** | 72.17 ± 4.88 | **76.95 ± 5.58** | **74.51 ± 4.84** | 85 |

ours) yield larger performance gains, benefiting from the more flexible convolutional mechanism in GNNs comparing with CNN. The convolution on GNN is not limited to spatial structure. Thus, more suitable topological learning for our diagnostic tasks. Compared with the graph-based methods, our proposed TPGNN significantly improved the performance with an accuracy of 72.5%, F1-score of 74.5%, precision of 72.2%, recall of 77.0%, which outperforms other comparative methods. The main reason may be that TPGNN combines topological information at three different scales to provide a more comprehensive understanding of the brain states.

## 3.2    Ablation Analysis

**Table 2.** Results summary of ablation analysis on the key components of TPGNN. G, C, R stand for global scale, community scale, and ROI scale respectively. × stands for the removed branch.

| G | C | R | Accuracy (%) | Precision (%) | Recall (%) | F1-score (%) | Param. (k) |
|---|---|---|---|---|---|---|---|
| × |   |   | 70.41 ± 4.29 | 67.88 ± 4.27 | 74.98 ± 5.12 | 71.87 ± 5.70 | 62 |
|   | × |   | 70.93 ± 3.77 | 69.28 ± 4.96 | 73.55 ± 5.73 | 72.92 ± 4.60 | 50 |
|   |   | × | 69.50 ± 4.83 | 71.26 ± 4.78 | 68.70 ± 5.52 | 70.80 ± 5.68 | 60 |
| ✓ | ✓ | ✓ | **72.49 ± 3.65** | **72.17 ± 4.88** | **76.95 ± 5.58** | **74.51 ± 4.84** | 85 |

Ablation analysis is performed to explore and validate our designed triple-pooling mechanism in TPGNN. We remove each branch in turn (Global, Community, ROI), and leave the other structure unchanged, to perform ASD classification tasks. The comparing results of these aspects are summarized in Table 2. It can be observed that when one of these branches is abandoned, the performance decreases significantly compared to the original TPGNN that included all scales, indicating that topological information of different scales is necessary for the analysis of BFC. The effect of the combination of global scale and ROI scale is better than that of the other two-scale combined methods, indicating that the information on the most refined scale and the coarsest scale is more critical for the topological learning of BFC. In the end, the classification performance of combining all scale information is the best, indicating that each designed pooling layer is effective for constructing and learning the corresponding scale and that different scales provide complementary information for BFC topology learning from their perspective. Our experiment further illustrates the important value of learning from the inherent scale properties of the brain.

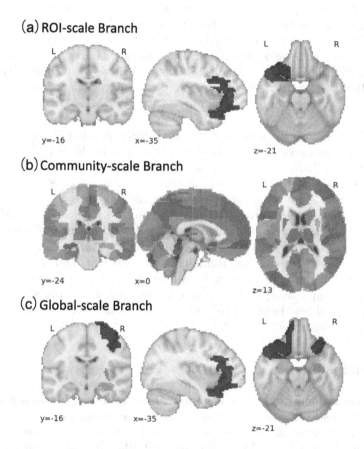

**Fig. 3.** The most contributing features selected by different pooling branches for ASD diagnostic tasks. (a) The eight most frequently selected regions by the ROI-topK pooling layer in the ROI-scale branch. (b) The community clustering pattern learned from the hierarchical pooling layer in the community-scale branch. (c) The twenty most frequently selected regions by the global pooling layer in the global-scale branch. Different colors denote different communities or regions.

### 3.3 Biomarkers Discovery

An interpretable network model is essential for investigating potential biomarkers of disease. Fortunately, the core of the pooling mechanism is to extract effective low-dimensional features from high-dimensional features. In TPGNN, each pooling branch can provide biomarkers that contribute to classification from its own perspective. For the global-scale branch, global max pooling selects the nodes with the greatest contribution to each input feature channel as a task-related global descriptor. The nodes with a larger frequency of being selected indicate having a weightier contribution to the ASD diagnosis task. For the community-scale branch, the learned embedding matrix $S^{(k)}$ provides the probability that each node belongs to a coarse community. The community clustering pattern of the whole brain under task induction will be got to benefit the task. For the ROI-scale branch, the ROI-topK pooling layer adaptively selects the most top-$k$ relevant brain regions for our task by $x_\mathrm{v}p/\|p\|$ as potential biomarkers.

Figure 3 visualizes the most contributing features selected by the three types of pooling mechanisms. It can be observed that most of the eight brain regions selected by the ROI-scale pooling are included in the twenty brain regions selected by the global-scale pooling, indicating that they have all carried out efficient feature learning and retrieval under the guidance of the task. Many previous studies have also reported these regions are implicated in ASDs. For example, the lateral prefrontal cortex and the lateral dorsal prefrontal cortex play an important role in a vital influence in multiple areas of child development, such as social cognition, communicative behavior, and moral behavior [15]. The superior parietal lobule is known to be involved in spatial orientation and receives many visual inputs as well as sensory input from one's hand, which may play a vital role in repetitive behavior and restricted interest symptoms found in individuals with ASD [27].

## 4    Conclusion

BFC has been widely used in the diagnosis of brain disorders. In this paper, motivated by the increasing evidence from neuroscience research that BFC presents a multi-scale topological structure, we proposed a multi-scale topological structure learning model called TPGNN to conduct BFC analysis at three different scales from coarse to fine for diagnosis of ASD. Specifically, a triple-pooling architecture is realized with three different branches, each branch with a different pooling layer is designed to achieve different-scale topological learning of BFC in a corresponding one-to-one architecture. The topological structure construction at different scales is done adaptively in a task-driven process, without any manual processing. Experimental results show that TPGNN effectively extracts task-related features on different scales from rs-fMRI data and improves the diagnostic performance of ASD over other methods. Moreover, the strong interpretability of TPGNN facilitates us to obtain the potential biomarkers of different scales to identify ASDs from HCs. Our work indicates that the multi-scale topological learning strategy of BFC shows great potential in the diagnosis of brain disorders and may provide insights for better understanding the mechanisms of brain function to move toward a more biologically meaningful brain connectome.

# References

1. Ahmedt-Aristizabal, D., Armin, M.A., Denman, S., Fookes, C., Petersson, L.: Graph-based deep learning for medical diagnosis and analysis: past, present and future. arXiv preprint arXiv:2105.13137 (2021)
2. Bassett, D.S., Sporns, O.: Network neuroscience. Nat. Neurosci. **20**(3), 353 (2017)
3. Betzel, R.F., Bassett, D.S.: Multi-scale brain networks. Neuroimage **160**, 73–83 (2017)
4. Büyükgök, D., Bayraktaroğlu, Z., Buker, H.S., Kulaksızoğlu, M.I.B., Gurvit, İH.: Resting-state fMRI analysis in apathetic Alzheimer's disease. Diagn. Interv. Radiol. **26**(4), 363 (2020)
5. Cangea, C., Veličković, P., Jovanović, N., Kipf, T., Liò, P.: Towards sparse hierarchical graph classifiers. arXiv preprint arXiv:1811.01287 (2018)
6. Davatzikos, C.: Machine learning in neuroimaging: progress and challenges. Neuroimage **197**, 652 (2019)
7. Dujardin, K., et al.: What can we learn from fMRI capture of visual hallucinations in Parkinson's disease? Brain Imaging Behav. **14**(2), 329–335 (2020)
8. Eslami, T., Mirjalili, V., Fong, A., Laird, A.R., Saeed, F.: ASD-DiagNet: a hybrid learning approach for detection of autism spectrum disorder using fMRI data. Front. Neuroinform. **13**, 70 (2019)
9. Feng, C., et al.: Prediction of trust propensity from intrinsic brain morphology and functional connectome. Hum. Brain Mapp. **42**(1), 175–191 (2021)
10. Fey, M., Lenssen, J.E.: Fast graph representation learning with PyTorch Geometric. In: ICLR Workshop on Representation Learning on Graphs and Manifolds (2019)
11. Finn, E.S., et al.: Functional connectome fingerprinting: identifying individuals using patterns of brain connectivity. Nat. Neurosci. **18**(11), 1664–1671 (2015)
12. Fornito, A., Zalesky, A., Breakspear, M.: The connectomics of brain disorders. Nat. Rev. Neurosci. **16**(3), 159 (2015)
13. Gao, H., Ji, S.: Graph U-Nets. In: International Conference on Machine Learning, pp. 2083–2092. PMLR (2019)
14. Hamilton, W., Ying, Z., Leskovec, J.: Inductive representation learning on large graphs. In: Advances in Neural Information Processing Systems, pp. 1024–1034 (2017)
15. Jun, E., Kang, E., Choi, J., Suk, H.I.: Modeling regional dynamics in low-frequency fluctuation and its application to autism spectrum disorder diagnosis. Neuroimage **184**, 669–686 (2019)
16. Khalid, A., Kim, B.S., Chung, M.K., Ye, J.C., Jeon, D.: Tracing the evolution of multi-scale functional networks in a mouse model of depression using persistent brain network homology. Neuroimage **101**, 351–363 (2014)
17. Khosla, M., Jamison, K., Ngo, G.H., Kuceyeski, A., Sabuncu, M.R.: Machine learning in resting-state fMRI analysis. Magn. Reson. Imaging **64**, 101–121 (2019)
18. Ktena, S.I., et al.: Metric learning with spectral graph convolutions on brain connectivity networks. Neuroimage **169**, 431–442 (2018)

19. Livingston, L.A., Colvert, E., Team, S.R.S., Bolton, P., Happé, F.: Good social skills despite poor theory of mind: exploring compensation in autism spectrum disorder. J. Child Psychol. Psychiatry **60**(1), 102–110 (2019)

20. Morris, C., et al.: Weisfeiler and leman go neural: higher-order graph neural networks. In: Proceedings of the AAAI Conference on Artificial Intelligence, vol. 33, pp. 4602–4609 (2019)

21. Murphy, K., Birn, R.M., Handwerker, D.A., Jones, T.B., Bandettini, P.A.: The impact of global signal regression on resting state correlations: are anti-correlated networks introduced? Neuroimage **44**(3), 893–905 (2009)

22. Parisot, S., et al.: Disease prediction using graph convolutional networks: application to autism spectrum disorder and Alzheimer's disease. Med. Image Anal. **48**, 117–130 (2018)

23. Qi, C.R., Su, H., Mo, K., Guibas, L.J.: PointNet: deep learning on point sets for 3D classification and segmentation. In: Proceedings of the IEEE Conference on Computer Vision and Pattern Recognition, pp. 652–660 (2017)

24. Raichle, M.E.: The brain's default mode network. Annu. Rev. Neurosci. **38**, 433–447 (2015)

25. Sarwar, T., Seguin, C., Ramamohanarao, K., Zalesky, A.: Towards deep learning for connectome mapping: a block decomposition framework. Neuroimage **212**, 116654 (2020)

26. Thomas Yeo, B., et al.: The organization of the human cerebral cortex estimated by intrinsic functional connectivity. J. Neurophysiol. **106**(3), 1125–1165 (2011)

27. Travers, B.G., Kana, R.K., Klinger, L.G., Klein, C.L., Klinger, M.R.: Motor learning in individuals with autism spectrum disorder: activation in superior parietal lobule related to learning and repetitive behaviors. Autism Res. **8**(1), 38–51 (2015)

28. Xu, T., Zhang, H., Huang, X., Zhang, S., Metaxas, D.N.: Multimodal deep learning for cervical dysplasia diagnosis. In: Ourselin, S., Joskowicz, L., Sabuncu, M.R., Unal, G., Wells, W. (eds.) MICCAI 2016. LNCS, vol. 9901, pp. 115–123. Springer, Cham (2016). https://doi.org/10.1007/978-3-319-46723-8_14

29. Yang, H., et al.: Interpretable multimodality embedding of cerebral cortex using attention graph network for identifying bipolar disorder. In: Shen, D., et al. (eds.) MICCAI 2019. LNCS, vol. 11766, pp. 799–807. Springer, Cham (2019). https://doi.org/10.1007/978-3-030-32248-9_89

30. Ying, Z., You, J., Morris, C., Ren, X., Hamilton, W., Leskovec, J.: Hierarchical graph representation learning with differentiable pooling. In: Advances in Neural Information Processing Systems, pp. 4800–4810 (2018)

31. Zhan, X., Yu, R.: A window into the brain: advances in psychiatric fMRI. BioMed Res. Int. **2015**, 542467 (2015)

32. Zhang, D., Chen, B., Chong, J., Li, S.: Weakly-supervised teacher-student network for liver tumor segmentation from non-enhanced images. Med. Image Anal. **70**, 102005 (2021)

33. Zhang, Y., Tetrel, L., Thirion, B., Bellec, P.: Functional annotation of human cognitive states using deep graph convolution. Neuroimage **231**, 117847 (2021)

34. Zhu, Z., Zhen, Z., Wu, X., Li, S.: Estimating functional connectivity by integration of inherent brain function activity pattern priors. IEEE/ACM Trans. Comput. Biol. Bioinform. (2020, early access)

# HierarIK: Hierarchical Inverse Kinematics Solver for Human Body and Hand Pose Estimation

Xinyu Yi, Yuxiao Zhou, and Feng Xu[✉]

BNRist and School of Software, Tsinghua University, Beijing, China
yixy20@mails.tsinghua.edu.cn, xufeng2003@gmail.com

**Abstract.** Estimating the 3D positions of the body and hand joints (keypoints) of human is widely investigated in computer vision. However, without knowing the joint rotations, the joint positions only cannot fully determine the body and hand motions and cannot drive a full geometry model of human to move. Thus the well-known Inverse Kinematics (IK) problem is proposed to solve for the full joint rotations based on some position constraints of the keypoints. Due to the noisy keypoint estimation and the ambiguity in twist rotations (Twist rotations cannot be solved by keypoint positions as they do not lead to position changes of any keypoint.), unnatural poses and jitters are commonly seen in current IK results. In this paper, we present a novel real-time IK solver where a deep multi-stage neural network takes the hierarchy of the human kinematic tree into consideration to robustly solve for the joint rotations in a depth-wise manner. Qualitative and quantitative results show the superiority of our hierarchical IK solver over the optimization-based methods and the solutions based on fully-connected neural networks.

**Keywords:** Inverse Kinematics · Pose estimation · Deep learning

## 1 Introduction

Human pose estimation from visual inputs is a widely-studied task with a variety of applications such as AR/VR, gaming, and sports. Recently, many works have explored estimating 3D keypoints from images [4,12,15,21,22,27]. However, such sparse keypoint coordinates are not sufficient for dense surface reconstruction. To obtain dense meshes, many works leverage statistical parametric human models [1,23,31] and regress the parameters for model-based mesh reconstruction. Estimating pose parameters, i.e. rotations of each joint, from keypoint positions, is known as solving the Inverse Kinematics (IK) problem. This task is previously addressed using iterative optimization [14,26,28–30,33,36]. Nevertheless, keypoint positions are often wrongly predicted in practice scenarios. In such cases, optimization-based IK approaches tend to give unnatural results due to the lack of prior knowledge on human poses. Solving IK problems accurately and robustly is of considerable importance in the area of human pose estimation.

This work was supported by the National Key R&D Program of China 2018YFA0704000, the NSFC (No.61822111, 61727808) and Beijing Natural Science Foundation (JQ19015).

L. Fang et al. (Eds.): CICAI 2021, LNAI 13070, pp. 371–382, 2021.
https://doi.org/10.1007/978-3-030-93049-3_31

To this end, we present HierarIK, a hierarchical neural IK solver that explicitly exploits the joint correlation information. To effectively learn the pose prior, we propose to formulate the IK task into sequential subtasks according to the kinematic tree structure, where each subtask incorporates a neural network to regress the local rotations for all joints at the same depth on the kinematic tree. Specifically, we handle each subtask in a depth-increasing order, i.e. we solve the rotation for the root joint first and the end effectors last. Once a subtask is solved, the estimated rotations are immediately used to rotate the descendant keypoint vectors inversely before entering the next subtask. As a result, the hierarchical solver achieves better accuracy than one-stage networks and optimization-based approaches, and shows great robustness to noisy 3D position inputs.

The hierarchical design of our IK solution is based on the observation that, according to the kinematic tree structure, 1) joints at a smaller depth (e.g. the root joint) are more important as their rotations affect all the descendants, while 2) joints at a larger depth (e.g. the end effectors) have less influence on the skeleton pose, but they can somehow correct the errors accumulated in ancestors by rotating towards the target. So the ancestors' rotations should be determined earlier than the descendants'. Therefore, we do not take all the joints equally and solve them together in a single forward pass as in [40,41], but instead, we divide the joints according to their hierarchical depths and solve them in a level order, which alleviates the error accumulation and helps the motion constraints be separately learned for each level of joints. Experiments show that our method is super-robust to noise and can even correct the errors in the input, which suggests HierarIK effectively learns the pose prior. In summary, our contributions are:

- An IK solver for human bodies and hands which significantly outperforms commonly-used iterative optimal algorithms and fully-connected networks.
- A novel hierarchical network that exploits the kinematic tree structure which results in super robustness and a large reduction of cumulative errors.

## 2 Related Work

General Inverse Kinematics (IK) problems, aiming at solving the joint rotations to reproduce the known positions of an articulated chain, have long been studied and can be solved using iterative optimization. IK for body and hand pose estimation can also be solved with neural networks leveraging motion datasets. In this section, we introduce the previous efforts on solving the IK problem.

### 2.1 Optimization-Based Methods

The most commonly used methods to solve IK problems are based on optimization. Due to the differentiability of the forward kinematics function, a gradient-descent-like updating strategy is often used. Many researches focus on the calculation of the inverse of the Jacobi matrix. The very early works take the transpose [35] or use the Moore-Penrose inverse [5]. To cope with the singularity, some works propose to use damped least squares [8], selectively damped least squares [6] or

Levenberg-Marquardt algorithm [34]. Colome and Torras [9] propose a filter for the singular values of the Jacobi matrix to limit its conditioning. Besides, some non-gradient-based greedy approaches such as angle-based [7,25] and position-based [2,3] algorithms are often used when end effector positions are known. However, these methods rely much on initialization and suffer from severe local-minimum problems. Some works [11,37] propose to solve a convex relaxation of the original IK problem for global solutions. For human pose estimation, adding constraint terms also helps to narrow the solution space [14,26,28–30,33]. As some restrictions are too complex to model and it is impossible to exclude all the infeasible solutions, optimization-based methods are limited. On the contrary, we propose a fully data-driven method where the motion prior can be effectively learned by neural networks, resulting in natural estimations.

## 2.2   Learning-Based Methods

Solving IK tasks for pose estimation with learning-based methods has attracted much attention recently. Leveraging human pose datasets, neural networks can automatically learn the constraints of motions and solve IK problems fast and accurately. Usually, a constraint of the degree of freedom (DOF) is imposed on the predicted poses to restrict the range of the Inverse Kinematics function [10,38]. A deep neural network is proved robust to noisy inputs and able to reduce errors through IK solving [16,40,41]. Zhou et al. [39] demonstrate the discontinuity of commonly used 3D rotation representations such as axis-angles, Euler angles, and quaternions, and present a continuous 6D representation which is suitable for network outputs. There are also works that regress joint angles from images directly [13,19,20,32]. However, none of these approaches take the advantage of the hierarchical structure of human bodies and hands, which often leads to severe error accumulation (i.e. large errors in the end effector positions). In contrast, we propose to use a hierarchical neural network that explicitly encodes the kinematic tree structure for IK problems in the pose estimation tasks, resulting in superior accuracy and robustness.

## 3   Method

In this section, we present HierarIK, a hierarchical Inverse Kinematics solver that estimates joint rotations from 3D positions for human pose estimation tasks. In Sect. 3.1, we introduce the kinematic model and the rotation representation we use in this paper. In Sect. 3.2, we introduce the proposed solution, HierarIK, for any specific kinematic tree structure. Finally, in Sect. 3.3, we present the details in training HierarIK on human body and hand pose datasets. An example of HierarIK for human body pose estimation is shown in Fig. 1.

### 3.1   Preliminary

**Body and Hand Model.** We use the SMPL [23] and MANO [31] model as our body and hand model respectively. The corresponding kinematic tree structures

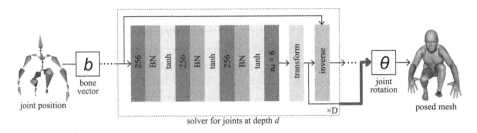

**Fig. 1.** HierarIK for human body pose estimation. We calculate bone vectors $b$ (position difference of each joint and its parent) from the input keypoints. Then we sequentially solve $D$ subtasks where $D$ is the height of the kinematic tree and $D = 9$ here for SMPL [23] body model. In the $d$th subtask, we estimate the local rotations represented by 6D [39] vectors for all $n_d$ joints at depth $d$ in the kinematic tree using a 4-layer neural network. Then we transform the results to rotation matrices and inversely rotate the corresponding descendant bone vectors. After solving $D$ subtasks sequentially, we get all joint rotations $\theta$ which can be used to drive the SMPL body model to move.

are shown in Fig. 2. We optionally add 5 external keypoints for each skeleton to deduce the rotations of the leaf joints. Both models share the same formulation defined as:

$$M(\boldsymbol{\beta}, \boldsymbol{\theta}) = W(\bar{\mathbf{T}} + B_S(\boldsymbol{\beta}) + B_P(\boldsymbol{\theta}), J(\boldsymbol{\beta}), \boldsymbol{\theta}, \mathcal{W}), \tag{1}$$

where $\bar{\mathbf{T}}$, $B_S$, and $B_P$ are the template mesh, the shape blendshape, and the pose blendshape respectively, $\boldsymbol{\beta}$ and $\boldsymbol{\theta}$ are shape and pose parameters respectively, $J$ is the joints, $\mathcal{W}$ is the blend weights, and $W(\cdot)$ is the linear blend skinning function. The mesh is used in our train data generation and result visualization.

**Rotation Representation.** We use 6D rotation representation [39] in our network output for better continuity. Compared with common representations such as quaternions, 6D is continuous and thus more suitable for neural networks, as demonstrated in [39]. Evaluations of rotation representations are in Sect. 4.2.

### 3.2 Hierarchical Inverse Kinematics Solver

We introduce our general IK solution in this section. IK problems are ill-posed as the same joint positions can be produced by more than one set of joint rotations. Usually, there are constraints on the skeleton movements, which can be difficult to model mathematically (e.g. human poses). Besides, the input positions may be noisy, making exact solutions infeasible. While the previous works [40,41] adopt neural networks to learn the direct mapping from joint positions to angles, we argue that it is significantly helpful to incorporate the hierarchy of the kinematic tree and solve IK in a multi-task manner. The overview of our method is shown in Fig. 1.

The input to our system is the joint positions $\{\boldsymbol{p}_j | j = 1, 2, \cdots, J\}$ where $J$ is the number of joints and $\boldsymbol{p}_j \in \mathbb{R}^3$ is the position of joint $j$. Leveraging the kinematic structure, we calculate the position difference of each joint and its parent in

| | Pelvis | | | | | | Wrist | | | |
|---|---|---|---|---|---|---|---|---|---|---|
| LHip | RHip | | Spine1 | | | Thumb0 | Index0 | Middle0 | Ring0 | Little0 |
| LKnee | RKnee | | Spine2 | | | Thumb1 | Index1 | Middle1 | Ring1 | Little1 |
| LAnkle | RAnkle | | Spine3 | | | Thumb2 | Index2 | Middle2 | Ring2 | Little2 |
| LFoot | RFoot | LCollar | Neck | RCollar | | Thumb3 | Index3 | Middle3 | Ring3 | Little3 |
| LToe | RToe | LShoulder | Head | RShoulder | | | | | | |
| | | LElbow | Head Top | RElbow | | | | | | |
| | | LWrist | | RWrist | | | | | | |
| | | LHand | | RHand | | | | | | |
| | | LFinger | | RFinger | | | | | | |

**Fig. 2.** The kinematic tree structure for SMPL body model (left) and MANO hand model (right). Root joints are shown in red. Additional keypoints are shown in blue. We add these keypoints to the IK input for the deduction of the leaf joints' rotations. (Color figure online)

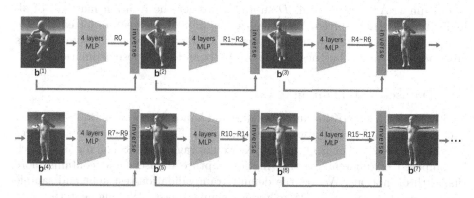

**Fig. 3.** Visualization of the input bone vectors for each subtask of HierarIK. We take the body IK as an example and visualize the bone vectors $b^{(d)}$ in SMPL meshes.

the kinematic tree, which is referred to as *bone vectors* $b = \{b_j | j = 2, 3, \cdots, J\}$ where $b_j = p_j - p_{\text{parent}(j)}$. This formulation encodes the joint relationships in the input and also facilitates the inverse operation of each subtask. To further exploit the kinematic tree structure, we group the joints by their depths in the tree, resulting in $D$ groups where the $d$th group contains $n_d$ joints at depth $d$. We denote the $m$th joint in the $d$th group as $j_m^{(d)}$ ($1 \leq m \leq n_d$). We formulate the IK task into corresponding $D$ subtasks which regress the rotations for each group sequentially in a depth-increasing order. Specifically, in the $d$th subtask, the input is bone vectors $b^{(d)}$ (and for the first one we have $b^{(1)} = b$). We first use a 4-layer neural network with batch normalization [17] and tanh activation to regress the local rotations for $n_d$ joints. Complex motion constraints for each group of joints can be effectively learned by these networks. Before entering the next subtask, we use the estimated rotations to inversely rotate the corresponding descendant bone vectors in $b^{(d)}$ to mitigate error accumulation, as shown in

---

**Algorithm 1:** The inverse operation of the $d$th subtask

---

**Input:** The input bone vectors $\boldsymbol{b}^{(d)}$ and the estimated joint rotation matrices
$\boldsymbol{R}_{j_1^{(d)}}, \boldsymbol{R}_{j_2^{(d)}}, \cdots, \boldsymbol{R}_{j_{n_d}^{(d)}}$.

**Output:** The input bone vectors for the next subtask $\boldsymbol{b}^{(d+1)}$.

1 **begin**

2     $\boldsymbol{b}^{(d+1)} \longleftarrow \boldsymbol{b}^{(d)}$

3     **for** $j_i^{(d)} \in \{j_1^{(d)}, j_2^{(d)}, \cdots, j_{n_d}^{(d)}\}$ **do**

4        **for** $j \in \text{descendant}(j_i^{(d)})$ **do**

5           $\boldsymbol{b}_j^{(d+1)} = \boldsymbol{R}_{j_i^{(d)}}^{-1} \boldsymbol{b}_j^{(d)}$

6        **end**

7     **end**

8 **end**

---

Algorithm 1. After solving all $D$ subtasks, we get the rotation matrices of all joints and the IK is solved. We visualize the input bone vectors for each stage in Fig. 3. As demonstrated in the experiments, our hierarchical design achieves great accuracy and robustness in human body and hand pose estimation.

### 3.3 Dataset and Training

We implement HierarIK for human bodies and hands (denoted as Body-HierarIK and Hand-HierarIK) for pose estimation tasks. The kinematic tree structures are shown in Fig. 2. For Body-HierarIK, we leverage AMASS [24] dataset which is composed of different existing motion capture datasets and contains more than 40 h of motions. We use the default train-validation-test split and sample the frames at a rate of 0.005 to reduce similar poses. We augment the data with random global rotations and normally distributed shape parameters with $\sigma_{\text{shape}} = 2$ for better robustness. The joint and the external keypoint positions are computed from the pose and shape parameters using Eq. 1. The bone vectors are directly calculated from the keypoint positions. We further add Gaussian noise with $\sigma_{\text{input}} = 0.04$ to the input bone vectors of each subtask to simulate real inputs which may be noisy. For Hand-HierarIK, we leverage the hand pose data in TCD-HandMocap of AMASS dataset, MANO [31] dataset, and FreiHAND [42] dataset. A similar augmentation with $\sigma_{\text{shape}} = 1.5$ and $\sigma_{\text{input}} = 0.004$ is applied during training and testing. For both tasks, we separately train each network with a batch size of 256 using an Adam [18] optimizer with a learning rate of $10^{-3}$. The training data is calculated by applying the inverse operations (see Algorithm 1) on the bone vectors using ground truth rotations, and the labels are acquired from ground truth poses. Both tasks use an L2 loss defined as:

$$\mathcal{L} = \|\boldsymbol{R}_{\text{pred}}^{(6\text{D})} - \boldsymbol{R}_{\text{GT}}^{(6\text{D})}\|_2, \tag{2}$$

where $\boldsymbol{R}_{\text{pred}}^{(6\text{D})}$ is the estimated local rotation in the 6D representation and $\boldsymbol{R}_{\text{GT}}^{(6\text{D})}$ is the corresponding ground truth.

# 4  Experiments

## 4.1  Data and Evaluation Metrics

We use the test split of AMASS dataset (including Transitions-Mocap and SSM-synced) for the evaluation of body IK, and the predetermined test split of the hand pose dataset (including MANO, TCD-HandMocap, and FreiHAND) for the evaluation of hand IK. We randomly initialize global rotations and shape parameters as similar in training. We use *joint error* as the metric which measures the mean distance error of all joints with the pelvis/middle0 (for body/hand) aligned, and *vertex error* which measures the mean distance error of all vertices of the estimated mesh also with the pelvis/middle0 joint aligned. The vertex coordinates are computed using Eq. 1 with the estimated poses and the mean shape. Note that twisting errors are reflected only in the vertex error.

## 4.2  Evaluations

**Rotation Representation.** To demonstrate the superiority of using the 6D rotation representation in IK tasks, we train the following two models for comparisons on solving body IK: *1)HierarIK-Q(uaternion)* where we use quaternions for all network outputs; *2)HierarIK-M(ixture)* where we use 6D for the first network (i.e. root rotation) and quaternions for the others. We use a similar loss as [41] for quaternions, which is defined as:

$$\mathcal{L} = \mathcal{L}_{l2} + \mathcal{L}_{\cos} + \mathcal{L}_{\text{norm}}. \tag{3}$$

Evaluation results on AMASS test split are shown in Table 1. HierarIK-Q performs significantly worse than HierarIK-M, indicating that 6D is necessary for the estimation of the root rotation due to its continuity property. Also, 6D performs a little better than quaternions on non-root joints as shown in the comparison between HierarIK-M and HierarIK. Due to the motion constraints of human bodies which limit the joint rotations to a contiguous space of quaternions, the accuracy improvements on non-root joints are small. Nevertheless, the continuous 6D still outperforms quaternions for the estimation of all joints.

**Training Noise.** To demonstrate the effectiveness of the training noise, we train HierarIK using different $\sigma_{\text{input}}$ ranging from 0 to 0.06 for bodies and 0 to 0.006 for hands. We show the estimated joint error under different input error expectations in Fig. 4. The results suggest that adding noise to the input during training is critical to improving robustness. Thus we use the exact model ($\sigma_{\text{input}} = 0$) only when solving IK with exact input as it is accurate enough, while we use $\sigma_{\text{input}} = 0.04$ for body and 0.004 for hand in practical IK tasks.

## 4.3  Comparisons

**Exact Input.** We compare HierarIK with commonly-used optimization- and learning-based methods to show our superiority. Concretely, we evaluate *1)IK-Optim(ize)* which is an optimization-based solution that minimizes the joint error

**Table 1.** Ablation study on rotation representations. We perform body IK on AMASS test split using different rotation representations. We adopt two settings where we train and evaluate the models using $\sigma_{\text{input}} = 0$ (i.e. solving IK with exact inputs) and $\sigma_{\text{input}} = 0.04$ (i.e. solving IK with noisy inputs, where the expected joint error in the input is 63.83 mm). The evaluations demonstrate the superiority of the 6D rotation representation [39] to quaternions in the IK tasks, which comes from its continuity property.

| | $\sigma_{\text{input}} = 0$ | | $\sigma_{\text{input}} = 0.04$ | |
|---|---|---|---|---|
| | Joint error (mm) | Vertex error (mm) | Joint error (mm) | Vertex error (mm) |
| HierarIK-Q | 224.10 | 244.39 | 263.27 | 293.49 |
| HierarIK-M | 8.63 | 10.66 | 60.50 | 72.05 |
| **HierarIK** | **7.08** | **9.15** | **58.62** | **69.72** |

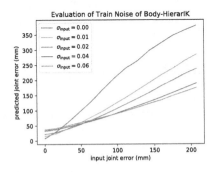 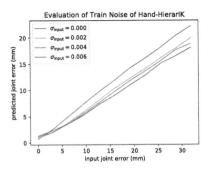

**Fig. 4.** Ablation study on training noise. We add Gaussian noise with different standard deviations $\sigma_{\text{input}}$ to the train data. We evaluate each model on noisy keypoints with different error expectations and plot the joint error curve. A lower slope indicates better robustness to input noise. Results show that training noise makes our model robust.

using the Levenberg-Marquardt algorithm; *2)IK-Optim(ize)-I(nitialize)* which is the same as IK-Optim but we initialize the pose with the ground truth global rotation (i.e. the root rotation is known); *3)MLP-L(ocal)* and *4)MLP-G(lobal)* which are deep multi-layer perceptions (MLPs) that directly regress local or global joint rotations from positions as in [41], except that we use 6D rather than quaternions as 6D performs significantly better (Sect. 4.2). We train and evaluate the MLPs and HierarIK on exact keypoint inputs. As shown in Table 2, HierarIK outperforms optimization-based and MLP solutions for both body and hand IK. We attribute our superiority to the hierarchical structure of our network, as we make full use of the joint correlations in the kinematic tree. The qualitative results in Fig. 5 also show that optimization-based methods suffer from local-minimum problems which lead to unnatural poses, and MLPs are often inaccurate due to the error accumulation from the root to the end effectors. Our hierarchical design helps to alleviate the problems and thus gives the best result. Please refer to our video for more qualitative comparison results.

**Table 2.** Quantitative comparison results for body and hand IK on exact inputs. HierarIK outperforms optimizing methods and MLPs due to the hierarchical design.

| | Body IK | | Hand IK | |
|---|---|---|---|---|
| | Joint error (mm) | Vertex error (mm) | Joint error (mm) | Vertex error (mm) |
| IK-Optim | 154.87 | 207.18 | 10.97 | 13.51 |
| IK-Optim-I | 14.40 | 23.11 | 1.91 | 3.17 |
| MLP-L | 32.83 | 39.35 | 2.16 | 2.73 |
| MLP-G | 22.33 | 22.73 | 2.20 | 2.64 |
| **HierarIK** | **7.08** | **9.15** | **0.57** | **0.72** |

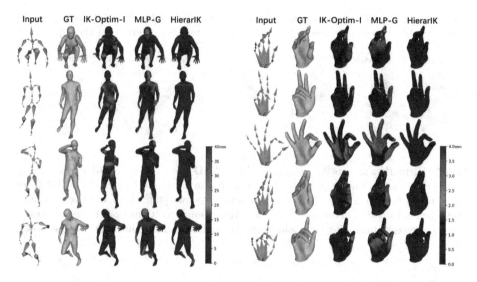

**Fig. 5.** Qualitative comparison results for body and hand IK. The input is exact joint positions. Color encodes the vertex error of the estimated pose. Front views are shown here but we should note that global rotations are not known except in IK-Optim-I.

**Noisy Input.** To demonstrate the robustness and accuracy of HierarIK, we compare learning-based methods on noisy inputs for body and hand IK tasks. As shown in Fig. 6, HierarIK outperforms the other methods and can reduce the errors in the input keypoints by IK solving. This is more evident in hand IK because there are more constraints for hand motions compared with body motions, and thus adding Gaussian noise to the hand joints may more easily result in wrong positions, which can be corrected by HierarIK. We show more cases where HierarIK corrects the errors in FreiHAND dataset by re-estimating the pose parameters from the ground truth poses, and estimates natural hand poses even from random noise (Fig. 7). We attribute this to the training noise and our multi-stage design which helps with the layer-wise learning of the pose prior. These results demonstrate our superior robustness to noisy inputs.

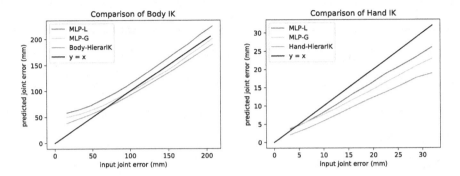

**Fig. 6.** Quantitative comparison results of solving body and hand IK with noisy inputs. The curve below $y = x$ means that the method corrects some errors in the input.

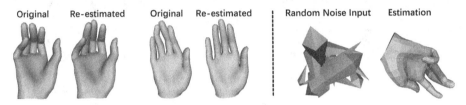

**Fig. 7.** HierarIK corrects the errors in FreiHAND dataset (left) and gives natural results from random noise (right). The original pose parameters in FreiHAND dataset are acquired by fitting a hand model from multiple views [42]. There are several unnatural poses in the original dataset. We compute keypoint positions from the pose parameters and re-estimate joint rotations using HierarIK, resulting in more natural poses.

## 5   Conclusion

This paper presents a novel data-driven Inverse Kinematics solver for human body and hand pose estimation tasks. As the commonly-used optimization-based methods rely on good initialization, noiseless input positions, and well-designed constraints which are difficult to achieve, we take the advantage of deep neural networks and learn the pose prior from body and hand pose datasets. To cope with the difficulty that the poses estimated by MLPs are often inaccurate due to the error accumulation from the root to the end effectors, we propose a novel hierarchical solution that exploits the correlations of the keypoints and estimates joint rotations sequentially according to the joint's depth in the kinematic tree. We use an inverse operation to broadcast the estimated joint rotation information to all its descendants, which helps with the estimation of local rotations and the mitigation of the error accumulation. The multi-stage design helps the model separately learn the motion constraints for joints at different depths. Experiments demonstrate our superiority to common optimization- and learning-based solutions: super accurate, and robust to noisy keypoint inputs.

# References

1. Anguelov, D., Srinivasan, P., Koller, D., Thrun, S., Rodgers, J., Davis, J.: SCAPE: shape completion and animation of people. ACM Trans. Graph. (TOG) **24**, 408–416 (2005)
2. Aristidou, A., Chrysanthou, Y., Lasenby, J.: Extending FABRIK with model constraints. Comput. Animat. Virtual Worlds **27**, 35–57 (2016)
3. Aristidou, A., Lasenby, J.: FABRIK: a fast, iterative solver for the inverse kinematics problem. Graph. Models **73**, 243–260 (2011)
4. Artacho, B., Savakis, A.: UniPose: unified human pose estimation in single images and videos, June 2020
5. Ben-Israel, A., Greville, T., Ben-Israel, A., Greville, T.N.E.: Generalized Inverses: Theory and Applications, ISBN 0-387-00293-6, January 2003
6. Buss, S., Kim, J.S.: Selectively damped least squares for inverse kinematics. J. Graph. Tools **10**, 37–49 (2005)
7. Canutescu, A.A., Dunbrack Jr, R.L.: Cyclic coordinate descent: a robotics algorithm for protein loop closure. Prot. Sci. **12**, 936–972 (2003)
8. Chiaverini, S., Siciliano, B., Egeland, O.: Review of the damped least-squares inverse kinematics with experiments on an industrial robot manipulator. IEEE Trans. Control Syst. Technol. **2**, 123–134 (1994)
9. Colome, A., Torras, C.: Closed-loop inverse kinematics for redundant robots: comparative assessment and two enhancements. IEEE/ASME Trans. Mechatron. **20**, 944–955 (2015)
10. Csiszar, A., Eilers, J., Verl, A.: On solving the inverse kinematics problem using neural networks, pp. 1–6, November 2017
11. Dai, H., Izatt, G., Tedrake, R.: Global inverse kinematics via mixed-integer convex optimization. Int. J. Robot. Res. **38**, 1420–1441 (2019)
12. Doosti, B., Naha, S., Mirbagheri, M., Crandall, D.: HOPE-Net: a graph-based model for hand-object pose estimation, pp. 6607–6616, June 2020
13. Habermann, M., Xu, W., Zollhofer, M., Pons-Moll, G., Theobalt, C.: DeepCap: monocular human performance capture using weak supervision, June 2020
14. Habermann, M., Xu, W., Zollhöfer, M., Pons-Moll, G., Theobalt, C.: LiveCap: real-time human performance capture from monocular video. ACM Trans. Graph. **38**, 1–17 (2019)
15. Habibie, I., Xu, W., Mehta, D., Pons-Moll, G., Theobalt, C.: In the wild human pose estimation using explicit 2D features and intermediate 3D representations, pp. 10897–10906 (2019)
16. Holden, D.: Robust solving of optical motion capture data by denoising. ACM Trans. Graph. (TOG) **37**(4), 1–12 (2018)
17. Ioffe, S., Szegedy, C.: Batch normalization: accelerating deep network training by reducing internal covariate shift (2015)
18. Kingma, D., Ba, J.: Adam: a method for stochastic optimization. In: International Conference on Learning Representations, December 2014
19. Kocabas, M., Athanasiou, N., Black, M.: VIBE: video inference for human body pose and shape estimation, pp. 5252–5262, June 2020
20. Kolotouros, N., Pavlakos, G., Black, M., Daniilidis, K.: Learning to reconstruct 3D human pose and shape via model-fitting in the loop, pp. 2252–2261, October 2019
21. Kundu, J., Seth, S., Jampani, V., Rakesh, M., Babu, R., Chakraborty, A.: Self-supervised 3D human pose estimation via part guided novel image synthesis, pp. 6151–6161, June 2020

22. Li, S., Ke, L., Pratama, K., Tai, Y.W., Tang, C.K., Cheng, K.T.: Cascaded deep monocular 3D human pose estimation with evolutionary training data, June 2020
23. Loper, M., Mahmood, N., Romero, J., Pons-Moll, G., Black, M.: SMPL: a skinned multi-person linear model, vol. 34, November 2015
24. Mahmood, N., Ghorbani, N., Troje, N., Pons-Moll, G., Black, M.: AMASS: archive of motion capture as surface shapes, pp. 5441–5450, October 2019
25. Martin, A., Barrientos, A., del Cerro, J.: The natural-CCD algorithm, a novel method to solve the inverse kinematics of hyper-redundant and soft robots. Soft Robot. **5**, 242–257 (2018)
26. Mehta, D., et al.: XNect: real-time multi-person 3D motion capture with a single RGB camera. ACM Trans. Graph. **39**, 82 (2020)
27. Mehta, D., et al.: Single-shot multi-person 3d pose estimation from monocular RGB, pp. 120–130, September 2018
28. Mehta, D., et al.: VNect: real-time 3D human pose estimation with a single RGB camera. ACM Trans. Graph. **36**, 1–14 (2017)
29. Mueller, F., et al.: GANerated hands for real-time 3D hand tracking from monocular RGB, pp. 49–59 (2018)
30. Mueller, F., Mehta, D., Sotnychenko, O., Sridhar, S., Casas, D., Theobalt, C.: Real-time hand tracking under occlusion from an egocentric RGB-D sensor, October 2017
31. Romero, J., Tzionas, D., Black, M.: Embodied hands: modeling and capturing hands and bodies together. ACM Trans. Graph. **36**, 1–17 (2017)
32. Shi, M., et al.: MotioNet: 3D human motion reconstruction from monocular video with skeleton consistency (2020)
33. Shimada, S., Golyanik, V., Xu, W., Theobalt, C.: PhysCap: physically plausible monocular 3D motion capture in real time. ACM Trans. Graph. **39**, 1–16 (2020)
34. Sugihara, T.: Solvability-unconcerned inverse kinematics by the Levenberg-Marquardt method. IEEE Trans. Robot. **27**(5), 984–991 (2011)
35. Wolovich, W., Elliott, H.: A computational technique for inverse kinematics. In: Proceedings of the IEEE Conference on Decision and Control 23, January 1985
36. Xiang, D., Joo, H., Sheikh, Y.: Monocular total capture: posing face, body, and hands in the wild (2018)
37. Yenamandra, T., Bernard, F., Wang, J., Mueller, F., Theobalt, C.: Convex optimisation for inverse kinematics, pp. 318–327, September 2019
38. Zhou, X., Sun, X., Zhang, W., Liang, S., Wei, Y.: Deep kinematic pose regression, September 2016
39. Zhou, Y., Barnes, C., Lu, J., Yang, J., Li, H.: On the continuity of rotation representations in neural networks, pp. 5738–5746, June 2019
40. Zhou, Y., Habermann, M., Habibie, I., Tewari, A., Theobalt, C., Xu, F.: Monocular real-time full body capture with inter-part correlations, December 2020
41. Zhou, Y., Habermann, M., Xu, W., Habibie, I., Theobalt, C., Xu, F.: Monocular real-time hand shape and motion capture using multi-modal data, pp. 5345–5354, June 2020
42. Zimmermann, C., Ceylan, D., Yang, J., Russell, B., Argus, M., Brox, T.: FreiHAND: a dataset for markerless capture of hand pose and shape from single RGB images, pp. 813–822, October 2019

# A Novel Conditional Knowledge Graph Representation and Construction

Tingyue Zheng, Ziqiang Xu, Yufan Li, Yuan Zhao, Bin Wang$^{(\boxtimes)}$,
and Xiaochun Yang

Northeastern University, Shenyang 110819, China
{binwang,yangxc}@mail.neu.edu.cn

**Abstract.** Many statements in the massive scientific text data are in the form of conditional sentences. Conditions are of great importance to facts. Existing conditional knowledge graphs have introduced condition triples, but ignore the latent semantic relations between fact and condition triples and the logical relationships among condition triples. To address these issues, we propose a novel conditional knowledge graph representation, which is a nested hierarchical triple. We design a new extraction strategy that employs a text hierarchy parsing module to extract the semantic relations between facts and conditions and a triple extraction module to extract fact and condition triples. Moreover, we provide a corresponding knowledge storage scheme which can store conditional knowledge. Experimental results on our constructed conditional dataset show that our model can not only capture semantic relations between fact and condition triples as well as logical relationships among condition triples, but also significantly improve the accuracy of triple extractions compared to baselines.

**Keywords:** Conditional knowledge graph · Knowledge representation · Triple extraction · Knowledge storage

## 1 Introduction

In recent years, knowledge graph has gradually become the core technology driving the development of artificial intelligence, which plays a vital role in various applications, such as question answering [16] and information retrieval. We observe that there are a certain percentage of facts or truths in the form of conditional sentences in the vast amount of scientific text data. We conduct statistics on scientific literature in several fields. The proportions of conditional sentences in biomedical, chemistry, mathematics and computer science are 25.6%, 20.5%,

The work is partially supported by the National Key Research and Development Program of China (2020YFB1707900), National Natural Science Foundation of China (62072088), Ten Thousand Talent Program (ZX20200035), and Liaoning Distinguished Professor (XLYC1902057).

L. Fang et al. (Eds.): CICAI 2021, LNAI 13070, pp. 383–394, 2021.
https://doi.org/10.1007/978-3-030-93049-3_32

384    T. Zheng et al.

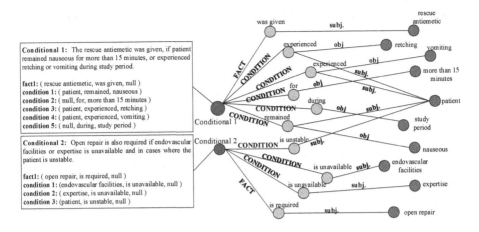

**Fig. 1.** The extractions generated by MIMO.

22.1% and 23.3% respectively. Traditional knowledge graphs represent knowledge as entity-level semantic networks. Their constructions only rely on plain triples extracted from text without distinguishing whether they are condition or fact triples. Consequently, important conditional information is lost, and the conditional semantics expressed by conditional sentences cannot be formalized completely and accurately. Furthermore, this limits the expression ability of knowledge graph, potentially affecting the exploration of downstream tasks, such as knowledge reasoning.

To the best of our knowledge, Jiang et al. [7] are the first to attempt to extract conditional information from text. They propose a MIMO model to extract fact and condition tuples, but ignore several issues. First, MIMO cannot clearly distinguish the mutual constraint relations between tuples. Take the first sentence in Fig. 1 as an example. The prepositional phrase "for more than 15 min" and "during study period" modify "patient remained nauseous" and "patient experienced retching or vomiting", respectively. However, following its extractions, we cannot conclude that condition 2 is a further constraint on condition 1, and condition 5 is a specific constraint on condition 3 and 4. Secondly, it does not consider the specific logical relations between condition tuples, such as "AND"/"OR", which have different meanings. For instance, from MIMO's extractions for the conditional 2 in Fig. 1, we cannot judge the logical relations between the three condition tuples. The real meaning of the conditional is that the fact will be valid under the existence of both condition 1 and 3 or both condition 2 and 3. To solve these problems, we intend to construct a richer knowledge graph based on the characteristics of existing data.

The construction of conditional knowledge graph is extremely challenging. First of all, conditional sentences have complex structures. To capture and formalize the conditional semantic information accurately, we need to analyze and demonstrate conditionals in depth from the perspective of morphology and syntax in linguistics. Secondly, we expect to extract not only fact and condition triples, but also rich semantic relations between triples, so we need to reckon with how to trade off both as well as possible. Finally, it has a significance to store and manage the extracted results. To be specific, during the construction process of conditional knowledge graph, we are confronted with the following three challenges:

- There are rich hierarchical semantic relations between triples which are beyond the capabilities of the current triple representation. It is a great challenge to express the hierarchical semantic information of conditional sentences more clearly without abandoning the advantages of the original triple representation. Therefore, we propose a novel conditional knowledge graph representation, which is in the form of nested hierarchical triples. It can represent entity-level and triple-level semantic relations simultaneously.
- In order to construct such a conditional knowledge graph, how to extract the fact and condition triples and the semantic relations is also a difficulty. We design a textual structure hierarchical parsing module to derive the hierarchical structure of conditional sentences. And then we extract fact and condition triples using the OpenIE method based on sequence labeling.
- Knowledge storage and management is necessary to facilitate the downstream application. Following our proposed logical model of conditional knowledge graph, knowledge structure to be stored has changed to some extent. How to build the mapping of knowledge logical representation to the physical storage based on the existing storage model is also crucial. Thus we propose a knowledge storage scheme to fit the proposed representation model based on the relational database to ensure the portability of knowledge.

Experiments show that our approach achieves significant and consistent improvements over other benchmarks on our proposed conditional dataset, and captures additional semantic relations between fact and condition triples. We apply our proposed approach to a great number of scientific literature to construct a conditional knowledge graph.

## 2   Conditional Knowledge Graph Representation

Triple is the basic unit of knowledge representation in the construction of knowledge graph. It has a simple structure and strong knowledge expression ability. Therefore, to keep the advantages of traditional triple representation and address the issues of existing works, we extend and improve the classical triple representation model, and propose a novel conditional knowledge graph representation to express conditional sentences more accurately and completely. Our knowledge representation not only uses triples to describe the semantic relationships at the entity level, but also extends to the deep semantic connections at the triple level.

More specifically, we construct a nested triple to formalize the representation of conditional sentences. The basic units of conditional knowledge graph representation are defined as the following three types:

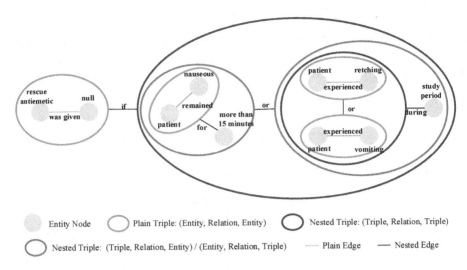

**Fig. 2.** The nested triple representation of conditional knowledge graph. Any type of triple can also serve as the head or tail node of a triple besides entities.

1) (Entity, Relation, Entity)
2) (Triple, Relation, Triple)
3) (Triple, Relation, Entity)/(Entity, Relation, Triple)

We define the first type as a plain triple, and the last two types as a nested triple. Different from the traditional knowledge representation, the head and tail node in the nested triple can be not only an entity, but also a triple. Or say, a triple can exist as a member of another triple, thus forming a nested triple representation. Figure 2 illustrates our proposed conditional knowledge graph representation more intuitively.

Take the conditional 1 in Fig. 1 for instance:

*"The rescue antiemetic was given, if patient remained nauseous for more than 15 min, or experienced retching or vomiting during study period."*

The expected knowledge representation in our method is as follows:

((rescue antiemetic, was given, null), if, (((patient, remained, nauseous), for, more than 15 min), or, (((patient, experienced, retching), or, (patient, experienced, vomiting)), during, study period))).

The whole is a nested triple with the condition subordinator "if" as the relation, where the head node is a fact triple belonging to type 1, and the tail node is the second type of nested condition triple that makes the fact triple valid. For the tail node, its head and tail node are all nested triples belonging to type 3. They are connected by the coordinate conjunction "or", which means that any one of the two condition triples is able to make the fact triple valid. The triple ((patient, experienced, retching), or, (patient, experienced, vomiting)) belongs to type 2, which contains two plain condition triples. From the above analysis, it is easy to know that the fact in conditional 1 has three sufficient conditions,

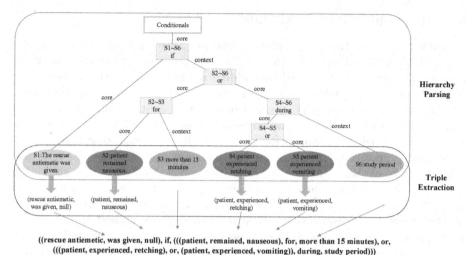

**Fig. 3.** Our approach for the nested triple extraction. The hierarchy parsing module constructs a semantic hierarchy tree to generate fact units (the orange leaf node), condition units (the blue leaf node) and supplementary units (the green leaf node) as well as extracts the semantic relations (the intermediate nodes) between facts and conditions. The triple extraction module is used to extract fact and condition triples. (Color figure online)

and is valid as long as any of them exists. Apparently, our proposed idea can effectively represent the complex semantics of conditional sentences.

## 3   Extraction Method for Nested Triples

In this section, the extraction strategy of fact and condition triples will be introduced in detail. Our approach aims to extract fact and condition triples and semantic hierarchical relations between triples from conditional sentences. Firstly, we operate text hierarchy semantic parsing for conditional sentences to construct a semantic hierarchical parse tree. The leaf nodes are a group of text units $\{c_1, \ldots, c_m, f_1, \ldots, f_n, s_1, \ldots, s_r\}$. We classify them into three types, $c_i$ $(i \in \{1, \ldots, m\})$ is a condition unit, $f_j$ $(j \in \{1, \ldots, n\})$ is a fact unit, and $s_k$ $(k \in \{1, \ldots, r\})$ is a supplementary unit. Notice that both the condition unit and the fact unit are simple sentences with a subject-predicate structure. The noun phrase structure is the form of the supplementary unit, which can be null. We extract triples from the fact and condition units to obtain the relevant triples. Finally, the triples are substituted into the leaves of the parse tree, and the tree is linearized into a hierarchical nested combination of triples. As shown in the Fig. 3, our model consists of two parts: hierarchy parsing module and triple extraction module. These modules are described in more detail in the following sections.

## 3.1   Hierarchical Parsing Module

In order to capture the deep semantic relations between facts and conditions, inspired by rhetorical structure theory (RST) in the field of linguistics, we make conditional sentences structured by building a semantic hierarchical parse tree. Naturally, we can derive local semantic connections between facts and conditions.

From a linguistic perspective, we conduct an in-depth linguistic analysis of conditional sentences. Conditional adverbial clauses, coordinate clauses, coordinate phrases (including coordinate verbs, coordinate nouns, coordinate adjectives) and prepositional phrases are the main scenarios that we need to deal with. In terms of these sentence patterns, we apply a small set of transformation rules [11] designed based on syntax driven for pattern matching and improve on them. Niklaus et al. [11] transform the prepositional phrases into simple sentences and fail to decompose coordinate relations. Instead of this method, we extract the noun phrase from the prepositional phrase as a stand-alone entity node to form nested triples and ensure the extensibility of knowledge representation. We retain the cue connectives with logical attributes (such as if/for/during/...) in conditional sentences as intermediate nodes of the parse tree to connect related text units to more clearly express the relations between the three types of units. In addition, we make further improvement on the transformation rules related to coordinate phrases, so that it can effectively split the coordinate components. We take the whole conditional sentence as the root node of the parse tree, and apply the syntactic transformation patterns recursively from top to bottom to simplify the sentence until we gain a set of basic text units that no pattern matches any more, which will be used as the input of the triple extraction task.

## 3.2   Triple Extraction Module

The purpose of this module is to extract triples from text units obtained by hierarchy parsing. The supplementary unit is generally a short noun phrase, which serves as an entity of a triple directly. The fact and the condition unit is a simple sentence, which serves as the input of the extraction model. Due to the simplification of input, a single triple can be extracted from each fact or condition unit, so the model does not need to consider the problem that the sentence may contain overlapping or multiple triples. We simplify the multi-input multi-output sequence labeling framework in [7] to multi-input single-output model. We use seven kinds of labels [6] to mark subject, predicate and object in the triple.

## 4   A Storage Schema for Conditional Knowledge Graph

Knowledge storage refers to how the acquired triples and schemas are stored in the computer. At present, relational database and graph database are used to store knowledge. In the graph database storage method, one entity of a triple corresponds to one vertex in the graph, which cannot store our representation because our triple nodes are not just entities, but triples. Therefore, we choose the storage scheme based on relational database which is widely used now.

fact_triple

| fact_id | subject | predicate | object |
|---|---|---|---|
| f1 | rescue antiemetic | was given | null |

condition_triple

| condition_id | subject | predicate | object |
|---|---|---|---|
| c1 | patient | remained | nauseous |
| c2 | patient | experienced | retching |
| c3 | patient | experienced | vomiting |

nested_triple

| nested_id | e_sub | f_sub | c_sub | n_sub | predicate | e_obj | f_obj | c_obj | n_obj |
|---|---|---|---|---|---|---|---|---|---|
| n1 | | | c1 | | for | more than 15 minutes | | | |
| n2 | | | c2 | | or | | | c3 | |
| n3 | | | | n2 | during | study period | | | |
| n4 | | | | n1 | or | | | | n3 |
| n5 | | f1 | | | if | | | | n4 |

**Fig. 4.** The physical storage scheme for nested triples. The `fact_triple` table and the `condition_triple` table store the plain fact and condition triples, respectively. The `nested_triple` table stores triples that their subject and object may be plain/nested triples, where the field ending with `sub` represents the subject of the triple, and the field ending with `obj` represents the object. The prefix of these fields indicates the type of node, for example, `e` denotes the entity, `f` and `c` denote the plain fact and condition triple, respectively, and `n` denotes the nested triple.

We carefully design the table schema, which can not only store the representation of nested triples, but also be compatible with traditional triples. As shown in Fig. 4, we design three tables to storage our knowledge. The `fact_triple` table and `condition_triple` table correspond to the first type of representation method, so traditional triples can also be stored in this way. The type of the head and tail nodes in nested triples is uncertain. To solve this uncertainty and preserve the hierarchical semantic information between triples, we provide a `nested_triple` table with fields representing triple node type. Only three columns in each row of the table are non-null, forming in a nested triple, and the predicate column represents the relation between the subject and object. Since a nested triple can be used as a column in other rows, multi-level nested triples can be stored in a relational database.

The specific storage algorithm is shown in Algorithm 1. We perform different operations depending on whether there is any condition connective $cc$ in the extraction. If no $cc$, the extraction is only a plain/nested fact triple, then we can insert it into the database through the recursive function $insertToDB$. Its termination condition is that the triple is no longer a nested triple, then we insert it into the `fact_triple` or `condition_triple` table according to the fact/condition type. Otherwise, we split the nested triple to get the corresponding subject, predicate and object, recursively execute $insertToDB$ on subject and object, and return the id of the triple in the table as the subject or object of the previous recursion to implement the insertion of the nested triple. If there is $cc$ in the extraction, we split it to obtain the plain/nested fact and condition triple according to $cc$. Then we apply $insertToDB$ on them and obtain the corresponding id of each triple in the table. Finally, the ids of the fact and condition triple combined with $cc$ are inserted into the `nested_triple` table.

---

**Algorithm 1:** Storage.

---

**Input:** extraction result $e$; condition connective $cc$
**Output:** the id of $e$ in a relation table

1 **if** $cc == null$ **then**
2     call $insertToDB(e, \text{``}fact\text{''})$ to insert the plain/nested fact triple $e$ into database;
3 **else**
4     split $e$ to acquire the plain/nested fact triple $f$ and condition triple $c$ according to $cc$;
5     call $insertToDB(f, \text{``}fact\text{''})$ to insert the plain/nested fact triple into database and get $tripleId(f)$;
6     call $insertToDB(c, \text{``}condition\text{''})$ to insert the plain/nested condition triple into database and get $tripleId(c)$;
7     Insert triple $(tripleId(f), cc, tripleId(c))$ into `nested_triple`;

8 **return** $tripleId(e)$;

---

## 5 Experiments

### 5.1 Experimental Setting

*Dataset:* We build a new dataset for the purpose of the joint extraction of fact and condition triples and their relations, named Biomedical Conditional Dataset (BioCD). Three participants manually annotated triples from conditional sentences with condition connectives such as "if", "unless" and so on, which are extracted from the abstracts of 81770 papers on Pubmed. BioCD contains 420 sentences, divided into 270 sentences for training (64%), 50 sentences for development (12%) and 100 sentences for test (24%).

*Comparison Methods:* We choose several advanced baseline methods to compare with our method, including the traditional rule-based OpenIE system Stanford OpenIE [2], sequence-labeling based model AllenNLP [14] and MIMO [7], as well as IMOJIE [10] based on sequence-generation. We use open-source implementations of these methods for experimental comparisons on our dataset.

*Evaluation Metrics:* We evaluate the performance of the above methods on two tasks. (i) triple unit extraction and (ii) triple extraction. For triple unit extraction, following Jiang et al. [7], we also use pair comparisons to match the constituent units (including subject, predicate, and object) of the extracted triples and ground-truth triples respectively to evaluate the accuracy of the triple units. For triple extraction, we undertake a pairwise comparison of the entire triple to evaluate the performance of the proposed approach more strictly.

### 5.2 Results Analysis

Table 1 reports the evaluation of all methods on triple unit extraction and triple extraction. The performance of our model consistently outperforms those

**Table 1.** Comparison of existing methods and our proposed model on triple unit extraction and triple extraction in the BioCD dataset. Higher score performs better.

| Baseline methods | Triple unit extraction | | | Triple extraction | | |
|---|---|---|---|---|---|---|
| | Prec.(%) | Rec.(%) | F1(%) | Prec.(%) | Rec.(%) | F1(%) |
| Stanford OpenIE | 23.47 | 25.71 | 24.54 | 1.62 | 2.67 | 2.01 |
| Allennlp OpenIE | 46.00 | 45.60 | 45.80 | 22.83 | 20.25 | 21.46 |
| IMoJIE | 61.25 | 57.76 | 59.45 | 33.67 | 32.12 | 32.87 |
| MIMO (BERT based) | 84.59 | 81.73 | 83.13 | 65.58 | 65.25 | 65.42 |
| Ours | **86.97** | **84.08** | **85.50** | **76.85** | **73.88** | **75.34** |

baseline methods on the two triple accuracy evaluation tasks. Compared with BERT-based MIMO, our model improves F1 score relatively by 2.9% on triple unit extraction and by 15.2% on triple extraction. Our model goes far beyond previous rule-based systems and neural approaches based on sequence labeling or generation. This suggests that it is useful to parse hierarchical structure of the conditional sentence in advance to get text units, and then extract triples, which reduces the difficulty of neural extraction model in learning the complex structure of conditional sentences. Moreover, it is observed that the results on triple extraction are relatively lower compared with those on triple unit extraction. Because our criterion for judging the correctness of a triple is that all the components of the triple are correct, and it is more difficult to extract the entire triple. Compared to other baselines, the performance of our model on triple extraction has a less decline than on triple unit extraction.

In addition to extracting fact and condition triples, our method also captures the deep semantic relations between triples, such as conditional and logical relations. However, the evaluation methods currently for OpenIE only handle the accuracy of triples and cannot evaluate the hierarchical dependencies between triples. So we manually evaluate the semantic hierarchical relations of 100 conditional sentences from the test set. As shown in Fig. 5, we undertake manual analysis on the final nested triples from the conditional sentences to check whether they are equivalent to the original meanings expressed by the conditional sentences and have the same hierarchical structure. We achieve 68% accuracy.

# 6  Related Work

*Conditional Knowledge Graph:* Conditional sentences exist widely in scientific literature. Most of the existing knowledge graphs ignore the influence of conditions. To address this issue, [8] proposes a representation and construction model of scientific conditional knowledge graph. But it cannot handle the overlapping tuple problem. [7] solves this problem on the basis. However, it cannot derive the relations between fact and condition triples.

---

**Conditional:** Atrial extrasystoles as such do not require any treatment unless they are accompanied by atrial fibrillation.
**Nested triple:** ((atrial extrasystoles as such, do not require, any treatment), unless, (atrial extrasystoles as such, are accompanied by, atrial fibrillation))   ✓

---

**Conditional:** If symptomatic treatment fails, pharyngeal airway obstruction is possible and a tonsillectomy may be necessary.
**Nested triple:** (((pharyngeal airway obstruction, is possible, null), if, (symptomatic treatment, fails, null)), and, (a tonsillectomy, may be necessary, null))   ✗

---

**Conditional:** Parathyroid tissue can be successfully autotransplanted and can be even allotransplanted if the host is immunosuppressed.
**Nested triple:** ((parathyroid tissue, can be successfully autotransplanted, null), and, ((parathyroid tissue, can be even allotransplanted, null), if, (the host, is immunosuppressed, null)))   ✓

---

**Fig. 5.** Examples for manual evaluation.

*Information Extraction:* Information extraction is an essential step in the construction of knowledge graph. Recent works [4,13,15,17,21–23] apply neural networks to extract pre-defined relations between entities. Open information extraction (OpenIE) extracts triples from text, which is not limited to pre-defined relations. The popular OpenIE methods mainly include two categories. [3,10] use sequence-to-sequence learning based on neural network. [9,14] learn to identify triples by tagging each word that composes the subject, predicate or object.

*Knowledge Graph Storage:* There are two main storage methods of knowledge graph, relational database based storage and graph based storage. [1,5,12,19,20] first build tables in a relational database, then map triples to records and store them in the database. Neo4j is a storage scheme based on attribute graph [18] and gStore is a method for RDF graph [24].

## 7   Conclusion

In this paper, we propose a novel knowledge graph representation, which is a nested triple structure. We implement a new extraction strategy for fact and condition triples which contains a text hierarchy parsing module to transform conditional sentences into a set of semantically related text units and an extraction module to extract fact and condition triples from text units. Our model can better structure conditional sentences and solve the problem that the existing conditional knowledge graph loses the deep semantic relations between facts and conditions. The proposed knowledge storage scheme can store fact and condition knowledge as well as their semantic hierarchical relations simultaneously. In the experiments, we show that our model significantly and consistently outperforms the state-of-the-art models in fact and condition triple extraction while capturing the relations between the triples.

# References

1. Abadi, D.J., Marcus, A., Madden, S.R., Hollenbach, K.: SW-store: a vertically partitioned DBMS for semantic web data management. VLDB J. **18**(2), 385–406 (2009)
2. Angeli, G., Premkumar, M.J.J., Manning, C.D.: Leveraging linguistic structure for open domain information extraction. In: Proceedings of the 53rd Annual Meeting of the Association for Computational Linguistics and the 7th International Joint Conference on Natural Language Processing (vol. 1: Long Papers), pp. 344–354. Association for Computational Linguistics, Beijing, China (2015)
3. Cui, L., Wei, F., Zhou, M.: Neural open information extraction. In: Proceedings of the 56th Annual Meeting of the Association for Computational Linguistics, (vol. 2: Short Papers), pp. 407–413. Association for Computational Linguistics, Melbourne, Australia (2018)
4. Guo, Z., Nan, G., LU, W., Cohen, S.B.: Learning latent forests for medical relation extraction. In: Proceedings of the Twenty-Ninth International Joint Conference on Artificial Intelligence, IJCAI-20, International Joint Conferences on Artificial Intelligence Organization, Virtual, Japan, pp. 3651–3657 (2020)
5. Harris, S., Gibbins, N.: 3store: efficient bulk RDF storage. In: Proceedings of the 1st International Workshop on Practical and Scalable Semantic Systems, pp. 81–95, Sanibel Island, Florida, USA (2004)
6. Hohenecker, P., Mtumbuka, F., Kocijan, V., Lukasiewicz, T.: Systematic comparison of neural architectures and training approaches for open information extraction. In: Proceedings of the 2020 Conference on Empirical Methods in Natural Language Processing (EMNLP), pp. 8554–8565. Association for Computational Linguistics, Online (2020)
7. Jiang, T., Zhao, T., Qin, B., Liu, T., Chawla, N., Jiang, M.: Multi-input multi-output sequence labeling for joint extraction of fact and condition tuples from scientific text. In: Proceedings of the 2019 Conference on Empirical Methods in Natural Language Processing and the 9th International Joint Conference on Natural Language Processing (EMNLP-IJCNLP), pp. 302–312. Association for Computational Linguistics, Hong Kong, China (2019)
8. Jiang, T., Zhao, T., Qin, B., Liu, T., Chawla, N.V., Jiang, M.: The role of "condition": a novel scientific knowledge graph representation and construction model. In: Proceedings of the 25th ACM SIGKDD International Conference on Knowledge Discovery & Data Mining, pp. 1634–1642. Association for Computing Machinery (2019)
9. Kolluru, K., Adlakha, V., Aggarwal, S., Mausam, Chakrabarti, S.: OpenIE6: iterative grid labeling and coordination analysis for open information extraction. In: Proceedings of the 2020 Conference on Empirical Methods in Natural Language Processing (EMNLP), pp. 3748–3761. Association for Computational Linguistics, Online (2020)
10. Kolluru, K., Aggarwal, S., Rathore, V., Mausam, Chakrabarti, S.: IMoJIE: iterative memory-based joint open information extraction. In: Proceedings of the 58th Annual Meeting of the Association for Computational Linguistics, pp. 5871–5886. Association for Computational Linguistics, Online (2020)
11. Niklaus, C., Cetto, M., Freitas, A., Handschuh, S.: Transforming complex sentences into a semantic hierarchy. In: Proceedings of the 57th Annual Meeting of the Association for Computational Linguistics, pp. 3415–3427. Association for Computational Linguistics, Florence, Italy (2019)

12. Pan, Z., Heflin, J.: DLDB: extending relational databases to support semantic web queries. In: Proceedings of the the 1st International Workshop on Practical and Scalable Semantic Systems, pp. 109–113, Sanibel Island, Florida, USA (2004)

13. Song, L., Zhang, Y., Gildea, D., Yu, M., Wang, Z., Su, J.: Leveraging dependency forest for neural medical relation extraction. In: Proceedings of the 2019 Conference on Empirical Methods in Natural Language Processing and the 9th International Joint Conference on Natural Language Processing (EMNLP-IJCNLP), pp. 208–218. Association for Computational Linguistics, Hong Kong, China (2019)

14. Stanovsky, G., Michael, J., Zettlemoyer, L., Dagan, I.: Supervised open information extraction. In: Proceedings of the 2018 Conference of the North American Chapter of the Association for Computational Linguistics: Human Language Technologies, vol. 1 (Long Papers), pp. 885–895. Association for Computational Linguistics, New Orleans, Louisiana (2018)

15. Sun, C., et al.: Chemical-protein interaction extraction via gaussian probability distribution and external biomedical knowledge. Bioinformatics **36**(15), 4323–4330 (2020)

16. Tong, P., Zhang, Q., Yao, J.: Leveraging domain context for question answering over knowledge graph. Data Sci. Eng. **4**(4), 323–335 (2019). https://doi.org/10.1007/s41019-019-00109-w

17. Wawrzinek, J., Pinto, J.M.G., Wiehr, O., Balke, W.T.: Exploiting latent semantic subspaces to derive associations for specific pharmaceutical semantics. Data Sci. Eng. **5**, 333–345 (2020)

18. Webber, J.: A programmatic introduction to Neo4j. In: Proceedings of the 3rd Annual Conference on Systems, Programming, and Applications: Software for Humanity, pp. 217–218. Association for Computing Machinery, New York, NY, USA (2012)

19. Weiss, C., Karras, P., Bernstein, A.: Hexastore: sextuple indexing for semantic web data management. Proc. VLDB Endowment **1**(1), 1008–1019 (2008)

20. Wilkinson, K.: Jena property table implementation. In: Proceedings of the 2nd International Workshop on Scalable Semantic Web Knowledge Base Systems, pp. 35–46, Athens, Georgia, USA (2006)

21. Zheng, W., et al.: An attention-based effective neural model for drug-drug interactions extraction. BMC Bioinform. **18**(1), 1–11 (2017)

22. Zhou, H., Liu, Z., Ning, S., Lang, C., Lin, Y., Du, L.: Knowledge-aware attention network for protein-protein interaction extraction. J. Biomed. Inform. **96**, 103234 (2019)

23. Zhou, H., et al.: Leveraging prior knowledge for protein-protein interaction extraction with memory network. Database **18** (2018)

24. Zou, L., Özsu, M.T.: Graph-based RDF data management. Data Sci. Eng. **2**, 56–70 (2017)

# Unlocking the Potential of MAPPO with Asynchronous Optimization

Wei Fu[1,2], Chao Yu[2], Yunfei Li[1,3], and Yi Wu[1,3(✉)]

[1] Shanghai Qizhi Institute, Shanghai, China
fuw17@tsinghua.org.cn
[2] Department of Electronics Engineering, Tsinghua University, Beijing, China
[3] Institute for Interdisciplinary Information Sciences,
Tsinghua University, Beijing, China

**Abstract.** It almost reaches a consensus that off-policy algorithms dominated research benchmarks of multi-agent reinforcement learning, while recent work [34] demonstrates that on-policy MARL algorithm, Multi-Agent Proximal Policy Optimization (MAPPO), can also attain comparable performance. In this paper, we propose a training framework based on MAPPO, named *async-MAPPO*, which supports scalable asynchronous training. We further re-examine async-MAPPO in StarCraftII micromanagement domain and obtain state-of-the-art performances on several hard and super-hard maps. Finally, we analyze three experimental phenomena and provide hypotheses behind the performance improvement of async-MAPPO.

**Keywords:** Multi-agent reinforcement learning · Asynchronous training · Distributed computing

## 1 Introduction

Recent research progress of multi-agent systems, such as AlphaStar [29], OpenAI Five [20] and hide-and-seek agents [1], indicate the general effectiveness and promising prospect of Multi-Agent Reinforcement Learning (MARL) in building intelligent agents that can behave cooperatively or competitively. It has been a growing trend to design and improve MARL algorithms [6,12,23,32], and to apply MARL to diverse applications, such as full Multiplayer Online Battle Arena (MOBA) game [33], autonomous driving [26] and social dilemmas [14].

Off-policy and value decomposition-based MARL algorithms have been preferred by researchers in recent years [27,30,31] since they are thought to be more sample efficient than on-policy ones. However, a recent work [34] demonstrates that with minimal hyperparameter tuning and restricted representation power, Multi-Agent Proximal Policy Optimization (MAPPO), i.e., PPO with centralized value function and decentralized policy, can match or surpass the performance of strong off-policy baselines on 3 categories of cooperative multi-agent benchmarks: Multi-agent Particle Environments (MPE) [17,18],

© Springer Nature Switzerland AG 2021
L. Fang et al. (Eds.): CICAI 2021, LNAI 13070, pp. 395–407, 2021.
https://doi.org/10.1007/978-3-030-93049-3_33

StarCraftII Multi-Agent Challenge (SMAC) [24] and Hanabi challenge [2]. The success of MAPPO indicates that on-policy multi-agent actor-critic algorithms are surprisingly effective and have great potential for MARL applications.

Even though the performance of MAPPO is impressive, there are some deficiencies in the original implementation[1] and experiments.

- **The original implementation of MAPPO is in a serial manner.** The agent sequentially collects data through environment interaction (referred to *rollout* stage) and then uses collected data for optimization (referred to *learning* stage). Rollout and learning need to wait for the completion of the other to enter the next round. If the data to be generated and consumed is vast, both rollout and learning require a longer time to complete and wait, which doubly increases training time and makes large-batch training unendurable.
- **MAPPO still requires carefully selected network architectures and moderate hyperparameter tuning on several maps.** To be more specific, Convolutional Neural Network (CNN) with frame-stacking is used on SMAC maps *3s_vs_4z* and *3s_vs_5z*, while the network architecture on other maps is either Multi-Layer Perceptron (MLP) or MLP with GRU [3]. In terms of hyperparameter tuning, uniquely different mini-batch numbers and initialization gain of the last action layer are utilized on the *MMM2* map.

Can we ameliorate the hyperparameter sensitiveness of MAPPO and accelerate the training procedure simultaneously? Authors of [1] found that batch size plays an imperative role in hide-and-seek training: larger batch size leads to faster convergence and even better sample efficiency, while training with a small batch may never converge. Besides, the aforementioned large-scale MARL applications, such as OpenAI Five and hide-and-seek agent, conformably utilize distributed RL system for fast data collection and large batch for stable training. This acquiescent agreement makes us wonder whether asynchronous training with a large batch is the key element to enhance original MAPPO and to unlock its full potential.

In this paper, we propose ***async-MAPPO***, a MARL framework that integrates MAPPO and the refined SEED (Scalable, EfficiEnt Deep-RL system [4]) architecture. We re-examine async-MAPPO on several hard and super-hard maps in SMAC domain and find that MAPPO algorithm can attain better final performance when training with asynchronous optimization and large batch. Notably, the final performance surpasses all the results reported in [34] and establishes a new state-of-the-art (SOTA). Finally, three experimental phenomena are proposed and discussed, through which we provide enlightenment about the reason behind the performance boost of async-MAPPO.

The contributions of this paper are summarized as follows:

1. We propose *async-MAPPO*, a scalable asynchronous training framework which integrates a refined SEED architecture with MAPPO.
2. We show that async-MAPPO can achieve SOTA performance on several hard and super-hard maps in SMAC domain with significantly faster training speed by tuning only one hyperparameter.

---

[1] https://github.com/marlbenchmark/on-policy.

3. We formulate hypotheses about the effectiveness of large-batch training based on the empirical results.

## 2   Related Works

Modern multi-agent deep reinforcement learning algorithms mostly follow the paradigm of Centralized-Training-with-Decentralized-Execution (CTDE [7,8]). Under CTDE paradigm, each agent independently behaves using its policy, while policies are jointly trained given global environment information. Popular MARL algorithms adopting CTDE paradigm can be roughly divided into off-policy and on-policy branches. Off-policy branch includes multi-agent actor-critic algorithms, such as MADDPG [17] and COMA [6], and value-decomposition based algorithms, such as QMIX [23], ROMA [31] and RODE [32]. On-policy branch typically includes MAPPO [34]. While off-policy MARL algorithms attract more attention and are nearly exhaustively developed, on-policy MARL algorithms are rarely studied. However, surprisingly, they are empirically promising and worth further improving [34].

Large-scale MARL projects are always supported by an efficient RL system, such as Rapid framework used in OpenAI Five [20]. RL system design was an early research focus [16]. To address the problem of iterative waiting in serial implementation, IMPALA [5] adopts a scalable actor-learner architecture, where each actor is placed on a CPU core and manages the whole rollout procedure independently, and the learner collects data generated by actors and optimizes model parameters on GPUs. To further utilize GPU/TPU resources in a cluster, SEED [4] decomposes the rollout procedure of IMPALA into a client-server mode. Every client steps through local environments on CPU, and issues action requests to the inference server, while inference server batches observations received from clients and provides action inference on a GPU or TPU. Providing the scalability, efficiency, and resource utilization requirement, SEED is currently one of the best architecture backbones in building a distributed RL system, whose modification has been applied in recent works [33].

Performance improvement with asynchronous training was reported previously in the domain of single-agent RL. Asynchronous training can both facilitate exploration and allow wider hyperparameter sweeping by setting different hyperparameters in every worker node [11]. When integrated with recurrent neural networks and distributed training, deep Q network can achieve state-of-the-art performance on most Atari games and outperform normal Q-learning-based algorithm by a large margin [13]. Considering the above results in single-agent benchmarks and the successful applications of PPO in large-scale multi-agent projects, we could expect a performance boost of MAPPO when combining it with an efficient asynchronous RL system in multi-agent benchmarks.

## 3   Preliminaries

We consider a decentralized partially observable Markov decision process (Dec-POMDP) [19] with shared rewards among agents defined by the tuple $G = \langle I, S, A, P, R, \Omega, O, n, \gamma \rangle$. $I$ defines the set of agents and $n$ is the number of agents.

---

**Algorithm 1.** Trainer Process of MAPPO with Asynchronous Optimization

---
**Input:** Parameters $\theta_0$, $\psi_0$, replay buffer $\mathcal{D}$, parameter queue $\mathcal{Q}$, learning rate $\alpha$

1: Send $\theta_0$ and $\psi_0$ into $\mathcal{Q}$
2: Launch rollout processes
3: **for** update iteration $i = 1, 2, \ldots$ **do**
4:     wait until there's enough data in $\mathcal{D}$ for optimization
5:     fetch data batch from $\mathcal{D}$
6:     compute $A_{\psi_i}^\pi$ using GAE [25] with PopArt [10] denormalization
7:     compute $V_{\text{target}}^\pi$ based on $A_\psi^\pi$
8:     use $V_{\text{target}}^\pi$ to update PopArt parameters
9:     compute loss $-J_\pi(\theta_i)$ and $-J_{V^\pi}(\psi_i)$ according to Equation (1) and (2)
10:     $\theta_{i+1} \leftarrow \theta i + \alpha \nabla J_\pi(\theta)|_{\theta=\theta_i}$
11:     $\psi_{i+1} \leftarrow \psi_i + \alpha \nabla J_{V^\pi}(\psi_i)|_{\psi=\psi_i}$
12:     Send $\theta_{i+1}$ and $\psi_{i+1}$ into $\mathcal{Q}$
13:     $i \leftarrow i + 1$
14: **return** policy $\pi_\theta$

---

$\gamma$ is the discount factor. $S$ is the support set of true state in the environment. At each timestep, agent $i$ receives an observation $o_i$ drawn from the observation function $O$, i.e., $o_i = O(s, i) \in \Omega$. After receiving an observation, agent $i$ infers an available action $a_i \in A$ to execute. A shared reward $R(s, \boldsymbol{a})$ is received by all agents once a joint action $\boldsymbol{a} \in A^n$ is formulated and a transition is triggered according to the transition function $P(s'|s, \boldsymbol{a})$. $\tau_i \in T \equiv (S \times A^n)^t$ denotes the trajectory of agent $i$ of the elapsed $t$ timesteps in one episode.

MAPPO follows the CTDE paradigm, where each agent learns a shared policy $\pi_\theta(\cdot|\tau_i) : (\Omega \times A)^* \to [0, 1]$ parameterized by $\theta$ conditioned on local history observation, and a centralized value function of current policy $\pi$, $V_\psi^\pi(\boldsymbol{\tau}) : S^* \to \mathbb{R}$ parameterized by $\psi$ conditioned on global history states. Here $X^*$ means the Cartesian product of set $X$ in an arbitrary number of timesteps. Note that trajectory $\boldsymbol{\tau}$ together with observation function $O$ contains the history information of both local observation and global state. Hence we use $\boldsymbol{\tau}$ to denote the input of both policy and value function, while the actual input may not be the same.

## 4  Async-MAPPO

### 4.1  MAPPO Algorithm

Similar as single-agent PPO, MAPPO simultaneously learns a shared policy $\pi_\theta(\cdot|\tau_i)$ and a centralized value function $V_\psi(\boldsymbol{\tau}) = \mathbb{E}_{(s_t, \boldsymbol{a}_t) \sim T}[\sum_{t=0}^\infty R(s_t, \boldsymbol{a}_t)]$ by optimizing the following objective:

$$J_\pi(\theta) = \sum_{i=1}^n \mathbb{E}_{(\tau_t^i, a_t^i) \sim G}\left[\min\left(\text{clip}\left(\frac{\pi_\theta(a_t^i|\tau_t^i)}{\pi_{\theta_{old}}(a_t^i|\tau_t^i)}\right)A_\psi^\pi(\tau_t^i, a_t^i), \frac{\pi_\theta(a_t^i|\tau_t^i)}{\pi_{\theta_{old}}(a_t^i|\tau_t^i)}A_\psi^\pi(\tau_t^i, a_t^i)\right)\right] \quad (1)$$

$$J_{V^\pi}(\psi) = -\sum_{i=1}^n \mathbb{E}_{\boldsymbol{\tau} \sim G}\left[V_{\text{target}}^\pi(\boldsymbol{\tau}) - V_\psi^\pi(\boldsymbol{\tau})\right]^2 \quad (2)$$

---

**Algorithm 2.** Client
___
**Input:** total $k*m$ environments in $m$ groups, remote reference of inference server $\mathcal{S}$
1: **for** group $g = 1, \ldots, m$ **do**
2:     **for** environment $e$ in $g$ **do**
3:         $o, o_{\text{share}}, a_{\text{avail}} = e.\text{reset}()$
4:     Batch $o$, $o_{\text{share}}$ and $a_{\text{avail}}$ into vector
5:     invoke RPC $\mathcal{S}.\text{select\_action}(o, o_{\text{share}}, a_{\text{avail}})$
6: **while** True **do**
7:     **for** group $g = 1, \ldots, m$ **do**
8:         wait for action response $a$ from $\mathcal{S}$
9:         **for** for environment $e$ in $g$ **do**
10:             $o, o_{\text{share}}, a_{\text{avail}}, r, d = e.\text{step}(a)$
11:         Batch $o$, $o_{\text{share}}$, $a_{\text{avail}}$, $r$, $d$ into vector
12:         invoke RPC $\mathcal{S}.\text{select\_action}(o, o_{\text{share}}, a_{\text{avail}}, r, d)$
___

In the above equations, $A_\psi^\pi(\tau_t^i, a_t^i)$ denotes the advantage function [28] and $V_{\text{target}}^\pi(\tau)$ denotes the target value computed by Generalized Advantage Estimation (GAE [25]). Pseudocode and algorithmic details of trainer process can be found in Algorithm 1.

## 4.2  Refined SEED Architecture

As mentioned in 2, SEED [4] is a high-performance distributed RL architecture built upon the gRPC package, which decomposes the rollout stage into client and server calls to fully utilize GPU/TPU resources in a cluster. Although it is scalable and cost-effective, an obvious flaw is that, in the aspect of a remote environment client, after issuing a request to the server, it must wait until the completion of inference and data transportation to continue the next step of environment simulation. This indicates that CPU resources may not be fully utilized due to the idle time of clients. To address this issue and remove the potential bottleneck, similar to the system designed by [22], we propose to store a vector of environments in each client, split the environments into multiple parts, e.g., 2 parts, and alternate stepping across them, which is referred to as *Multiple-Buffered Sampling.*

To be more specific, a client possesses $m$ environment splits, and an environment split is composed of $k$ environments. The simulation of each environment in the same split is executed sequentially in a *for* loop. After the request initiation of any environment split, the client will keep stepping through the other environment splits instead of waiting for the response. For example, while the second environment split is stepped through, the actions of the first split are computed on the inference server. Hence, clients can attain full CPU utilization once $k$ is correctly chosen such that the time of inference and data transportation can be overlapped by simulation of other environment splits. Graphical illustration can be found in Fig. 1.

**Algorithm 3.** Inference Server Invocation

**Input:** replay buffer $\mathcal{D}$, parameter queue $\mathcal{Q}$, inference model $\pi$ and $V$, inference batch size $B$, observation $o$, centralized observation $o_{\text{share}}$, available action $a_{\text{avail}}$, reward $r$, termination indicator $d$

1: **if** there's new parameter $\theta'$, $\psi'$ in $\mathcal{Q}$ **then**
2:     $\theta \leftarrow \theta', \psi \leftarrow \psi'$
3: Store $o$, $o_{\text{share}}$, $a_{\text{avail}}$, $r$, $d$ into $\mathcal{D}$
4: batching count $\leftarrow$ batching count + 1
5: Set callback object $a$ to be the slice of $a^{\text{batch}}$
6: **if** batching count $\geq B$ **then**
7:     Get batched $o^{\text{batch}}$, $o^{\text{batch}}_{\text{share}}$, $a^{\text{batch}}_{\text{avail}}$ from $\mathcal{D}$
8:     Get hidden state $h^{\text{batch}}$ from $\mathcal{D}$
9:     $a^{\text{batch}}, \hat{h}^{\text{batch}} = \text{model.inference}(o^{\text{batch}}, o^{\text{batch}}_{\text{share}}, a^{\text{batch}}_{\text{avail}}, h^{\text{batch}})$
10:    Store $a^{\text{batch}}, \hat{h}^{\text{batch}}$ into $\mathcal{D}$
11:    batching count $\leftarrow 0$
12:    Trigger callback function on $a$
13: **return** $a$

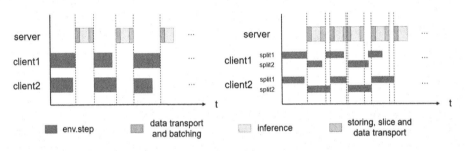

**Fig. 1.** Graphical illustration of *Multiple-Buffered Sampling*. (Left) original SEED sampling architecture. Servers wait for client requests, batch data, conduct inference, store inference outputs, and then send action slices back to clients. Clients wait for the response from servers after issuing a request. Idle time of client CPU is non-negligible. (Right) *Multiple-Buffered Sampling*. Clients alternate among $m = 2$ environment splits and step through them. Overlapping between client CPU and server GPU makes full use of computation resources.

### 4.3   Implementation Details

Communication between clients and the inference server is supported by *PyTorch* [21] *torch.distributed* [15] package. Remote reference of inference server is kept by several clients, through which clients can invoke remote procedure calls (RPC) to request actions from the server. After receiving a request from any client, the server stores observations returned in the previous environment step into the replay buffer and returns a *torch.futures.Future* object immediately. Once a certain number of clients invoke RPCs, the server will fetch a data batch from the buffer, conduct inference, and save data returned by inference back into the buffer. At this point, slicing callback function chained in *torch.futures.Future* object is triggered. Finally, clients receive the corresponding slice of the action batch as a response and start a new round of environment steps. Detailed algorithm description of client and server is illustrated in Algorithm 2 and Algorithm 3 respectively.

Replay buffer is based on *NumPy* library [9] and shared memory in Python *multiprocessing* package, i.e., every data segmentation is a *NumPy* array in shared memory, such that the learner can benefit from zero-copy communication when requesting for a data batch stored by inference servers. During optimization, the learner converts a data batch in replay buffer into *PyTorch Tensor* [21], loads it into GPU memory, and optimizes model parameters. After every full update iteration (several PPO update epochs), the updated parameter is pushed into a queue, through which servers synchronize the local model with the latest one to ensure that the rollout procedure is sufficiently on-policy.

## 5  Experiment

In this section, we examine both system-level and algorithm-level performance of async-MAPPO. In Sect. 5.1, experiments concerning the effect of *Multiple-Buffered Sampling* and scalability are conducted, which show that refined SEED architecture provides higher system throughput than the original one and still scales well. Algorithm performance is measured in Sect. 5.2 on selected SMAC maps. These results meet our expectation of performance enhancement with asynchronous optimization and a large batch. We analyze the reason behind algorithm performance gain and formulate two hypotheses derived from experimental observations in Sect. 5.3.

### 5.1  System-level Evaluation

We examine the scalability of the refined SEED architecture and the influence of refinement in terms of system throughput, i.e., collected environment Frames Per Second (FPS), on Hanabi learning environment [2]. Table 1 demonstrates the numerical results tested on a single machine and a cluster. In both systems, the first 3 GPUs are used for optimization and the last one is used for rollout inference, where four inference servers are initialized. Detailed hardware description of System #1 and System #2 can be found in the caption of Table 1.

On the one hand, from the first three rows of system #2, we can see that refined SEED architecture can be employed to a cluster with near-linear scaling, i.e., system throughput improves linearly as the number of actors increases, which shows promising scalability. On the other hand, by setting environment from single split into double splits, system throughput is improved in both local and distributed settings (first two rows of system #1, last two rows of system #2), which verifies the necessity of *Multiple-Buffered Sampling* in SEED architecture. If environment splits are further increased (last row of system #1), system throughput will be hurt because there exists a group of environments that are neither stepped through in clients nor waiting for action response from servers. Empirically, splitting environments into two groups is the best practice.

**Table 1.** System throughput measurement results and corresponding experiment configuration. System # 1 is a laboratory-level server machine with one physical CPU of 64 cores, 128 GB memory, and 4 NVIDIA 2080Ti GPUs. System #2 is an Ali-Cloud cluster, whose head node is a GPU machine with 48 cores and 4 NVIDIA V100 GPUs and worker nodes are homogeneous CPU machines with 104 virtual cores. Rounded average FPS of 3 independent runs across 30 s is presented. The refinement of SEED architecture improves system throughput, with which SEED architecture still scales well.

| System | #Actors | #Envs per actor | #Env splits | FPS |
|--------|---------|-----------------|-------------|------|
| #1 | 48 | 80 | 1 | 13.5k |
|    |    |    | 2 | 17.5k |
|    |    |    | 3 | 16.3k |
| #2 | 32 | 128 | 2 | 11.1k |
|    | 64 |    |   | 22.0k |
|    | 128 |   |   | 44.5k |
|    |    |    | 1 | 30.6k |

## 5.2   Algorithm-level Evaluation

**Algorithmic Details.** Experiments in this section are conducted on System #1 described in the previous section. Only 1 GPU is utilized for both rollout inference and network optimization in consistent with [34]. The number of environment splits is fixed to 2, while the number of actors and environments in each actor varies across selected environments. For StarCraftII environment, the more agents are in the map, the lower FPS and the fewer total environments a single machine can support. The same setting on maps with fewer agents (e.g. *3s_vs_5z*), which initializes a relatively large number of environments, can not be adopted on maps with plenty of agents (e.g. *27m_vs_30m*), otherwise, the game will cause a memory overflow problem. Hence, we report the batch size magnification and FPS acceleration factor instead of specific system configuration and actual FPS. We believe this substitution will make the conclusion more clear.

Two separate networks with the same recurrent structure as [34] are maintained for policy $\pi_\theta$ and value function $V_\psi$ respectively. Hyperparameters, including hidden size, learning rate, and so on, are mostly the same as [34] except for reuse times of each data batch, which is presented in Table 2 in details. We follow the suggestions proposed by [34] and include all the recommended tricks in async-MAPPO, including agent-specific global state, training data usage, value normalization (PopArt) [10], action masking, and death masking, since they are all found to be critical to MAPPO's practical performance. We directly modify the codebase released with [34] for maximum consistency. If not specified, all evaluation procedure is the same as that reported in [34]. To better distinguish original MAPPO implementation from async-MAPPO, we refer to it as *serial-MAPPO* in the remaining part of this section.

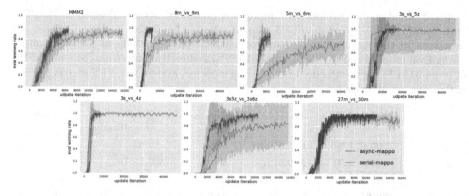

**Fig. 2.** Learning curves of serial-MAPPO and async-MAPPO on selected SMAC maps. Results of serial-MAPPO is taken from [34] in communication with authors. The Y-axis is the median evaluation winning rate throughout the training procedure. The shaded area indicates the standard deviation of different runs with different random seeds. The X-axis is the number of update iterations. (#update iteration = #total environment steps × sample reuse/batch size)

**StarCraftII Multi-Agent Challenge.** Because all the features and tricks of serial-MAPPO are preserved, we expect *no performance drop* in async-MAPPO given that correctly tuned hyperparameters for asynchronous training. Therefore, we omit experiments on easy SMAC maps where the final performance of serial-MAPPO can not be further improved. SMAC maps that meet any of the following conditions are considered:

(1) Serial-MAPPO can not achieve SOTA performance. (*8m_vs_9m, 3s_vs_5z, 27m_vs_30m, 3s5z_vs_3s6z*);
(2) Even though serial-MAPPO achieves SOTA performance, there is plenty of room to improve. (*5m_vs_6m*);
(3) There is uniquely different hyperparameter or network architecture selection in serial-MAPPO. (*MMM2, 3s_vs_4z, 3s_vs_5z*)

Figure 2 demonstrates the median evaluation winning rate and the corresponding standard deviation across the training process. Final evaluation performance, sample reuse[2] times, FPS speedup and batch size magnification are presented in Table 2. When training with asynchronous optimization and large batch, the performance of async-MAPPO matches or exceeds serial-MAPPO, QMIX, and RODE examined in [34], either of which achieved SOTA results on these maps. Besides, training is significantly accelerated compared with serial-MAPPO for about 4~5 times in terms of wall clock speed.

## 5.3 Discussion

We provide three experimental phenomena and corresponding analyses in async-MAPPO practice.

---

[2] Referred to as *ppo_epoch* in [34].

**Table 2.** Final median evaluation winning rate (standard deviation) of async-MAPPO and serial-MAPPO. Speedup in FPS, magnification in batch size, and sample reuse are presented in the right columns. Runs of async-MAPPO include at least 3 random seeds. Evaluation winning rate of serial-MAPPO is directly taken from [34], while FPS of serial-MAPPO is re-measured in the same machine, system #1. FPS metric applied is the same as in Table 1.

| Map | async-MAPPO | serial-MAPPO | FPS Speedup | Batchsize magnification | Sample reuse |
|---|---|---|---|---|---|
| *3s_vs_4z* | **100.0(1.5)** | **100.0(0.9)** | 5.50x | 7.5× | 10 |
| *3s_vs_5z* | **100.0(2.3)** | 96.9(37.5) | 5.84× | 7.5× | 15 |
| *5m_vs_6m* | **90.6(3.1)** | 75.0(18.2) | 5.22× | 7.5× | 10 |
| *8m_vs_9m* | **96.8(1.5)** | 87.5(4.0) | 4.66× | 7.5× | 10 |
| *27m_vs_30m* | **98.4(3.3)** | 93.8(2.4) | 3.91× | 2× | 5 |
| *3s5z_vs_3s6z* | **96.8(3.3)** | 84.4(34.0) | 5.20× | 7.5× | 10 |
| *MMM2* | **96.8(1.1)** | 90.6(2.8) | 4.96× | 6× | 5 |

**Async-MAPPO Can Endure More Reuse Times.** Note that sample reuse reported in Table 2 is greater or equal to that in [34]. First, a large batch reduces the variance of episode returns and improves the precision of value function and advantage estimation. Second, with more diverse experiences in a data batch, the expectation in Eq. 1, i.e., off-policy correction of advantage function, is also more accurate. Consequently, policy improvement direction is more accurate and off-policy correction is not prone to diverge, which is the reason why more reuse times can be applied on the same data batch in async-MAPPO.

**Async-MAPPO Requires Fewer Update Iterations.** Even though sample reuse and total environment steps of async-MAPPO may be larger, the update iterations are fewer than serial-MAPPO, as shown in Fig. 2. In the context of advantage normalization, a large batch reduces the variance of advantage estimation and increases the scale of normalized advantage (because the divisor is reduced) while maintaining high accuracy of mean advantage estimation. This means async-MAPPO may step further in the correct direction when using the same learning rate as serial-MAPPO, which explains fewer update iterations are required.

**Async-MAPPO Is Less Sensitive to Network Architectures and Hyperparameters.** To attain better results, on *MMM2*, serial-MAPPO splits one data batch into two mini-bathes to escape local optima, while async-MAPPO uses the whole data batch. For a similar reason, CNN with frame-stacking was used in maps *3s_vs_4z* and *3s_vs_5z*, while async-MAPPO uses a universal MLP+GRU network architecture. However, async-MAPPO achieves even better performance on these maps. We reckon the analysis in the previous two phenomena also applies in that accurate policy evaluation and fast policy improvement help escaping local optima, causing hyperparameter and network architecture less imperative.

## 6 Conclusion

In this work, we propose async-MAPPO, the integration of MAPPO algorithm, and refined SEED architecture. Async-MAPPO has promising system-level and algorithm-level performance and establishes a new SOTA result on selected hard and super-hard SMAC maps. We formulate hypotheses about the reason behind based on experimental phenomena. We conjecture that training with asynchronous optimization and a large batch is possibly a generally beneficial choice to use MAPPO. Systematic and theoretical verification of these hypotheses remains in future works.

## References

1. Baker, B., et al.: Emergent tool use from multi-agent autocurricula. arXiv preprint arXiv:1909.07528 (2019)
2. Bard, N., et al.: The Hanabi challenge: a new frontier for AI research. Artif. Intell. **280**, 103216 (2020)
3. Chung, J., Gulcehre, C., Cho, K., Bengio, Y.: Empirical evaluation of gated recurrent neural networks on sequence modeling. arXiv preprint arXiv:1412.3555 (2014)
4. Espeholt, L., Marinier, R., Stanczyk, P., Wang, K., Michalski, M.: Seed rl: Scalable and efficient deep-rl with accelerated central inference. arXiv preprint arXiv:1910.06591 (2019)
5. Espeholt, L., et al.: Impala: Scalable distributed deep-rl with importance weighted actor-learner architectures. In: International Conference on Machine Learning, pp. 1407–1416. PMLR (2018)
6. Foerster, J., Farquhar, G., Afouras, T., Nardelli, N., Whiteson, S.: Counterfactual multi-agent policy gradients. In: Proceedings of the AAAI Conference on Artificial Intelligence, vol. 32 (2018)
7. Foerster, J.N., Assael, Y.M., De Freitas, N., Whiteson, S.: Learning to communicate with deep multi-agent reinforcement learning. arXiv preprint arXiv:1605.06676 (2016)
8. Gupta, J.K., Egorov, M., Kochenderfer, M.: Cooperative multi-agent control using deep reinforcement learning. In: Sukthankar, G., Rodriguez-Aguilar, J.A. (eds.) AAMAS 2017. LNCS (LNAI), vol. 10642, pp. 66–83. Springer, Cham (2017). https://doi.org/10.1007/978-3-319-71682-4_5
9. Harris, C.R., et al.: Array programming with numPy. Nature **585**(7825), 357–362 (2020)
10. Hessel, M., Soyer, H., Espeholt, L., Czarnecki, W., Schmitt, S., van Hasselt, H.: Multi-task deep reinforcement learning with popart. In: Proceedings of the AAAI Conference on Artificial Intelligence, vol. 33, pp. 3796–3803 (2019)
11. Horgan, D., et al.: Distributed prioritized experience replay. arXiv preprint arXiv:1803.00933 (2018)
12. Hu, H., Foerster, J.N.: Simplified action decoder for deep multi-agent reinforcement learning. arXiv preprint arXiv:1912.02288 (2019)

13. Kapturowski, S., Ostrovski, G., Quan, J., Munos, R., Dabney, W.: Recurrent experience replay in distributed reinforcement learning. In: International Conference on Learning Representations (2018)
14. Leibo, J.Z., Zambaldi, V., Lanctot, M., Marecki, J., Graepel, T.: Multi-agent reinforcement learning in sequential social dilemmas. arXiv preprint arXiv:1702.03037 (2017)
15. Li, S., et al.: Pytorch distributed: Experiences on accelerating data parallel training. arXiv preprint arXiv:2006.15704 (2020)
16. Li, Y., Schuurmans, D.: MapReduce for parallel reinforcement learning. In: Sanner, S., Hutter, M. (eds.) EWRL 2011. LNCS (LNAI), vol. 7188, pp. 309–320. Springer, Heidelberg (2012). https://doi.org/10.1007/978-3-642-29946-9_30
17. Lowe, R., Wu, Y., Tamar, A., Harb, J., Abbeel, P., Mordatch, I.: Multi-agent actor-critic for mixed cooperative-competitive environments. In: Neural Information Processing Systems (NIPS) (2017)
18. Mordatch, I., Abbeel, P.: Emergence of grounded compositional language in multi-agent populations. arXiv preprint arXiv:1703.04908 (2017)
19. Oliehoek, F.A., Amato, C.: A Concise Introduction to Decentralized POMDPs. Springer, Cham (2016). https://doi.org/10.1007/978-3-319-28929-8
20. OpenAI: Openai five. https://blog.openai.com/openai-five/ (2018)
21. Paszke, A., et al.: Pytorch: An imperative style, high-performance deep learning library. arXiv preprint arXiv:1912.01703 (2019)
22. Petrenko, A., Huang, Z., Kumar, T., Sukhatme, G., Koltun, V.: Sample factory: Egocentric 3d control from pixels at 100000 fps with asynchronous reinforcement learning. In: International Conference on Machine Learning, pp. 7652–7662. PMLR (2020)
23. Rashid, T., Samvelyan, M., Schroeder, C., Farquhar, G., Foerster, J., Whiteson, S.: Qmix: monotonic value function factorisation for deep multi-agent reinforcement learning. In: International Conference on Machine Learning, pp. 4295–4304. PMLR (2018)
24. Samvelyan, M., et al.: The StarCraft Multi-Agent Challenge. CoRR abs/1902.04043 (2019)
25. Schulman, J., Moritz, P., Levine, S., Jordan, M., Abbeel, P.: High-dimensional continuous control using generalized advantage estimation. arXiv preprint arXiv:1506.02438 (2015)
26. Shalev-Shwartz, S., Shammah, S., Shashua, A.: Safe, multi-agent, reinforcement learning for autonomous driving. arXiv preprint arXiv:1610.03295 (2016)
27. Son, K., Kim, D., Kang, W.J., Hostallero, D.E., Yi, Y.: Qtran: learning to factorize with transformation for cooperative multi-agent reinforcement learning. In: International Conference on Machine Learning, pp. 5887–5896. PMLR (2019)
28. Sutton, R.S., Barto, A.G.: Reinforcement Learning: An introduction. MIT Press, Cambridge (2018)
29. Vinyals, O., et al.: Grandmaster level in StarCraft ii using multi-agent reinforcement learning. Nature **575**(7782), 350–354 (2019)
30. Wang, J., Ren, Z., Liu, T., Yu, Y., Zhang, C.: Qplex: Duplex dueling multi-agent q-learning. arXiv preprint arXiv:2008.01062 (2020)
31. Wang, T., Dong, H., Lesser, V., Zhang, C.: Roma: multi-agent reinforcement learning with emergent roles. In: Proceedings of the 37th International Conference on Machine Learning (2020)
32. Wang, T., Gupta, T., Mahajan, A., Peng, B., Whiteson, S., Zhang, C.: Rode: Learning roles to decompose multi-agent tasks. arXiv preprint arXiv:2010.01523 (2020)

33. Ye, D., et al.: Towards playing full moba games with deep reinforcement learning. arXiv preprint arXiv:2011.12692 (2020)
34. Yu, C., Velu, A., Vinitsky, E., Wang, Y., Bayen, A., Wu, Y.: The surprising effectiveness of mappo in cooperative, multi-agent games. arXiv preprint arXiv:2103.01955 (2021)

# A Random Opposition-Based Sparrow Search Algorithm for Path Planning Problem

Guangjian Zhang and Enhao Zhang[✉]

Chongqing University of Technology, Chongqing, China
52192325122@2019.cqut.edu.cn

**Abstract.** In this paper, an improved intelligence algorithm is proposed for path planning problem. The algorithm is based on Sparrow Search Algorithm and is combined with Random Opposition-based Learning and linear decreasing strategy, named ROSSA. The mobile robot path planning problem can be mathematically transformed into an optimization problem, which can be solved by intelligent optimization algorithms. With this consideration, an SSA-based optimization algorithm is proposed. Random opposition-based learning increases the diversity of the population and enhances the exploration ability of the algorithm; the linear decreasing strategy balances the ability of the algorithm to explore globally and exploit locally by adjusting the algorithm parameters. Meanwhile, the Bezier curve satisfies the requirement of path smoothness for the robot path planning problem. The superiority of the proposed algorithm is verified by conducting experiments with three standard algorithms for 11 benchmark test functions, and some comparison experiments on the path planning problem with PSO and SSA to confirm that the proposed algorithm can find a safe and optimal path in the mobile robot path planning problem.

**Keywords:** Path planning · Sparrow search algorithm · Opposition-based learning · Bezier curve

## 1 Introduction

Robot path planning is a very important part in the field of robotics, because it gives robots the ability to move, so that it can handle a variety of tasks that need to move between two points [1].

Given the start and goal position for robot in a 2D environment with static obstacles, the goal of path planning is to search for an optimal or suboptimal collision-free path so that robots can move from the start point to the target point without collision with obstacles [2]. Based on the mastery of the environment, path planning can be divided into global path planning and local path planning.

There has been lots of research on motion planning since the pioneering work presented by N. J. Nilsson in late 1960 s [3]. Thus far, various motion planning

© Springer Nature Switzerland AG 2021
L. Fang et al. (Eds.): CICAI 2021, LNAI 13070, pp. 408–418, 2021.
https://doi.org/10.1007/978-3-030-93049-3_34

algorithms have been presented such as Probabilistic Roadmaps [4,5], Rapidly Exploring Random Trees [6,7], and Potential Fields [8,9], etc. These algorithms can be divided into deterministic and undeterministic algorithms. Deterministic algorithms must find the optimal solution when the problem has an optimal solution, otherwise they return information that there is no optimal solution. However, as the size of the problem becomes more complex, the complexity of modeling the problem and the amount of computation required by the algorithm grows exponentially. Besides, since practical engineering problems usually have many locally optimal solutions, it is difficult for these deterministic algorithms to cope with increasingly difficult problems. Unlike deterministic algorithms, meta-heuristic algorithms can find an approximate solution in case the exact solution cannot be found. This can significantly reduce the amount of computation. Also, meta-heuristic algorithms introduce stochasticity, which gives it the ability to get rid of the local optimum problem. These advantages provide important implications for metaheuristic algorithms to solve global optimization problems. In the past decades, researchers have proposed various Swarm intelligence algorithms, including: Particle Swarm Optimization Algorithm [10], Krill Herd Optimization Algorithm [11], Beetle Antenna Search Algorithm [12], etc. The sparrow search algorithm [13] is a novel metaheuristic optimization algorithm recently proposed with faster convergence, fewer control parameters, and simpler computation, but like other swarm intelligence algorithms, it tends to converge early when solving complex optimization problems, thus falling into the local optima.

In this paper, a novel SSA-based path planning algorithm is proposed. The algorithm incorporates random opposition-based learning strategy and linear decreasing mechanism and is utilized to optimize the control points of Bezier curve, which is used to generate an optimal feasible path. The Bezier curve requires only a few control points to generate a smooth curve, which makes the dimension of the path planning problem not increase exponentially with the complexity of the environment and greatly reduces the complexity of the path planning problem. The superiority of the proposed algorithm is verified by benchmark function experiments, and the smooth optimal path of the robot is designed more stably in the contrast experiments of the path planning problem.

The remaining of the article is arranged as follows: Sect. 2 explains basic SSA algorithm. Section 3 presents the proposed algorithm ROSSA. The description of robot path planning problem and Bezier curve are discussed and contrast experiments with PSO and SSA are conducted in Sect. 4. Finally, Sect. 5 gives the conclusion.

## 2  Sparrow Search Algorithm(SSA)

SSA is a novel swarm intelligence-based optimization algorithm inspired by the foraging and anti-predatory behaviors of a sparrow population. It has three phases: producer phase, scrounger phase and scouter phase. The key steps of the SSA algorithm are following:

Initialization: First of all, SSA initializes all the parameters and random population of sparrow as follows:

$$X_{i,j} = rand \times (UB_j - LB_j) \tag{1}$$

where $i = 1, 2, ..., pop$, $j = 1, 2, ..., dim$. $LB_j$ and $UB_j$ are lower and upper bounds of search spaceseparately. $rand \in (0, 1)$ is a random number.

Producer phase: After initialization, sparrows in the top 10%–20% fitness values (producers) start to search for a better solution in the search space. In this phase, the location of the sparrow is updated by

$$X_{i,j}^{t+1} = \begin{cases} X_{i,j}^t \cdot \exp\left(\frac{-i}{\alpha \cdot T}\right) & if\, R_2 \leq ST \\ X_{i,j}^t + Q \cdot L & if\, R_2 \geq ST \end{cases} \tag{2}$$

where $t$ represents the current iteration. $\alpha$ is a random number in the range $(0, 1)$. $T$ is the max iteration. $Q$ obeys normal distribution. $R_2 \in [0, 1]$, $ST \in [0.5, 1]$ are the alarm value and the safety threshold respectively.

Scrounger phase: The rest of population are called scroungers. The movement of scrounger individuals can be defined as:

$$X_{i,j}^{t+1} = \begin{cases} Q \cdot \exp\left(\frac{x_{worst}^t - X_{i,j}^t}{i^2}\right) & i > n/2 \\ X_p^t + \left|X_{ij}^t - X_p^t\right| \cdot A^+ \cdot L & otherwise \end{cases} \tag{3}$$

where $X_p$ is the best position occupied by the producer. $X_{worst}$ denotes the current global worst location. $A$ represents a matrix of $1 \times d$ for which each element inside is randomly assigned 1 or –1, and $A^+ = A^T \left(AA^T\right)^{-1}$.

Scout phase: Randomly select 10%–20% of population as scout. The update formula of scout is described as follows:

$$X_{i,j}^{t+1} = \begin{cases} X_{best}^t + \beta \cdot \left|X_{i,j}^t - X_{best}^t\right| & f_i > f_g \\ X_{i,j}^t + K \cdot \left(\frac{|x_{i,j}^t - X_{worst}^t|}{(f_i - f_{worst}) + \varepsilon}\right) & f_i = f_g \end{cases} \tag{4}$$

where $X_{best}$ is the current global optimal location. $b$ is the step control parameter that obeys a normal distribution. $K \in [-1, 1]$ is a random number. $f_i$ represents the fitness value of sparrow $i$. $f_g$ and $f_w$ denote the current global best and worst fitness values, respectively. $\varepsilon$ is a constant used to avoid the denominator being 0.

## 3    Improvement

### 3.1    Opposition-based Learning(OBL)

OBL was first proposed by Tizhoosh [14], and a large amount of variants of opposition-based learning were proposed, such as quasi-opposition [15], quasi-reflection [16], comprehensive opposition [17], etc. Studies showed that considering both random outcomes and their opposite results is more advantageous than considering only random results [18]. The concept of opposition-based learning is based on opposite numbers. It is expressed as follows: Let $x \in [a, b]$ be a real number. Then its opposite number, $\breve{x}$, is given by following equation:

$$\breve{x} = a + b - x \tag{5}$$

In higher dimensional space, the extended definition is defined as follows:

Let $x(x_1, \ldots, x_d)$ be a point in d-dimensional space and $x_i \in [a_i, b_i]$, $i = 1, 2, \ldots, d$. The opposite point of x, $\breve{x}(\breve{x}_1, \ldots, \breve{x}_d)$, can be expressed as:

$$\breve{x}_i = a_i + b_i - x_i \tag{6}$$

Different from basic opposite point, this paper uses a variant strategy called random opposite point [19], which is defined by:

$$\breve{x}_i = \begin{cases} a_i + b_i - x_i & rand \geq R \\ x_i & otherwise \end{cases} \tag{7}$$

It is reported that by this reverse strategy there are more possible positions than the base reverse strategy, further increasing the diversity of the population.

## 3.2   Random Opposition-based Sparrow Search Algorithm(ROSSA)

SSA has the disadvantage that when the search is close to the global optimum, the population diversity decreases and it is easy to fall into the local optimum solution [20]. This paper uses random opposition-based strategy to improve SSA. First, a random OBL strategy is used to generate the opposite initial solution when initializing the population, and an elite strategy is used to select better individuals from the initial population and the opposite initial population to form the final initial population. This gives the algorithm an advantage at the beginning. Meanwhile, the producers in SSA searches the whole search space, and random opposition-based strategy can effectively increase the population diversity and optimize the global search ability.

Both producers and scouters in SSA can enhance the global exploration ability of the algorithm, but their proportion is fixed, which does not balance well between global exploration and local exploitation in the first and second stages of SSA algorithm. Therefore, this paper adopts a linear decreasing strategy to control the number of both producers and scouters, which is beneficial to the convergence of the algorithm. The decreasing formula is as follows.

$$p = p_{\max} - (p_{\max} - p_{\min}) \cdot \frac{t}{T} \tag{8}$$

where $p$ is the proportion of producers and scouters, $p_{max}$ and $p_{min}$ denote the maximum and minimum number of $p$. In this paper, the maximum and minimum values of both are taken as 0.4 and 0.1, respectively.

The main flow of ROSSA is shown in Algorithm 1.

---

**Algorithm 1:** Framework of ROSSA

---

**input** : $T$: the maximum iterations,

　　　　 $pop$: the size of population,

　　　　 $num_p$: the number of producers,

　　　　 $num_s$: the number of scouters,

　　　　 $ST$: the threshold of alert value

**output**: $X_{best}$, $f_{best}$

1 Initialize the population and opposition population;

2 Sort the population by fitness and retain the pop individuals with better fitness values;

3 **while** $t < T$ **do**

4 　| Calculate $num_p$ with equation 8;

5 　| **for** $i = 1 : num_p$ **do**

6 　| 　| update the producers' location with Equation 7;

7 　| **end**

8 　| **for** $i = (num_p + 1) : pop$ **do**

9 　| 　| update the scroungers' location with Equation 3;

10 　| **end**

11 　| Calculate $num_s$ with equation 8;

12 　| **for** $i = 1 : num_s$ **do**

13 　| 　| update the scouters' location with Equation 4;

14 　| **end**

15 　| t=t+1;

16 **end**

17 return $X_{best}$, $f_{best}$;

---

### 3.3 Benchmark Test

To verify the advancedness of the proposed algorithm, PSO, KH, SSA and ROSSA are used to solve these test functions, which are shown in Table 1. In the tests, the population size is set to 30, the total number of iterations is set to 500, and the dimension of each test function are 30. The other properties of the functions are shown in the following table. 30 simulation experiments were conducted for each test function separately, and the mean and variance obtained from 30 experiments were counted as shown in Table 2.

Among them, F1–F6 are unimodal test functions, which are mainly used to test the exploitation ability of the algorithm. The results of these 6 unimodal functions show that ROSSA has the best effect of finding the best solution for unimodal functions, and can obtain the global optimal solution to all of these 6 functions, and the stability of ROSSA is better than other 3 algorithms.

F7–F11 are multimodal test functions, which have multiple local optimal solutions, and the intelligent optimization algorithm is easy to fall into the local optimum when solving, so the multimodal test functions are mainly used to test the exploration ability of the algorithm. In solving F7, all four algorithms have unsatisfactory results for this function on average, but SSA and ROSSA can

**Table 1.** benchmark functions

| Unimodal functions | Range | fmin |
|---|---|---|
| Sphere | [−100,100] | 0 |
| Schwefel 2.22 | [−10,10] | 0 |
| Schwefel 1.2 | [−100,100] | 0 |
| Schwefel 2.21 | [−100,100] | 0 |
| Rosenbrock | [−30,30] | 0 |
| Step | [−100,100] | 0 |
| Multimodal functions | Range | fmin |
| Schwefel 2.26 | [−500,500] | 0 |
| Rastrigin | [−5.12,5.12] | 0 |
| Ackley | [−32,32] | 0 |
| Griewank | [−600,600] | 0 |
| Penalty | [−50,50] | 0 |

**Table 2.** Test results

| Function | Values | PSO | KH | SSA | ROSSA |
|---|---|---|---|---|---|
| Sphere | Ave | 1.139900e+03 | 1.834809e+03 | 6.655620e-66 | 0 |
|  | Best | 7.770424e+01 | 1.366952e+03 | 0 | 0 |
|  | Var | 6.449311e+06 | 7.730924e+04 | 7.504269e-130 | 0 |
| Schwefel 2.22 | Ave | 6.683219e+01 | 1.610389e+10 | 4.019652e-38 | 0 |
|  | Best | 3.229506e+01 | 7.338242e+06 | 0 | 0 |
|  | Var | 4.282651e+02 | 2.609927e+21 | 4.826645e-74 | 0 |
| Schwefel 1.2 | Ave | 1.989746e+04 | 5.264767e+03 | 3.469262e-41 | 0 |
|  | Best | 9.569614e+03 | 2.534085e+03 | 0 | 0 |
|  | Var | 2.610611e+07 | 2.632828e+06 | 3.610733e-80 | 0 |
| Schwefel 2.21 | Ave | 2.519220e+01 | 1.506538e+00 | 2.556319e-27 | 0 |
|  | Best | 1.836057e+01 | 1.066817e+00 | 0 | 0 |
|  | Var | 1.893842e+01 | 8.679998e-02 | 1.960410e-52 | 0 |
| Rosenbrock | Ave | 9.918774e+03 | 5.557089e+05 | 1.722730e-05 | 1.349896e-09 |
|  | Best | 2.237782e+02 | 2.341405e+05 | 0 | 0 |
|  | Var | 6.563432e+08 | 6.284255e+10 | 4.127078e-09 | 5.466655e-17 |
| Step | Ave | 1.876022e+00 | 4.683979e+02 | 7.875731e-08 | 2.208193e-11 |
|  | Best | 3.994384e-01 | 3.481912e+02 | 0 | 0 |
|  | Var | 4.533699e+00 | 3.435012e+03 | 3.706890e-14 | 1.462835e-20 |
| Schwefel 2.26 | Ave | 2.143348e+03 | 4.225624e+03 | 3.038711e+03 | 1.908453e+03 |
|  | Best | 9.521574e+02 | 2.625807e+03 | 3.818270e-04 | 3.818270e-04 |
|  | Var | 3.464213e+05 | 6.830527e+05 | 9.286784e+06 | 3.112237e+06 |
| Rastrigin | Ave | 1.185528e+02 | 3.808714e+01 | 0 | 0 |
|  | Best | 6.077383e+01 | 2.178187e+01 | 0 | 0 |
|  | Var | 1.086724e+03 | 1.342092e+02 | 0 | 0 |
| Ackley | Ave | 1.406729e+00 | 1.891097e+00 | 8.881784e-16 | 8.881784e-16 |
|  | Best | 2.128831e-01 | 1.439818e+00 | 8.881784e-16 | 8.881784e-16 |
|  | Var | 5.047473e-01 | 4.622237e-02 | 0 | 0 |
| Griewank | Ave | 8.646091e-01 | 5.004274e+00 | 0 | 0 |
|  | Best | 3.867774e-01 | 3.705478e+00 | 0 | 0 |
|  | Var | 4.332038e-02 | 5.003298e-01 | 0 | 0 |
| Penalty | Ave | 5.049397e+00 | 5.059365e+04 | 3.214013e-09 | 6.214156e-10 |
|  | Best | 1.200169e+00 | 2.353369e+03 | 1.570545e-32 | 1.570545e-32 |
|  | Var | 7.984803e+00 | 3.192327e+09 | 7.707496e-17 | 1.124535e-17 |

explore better positions; in solving F8, F9 and F10, SSA and ROSSA outperform the other two algorithms; in solving F11, both SSA and ROSSA have the ability to find excellent solutions, but ROSSA has a slight advantage. In summary, ROSSA performs better than the other three algorithms in the benchmark function experiments.

Figure 1 shows the convergence curves of the partial functions of each algorithm. The horizontal axis represents the number of update generations and the vertical axis represents the log of the fitness value. It can be seen that, ROSSA has better convergence speed, accuracy and stability.

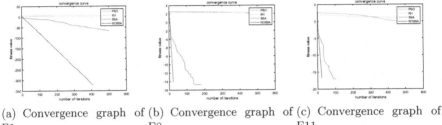

(a) Convergence graph of F1.  (b) Convergence graph of F8.  (c) Convergence graph of F11.

**Fig. 1.** Convergence curves of partial functions: (a) F1, (b) F8, (c) F11.

## 4    ROSSA for Path Planning Problems

### 4.1    Problem Description

In this paper, ROSSA is used to solve the robot path planning problem. The target environment is a two-dimensional plane with static obstacles. Each individual in the algorithm denotes a path, represented by $N$ control points as $p[p_1, p_2, \ldots, p_N]$, where p1 is the starting point and $p_N$ is the end point. In the implementation, the SSA individuals are represented as $[x_2, y_2, x_3, y_3, \ldots, x_{(N-1)}, y_{(N-1)}]$ for coding. In the actual environment, obstacles have various shapes. In this paper, for the simplification of the environment model, the circumcircle of the obstacle is used to simplify modeling.

### 4.2    Bezier Curve

Bezier curve was first proposed by engineer P.E. Bezier [21] and is widely used in practices such as computer graphics and mechanical design [22]. Bezier curve is generated by a series of control points and these points are not on the curve except for the start and end points. Given a set of control points $P_0, \ldots, P_n$, the corresponding Bezier curve can be expressed as

$$P(t) = \sum_{i=0}^{n} P_i B_{i,n}(t), \quad t \in [0, 1] \tag{9}$$

where $t$ is the normalized time variable, $B_{i,n}$ is the Bernstein basis polynomials, which represents the base function in the expression of a Bezier curve:

$$B_{i,n}(t) = C_n^i t^i (1-t)^{n-i} \quad i = 0, 1, \ldots n \tag{10}$$

In this way, a smooth curve can be created with only a small number of control points.

### 4.3  Fitness Function

The purpose of this paper is to find an optimal path for the robot that satisfies the constraints, where the constraints include (1) feasibility, (2) optimality, and (3) safety.

1. Feasibility

   Feasibility is the most important goal of path planning. If the path collides with an obstacle, the fitness should be large, which is set here to 10000:

$$f_{feasible} = \begin{cases} 0 & if \quad feasible \\ 10000 & otherwise \end{cases} \tag{11}$$

2. Shortest distance

   The second target is to minimize the length of the solution generated by the algorithm. For simplicity, we choose 100 points on the path and calculate the Euclidean distance between two adjacent points:

$$f_{length} = \sum_{i=1}^{n-1} \|p_{i+1} - p_i\| \tag{12}$$

3. Safety
   An excellent path should be as far away from obstacles as possible. If the distance between the path and the obstacle is less than the safe distance, $d_{safe}$, it will be penalized:

$$f_{safe}(o_j) = \begin{cases} \left(1 - \frac{d_{min}(o_j)}{d_{safe}(o_j)}\right)^2 & if \quad d_{min}(o_j) \le d_{safe}(o_j) \\ 0 & otherwise \end{cases} \tag{13}$$

$$f_{risk} = \max(f_{safe}) \tag{14}$$

where $d_{min}(o_j)$ means the minimum distance of the path from the obstacle $j$. $d_{safe}(o_j)$ can be expressed as follow:

$$d_{safe}(o_j) = kr_{o_j} \tag{15}$$

where $k$ indicates the scale factor and $r_{o_j}$ denotes the radius of obstacle $j$.

Considering the above factors, the fitness function of the robot path planning problem can be expressed as:

$$f = f_{feasible} + w_1 * f_{length} + w_2 * f_{risk} \tag{16}$$

## 4.4   Comparision

The parameters of the path planning model are set as follows: the map is $500 \times 500$, as shown in Fig. 2. The number of control points is 5, the robot moves from (10,10) to (490,490).

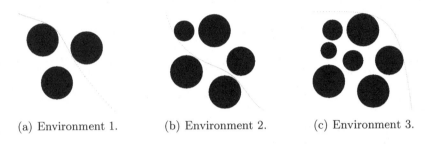

(a) Environment 1.          (b) Environment 2.          (c) Environment 3.

**Fig. 2.** Environments.

The objective function of this path planning model is solved using PSO, SSA and ROSSA respectively to obtain the desired paths. The population size is set to 30, each individual is a path, the maximum number of iterations is 500, and 30 simulation experiments are conducted. Figure 3 shows the convergence curves of the three algorithms and related data are shown in Table 3.

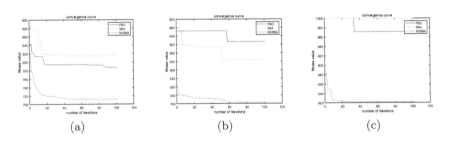

(a)                          (b)                          (c)

**Fig. 3.** Convergence curve of each environment.

The comparison in Fig. 3 and Table 3 show that the ROSSA algorithm out-performs the PSO and SSA algorithms for the path planning problem. As can be seen from Fig. 3, due to the opposition-based initialization of ROSSA, the initial solution of ROSSA is in a more optimal position. Meanwhile, the rapid convergence to the better position and the continuous approximation to the optimum can stabilize the convergence to the optimal value. Table 3 shows that the average and minimum fitness values obtained by the ROSSA algorithm are lower than those of the PSO and SSA algorithms, and that it is able to solve the path planning problem stably, resulting in a safe and feasible trajectory that is optimal and satisfies the constraints.

**Table 3.** Performance comparison of three algorithms

(a) Environment 1.

| Algorithm | Ave | Best | Var | Success rate |
|---|---|---|---|---|
| PSO | 7.679612e+02 | 7.160134e+02 | 7.450565e+03 | 90% |
| SSA | 8.636940e+02 | 7.040057e+02 | 2.140079e+04 | 50% |
| ROSSA | 7.145130e+02 | 7.023720e+02 | 9.443526e+01 | 100% |

(b) Environment 2.

| Algorithm | Ave | Best | Var | Success rate |
|---|---|---|---|---|
| PSO | 8.929376e+02 | 7.090549e+02 | 1.959787e+04 | 40% |
| SSA | 9.395579e+02 | 6.977466e+02 | 1.623664e+04 | 30% |
| ROSSA | 8.003072e+02 | 7.034846e+02 | 2.640106e+03 | 100% |

(c) Environment 3.

| Algorithm | Ave | Best | Var | Success rate |
|---|---|---|---|---|
| PSO | INF | INF | 0 | 0% |
| SSA | 9.714638e+02 | 9.146378e+02 | 8.143158e+03 | 10% |
| ROSSA | 8.216849e+02 | 8.152322e+02 | 5.115674e+01 | 100% |

## 5  Conclusions

In this paper, an improved SSA is used to solve the path planning problem. The random OBL strategy and a linear decreasing strategy are introduced into the basic SSA. These strategies are used to increase the population diversity, balance the local exploitation and global exploration ability of the algorithm, and avoid the algorithm from falling into local optimum. The results of benchmark function test show that the proposed algorithm has a significant improvement in the performance of convergence speed, accuracy and stability. The path planning simulation results show that the path planning based on ROSSA can effectively find the optimal path and steadily plan a feasible and efficient path.

## References

1. Reif, J.H.: Complexity of the mover's problem and generalizations. In: 20th Annual Symposium on Foundations of Computer Science (SFCS 1979), pp. 421–427. IEEE (1979)
2. Li, G., Chou, W.: Path planning for mobile robot using self-adaptive learning particle swarm optimization. Sci. China Inf. Sci. **61**(5), 1–18 (2017). https://doi.org/10.1007/s11432-016-9115-2
3. Nilsson, N.J.: A mobile automaton: An application of artificial intelligence techniques. Technical Report, Sri International Menlo Park Ca Artificial Intelligence Center (1969)
4. Kavraki, L.E., Svestka, P., Latombe, J.C., Overmars, M.H.: Probabilistic roadmaps for path planning in high-dimensional configuration spaces. IEEE Trans. Robot. Autom. **12**(4), 566–580 (1996)

5. Yan, F., Liu, Y.S., Xiao, J.Z.: Path planning in complex 3d environments using a probabilistic roadmap method. Int. J. Autom. Comput. **10**(6), 525–533 (2013)
6. LaValle, S.M.: Rapidly-exploring random trees: A new tool for path planning (1998)
7. Yuan, C., Liu, G., Zhang, W., Pan, X.: An efficient RRT cache method in dynamic environments for path planning. Robot. Auton. Syst. **131**, 103595 (2020)
8. Khatib, O.: Real-time obstacle avoidance for manipulators and mobile robots. In: Cox I.J., Wilfong G.T. (eds.) Autonomous robot vehicles, pp. 396–404. Springer, New York (1986). https://doi.org/10.1007/978-1-4613-8997-2_29
9. Sabudin, E.N., et al.: Improved potential field method for robot path planning with path pruning. In: Md Zain, Z., et al. (eds.) Proceedings of the 11th National Technical Seminar on Unmanned System Technology 2019. LNEE, vol. 666, pp. 113–127. Springer, Singapore (2021). https://doi.org/10.1007/978-981-15-5281-6_9
10. Li, X., Wu, D., He, J., Bashir, M., Liping, M.: An improved method of particle swarm optimization for path planning of mobile robot. J. Control Sci. Eng. **2020** (2020)
11. Rao, D.C., Kabat, M.R., Das, P.K., Jena, P.K.: Cooperative navigation planning of multiple mobile robots using improved krill herd. Arab. J. Sci. Eng. **43**(12), 7869–7891 (2018)
12. Zhang, B., Duan, Y., Zhang, Y., Wang, Y.: Particle swarm optimization algorithm based on beetle antennae search algorithm to solve path planning problem. In: 2020 IEEE 4th Information Technology, Networking, Electronic and Automation Control Conference (ITNEC), vol. 1, pp. 1586–1589. IEEE (2020)
13. Xue, J., Shen, B.: A novel swarm intelligence optimization approach: sparrow search algorithm. Syst. Sci. Control Eng. **8**(1), 22–34 (2020)
14. Tizhoosh, H.R.: Opposition-based learning: a new scheme for machine intelligence. In: International Conference on Computational Intelligence for Modelling, Control and Automation and International Conference on Intelligent Agents, Web Technologies and Internet Commerce (CIMCA-IAWTIC 2006), vol. 1, pp. 695–701. IEEE (2005)
15. Rahnamayan, S., Tizhoosh, H.R., Salama, M.M.: Quasi-oppositional differential evolution. In: 2007 IEEE Congress on Evolutionary Computation, pp. 2229–2236. IEEE (2007)
16. Ergezer, M., Simon, D., Du, D.: Oppositional biogeography-based optimization. In: 2009 IEEE International Conference on Systems, Man and Cybernetics, pp. 1009–1014. IEEE (2009)
17. Seif, Z., Ahmadi, M.B.: An opposition-based algorithm for function optimization. Eng. Appl. Artif. Intell. **37**, 293–306 (2015)
18. Rahnamayan, S., Tizhoosh, H.R., Salama, M.M.: Opposition-based differential evolution. IEEE Trans. Evol. Comput. **12**(1), 64–79 (2008)
19. Bairathi, D., Gopalani, D.: Random-opposition-based learning for computational intelligence. In: Tuba, M., Akashe, S., Joshi, A. (eds.) Information and Communication Technology for Sustainable Development. AISC, vol. 933, pp. 111–120. Springer, Singapore (2020). https://doi.org/10.1007/978-981-13-7166-0_11
20. Qinghua, M., Qiang, Z.: Improved sparrow algorithm combining cauchy mutation and opposition-based learning. J. Front. Comput. Sci. Technol. **15**, 1–12 (2020)
21. Farin, G.: Curves and Surfaces for Computer-Aided Geometric Design: a Practical Guide. Elsevier, Amsterdam (2014)
22. Song, B., Wang, Z., Zou, L.: An improved PSO algorithm for smooth path planning of mobile robots using continuous high-degree Bezier curve. Appl. Soft Comput. **100**, 106960 (2021)

# Communication-Efficient Federated Learning with Multi-layered Compressed Model Update and Dynamic Weighting Aggregation

Kaiyang Zhong$^{(\boxtimes)}$ and Guiquan Liu

School of Computer Science and Technology, University of Science
and Technology of China, Hefei 230026, China
972887209@qq.com, gqliu@ustc.edu.cn

**Abstract.** Federated learning (FL) aims to build a deep learning model based on distributed datasets. Different from traditional deep learning, federated learning does not need to centralize data from multi-party. Clients store datasets locally and train a central model through interaction with the central server. And the data privacy of clients could be preserved very well. However, sending a huge number of information to the central server will lead to huge communication overhead. This article presented two methods to solve the problem. The first is the enhanced federated learning technique with multi-layered compressed model. The second is the dynamic weighting aggregation algorithm considering the size of the dataset, local learning accuracy and the frequency of local model update for each client. The result of experiments demonstrates that the proposed framework with multi-layered compressed model and dynamic weighting aggregation performs better than the baseline algorithm in both accuracy and communication efficiency.

**Keywords:** Federated learning · Multi-layered compressed model · Aggregation · Huffman

## 1 Introduction

In the past few years, we have seen the rapid development of machine learning in the field of artificial intelligence, such as computer vision [3,19], natural language processing [22,26] and recommendation system [20,21]. The success of these machine learning technologies is based on a large amount of data. For example, the target detection system of Facebook company is trained by 350 million images from Instagram [18]. With the continuous development of society, people pay more attention to user privacy and data protection. In this case, the owners of some highly sensitive data (such as financial transaction data and health data) can only keep the data confidential strictly. This phenomenon is called isolated data islands [23], which brings great resistance to the data integration of artificial intelligence technology.

© Springer Nature Switzerland AG 2021
L. Fang et al. (Eds.): CICAI 2021, LNAI 13070, pp. 419–430, 2021.
https://doi.org/10.1007/978-3-030-93049-3_35

The reason why the traditional machine learning technology is restricted by the phenomenon of isolated data islands is that the training of the traditional machine learning model [2,12] needs to gather all the data together, and then unified by the learning model for training. A feasible method to solve the problem is that each client trains their own model, then each client communicates with each other on their own model, and finally gets a central model through model aggregation. The data of each client is only stored locally, so that no one can guess the privacy data content of each client.

Compared with the traditional machine learning model, federated learning [5,11,13] still has the challenge of communication bandwidth and learning accuracy. Nowadays, there are two strategies for improving communication efficiency. The first is the sketched updates [10]. Clients compute local updates and compress them locally. The compressed model parameters update is the unbiased estimation of the real parameters update. The second is the structured updates [8]. In the training process of federated model, the updating of model parameters is limited to the form that allows effective compression operation. For example, model parameters may be required to be sparse or low-order. But this method will reduce the learning accuracy to a certain extent.

To tackle the challenges mentioned above, this article proposed multi-layered compressed model update and dynamic weighting aggregation to reduce the communication cost and improve the accuracy rate. The main contributions of this article are as following.

First, the enhanced federated learning technique with multi-layered compressed model is proposed to reduce the communication costs. Layers of the network are categorized into feature layer and composition layer. Parameters of the different layers are compressed with Huffman coding and updated asynchronously.

Second, different from the traditional aggregation algorithm only considering the size of dataset, dynamic weighting aggregation algorithm considering the size of the dataset, local learning accuracy and the frequency of local model update for each client is proposed to improve the speed and stability of the convergence.

## 2   Related Work

### 2.1   Horizontal Federated Learning

Horizontal federated learning [6,28] is applied to scenarios where each client's dataset has the similar feature space and different sample space. The federated learning mentioned later is all horizontal federated learning. The detailed steps of federated learning are divided into server execution part and client update part.

Server Execution: First, the central server initializes the model parameters and broadcasts them to all clients. Second, the central server determines the number of participating clients and randomly selects the corresponding number of clients. Third, the central server collects the parameters uploaded by each client. Fourth, the server broadcasts the aggregated result to all clients.

Client Update: First, the clients get the aggregation result from the server. Second, split the dataset into the batches of certain size. Third, SGD [14] is performed on each batch of data to calculate the parameters. Fourth, return the parameters to the central server.

## 2.2 Federated Learning Algorithm

Federated averaging algorithm (FedAvg) [15] is proposed by Google to obtain a central prediction model of Google's Gboard app. The aggregated model weights in FedAvg is decided by the size of each client's local dataset. It's showed as following.

$$w_{t+1} \longleftarrow \sum_{k=1}^{K} \frac{n_k}{n} * w^k \tag{1}$$

$n$ is the total number of the dataset of all clients which participates in the $t_{th}$ federated learning. $n_k$ is the number of the dataset which the $k_{th}$ client has. $w^k$ is the model parameters which learned from the $k_{th}$ client. The larger size of dataset that the client uses to train the local model, the more weight that the client has in the aggregation.

## 2.3 Temporally Weighted Aggregation

In federated learning, the training datasets of each client may change all the time. Therefore, clients whose local datasets had changed recently should have more weight in the aggregation. Different from FedAvg only considering the size of dataset, temporally weighted aggregation algorithm [1] takes into account of both the size of dataset and the latest learning round of the client. It's showed as following.

$$w_{t+1} \longleftarrow \sum_{k=1}^{K} \frac{n_k}{n} * (\frac{e}{2})^{-(t-timestamp^k)} * w^k \tag{2}$$

$t$ is the label of the current round. $timestamp^k$ is the label of the round in which the newest local model parameters of the client was updated. $w^k$ is the model parameters which learned from the $k_{th}$ client. Temporally weighted aggregation algorithm is proposed to increase the communication efficiency of federated learning. It enables faster convergence of the training accuracy compared to FedAvg.

# 3 Proposed Method

## 3.1 Multi-Layered Model Update

In the implementation of traditional model aggregation, each client needs to send complete model parameter updates to the central server in each central model training round. Because the modern DNN model usually has millions of parameters, sending so many values to the central server will lead to huge communication overhead. And the overhead will also increase with the number of clients and iterations.

The idea of multi-layered model update is inspired by the interpretability of deep neural network model [16]. Different from the design of logical regression [7] and decision tree [25] which is easy for people to understand, deep neural network can fit highly complex data with a large number of parameters but how to explain this is very difficult. We can roughly assume that the lower-level layers in the deep neural network learn the basic features and the higher-level layers in the deep neural network learn the feature composition laws in specific datasets. In other words, when the training datasets does not change greatly, the model learns more basic features through the new training data. The relevant practice has also proved that the lower-level layers in the deep neural network change more frequently than the higher-level layers in the later stage of the training process.

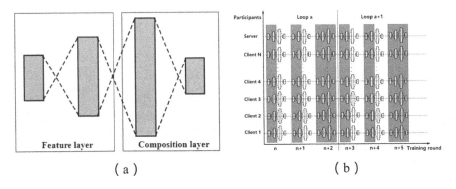

**Fig. 1.** The classification of feature layer and composition layer. (b) The diagram of multi-layered model update strategy.

The simple diagram of deep neural network is showed in Fig. 1(a). The Layers of deep neural network are categorized into feature layer and composition layer. The feature layer learns the basic features and the composition layer learns feature composition laws in specific datasets. The composition layer usually has more parameters than the feature layer because of the complexity of the feature composition laws. In the proposed multi-layered model update strategy, parameters in the feature layer will be updated more frequently than that in the composition layer. The process of federated learning consists of a large number of training rounds. We record three training rounds as one loop. In each loop, only parameters of the feature layer are updated during the first two training rounds both in local clients and the central server. And only in the last training round, parameters of both the feature layer and the composition layer are updated synchronously.

An example is given in Fig. 1(b) to show how the multi-layered model update strategy works. In this example, there are a large number of clients whose labels are from 1 to N and a central server. The process of the federated learning consists of six training rounds (n, n + 1, ..., n + 5). Point (n, client1) means client1 is participating in training the central server during the stage of training round 'n'. The parts surrounded by gray rectangles mean they will be transmitted to the central server to participate in the model aggregation. The training rounds are categorized into loop 'a' which consists of training round from 'n' to 'n + 2' and loop 'a + 1' which consists of training round from 'n + 3' to 'n + 5'. It can be seen from the Fig. 1(b) that parameters of the composition layer are only updated in the training round 'n + 2' and 'n + 5' while parameters of the feature layer are updated in the whole process of the federated learning. As a result, the number of parameters which should have been transmitted is reduced greatly.

## 3.2 Dynamic Weighting Aggregation

Dynamic weighting aggregation algorithm is proposed as a more mature and comprehensive model aggregation algorithm to improve the speed and stability of the convergence. It considers the size of the dataset, local learning accuracy and the frequency of local model update for each client. And it adopts a more reasonable way of selecting participants rather than random selection. The specific design of the dynamic weighting aggregation is as following.

Local Accuracy: Clients have a test dataset locally. The local accuracy of participants will affect their weight when participating in model aggregation. The local accuracy of each participant is recorded as $a_k$. $a$ is the sum of the local accuracy for all participants.

Select Participants: In traditional model aggregation algorithm, the server uses the random selection method while choosing the federated learning participants. In order to motivate more participants to join the federated learning framework and contribute more, we optimize the process of participants selection. All clients are ranked according to a reliable index. The index is calculated by the size of the datasets and the local accuracy. Then select the appropriate number of participants according to the index from large to small. $q$ represents the proportion of the local accuracy in the index.

$$index \longleftarrow q * \frac{a_i}{a} + (1 - q) * \frac{n_i}{n} \qquad (3)$$

Frequency of Local Model Update: When the local datasets of participants do not change for a long time, it has no effect on the training of the central model. The frequency of local model update is the proportion of the number of local dataset updates in the total number of training rounds. It's recorded as $f_k$. The central server will record the number of local dataset updates and the total number of training rounds for each client. However, if the frequency is directly used in model aggregation without modification, the influence of the frequency in model aggregation will be too large because some frequencies may approach zero. So, we normalize the frequency between 0.5 and 1 as following. $f$ is the sum of the update frequency for all participants.

$$f_k \longleftarrow (f_k + 1)/2 \tag{4}$$

Dynamic Weighting Aggregation: First, the server selects the optimal participants according to the size of the dataset and local learning accuracy. Then the server execution in Sect. 2.1 is performed for each participant. After the server receiving information from all participants, it will update the number of local dataset updates and the total number of training rounds for each client. At last, the dynamic weighting aggregation algorithm based on the size of the dataset, local learning accuracy and the frequency of local model update for each participant is performed as following. $\alpha$, $\beta$, $\gamma$ is the proportion of the three parts in the model aggregation.

$$w_{t+1} \longleftarrow \sum_{k=1}^{K} (\alpha * \frac{n_k}{n} * (\frac{e}{2})^{-(t-timestamp^k)} + \beta * \frac{a_k}{a} + \gamma * \frac{f_k}{f}) * w^k \tag{5}$$

### 3.3   Model Compression

As mentioned in Sect. 3.1, different layers of the deep neural networks are categorized into feature layer and composition layer and they are updated asynchronously. In order to further reduce the communication overhead of federated learning, the model is compressed by pruning [27] and Huffman coding [9].

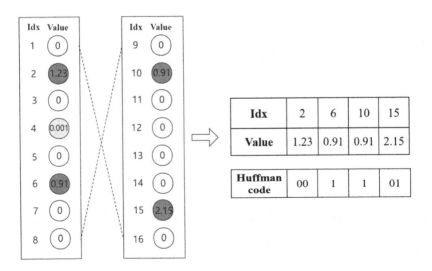

**Fig. 2.** The simple diagram of pruning and Huffman coding.

Pruning: Pruning is a method of deleting some calculation costs with low benefits. Deep neural networks usually have millions of parameters. But the value of zero accounts for a large part of these parameters. Multiplying zero by one of the input values will come to zero which takes up storage resources and brings low benefits. So, the weights which are close to zero can be cut off while compressing the model.

An example of pruning with two-layers model is given in Fig. 2. The model is stored in the program as the form of Idx:Value. In the proposed pruning method, parameters with absolute value less than or equal to 0.001 are cut off. As shown in Fig. 2, the two-layers model is stored as 2:1.23, 6:0.91, 10:0.91 and 15:2.15 in the program.

Huffman Coding: Huffman code can represent complex numbers with short code value. As the example shown in Fig. 2, the two-layers model in the program has become 2:00, 6:1, 10:1, 15:01 in the form of Idx:Huffmancode. The original average code length is 4 bit/sym. After Huffman coding, the average code length is 1.5 bit/sym. It can be clearly seen that the storage space utilization can be improved greatly after the proposed method of pruning and Huffman coding.

## 3.4   Framework

The framework of the proposed federated learning with multi-layered compressed model update and dynamic weighting aggregation is given in Fig. 3.

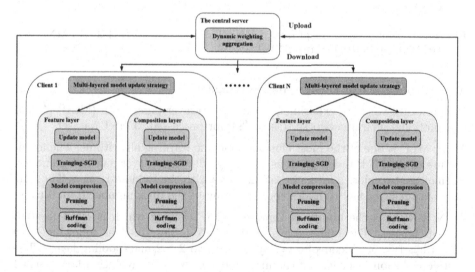

**Fig. 3.** Federated learning with multi-layered compressed model update and dynamic weighting aggregation.

## 4   Experiments

### 4.1   Dataset and Dataset Processing

Commercial data is not available for our experiments due to its privacy. So, the MNIST dataset [17] is used in the experiment. The MNIST dataset is a classic dataset in the field of machine learning. It consists of 60000 training samples and 10000 test samples. Each sample is a 28 * 28 pixel gray handwritten digital image. Each picture represents a number from 0 to 9.

To make the framework have better robustness, we preprocess the MNIST dataset to make it satisfied the requirements of Non-IId data [29], unbalanced data and large number of clients. As shown in Fig. 4, We have five processed datasets with different distributions. The X-axis represents the Id of clients. The Y-axis represents the label of digit number. The Z-axis represents the number of samples. An example is given in Fig. 4(a), point (clien6, digit8, 991) means the client6 has 991 samples of digit number 8.

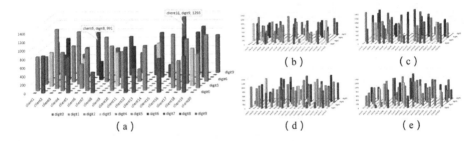

**Fig. 4.** 3-D column charts of processed MNIST dataset. (a) MNIST-1. (b) MNIST-2. (c) MNIST-3. (d) MNIST-4. (e) MNIST-5.

### 4.2   Experimental Design and Settings

We perform two set of experiments using multi-layer perceptron (MLP) [4] and convolutional neural networks (CNN) [24] for processed MNIST dataset.

MLP: MLP is one of the most classical machine learning algorithms. The architecture of the MLP is described as following. The input layer has 28 * 28 nodes. Both of the two hidden layers have 512 nodes with the Relu activation. The softmax output layer has 10 nodes corresponding to the digit number from 0 to 9.

CNN: CNN has the ability of representation learning which is often used to process images. The architecture of the CNN is described as following. The model has two 5 * 5 convolution layers and two 2 * 2 max-pooling layers. The first convolution layer has 32 channels and the second convolution layer has 64 channels. After that, a fully connected layer with 512 nodes and Relu activation will follow. At last, the softmax output layer has 10 nodes corresponding to the digit number from 0 to 9.

The parameters of $\alpha$, $\beta$ and $\gamma$ in dynamic weighting aggregation are set as 1/2, 1/4 and 1/4.

## 4.3   Results and Analysis

As mentioned above, experiments of MLP and CNN are designed on the dataset from MNIST-1 to MNIST-5. The baseline algorithm is FedAvg. The comparative algorithms are federated learning with only dynamic weighting aggregation (DW_FedAvg) and federated learning with multi-layered compressed model update and dynamic weighting aggregation (MCDW_FedAvg). We use accuracy (Acc) and communication cost (C.Cost) [20] as the metrics. Accuracy is the accuracy when the model has converged. Communication cost is the time required when the accuracy of the model reaches 90%. The result of comparative experiments is showed as following. In each experiment, the communication cost of MCDW_FedAvg is termed as 1. The communication costs of other algorithms are compared to that of MCDW_FedAvg (Table 1).

**Table 1.** The result of the comparative experiments.

| Model | Dataset | FedAvg | | DW_FedAvg | | MCDW_FedAvg | |
|---|---|---|---|---|---|---|---|
| | | Round(Acc) | C.Cost | Round(Acc) | C.Cost | Round(Acc) | C.Cost |
| MLP | MNIST-1 | 103(95.2%) | 4.89 | 79(95.8%) | 1.37 | 135(93.4%) | 1 |
| MLP | MNIST-2 | 121(95.1%) | 4.72 | 92(95.5%) | 1.49 | 143(95.2%) | 1 |
| MLP | MNIST-3 | 97(95.6%) | 7.34 | 67(96.1%) | 1.42 | 132(94.1%) | 1 |
| MLP | MNIST-4 | 142(95.1%) | 2.42 | 103(94.7%) | 2.17 | 189(93.5%) | 1 |
| MLP | MNIST-5 | 114(95.2%) | 5.66 | 96(95.5%) | 1.51 | 157(93.3%) | 1 |
| CNN | MNIST-1 | 201(96.1%) | 3.25 | 114(96.6%) | 0.89 | 249(93.4%) | 1 |
| CNN | MNIST-2 | 187(96.5%) | 4.92 | 98(97.1%) | 3.18 | 194(94.3%) | 1 |
| CNN | MNIST-3 | 225(95.7%) | 3.81 | 136(96.2%) | 1.72 | 285(95.8%) | 1 |
| CNN | MNIST-4 | 198(96.1%) | 5.14 | 131(96.4%) | 2.13 | 238(93.8%) | 1 |
| CNN | MNIST-5 | 210(96.2%) | 5.76 | 144(96.3%) | 1.53 | 251(94.1%) | 1 |

It can be seen that, MCDW_FedAvg performs better than FedAvg in communication cost. MCDW_FedAvg uses the lossy model compression method, so its accuracy would be lower. But with more communication rounds, MCDW_FedAvg could perform better than FedAvg in accuracy for certain experiments, such as the MLP on MNIST-2 and the CNN on MNIST-3. DW_FedAvg performs better than FedAvg in both accuracy and communication cost. But it is inferior to MCDW_FedAvg in improving communication efficiency.

We also design ablation experiments on the proportion of composition layer in all the networks. The proportion is set as 1/2 and 1/4 on the MLP in MNIST-1. The result is showed in Fig. 5. It can be seen that the training could converge in less communication rounds if the proportion is lower. But each communication round would take more time when the proportion is set as 1/4 because a larger number of parameters should be transmitted in each communication round.

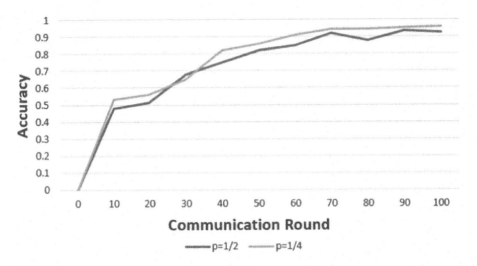

**Fig. 5.** Experiments on proportion of composition layer.

## 5  Conclusion and Future Work

The article proposed federated learning framework with multi-layered compressed model update and dynamic weighting aggregation to reduce the communication cost and improve the accuracy rate. Comparative experiments of the MLP and CNN on five processed MNIST dataset demonstrate that the proposed method could perform better than the baseline algorithm in both accuracy and communication cost.

This article uses the lossy compression method to reduce the communication cost at the expense of accuracy. In the future, we are going to develop new federated learning algorithms to further improve the learning accuracy and reduce communication cost.

## References

1. Chen, Y., Sun, X., Jin, Y.: Communication-efficient federated deep learning with asynchronous model update and temporally weighted aggregation. IEEE Trans. Neural Netw. Learn. Syst. **31**(10), 4229–4238 (2019)
2. Eykholt, K., et al.: Robust physical-world attacks on deep learning models. In: 2018 IEEE/CVF Conference on Computer Vision and Pattern Recognition (CVPR) (2018)
3. Fan, Q., Zhuo, W., Tang, C.K., Tai, Y.W.: Few-shot object detection with attention-RPN and multi-relation detector. In: 2020 IEEE/CVF Conference on Computer Vision and Pattern Recognition (CVPR) (2020)
4. Garg, G., Kumar, D., ArvinderPal, Sonker, Y., Garg, R.: A hybrid MLP-SVM model for classification using spatial-spectral features on hyper-spectral images (2021)

5. Hao, M., Li, H., Luo, X., Xu, G., Liu, S.: Efficient and privacy-enhanced federated learning for industrial artificial intelligence. IEEE Trans. Industr. Inf. **16**(10), 6532–6542 (2019)
6. Hardy, S., Henecka, W., Ivey-Law, H., Nock, R., Patrini, G., Smith, G., Thorne, B.: Private federated learning on vertically partitioned data via entity resolution and additively homomorphic encryption (2017)
7. Huang, X., Chu, F.: Risk grade evaluation model of Goaf based on logical regression and clustering algorithm. Metal Mine **48**(08), 179 (2019)
8. Kamp, M., et al.: Efficient decentralized deep learning by dynamic model averaging. In: Berlingerio, M., Bonchi, F., Gärtner, T., Hurley, N., Ifrim, G. (eds.) ECML PKDD 2018. LNCS (LNAI), vol. 11051, pp. 393–409. Springer, Cham (2019). https://doi.org/10.1007/978-3-030-10925-7_24
9. Klein, S.T., Radoszewski, J., Serebro, T.C., Shapira, D.: Optimal skeleton and reduced Huffman trees. Theoret. Comput. Sci. **852**(4), 157–171 (2021)
10. Konen, J., Mcmahan, H.B., Yu, F.X., Richtárik, P., Bacon, D.: Federated learning: strategies for improving communication efficiency (2016)
11. Li, T., Sahu, A.K., Talwalkar, A., Smith, V.: Federated learning: challenges, methods, and future directions. IEEE Signal Process. Mag. **37**(3), 50–60 (2020)
12. Lin, K., Lu, J., Chen, C.S., Zhou, J., Sun, M.T.: Unsupervised deep learning of compact binary descriptors. IEEE Trans. Pattern Anal. Mach. Intell. **41**(6), 1501–1514 (2019)
13. Lu, Y., Huang, X., Dai, Y., Maharjan, S., Zhang, Y.: Blockchain and federated learning for privacy-preserved data sharing in industrial iot. IEEE Trans. Industr. Inf. **16**(6), 4177–4186 (2019)
14. Luo, L., Xiong, Y., Liu, Y., Sun, X.: Adaptive gradient methods with dynamic bound of learning rate (2019)
15. Mcmahan, H.B., Moore, E., Ramage, D., Hampson, S., Arcas, B.: Communication-efficient learning of deep networks from decentralized data (2016)
16. Miller, T.: Explanation in artificial intelligence: Insights from the social sciences. Artif. Intell. **267**, 1–38 (2017)
17. Or, A., Ea, B.: Dropout regularization in hierarchical mixture of experts. Neurocomputing **419**, 148–156 (2021)
18. Reece, A.G., Danforth, C.M.: Instagram photos reveal predictive markers of depression. EPJ Data Sci. **6**(1), 15 (2016). https://doi.org/10.1140/epjds/s13688-017-0118-4
19. Shi, W., Rajkumar, R.: Point-GNN: Graph neural network for 3d object detection in a point cloud. In: IEEE (2020)
20. Song, Q., Cheng, D., Zhou, H., Yang, J., Hu, X.: Towards automated neural interaction discovery for click-through rate prediction. In: ACM (2020)
21. Sun, J., Guo, W., Zhang, D., Zhang, Y., Coates, M.: A framework for recommending accurate and diverse items using bayesian graph convolutional neural networks. In: KDD 2020: the 26th ACM SIGKDD Conference on Knowledge Discovery and Data Mining (2020)
22. Xia, R., Ding, Z.: Emotion-cause pair extraction: a new task to emotion analysis in texts. In: Proceedings of the 57th Annual Meeting of the Association for Computational Linguistics (2019)
23. Yang, Q., Liu, Y., Chen, T., Tong, Y.: Federated machine learning: concept and applications. ACM Trans. Intell. Syst. Technol. **10**(2), 1–19 (2019)
24. Young, S.I., Zhe, W., Taubman, D., Girod, B.: Transform quantization for cnn compression. IEEE Trans. Pattern Anal. Mach. Intell. **99**, 1–13 (2020)

25. Yu, C.S., et al.: Predicting metabolic syndrome with machine learning models using a decision tree algorithm: retrospective cohort study. JMIR Med. Inf. **8**(3), e17110 (2020)
26. Zhang, W., Feng, Y., Meng, F., You, D., Liu, Q.: Bridging the gap between training and inference for neural machine translation (2019)
27. Zhang, Y., Zhu, B., Ma, Q., Wang, H.: Effects of gradient optimizer on model pruning. In: IOP Conference Series: Materials Science and Engineering, vol. 711, p. 012095 (2020)
28. Zhao, J., Zhu, X., Wang, J., Xiao, J.: Efficient client contribution evaluation for horizontal federated learning (2021)
29. Zhao, Y., Li, M., Lai, L., Suda, N., Civin, D., Chandra, V.: Federated learning with non-iid data (2018)

# Author Index

An, Pei   I-408
An, Yanqing   I-712

Bai, Jing   I-638
Bao, Feng   II-91
Bao, Wei   I-650
Bi, Zhen   II-215

Cai, Tao   I-528
Cao, Shiyun   II-285
Cao, Wei   I-783
Cao, Xun   I-311
Cen, Jiajing   I-408
Chen, Changzu   I-89
Chen, Dewei   I-323
Chen, Enhong   I-111, I-783, II-203
Chen, Gaojie   I-408
Chen, Huajun   II-215
Chen, Jing   I-51
Chen, Jinyong   I-65
Chen, Jun   II-251
Chen, Junji   I-323
Chen, Long   II-164
Chen, Shaofei   I-51
Chen, Shuangye   II-79
Chen, Xiang   II-215
Chen, Xun   I-149
Chen, Yanmin   I-712
Cheng, Qimin   I-136
Cheng, Shuli   I-469, I-771
Chu, Boce   I-65

Dai, Hongying   I-323
Deng, Bin   II-103
Deng, Qiang   II-273
Deng, Shumin   II-215
Deng, Sinuo   I-553
Di, Huijun   I-528
Ding, Guiguang   I-39
Ding, Nai   I-444
Ding, Xin   I-662
Ding, Zhiyong   I-759
Dong, Jingjing   I-613
Dong, Ruihai   I-553

Du, Jiangnan   II-251
Du, Junping   I-724, I-735
Du, Sidan   I-456
Du, Yichao   I-111
Duan, Lijuan   II-55
Duan, Ruixue   II-227

Fan, Lei   II-103
Fan, Xin   I-252
Fan, Xiuyi   II-16
Fang, Baofu   II-334
Feng, Guorui   I-565
Feng, Tianshuo   II-127
Fu, Bo   II-152
Fu, Wei   II-395

Gan, Deqiao   I-136
Gao, Feng   I-65
Gao, Hao   I-432
Gao, Xiang   II-323
Gao, Yue   I-123
Gong, Peixian   I-158
Gong, Yongchang   I-469
Gu, Jing   I-384
Gu, Tiankai   I-15
Guan, Bochen   I-227
Guo, Jingwen   I-77
Guo, Qi   I-65
Guo, Wenyi   II-67
Guo, Yike   I-747

Han, Jungong   I-39
Han, Zhiwei   II-3
Han, Zhongyi   II-347
Hang, Cheng   I-516
He, L.   I-674
He, Qishan   I-191
He, Rundong   II-347
He, Weidong   I-712, I-783
He, Xueying   II-347
He, Yiming   I-540
Hu, Bo   I-359
Hu, Junlin   I-481
Hu, Menghan   II-67

Hu, Mingzhe I-89
Hu, Tianying II-299
Hu, Wei I-540
Hu, Wenjin I-553
Hu, Xinyi II-3
Hu, Yan I-136
Hu, Zhenzhen I-51, I-613, I-626
Hu, Zhongyi I-89, I-347
Huang, Bangbo I-301
Huang, Bin I-492
Huang, Haiyan I-136
Huang, Mei I-575
Huang, Meiyu I-650
Huang, Ran I-674
Huang, Ruqi I-371
Huang, Weibo I-77
Huang, Xiaosong I-136
Huang, Yipo I-359
Huang, Zhangjin I-202

Ji, Mengqi I-371
Ji, Xiang I-51
Jia, Yunde I-528
Jiang, Wenlan I-180
Jiao, Licheng I-384
Jin, Shan I-89, I-347

Kong, Qi II-188
Kong, Xiangwei II-39
Kou, Feifei I-724, I-735
Kuang, Gangyao I-191
Kuang, Kun I-689

Lang, Xianglong I-504
Li, Baojuan I-123
Li, Cuijin I-323
Li, Daoming I-700
Li, Fan II-115
Li, Feng I-65
Li, Hongjun I-276
Li, Jianfeng II-251
Li, Ju I-101
Li, Leida I-359
Li, Ming I-456
Li, Peng I-759
Li, Qingli II-67
Li, Shuo II-359
Li, Wei II-323
Li, Weiyi I-227
Li, Xiaojing I-168

Li, Yang I-456, II-239
Li, Yongming I-469, I-771
Li, Yufan II-383
Li, Yunfei II-395
Li, Zhongnian I-213, II-27
Li, Zi I-264
Li, Zuoyong I-347
Liang, Junxiong I-408
Liang, Meiyu I-724, I-735
Liu, Aiping I-149
Liu, Di I-589
Liu, Fang I-227
Liu, Guiquan II-419
Liu, Hong I-77
Liu, Jiahao I-700
Liu, Jian I-123
Liu, Jinhui I-289
Liu, Li I-191
Liu, Mengfan I-27
Liu, Qi I-712, I-783
Liu, Qun I-747
Liu, Risheng I-240, I-252, I-264
Liu, Siyuan II-16
Liu, Yanli I-227
Liu, Ye I-712
Liu, Zhaoxin I-301
Liu, Zhen-Tao II-261
Liu, Zhigui I-420
Liu, Zhouyong II-3
Lu, Ke I-492
Lu, Ming I-311
Lu, Qiang I-3
Lu, Zhaofeng II-152
Lu, Zhisheng I-77
Luo, Junren I-51
Luo, Zhenzhen I-89, I-347
Luo, Zhongxuan I-252, I-264
Lv, Chunpu I-180
Lv, Guangyi II-203
Lv, Kai I-674
Lv, Xun I-432

Ma, Jianhui I-783
Ma, Jie I-408
Ma, Long I-252, I-444
Ma, Qianxia II-176
Mei, Jie I-674
Min, Xiongkuo I-3
Mo, Hongwei II-311

Ni, Yihua   I-601
Nie, Jianhui   I-432
Nie, Xiushan   II-347

Pan, Kang   I-601
Pang, Wei   II-227
Peng, Chenglei   I-456

Qian, Jianjun   I-396
Qiao, Yuanhua   II-55
Quan, Shuxue   I-227

Rehman, Abdul   II-261

Sethares, William A.   I-227
Shan, Shouping   I-384
Shang, Jingjie   I-444
Shang, Xiaoke   I-444
Shao, Feifei   II-164
Shao, Jingjing   I-89
Shao, Liyuan   I-136
Shao, Pengyang   I-27
Shi, Ge   I-553
Shi, Guangming   I-301
Shi, Wenchuan   I-771
Si, Pengda   II-273
Song, Jie   I-724
Song, Qianqian   II-39
Sun, Jianguo   I-101
Sun, Jianing   I-240
Sun, Yuqing   II-239

Tan, Yanhao   I-492
Tang, Baoqing   I-601
Tang, Jingyu   II-67
Tang, Xu   I-384
Tang, Zuoqi   II-164
Tao, Hanqing   I-783
Tian, Hui   I-481
Tian, Ye   I-101

Wang, Baoyun   I-516
Wang, Bin   I-168, I-371, II-383
Wang, Boyu   II-359
Wang, Chaokun   I-15
Wang, Chunyu   I-158
Wang, Duo   II-176
Wang, Guoyin   I-747
Wang, Hao   I-626, II-334
Wang, Hongqiang   II-103

Wang, Huangang   I-180, I-700
Wang, Jiayu   I-276
Wang, Jing   I-420
Wang, Liejun   I-469, I-771, II-115
Wang, Meirui   I-65
Wang, Meng   II-203
Wang, Pengpai   I-213
Wang, Qi   I-504
Wang, Qinghu   I-396
Wang, Quan   I-565
Wang, Shengjin   I-371
Wang, Suhong   II-251
Wang, Tao   I-3
Wang, Ti   I-77
Wang, Xiao   I-420
Wang, Yincan   II-334
Wang, Yiru   II-273
Wang, Yuan   I-432
Wang, Zhibo   I-311
Wang, Ziming   II-39
Welch, William J.   I-662
Wen, Zhiqing   II-323
Wong, Waikeung   I-396
Wu, Chao   II-164
Wu, Cheng   I-15
Wu, Fei   I-689
Wu, Jinjian   I-301
Wu, Le   I-149
Wu, Lifang   I-504, I-553
Wu, Qi   I-89
Wu, Yi   II-395
Wu, Yue   II-67
Wu, Yuwei   I-528

Xianbo, Deng   I-136
Xiang, Liuyu   I-39
Xiang, Xueshuang   I-650
Xiang, Ye   I-504
Xiao, Jun   II-164
Xiao, Lei   I-89
Xiao, Zhifeng   II-251
Xie, Haibin   I-759
Xie, Xin   II-215
Xin, Fan   I-240, I-264
Xiong, Hui   I-111
Xu, Fan   II-55
Xu, Feiyi   I-432
Xu, Feng   I-311, I-432, II-371
Xu, Hao   I-638
Xu, Hongteng   II-140

Xu, Jiahui    I-51
Xu, Jin    II-273
Xu, Jin-Meng    II-261
Xu, Kai    II-79
Xu, Mengting    II-27
Xu, Mingying    I-724, I-735
Xu, Qinwen    I-227
Xu, Tong    I-111
Xu, Xin    II-127, II-188
Xu, Yao    I-650
Xu, Yifang    I-456
Xu, Yuan    I-136
Xu, Ziqiang    II-383
Xue, Bo    I-149
Xue, Jian    I-492
Xue, Xuqian    I-420
Xue, Zhe    I-724, I-735

Yan, Shengye    I-601
Yang, Jian    I-396
Yang, Le    II-115
Yang, Luxi    II-3
Yang, Qi    II-103
Yang, Wenzhong    I-575
Yang, Xiaochun    I-168, II-383
Yang, Yongjia    I-601
Yang, Yujiu    II-273, II-285
Yao, Chang    II-67
Ye, Hongbin    II-215
Ye, Yunan    II-164
Yi, Xinyu    II-371
Yin, Guisheng    I-101
Yin, Yilong    II-347
Yu, Chao    II-395
Yu, Xin    I-289
Yuan, Liang    I-674
Yuan, Ming    I-747
Yuan, Senchao    I-712
Yuan, Shen    II-140
Yuan, Weilin    I-51
Yuan, Xin    I-335
Yue, Linan    I-712

Zeng, Yang    II-103
Zhai, Guangtao    I-3
Zhang, Daoqiang    I-213, II-27
Zhang, Enhao    II-408
Zhang, Guangjian    II-408
Zhang, Heng    I-553
Zhang, Jiaao    I-240
Zhang, Jinnian    I-227

Zhang, Kai    I-712
Zhang, Kun    I-27, I-783, II-203
Zhang, Le    I-111
Zhang, Liangliang    II-188
Zhang, Liguo    I-101
Zhang, Lihua    I-158
Zhang, Ming    II-176
Zhang, Ningyu    II-215
Zhang, Qiong    I-662
Zhang, Shaomin    I-444
Zhang, Tao    I-180, II-27, II-176
Zhang, Wanpeng    I-51
Zhang, Xiangrong    I-384
Zhang, Xiaochuan    II-127
Zhang, Xiaodong    I-252
Zhang, Xiaohong    I-168
Zhang, Xiao-Ping    I-3, II-67
Zhang, Xiaoqian    I-420
Zhang, Xifeng    II-311
Zhang, Xinglong    II-127
Zhang, Xu    I-149
Zhang, Yan    I-589
Zhang, Yu    I-589, II-347
Zhang, Zhe    I-111
Zhang, Zhengming    II-3
Zhang, Zizhao    I-123
Zhang, Zonghui    I-202
Zhao, Bo    I-516
Zhao, Lili    I-712
Zhao, Lingjun    I-191
Zhao, Yan    I-589
Zhao, Yang    I-492
Zhao, Yaping    I-371
Zhao, Yuan    II-383
Zheng, Changwei    I-735
Zheng, Tingyue    II-383
Zheng, Ziwei    II-115
Zhong, Bin    II-273
Zhong, Kaiyang    II-419
Zhong, Wei    I-252
Zhou, Yuanen    I-613
Zhou, Yueying    I-213
Zhou, Yuxiao    II-371
Zhu, Hancheng    I-359
Zhu, Yucheng    I-3
Zhu, Zhiyuan    II-359
Ziyu, Zhao    I-689
Zou, Lu    I-202
Zou, Mianlu    I-89
Zou, Xingxing    I-396